Advances in Intelligent Systems and Computing

Volume 701

Series editor

Janusz Kacprzyk, Polish Academy of Sciences, Warsaw, Poland
e-mail: kacprzyk@ibspan.waw.pl

The series "Advances in Intelligent Systems and Computing" contains publications on theory, applications, and design methods of Intelligent Systems and Intelligent Computing. Virtually all disciplines such as engineering, natural sciences, computer and information science, ICT, economics, business, e-commerce, environment, healthcare, life science are covered. The list of topics spans all the areas of modern intelligent systems and computing such as: computational intelligence, soft computing including neural networks, fuzzy systems, evolutionary computing and the fusion of these paradigms, social intelligence, ambient intelligence, computational neuroscience, artificial life, virtual worlds and society, cognitive science and systems, Perception and Vision, DNA and immune based systems, self-organizing and adaptive systems, e-Learning and teaching, human-centered and human-centric computing, recommender systems, intelligent control, robotics and mechatronics including human-machine teaming, knowledge-based paradigms, learning paradigms, machine ethics, intelligent data analysis, knowledge management, intelligent agents, intelligent decision making and support, intelligent network security, trust management, interactive entertainment, Web intelligence and multimedia.

The publications within "Advances in Intelligent Systems and Computing" are primarily proceedings of important conferences, symposia and congresses. They cover significant recent developments in the field, both of a foundational and applicable character. An important characteristic feature of the series is the short publication time and world-wide distribution. This permits a rapid and broad dissemination of research results.

Advisory Board

Chairman

Nikhil R. Pal, Indian Statistical Institute, Kolkata, India
e-mail: nikhil@isical.ac.in

Members

Rafael Bello Perez, Universidad Central "Marta Abreu" de Las Villas, Santa Clara, Cuba
e-mail: rbellop@uclv.edu.cu

Emilio S. Corchado, University of Salamanca, Salamanca, Spain
e-mail: escorchado@usal.es

Hani Hagras, University of Essex, Colchester, UK
e-mail: hani@essex.ac.uk

László T. Kóczy, Széchenyi István University, Győr, Hungary
e-mail: koczy@sze.hu

Vladik Kreinovich, University of Texas at El Paso, El Paso, USA
e-mail: vladik@utep.edu

Chin-Teng Lin, National Chiao Tung University, Hsinchu, Taiwan
e-mail: ctlin@mail.nctu.edu.tw

Jie Lu, University of Technology, Sydney, Australia
e-mail: Jie.Lu@uts.edu.au

Patricia Melin, Tijuana Institute of Technology, Tijuana, Mexico
e-mail: epmelin@hafsamx.org

Nadia Nedjah, State University of Rio de Janeiro, Rio de Janeiro, Brazil
e-mail: nadia@eng.uerj.br

Ngoc Thanh Nguyen, Wroclaw University of Technology, Wroclaw, Poland
e-mail: Ngoc-Thanh.Nguyen@pwr.edu.pl

Jun Wang, The Chinese University of Hong Kong, Shatin, Hong Kong
e-mail: jwang@mae.cuhk.edu.hk

More information about this series at http://www.springer.com/series/11156

Suresh Chandra Satapathy
Joao Manuel R. S. Tavares
Vikrant Bhateja · J. R. Mohanty
Editors

Information and Decision Sciences

Proceedings of the 6th International
Conference on FICTA

 Springer

Editors

Suresh Chandra Satapathy
Department of Computer Science
  and Engineering
PVP Siddhartha Institute of Technology
Vijayawada, Andhra Pradesh
India

Joao Manuel R. S. Tavares
Departamento de Engenharia Mecânica
Universidade do Porto
Porto
Portugal

Vikrant Bhateja
Department of Electronics
  and Communication Engineering
SRMGPC
Lucknow, Uttar Pradesh
India

J. R. Mohanty
School of Computer Application
KIIT University
Bhubaneswar, Odisha
India

ISSN 2194-5357   ISSN 2194-5365   (electronic)
Advances in Intelligent Systems and Computing
ISBN 978-981-10-7562-9   ISBN 978-981-10-7563-6   (eBook)
https://doi.org/10.1007/978-981-10-7563-6

Library of Congress Control Number: 2018930369

© Springer Nature Singapore Pte Ltd. 2018
This work is subject to copyright. All rights are reserved by the Publisher, whether the whole or part of the material is concerned, specifically the rights of translation, reprinting, reuse of illustrations, recitation, broadcasting, reproduction on microfilms or in any other physical way, and transmission or information storage and retrieval, electronic adaptation, computer software, or by similar or dissimilar methodology now known or hereafter developed.
The use of general descriptive names, registered names, trademarks, service marks, etc. in this publication does not imply, even in the absence of a specific statement, that such names are exempt from the relevant protective laws and regulations and therefore free for general use.
The publisher, the authors and the editors are safe to assume that the advice and information in this book are believed to be true and accurate at the date of publication. Neither the publisher nor the authors or the editors give a warranty, express or implied, with respect to the material contained herein or for any errors or omissions that may have been made. The publisher remains neutral with regard to jurisdictional claims in published maps and institutional affiliations.

Printed on acid-free paper

This Springer imprint is published by the registered company Springer Nature Singapore Pte Ltd. part of Springer Nature
The registered company address is: 152 Beach Road, #21-01/04 Gateway East, Singapore 189721, Singapore

Preface

This book is a collection of high-quality peer-reviewed research papers presented at the 6th International Conference on Frontiers of Intelligent Computing: Theory and Applications (FICTA-2017) held at School of Computer Applications, KIIT University, Bhubaneswar, Odisha, India, during October 14–15, 2017.

The idea of this conference series was conceived by few eminent professors and researchers from premier institutions of India. The first three editions of this conference: FICTA-2012, FICTA-2013, and FICTA-2014 were organized by Bhubaneswar Engineering College (BEC), Bhubaneswar, Odisha, India. Due to its popularity and wide visibilities in the entire country as well as abroad, the fourth edition FICTA-2015 has been organized by the prestigious NIT Durgapur, WB, India. The fifth edition FICTA-2016 was organized by KIIT University, Bhubaneswar, Odisha, India. All papers of past FICTA editions are published by Springer AISC series. Presently, FICTA-2017 is the sixth edition of this conference series which aims to bring together researchers, scientists, engineers, and practitioners to exchange and share their theories, methodologies, new ideas, experiences, applications in all areas of intelligent computing theories, and applications to various engineering disciplines like Computer Science, Electronics, Electrical, Mechanical, Biomedical Engineering.

FICTA-2017 had received a good number of submissions from the different areas relating to decision sciences, intelligent computing, and its applications. These papers have undergone a rigorous peer review process with the help of our program committee members and external reviewers (from the country as well as abroad). The review process has been very crucial with minimum 2 reviews each and in many cases 3–5 reviews along with due checks on similarity and content overlap as well. FICTA-2017 witnessed more than 300 papers including the main track as well as special sessions. The conference featured seven special sessions in various cutting-edge technologies of specialized focus which were organized and chaired by eminent professors. The total toll of papers received, included submissions received cross country along with 7 overseas countries. Out of this pool, only 131 papers were given acceptance and segregated as two different volumes for publication under the proceedings. This volume consists of 59 papers from diverse areas of information and decision sciences.

The conference featured many distinguished keynote addresses by eminent speakers like Dr. Siba K. Udgata (University of Hyderabad, Telangana, India) on Intelligent and Soft Sensor for Environment Monitoring; Dr. Goutam Sanyal (NIT Durgapur, WB, India) on Vision-based Biometric Features; Dr. Kamiya Khatter (Sr. Editorial Assistant, Springer Nature, India) on Author Services and Tools. These keynote lectures embraced a huge toll of an audience of students, faculties, budding researchers, as well as delegates.

We thank the General Chairs: Prof. Samaresh Mishra, Prof. Veena Goswami, KIIT University, Bhubaneswar, India, and Prof. Suresh Chandra Satapathy, PVPSIT, Vijayawada, India, for providing valuable guidelines and inspiration to overcome various difficulties in the process of organizing this conference.

We extend our heartfelt thanks to the Honorary Chairs of this conference: Dr. B. K. Panigrahi, IIT Delhi, and Dr. Swagatam Das, ISI, Kolkata, for being with us from the beginning to the end of this conference and without their support this conference could never have been successful.

We would also like to thank School of Computer Applications and Computer Engineering, KIIT University, Bhubaneswar, who jointly came forward to support us to organize sixth edition of this conference series. We are amazed to note the enthusiasm of all faculty, staff, and students of KIIT to organize the conference in such a professional way. Involvements of faculty coordinators and student volunteer are praiseworthy in every respect. We are confident that in the future too we would like to organize many more international-level conferences in this beautiful campus. We would also like to thank our sponsors for providing all the support and financial assistance.

We take this opportunity to thank authors of all submitted papers for their hard work, adherence to the deadlines, and patience with the review process. The quality of a refereed volume depends mainly on the expertise and dedication of the reviewers. We are indebted to the program committee members and external reviewers who not only produced excellent reviews but also did these in short time frames. We would also like to thank the participants of this conference, who have participated in the conference above all hardships. Finally, we would like to thank all the volunteers who spent tireless efforts in meeting the deadlines and arranging every detail to make sure that the conference can run smoothly. All the efforts are worth and would please us all, if the readers of this proceedings and participants of this conference found the papers and conference inspiring and enjoyable. Our sincere thanks to all press print and electronic media for their excellent coverage of this conference.

We take this opportunity to thank all Keynote Speakers, Track and Special Session Chairs for their excellent support to make FICTA-2017 a grand success.

Vijayawada, India   Dr. Suresh Chandra Satapathy
Porto, Portugal   Dr. Joao Manuel R. S. Tavares
Lucknow, India   Dr. Vikrant Bhateja
Bhubaneswar, India   Dr. J. R. Mohanty

Organization

Chief Patron

Achyuta Samanta, KISS and KIIT University

Patron

H. Mohanty, KIIT University

Advisory Committee

Sasmita Samanta, KIIT University
Ganga Bishnu Mund, KIIT University
Samaresh Mishra, KIIT University

Honorary Chairs

Swagatam Das, ISI, Kolkata
B. K. Panigrahi, IIT Delhi

General Chairs

Veena Goswami, KIIT University
Suresh Chandra Satapathy, PVPSIT, Vijayawada

Convener

Sachi Nandan Mohanty, KIIT University
Satya Ranjan Dash, KIIT University

Organizing Chairs

Sidharth Swarup Routaray, KIIT University
Manas Mukul, KIIT University

Publication Chair

Vikrant Bhateja, SRMGPC, Lucknow

Steering Committee

Suresh Chandra Satapathy, PVPSIT, Vijayawada
Vikrant Bhateja, SRMGPC, Lucknow
Siba K. Udgata, UoH, Hyderabad
Manas Kumar Sanyal, University of Kalyani
Nilanjan Dey, TICT, Kolkata
B. N. Biswal, BEC, Bhubaneswar

Editorial Board

Suresh Chandra Satapathy, PVPSIT, Vijayawada, India
Vikrant Bhateja, SRMGPC, Lucknow (UP), India
Dr. J. R. Mohanty, KIIT University
Prasant Kumar Pattnaik, KIIT University, Bhubaneswar, India
Joao Manuel R. S. Tavares, Universidade do Porto (FEUP), Porto, Portugal
Carlos Artemio Coello Coello, CINVESTAV-IPN, Mexico City, Mexico

Transport and Hospitality Chairs

Ramakant Parida, KIIT University
K. Singh, KIIT University
B. B. Dash, KIIT University

Session Management Chairs

Chinmay Mishra, KIIT University
Sudhanshu Sekhar Patra, KIIT University

Registration Chairs

Ajaya Jena, KIIT University
Utpal Dey, KIIT University
P. S. Pattanayak, KIIT University
Prachi Viyajeeta, KIIT University

Publicity Chairs

J. K. Mandal, University of Kalyani, Kolkata
Himanshu Das, KIIT University
R. K. Barik, KIIT University

Workshop Chairs

Manoj Mishra, KIIT University
Manas Kumar Rath, KIIT University

Track Chairs

Machine Learning Applications: Steven L. Fernandez, The University of Alabama, Birmingham, USA
Image Processing and Pattern Recognition: V. N. Manjunath Aradhya, SJCE, Mysore, India
Signals, Communication, and Microelectronics: A. K. Pandey, MIET, Mccrut (UP), India
Data Engineering: M. Ramakrishna Murty, ANITS, Visakhapatnam, India

Special Session Chairs

SS01: Computational Intelligence to Ecological Computing through Data Sciences: Tanupriya Choudhury and Praveen Kumar, Amity University, UP, India
SS02: Advances in Camera Based Document Recognition: V. N. Manjunath Aradhya and B. S. Harish, SJCE, Mysore, India
SS03: Applications of Computational Intelligence in Education and Academics: Viral Nagori, GLS University, Ahmedabad, India
SS04: Modern Intelligent Computing, Human Values and Professional Ethics for Engineering and Management: Hardeep Singh and B. P. Singh, FCET, Ferozepur, Punjab
SS05: Mathematical Modelling and Optimization: Deepika Garg, G. D. Goenka University, India, and Ozen Ozer, Kırklareli Üniversitesi, Turkey
SS06: Computer Vision and Image Processing: Synh Viet-Uyen Ha, Vietnam National University, Vietnam
SS07: Data Mining Applications in Network Security: Vinutha H. P., BIET, Karnataka, India, and Sagar B. M., RVCE, Bangalore, Karnataka, India

Technical Program Committee/International Reviewer Board

A. Govardhan, India
Aarti Singh, India
Almoataz Youssef Abdelaziz, Egypt
Amira A. Ashour, Egypt
Amulya Ratna Swain, India
Ankur Singh Bist, India
Athanasios V. Vasilakos, Athens
Banani Saha, India
Bhabani Shankar Prasad Mishra, India
B. Tirumala Rao, India
Carlos A. Coello, Mexico

Charan S. G., India
Chirag Arora, India
Chilukuri K. Mohan, USA
Chung Le, Vietnam
Dac-Nhuong Le, Vietnam
Delin Luo, China
Hai Bin Duan, China
Hai V. Pham, Vietnam
Heitor Silvério Lopes, Brazil
Igor Belykh, Russia
J. V. R. Murthy, India
K. Parsopoulos, Greece
Kamble Vaibhav Venkatrao, India
Kailash C. Patidar, South Africa
Koushik Majumder, India
Lalitha Bhaskari, India
Jeng-Shyang Pan, Taiwan
Juan Luis Fernández Martínez, California
Le Hoang Son, Vietnam
Leandro Dos Santos Coelho, Brazil
L. Perkin, USA
Lingfeng Wang, China
M. A. Abido, Saudi Arabia
Maurice Clerc, France
Meftah Boudjelal, Algeria
Monideepa Roy, India
Mukul Misra, India
Naeem Hanoon, Malaysia
Nikhil Bhargava, India
Oscar Castillo, Mexico
P. S. Avadhani, India
Rafael Stubs Parpinelli, Brazil
Ravi Subban, India
Roderich Gross, England
Saeid Nahavandi, Australia
Sankhadeep Chatterjee, India
Sanjay Sengupta, India
Santosh Kumar Swain, India
Saman Halgamuge, India
Sayan Chakraborty, India
Shabana Urooj, India
S. G. Ponnambalam, Malaysia
Srinivas Kota, Nebraska
Srinivas Sethi, India
Sumanth Yenduri, USA

Suberna Kumar, India
T. R. Dash, Cambodia
Vipin Tyagi, India
Vimal Mishra, India
Walid Barhoumi, Tunisia
X. Z. Gao, Finland
Ying Tan, China
Zong Woo Geem, USA
Monika Jain, India
Rahul Saxena, India
Vaishali Mohite, India
And many more …

Contents

Contents

About the Editors

Suresh Chandra Satapathy, Ph.D. is currently working as Professor and Head of the Department of CSE, PVPSIT, Vijayawada, India. He was the National Chairman Div-V (Educational and Research) of the Computer Society of India from 2015 to 2017. A Senior Member of IEEE, he has been instrumental in organizing more than 18 international conferences in India and has edited more than 30 books as a corresponding editor. He is highly active in research in the areas of Swarm Intelligence, Machine Learning, and Data Mining. He has developed a new optimization algorithm known as social group optimization (SGO) and authored more than 100 publications in reputed journals and conference proceedings. Currently, he serves on the editorial board of the journals IGI Global, Inderscience, and Growing Science.

Joao Manuel R. S. Tavares earned his Ph.D. in Electrical and Computer Engineering in 2001 and his postdoctoral degree in Mechanical Engineering in 2015. He is a Senior Researcher and Project Coordinator at the Instituto de Ciência e Inovação em Engenharia Mecânica e Engenharia Industrial (INEGI), Portugal, and an Associate Professor at the Faculdade de Engenharia da Universidade do Porto (FEUP), Portugal. He is the co-editor of more than 35 books, co-author of more than 550 articles in international and national journals and conferences, and holder of 3 international and 2 national patents. He is the co-founder and co-editor of the book series "Lecture Notes in Computational Vision and Biomechanics," founder and editor-in-chief of the journal "Computer Methods in Biomechanics and Biomedical Engineering: Imaging and Visualization," and co-founder and co-chair of the international conference series: CompIMAGE, ECCOMAS VipIMAGE, ICCEBS, and BioDental.

Vikrant Bhateja is an Associate Professor in the Department of ECE, SRMGPC, Lucknow, and also the Head of Academics and Quality Control at the same college. His areas of research include Digital Image and Video Processing, Computer Vision, Medical Imaging, Machine Learning, Pattern Analysis, and Recognition. He has authored more than 120 publications in various international journals and conference proceedings. He is an Associate Editor for the International Journal of Synthetic Emotions (IJSE) and International Journal of Ambient Computing and Intelligence (IJACI).

J. R. Mohanty is a Professor and Associate Dean School of Computer Applications at KIIT University, Bhubaneswar, India. He holds a Ph.D. (Computer Science) and has 20 years of experience teaching postgraduate students. His key strengths in research and academia include Database Management Systems, Operating Systems, Evolutionary Algorithms, Queueing Networks, and Cloud Computing. He has published a number of research papers in peer-reviewed international journals and conferences.

A New Approach for Authorship Attribution

P. Buddha Reddy, T. Raghunadha Reddy, M. Gopi Chand
and A. Venkannababu

Abstract Authorship attribution is a text classification technique, which is used to find the author of an unknown document by analyzing the documents of multiple authors. The accuracy of author identification mainly depends on the writing styles of the authors. Feature selection for differentiating the writing styles of the authors is one of the most important steps in the authorship attribution. Different researchers proposed a set of features like character, word, syntactic, semantic, structural, and readability features to predict the author of a unknown document. Few researchers used term weight measures in authorship attribution. Term weight measures have proven to be an effective way to improve the accuracy of text classification. The existing approaches in authorship attribution used the bag-of-words approach to represent the document vectors. In this work, a new approach is proposed, wherein the document weight is used to represent the document vector instead of using features or terms in the document. The experimentation is carried out on reviews corpus with various classifiers, and the results achieved for author attribution are prominent than most of the existing approaches.

Keywords Authorship attribution · Author prediction · Term weight measure
BOW approach

P. B. Reddy (✉) · T. R. Reddy · M. G. Chand
Department of IT, Vardhaman College of Engineering, Hyderabad, India
e-mail: buddhareddy.polepelli@gmail.com

T. R. Reddy
e-mail: raghu.sas@gmail.com

M. G. Chand
e-mail: gopi_merugu@yahoo.com

A. Venkannababu
Department of CSE, Sri Vasavi Engineering College, Tadepalligudem, Andhra Pradesh, India
e-mail: venkannababu.alamuru@gmail.com

© Springer Nature Singapore Pte Ltd. 2018
S. C. Satapathy et al. (eds.), *Information and Decision Sciences*,
Advances in Intelligent Systems and Computing 701,
https://doi.org/10.1007/978-981-10-7563-6_1

1   Introduction

The web is growing constantly with a huge amount of text mainly through blogs, twitter, reviews, and social media. Most of this text is written by various authors in different contexts. The availability of text challenges the researchers and information analysts to develop automated tools for information analysis. Authorship analysis is one such area attracted by several researchers to extract information from the anonymous text. Authorship analysis is a procedure of finding the authorship of a text by inspecting its characteristics. Authorship analysis is classified into two categories such as authorship identification and author profiling [1].

Authorship identification finds the authorship of a document. It is categorized into two classes namely, authorship attribution and authorship verification. Authorship attribution predicts the author of a given anonymous document by analyzing the documents of multiple authors [2]. Authorship verification finds whether the given document is written by a particular author or not by analyzing the documents of a single author [3]. Author profiling is a type of text classification task, which predicts profiling characteristics such as gender, age, occupation, native language, location, education, and personality traits of the authors by analyzing their writing styles [4].

Authorship attribution is used in several applications such as forensic analysis, security, and literary research. In forensic analysis, the suicide notes and property wills are analyzed whether the note or will is written by a correct person or not by analyzing the writing styles of suspected authors. The terrorist organizations send threatening mails; authorship attribution techniques are used to identify which terrorist organization send the mail or to confirm the mail whether it came from correct source or not. In literary research, some researchers try to claim the innovations of others without proper acknowledgement. The authorship attribution is used to identify author of a document by analyzing the writing styles of the various authors.

The authorship attribution approaches are divided into two categories such as profile-based approaches and instance-based approaches [1]. In this work, an instance-based approach is discussed. In the profile-based approach, the available training texts per author were concatenated to get a single text file. This single big file is used to extract the different properties of the author's style. In the instance-based approaches, the known authorship text sample is considered as an instance and every text sample as a unit.

This paper is planned as follows. Section 2 explains the related work done by the researchers in authorship attribution area. The existing model used by several researchers to represent a document is described in Sect. 3. In Sect. 4, the proposed model is explained with a term weight measure and document weight measure. The experimental results are analyzed in Sect. 5. Section 6 concludes this work and suggests the future work in authorship attribution.

2 Existing Work

Authorship attribution task is subdivided into two different subtasks: First, extraction of most discriminative features to differentiate the writing styles of the authors; and second, identifying the suitable classification algorithm to detect the most probable author of a test document [5].

The character trigrams, POS bigrams and trigrams, suffixes, word length, syntactic complexity and structure, and percentage of direct speech from the documents is extracted in Stefan Ruseti et al. [6] to represent the document vector. They experimented with Sequential Minimal Optimization (SMO) algorithm and obtained an accuracy of overall 77% in authorship identification. It was observed that the accuracy is increased when the application-specific features are added.

Ludovic Tanguy et al. experimented [7] with rich linguistic features such as frequency of character trigrams, contraction forms, phrasal verbs, lexical generosity and ambiguity, frequency of POS trigrams, syntactic dependencies, syntactic complexity, lexical cohesion, morphological complexity, lexical absolute frequency, punctuation and case, quotations, first/third person narrative and proper names, and maximum entropy machine learning tool. It was observed that the rich set of linguistic features perform well for author identification when compared to trigrams of characters and word frequencies.

Ludovic Tanguy et al. experimented [8] with linguistic features such as sub-word-level features, word-level features, sentence-level features, and message-level features to represent the document vectors. They used maximum entropy technique and machine learning algorithms such as rule-based learners and decision trees to evaluate the accuracy of author prediction. It was observed that maximum entropy technique achieved good accuracy than machine learning algorithms.

N. Akiva used [9] single vector representation that captures the presence or absence of common words in a text. They used SVM-Light classification algorithm to generate the classification model. They observed that the accuracy of author prediction is increased when binary BOW representation is used to represent the document vector and also observed that the accuracy is increased when the number of authors is increased in the training data.

In this work, the dataset was collected from www.amazon.com, and it contains 10 different authors' reviews on different products. The corpus is balanced in terms of number of documents in each author group, and each author group contains 400 reviews of each. The accuracy measure is used to evaluate the performance of the author prediction. Accuracy is the ratio of number of documents correctly predicted their author and the number of documents considered.

3 Existing Approach

Most of the authorship attribution approaches used the Bag-Of-Words (BOW) - approach to represent the document vector.

3.1 Bag-of-Words (BOW) Approach

Figure 1 shows the model of the BOW approach. In this approach, first collect the corpus. Then, preprocessing techniques are applied to the collected dataset. Extract

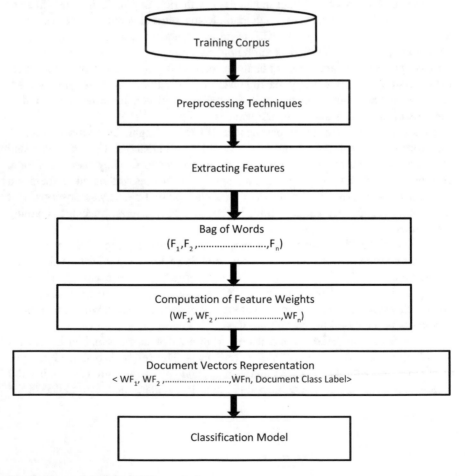

Fig. 1 The model of the BOW approach

the most frequent terms or features that are important to discriminate the writing styles of the authors from the modified dataset. Consider these terms or features as BOW. Every document in the dataset is represented by this BOW. Each value in the document vector is the weights of the BOW. Finally, the document vectors are used to generate classification model.

4 Proposed Author-Specific Document Weighted (ADW) Approach

Figure 2 depicts the model of proposed author-specific document weighted approach.

In the proposed approach first, collect the reviews corpus. Apply preprocessing techniques to the corpus of documents such as stop words removal and stemming. Extract frequent terms that occur at least 5 times in the updated corpus. Compute

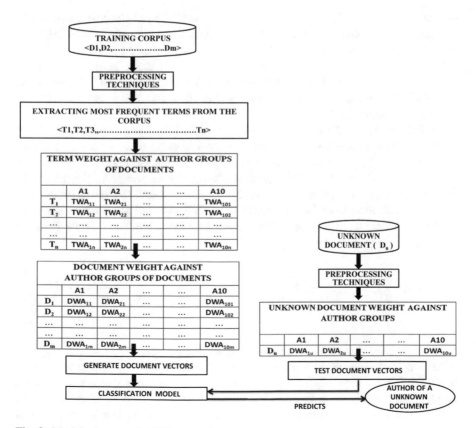

Fig. 2 Model of proposed ADW approach

term weights in each author group of documents using term weight measures. Document weights are determined for each author group by aggregating the weights of the terms in a document using document weight measure. Generate document vectors with document weights to build a classification model. Finally, the classification model is used to predict the author of an unknown document.

In this model, $\{A_1, A_2, \ldots A_q\}$ is a set of author groups, $\{D_1, D_2, \ldots D_m\}$ is a list of documents in the corpus, and $\{T_1, T_2, \ldots T_n\}$ is a list of vocabulary terms. TWA_{xy} is the weight of the term T_y weight in the author group A_x, and DWA_{pq} is the document D_q weight in the author group A_p. The profiles of an anonymous document are predicted using classification model. In this approach, identification of suitable weight measures for calculating term weight and document weight is important. Sections. 4.1 and 4.2 discuss about the weight measures used in this approach.

4.1   Term Weight Measure

In text classification, the text document is represented by the vector space model and predefined classes are assigned to documents through machine learning algorithms. In vector space model, the documents are represented as numerical feature vector, which consists of the weights of the terms that are extracted from the document. The basic problem in text classification is identification of appropriate term weight measure to calculate the weight of the terms that directly affects the classification accuracy.

In this work, the Discriminative Feature Selection Term Weight (DFSTW) measure [10] was identified from text categorization domain to select the best features to categorize the text. This measure is tested on our corpus to predict the characteristics of the authors. DFS measure assigns more weight to the terms that are having high average term frequency in class c_j and the terms with high occurrence rate in most of the documents of c_j. DFS measure assigns less weight to the terms that occurred in most of the documents in both c_j and $\overline{c_j}$. The DFSTW measure is represented in Eq. (1).

$$W(t_i, c_j) = \frac{tf(t_i, c_j)/df(t_i, c_j)}{tf(t_i, \overline{c_j})/df(t_i, \overline{c_j})} \times \frac{a_{ij}}{(a_{ij} + b_{ij})} \times \frac{a_{ij}}{(a_{ij} + c_{ij})} \times \left| \frac{a_{ij}}{(a_{ij} + b_{ij})} - \frac{c_{ij}}{(c_{ij} + d_{ij})} \right|$$

(1)

where $A = \{A_1, A_2, \ldots A_q\}$ is the set of author groups, $\{D_1, D_2, \ldots, D_m\}$ is a collection of documents in the corpus, and $V = \{t_1, t_2, \ldots, t_n\}$ is a collection of vocabulary terms for analysis. a_{ij} is the number of documents of class c_j which contains the term t_i, b_{ij} is the number of documents of class c_j which does not contain term t_i, c_{ij} is the number of documents which contains term t_i and does not belong to class c_j, and d_{ij} is the number of documents which does not contain term t_i and does not belong to class c_j.

4.2 Document Weight Measure

The proposed document weight measure as in Eq. (2) is used to calculate the weight of a document or corpus of each author group. This measure used the combination of term weights that are specific to document and specific to author group. The Term Frequency (TF) is used to compute term weights specific to a document, and term weight measures are used to determine the term weights specific to author group. The document weight computation considers the correlation between the terms in that document.

$$W_{d_{kj}} = \sum_{t_i \in d_k, d_k \in A_j} TF(t_i, d_k) * W_{t_{ij}}, \qquad (2)$$

where $W_{d_{kj}}$ is the weight of document d_k in the author group A_j and Wt_{ij} is the weight of a term t_i in the corpus of author group A_j.

5 Experimental Results

5.1 BOW Approach

Table 1 represents the accuracies of author prediction in BOW approach using various classifiers. In the BOW approach, unlike other classifiers, the Naive Bayes Multinomial classifier achieved an accuracy of 82.16% for the most frequent 8000 terms.

Table 1 Accuracies of author prediction when BOW approach is used

Classifier/Number of terms	Naive Bayes multinomial	Logistic	Random forest
1000	72.09	69.67	68.21
2000	74.29	70.71	69.45
3000	75.48	72.23	71.01
4000	76.31	74.49	71.13
5000	78.17	76.97	71.87
6000	79.87	77.02	73.51
7000	81.39	77.11	73.92
8000	82.16	78.82	75.83

Table 2 Accuracies of author prediction when ADW approach is used

Classifier/Number of terms	Naive Bayes multinomial	Logistic	Random forest
1000	88.21	85.55	82.79
2000	89.06	87.90	83.44
3000	91.12	88.31	84.74
4000	92.67	89.56	86.43
5000	93.71	91.81	87.39
6000	95.47	92.65	89.81
7000	96.23	94.32	91.38
8000	97.89	94.91	92.17

5.2 ADW Approach

The accuracies of author prediction in ADW approach using various classification algorithms are represented in Table 2. With the ADW approach, the Naïve Bayes Multinomial classifier obtained accuracy with 97.89% for the most frequent 8000 terms. When the number of terms is increased, the accuracy is also increased in all the classifiers.

The proposed ADW approach is evaluated and compared against the traditional BOW approach. This approach achieved better results than most of the existing approaches in authorship attribution. The Naïve Bayes Multinomial classifier produced more accurate results for author prediction with BOW and ADW approaches.

It is also notable that the most frequent 8000 terms as feature set, the proposed model generated good accuracies when compared with existing approaches. The Logistic and RandomForest classifiers achieve a good accuracy with 8000 frequent terms in both approaches.

6 Conclusions and Future Scope

In this work, the experimentation is carried out with most frequent terms with BOW model and proposed ADW model. The proposed model achieved an accuracy of 97.89% for author prediction when Naïve Bayes Multinomial classifier is used. The BOW approach obtained an accuracy of 82.16% for author prediction when Naïve Bayes Multinomial classifier is used. In both models, the most frequent 8000 terms obtained highest accuracy. In BOW approach, the terms are independently participated in the classification process but in proposed ADW model the terms are collaboratively in the form of document weight participated in the classification process. This is the main reason for obtaining good accuracies in the proposed model.

In our future work, it is planned to consider the domain characteristics and categorical features while computing a document weight. It is also planned to usage of semantic and syntactic structure of the language while assigning weights to the document.

References

1. Stamatatos, E.: A survey of modern authorship attribution methods. JASIST (2009)
2. Elayidom, M.S., Jose, C., Puthussery, A., Sasi, N.K.: Text classification for authorship attribution analysis. Advanc. Comput. Int. J. **4**(5) (2013)
3. Koppel, M., Schler, J., Bonchek-Dokow, E.: Measuring differentiability: Unmasking pseudonymous authors. J. Mach. Learn. Res. **8**, 1261–1276 (2007)
4. Koppel, M., Argamon, S., Shimoni, A.R.: Automatically categorizing written texts by author gender. Liter. Linguist. Comput. **17**(4), 401–412 (2002)
5. Juola, P.: Authorship attribution. Found. Trends Inf. Retr. **1**, 233–334 (2006)
6. Stefan, R., Traian, R.: Authorship identification using a reduced set of linguistic features—notebook for PAN at CLEF 2012. In: CLEF 2012 Evaluation Labs and Workshop, 17–20 September, Rome, Italy, September 2012. ISBN 978-88-904810-3-1. ISSN 2038-4963
7. Ludovic, T., Franck, S., Basilio, C., Nabil, H.: Authorship attribution: using rich linguistic features when training data is scarce. In: CLEF 2012 Evaluation Labs and Workshop, 17–20 September, Rome, Italy, September 2012. ISBN 978-88-904810-3-1. ISSN 2038-4963
8. Ludovic, T., Assaf, U., Basilio, C., Nabil, H., Franck, S.: A Multitude of Linguistically-rich Features for Authorship Attribution. CLEF 2011 Labs and Workshops, 19–22 September, Amsterdam, Netherlands, September 2011. ISBN 978-88-904810-1-7. ISSN 2038-4963
9. Navot, A.: Authorship and plagiarism detection using binary BOW features. In: CLEF 2012 Evaluation Labs and Workshop, 17–20 September, Rome, Italy, September 2012. ISBN 978-88-904810-3-1. ISSN 2038-4963
10. Wei, Z., Feng, Wu, Lap-Keung, C., Domenic, S., A discriminative and semantic feature selection method for text categorization. Int. J. Prod. Econom. Elsevier, 215–222 (2015)

Using Aadhaar for Continuous Test-Taker Presence Verification in Online Exams

N. Sethu Subramanian, Sankaran Narayanan, M. D. Soumya, Nitheeswar Jayakumar and Kamal Bijlani

Abstract In the context of Indian higher education, deployment of online exams for summative assessments is hampered by the lack of a reliable e-assessment system that can guarantee high degree of examination integrity. Current methods for maintaining integrity of online exams largely depend on proctor-based invigilation. This cumbersome method does not fully prevent test-taker impersonation. We propose leveraging the Aadhaar biometric data to verify the presence of the test-taker throughout the entire duration of the examination. Such a continuous presence verification technique is likely to be far more successful in preventing impersonation than the conventional methods. Our method can help eventually eliminate the need for a remote proctor. We identify a set of strategies to systematically validate the effectiveness of the proposed system.

Keywords LMS · Cloud · Exam cheating · Proctoring · e-Assessment
Multimodal biometrics

1 Introduction

Online assessments are gaining increased acceptance in higher education as the preferred appraisal scheme, gradually replacing the manual paper-and-pen assessment

N. S. Subramanian (✉) · S. Narayanan · M. D. Soumya · N. Jayakumar · K. Bijlani
Amrita e-Learning Research Lab (AERL), Amrita School of Engineering, Amrita Vishwa
Vidyapeetham, Amrita University, Amritapuri, India
e-mail: sethus@am.amrita.edu

S. Narayanan
e-mail: nsankaran@am.amrita.edu

M. D. Soumya
e-mail: soumyamd@am.amrita.edu

N. Jayakumar
e-mail: nitheeswar90@gmail.com

K. Bijlani
e-mail: kamal@amrita.edu

© Springer Nature Singapore Pte Ltd. 2018
S. C. Satapathy et al. (eds.), *Information and Decision Sciences*,
Advances in Intelligent Systems and Computing 701,
https://doi.org/10.1007/978-981-10-7563-6_2

format for both formative and summative assessments [1]. Summative evaluations, such as final exams, are also transitioning from written to a blended assessment model [2] composed of a mix of open- and close-ended questions. These unified evaluations could very well be conducted in a proctored online setting obviating the need for the pen-and-paper option. However, there is a wide perception that online exams are inherently insecure despite being proctored [3, 4]. This perception hampers widespread adoption or constrains institutions, to conduct online exams in controlled settings under strict uninterrupted vigilance. These burdensome options motivate the development of practical alternative measures to administer online examinations, untainted by cheating and concomitant security violations.

Test-taker validation, identity verification, and impersonation prevention are major security goals that need to be achieved to ensure the integrity of summative e-assessments [5]. These three criteria need to be satisfied in any solution so that consumers of reliable, vetted academic grading systems such as educators, employers, public and private sector enterprises, and government agencies are assured of the authenticity and integrity of the examination process. Conventional methods such as usernames and passwords, hall tickets, or even biometrics provide only a partial security solution by meeting the initial validation criteria. But these checks are incapable to certify that the test-taker has not subsequently switched during the exam. Thus, frequent reports of compromised exams due to student–student or student–proctor collusions form part of the daily news headlines. These occurrences can be due to empathy, blackmail, bribery, or coercion. Test-taker impersonation is thus a key threat that needs comprehensive attention.

Impersonation occurs when someone other than the designated individual seeks all access privileges duly granted to the vetted person, whose identity is being claimed by the impersonator. In the case of online exams, this usually occurs when test-takers reveal deliberately their security credentials to the impersonating person who can then complete the exam on their behalf. Continuous presence verification [6] has been proposed as a means to reaffirm test-taker integrity throughout the examination process. The presence of the test-taker is verified continuously using biometric features such as fingerprints or keystroke dynamics [7] for the entire duration of the examination. Several factors serve to inhibit the wide deployment of sophisticated security technology such as these. First among these is the lack of a central repository database to collect and store test-taker demographic and security profile data during the enrollment/registration phase. Given the globally distributed test-taker body in large educational institutions and learning centers, building and maintaining an aggregated, real-time updated, and verifiable database poses a significant logistical and engineering challenge. The other security challenge is safeguarding such a database against attacks by hackers and terrorists.

In this paper, we propose a simple, efficient, and cost-effective architectural approach to continuous test-taker presence verification in the mega-sized Indian educational context. Our concept incorporates the field-proven Aadhaar database [8, 9] with a collection of biometric data of over a billion Indian citizens. Mobile devices have facilitated increased awareness of biometrics, such as fingerprint scanning technology, especially among the younger citizens. Moreover, inexpensive USB-based fingerprint scanning devices are gaining wide adoption in a variety of commercial settings for use with Aadhaar and other biometric systems. The confluence of these factors renders the proposed architecture ideal for continuous test-taker presence verification in large-scale distributed e-assessment applications.

2 Related Work

Authentication of an initial user (i.e., test-taker) can be accomplished via several mainstream techniques [7]. Knowledge-proof methods require the user to demonstrate knowledge of a secret possessed only by that user such as usernames–passwords, security questions–answers, etc. As standalone authentication factors, these methods are susceptible to inter test-taker collusion. Possession proof methods require the user to possess a physical entity such as an ID card. These are generally used when the exam is conducted in a proctored environment such as a university examination hall or an accredited exam center but vulnerable as the card can be cloned. Biometric methods rely on the user's unique physical characteristics such as fingerprint [10] or iris recognition [11]. By far, biometric methods are the most reliable for initial authentication but they do not provide continuous presence verification by themselves.

Proctoring methods using a remote proctor observing test-takers via webcams [12] or centralized video cameras is a widely accepted practice to minimize the likelihood of malpractices in exam centers. This method is generally good at capturing abnormal activities in an examination center. However, these and other video-camera-based methods are storage-greedy and bandwidth-intensive [13] that do not scale well when there is large number of test-takers involved. Observing the test-taker's face via an individualized webcam by the remote proctor could be a reasonable method for continuous presence verification. Currently, this approach has a few problems that limits its practical effectiveness. First of all, it requires an alert proctor, undistracted for the entire duration of the exam. Second, each proctor is able to monitor only a few test-takers effectively. Lastly, there is a paucity of inexpensive off-the-shelf proctoring solutions that can be readily deployed.

Several methods have been proposed to reduce the dependency on remote proctor. Swathi et al. [14] propose a method to perform feature extraction for automated face understanding. Their work was geared to ensure presence of the examinee throughout the examination session, with the assumption that if the face is not present it could indicate malicious activity. Krishnamoorthy et al. [15] used image recognition to automatically recognize the test-taker's face. These measures remain under the domain of active research.

Flior et al. [7] proposed keystroke dynamics for biometric authentication. By detecting the keystroke typing patterns of the pre-authenticated users, a model is constructed to provide continuous presence verification. Although keystroke detection has been shown to have low error rates in lab conditions [5], its availability and technical maturity for mass-scale deployment such as the heterogeneous test-taker population in India is far from clear [16]. The same can be said for other types of advanced biometrics like Iris recognition, too expensive to be established on a mass-scale.

Fingerprint biometrics [10] are used ubiquitously all over India in various public and private sector enterprises like military, banking, etc. They are easy to use, have fast response time, and impose minimal technology training and deployment over-head. Most modern mobile phones are equipped with a built-in fingerprint sensor that can unlock a phone by recognition of the human fingerprints. USB-based fin-gerprint scanning devices are also available that scan fingerprints with greater details and higher levels of accuracy. Studies have shown that among the domain biometric options, the fingerprint biometric option is the least intrusive and most reliable [17]. Gil et al. [10] have proposed the design of a middleware that links up a fingerprint identification system with the university learning management system (LMS). Their solution requires a computer connected with a biometric reader for every test-taker.

Clarke et al. [13] outline a generic e-invigilation methodology that can utilize sev-eral methods to provide continuous presence verification. They envisage extensive use of facial recognition as it can be implemented naturally by placing a webcam on top of the test-taker's monitor. Their prototype performed fingerprint authentication by uploading fingerprints to a central server.

The goal of the present work is to transform the fingerprint scanning option into an efficient, practical method approach for continuous presence verification of test-takers (such as students) in e-assessment sessions.

3 Proposed Methodology

In this paper, we propose a simple architecture to enable continuous authentication by means of the Indian government's Aadhaar database [8, 9]. Our first observation is that the biometric data collection and management process is simplified significantly by the presence of the Aadhaar database. Aadhaar Web Services make available a portal interface that can be dedicated to biometric verification by third-party service providers.

Modern smartphones and laptop personal computers (PCs) are built-in with a fin-gerprint sensor that can be configured as a login mechanism. These built-in sensors are designed for single-user identity verification scenarios. Devices with these sen-sors do not allow export of the private biometric information beyond that device to protect the user's privacy. Thus, these sensors are of no use in public, multiuser identity verification settings. In contrast to fingerprint sensors, fingerprint scan-ning devices are designed for public, multiuser identify verification scenarios. These devices can be interfaced to a PC or a smartphone via USB. They are widely avail-able, inexpensive, and easy to train and deploy. Coupled with image recognition-based fingerprint detection technology, these scanning devices serve as an effective mechanism to identify fingerprints in large multiuser identity verification settings.

Test-taker information stored in a profile database (such as LMS in academic environments) is normally the starting point for exam authentication and launch of the exam sessions and storing of results. This is most commonly done using a Web Portal that the test-takers connect to from their Browsers. To bring the benefits of Biometrics to an academic environment, integration with such profile databases is necessary.

Integrating Aadhaar with student profile databases (such as Learning Management Systems) can be easily achieved via a cloud service. This architecture provides a seamless channel for large-scale dispensation of benefits of biometric authentication to the online exam environment.

3.1 Aadhaar Service Overview

Aadhaar identity management system was originally intended as a mechanism that can efficiently enable reliable large-scale delivery of various social services and entitlement programs. It has since become the largest biometric verification system in the world. Aadhaar service supports hundreds of millions of verification requests per day with sub-second latencies. Aadhaar exposes open services to help verify identity of the claimant using a combination of biometrics and demographic data. Aadhaar service is being used for fraud detection in near real time, in a variety of application domains.

The Aadhaar system architecture is shown in Fig. 1. The Authentication User Access Server (AUA) is an application-domain-specific server that receives requests for authentication from client agents and forward it to the Aadhaar Authentication Service (AAS). The authentication service returns a simple Boolean response to the identity verification request. The client agent supplies the biometrics data collected from the user along with the user information data to the AUA. The AUA responds with the Boolean answer received from the AAS verifying the claimant (whether "the user is who he/she claims to be").

3.2 System Architecture

Our system for continuous presence verification involves capturing test-taker fingerprints at random unpredictable intervals using a fingerprint scanning device. These fingerprints are processed by a cloud service. The cloud service acts as an AUA that can leverage Aadhaar service for biometrics verification. Further, it contains a broker component to read/write information from the profile database. This architecture allows seamless integration of fingerprint authentication to any standard online exam web portal. This architecture is shown in Fig. 2.

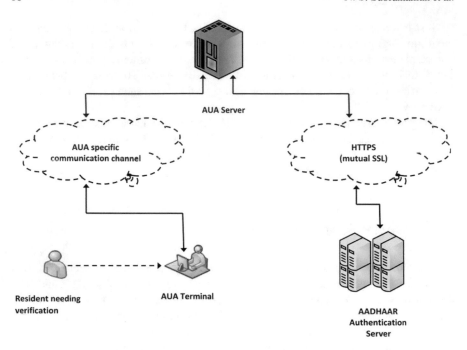

Fig. 1 Aadhaar system architecture

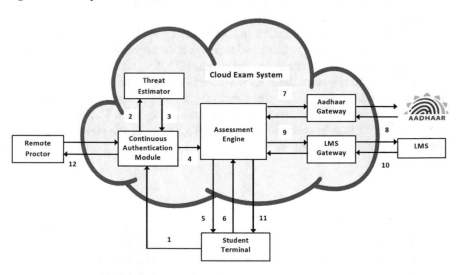

Fig. 2 Proposed system architecture

- Assessment Engine: Responsible for generating assessments that will be delivered to pre-authenticated and verified (by Aadhaar-LMS system) test-takers.
- Continuous Verification: Orchestrator for ensuring test-taker's continuous presence for the entire duration of the examination.
- Threat Estimator: Determines whether a re-verification request is necessary based on the client inputs.
- Aadhaar Gateway: Acts as the AUA middleware entity (ref. Fig. 1) responsible for managing the interaction with the Aadhaar cloud.
- LMS Gateway: Supplies the middleware necessary to connect to various test-taker profile databases (e.g., Moodle, A-VIEW [18], etc.).

The runtime interaction between the modules is as given below:

1. The video feed via webcam is fed into the continuous verification module by the client agent. Other parameters such as keystrokes or mouse movement pattern could also be used.
2. The Threat Estimator module builds a model of the test-taker's behavior obtained from the client agent. The threat estimator determines whether a re-verification request is needed. As explained before, in our current system, we have implemented this using a simple random interval to make the re-verification requests unpredictable.
3. When the continuous verification module determines re-verification is needed, it notifies the assessment engine to perform the actual verification step.
4. The user is prompted for a fingerprint scan by the client agent. In a proctored setting, we could use this step to alert the proctor to check the test-taker. The fingerprint scan method helps move toward a proctor-less examination solution.
5. The fingerprint data is passed back to the assessment engine by the client agent.
6. The assessment engine sends the obtained credential data to the Aadhaar Gateway acting as an AUA.
7. The Aadhaar Gateway makes a verification request to the Aadhaar Service (AAS). In the request, the test-taker's personal information obtained from the LMS is passed along with the biometric data collected for verification.
8. The return value is a Yes or No Boolean value which is passed back to the orchestrator.
9. Optionally, the assessment engine could notify the LMS about the re-verification and the results of re-verification.
10. The LMS system stores the supplied re-verification event as part of the test-taker's assessment history.
11. The client agent is notified of the result. If the re-verification request fails, and instructor/proctor alerting is issued for manual verification.

4 Conclusion

In this paper, we have proposed a continuous presence verification system that can be reliably implemented to guarantee high degree of examination integrity. By capturing test-taker biometrics at unpredictable intervals and validating them via the national Aadhaar service, we have shown that it is possible to reduce and eventually completely eliminate the dependency on remote proctor. Our system can be implemented using off-the-shelf components. We are currently working to validate our prototype in several university-level trials.

5 Future Work

Instead of using random unpredictable intervals, a threat model [6] could be built to decide when a re-verification is really necessary. This could help reduce distraction. Behavior of the system under miscalibrated biometric devices is an area that deserves closer attention. Offline mode that can work with partial unavailability of Aadhaar Service is another interesting area of study.

References

1. Perera-Diltz, D., Moe, J.: Formative and summative assessment in online education. J. Res. Innovat. Teach. **7**(1), 130–142 (2014)
2. Ardid, M., Gómez-Tejedor, J.A., Meseguer-Dueñas, J.M., Riera, J., Vidaurre, A.: Online exams for blended assessment. study of different application methodologies. Comput. Educat. **81**, 296–303 (2015)
3. Fask, A., Englander, F., Wang, Z.: Do online exams facilitate cheating? an experiment designed to separate possible cheating from the effect of the online test taking environment. J. Acad. Ethics **12**(2), 101–112 (2014)
4. Harmon, O.R., Lambrinos, J., Buffolino, J.: Assessment design and cheating risk in online instruction. Online J. Dist. Learn. Administ. **13**(3), n3 (2010)
5. Apampa, K.M., Wills, G., Argles, D.: User security issues in summative e-assessment security. Int. J. Digit. Soc. **1**(2), 1–13 (2010)
6. Al Solami, E., Boyd, C., Clark, A., Islam, A.K.: Continuous biometric authentication: Can it be more practical? In: 2010 12th IEEE International Conference on High Performance Computing and Communications (HPCC), pp. 647–652. IEEE (2010)
7. Flior, E., Kowalski, K.: Continuous biometric user authentication in online examinations. In: 2010 Seventh International Conference on Information Technology: New Generations (ITNG), pp. 488–492. IEEE (2010)
8. Aadhar Unique Identification Authority of India (uidai)
9. Varma, P.: Building an open identity platform for India. In: 2015 Asia-Pacific Software Engineering Conference (APSEC), pp. 3–3. IEEE (2015)
10. Gil, C., Díaz, G., Castro, M.: Fingerprint identification in lms and its empirical analysis of engineer students' views. In: 2010 IEEE Education Engineering (EDUCON), pp. 1729–1736. IEEE (2010)

11. Wildes, R.P.: Iris recognition: an emerging biometric technology. Proc. IEEE **85**(9), 1348–1363 (1997)
12. Kryterion Global Testing Solutions. http://www.kryteriononline.com
13. Clarke, N.L., Dowland, P., Furnell, S.M.: e-invigilator: a biometric-based supervision system for e-assessments. In: 2013 International Conference on Information Society (i-Society), pp. 238–242. IEEE (2013)
14. Prathish, S., Bijlani, K., et al.: An intelligent system for online exam monitoring. In: International Conference on Information Science (ICIS), pp. 138–143. IEEE (2016)
15. Krishnamoorthy, S., Soman, K.: Implementation and comparative study of image fusion algorithms. Int. J. Comput. Appl. **9**(2) (2010)
16. Gonzalez, N., Calot, E.P., Ierache, J.S.: A replication of two free text keystroke dynamics experiments under harsher conditions. In: 2016 International Conference of the Biometrics Special Interest Group (BIOSIG), pp. 1–6. IEEE (2016)
17. Asha, S., Chellappan, C.: Authentication of e-learners using multimodal biometric technology. In: International Symposium on Biometrics and Security Technologies, 2008. ISBAST 2008, pp. 1–6. IEEE (2008)
18. Subramanian, N.S., Anand, S., Bijlani, K.: Enhancing e-learning education with live interactive feedback system. In: Proceedings of the 2014 International Conference on Interdisciplinary Advances in Applied Computing. p. 53. ACM (2014)
19. Apampa, K.M., Wills, G., Argles, D.: An approach to presence verification in summative e-assessment security. In: 2010 International Conference on Information Society (i-Society), pp. 647–651. IEEE (2010)

Determining the Popularity of Political Parties Using Twitter Sentiment Analysis

Sujeet Sharma and Nisha P. Shetty

Abstract With the advancement in the Internet Technology, many people have started connecting to social networking websites and are using these microblogging websites to publically share their views on various issues such as politics, celebrity, or services like e-commerce. Twitter is one of those very popular microblogging website having 328 million of users around the world who posts 500 million of tweets per day to share their views. These tweets are rich source of opinionated User-Generated Content (UGC) that can be used for effective studies and can produce beneficial results. In this research, we have done Sentiment Analysis (SA) or Opinion Mining (OM) on user-generated tweets to get the reviews about major political parties and then used three algorithms, Support Vector Machine (SVM), Naïve Bayes Classifier, and k-Nearest Neighbor (k-NN), to determine the polarity of the tweet as positive, neutral, or negative, and finally based on these polarities we made a prediction of which party is likely to perform more better in the upcoming election.

Keywords Sentiment analysis · Opinion mining · Tokenization
Classification · Natural language processing (NLP)

1 Introduction

The rapid development in the Internet technology has also accelerated the usage of microblogging websites. There are a lot of such websites like Twitter, Facebook, Tumblr, etc. Using these services, people not only gets connected to their family and friends but also uses these services for sharing their views or opinion publically

S. Sharma (✉)
Sikkim Manipal Institute of Technology, Majitar, East Sikkim 737136, India
e-mail: sujeetsharma1107@gmail.com

N. P. Shetty
Manipal Institute of Technology, Manipal University, Manipal 576104, India
e-mail: nisha.pshetty@manipal.edu

© Springer Nature Singapore Pte Ltd. 2018
S. C. Satapathy et al. (eds.), *Information and Decision Sciences*,
Advances in Intelligent Systems and Computing 701,
https://doi.org/10.1007/978-981-10-7563-6_3

about various issues like any product, celebrity, or politics. During the election time, these microblogging websites get full of comments and reviews about the political parties and their leaders. Such reviews or comments are known as User-Generated Content (UGC) [1].

These UGCs are rich source of opinionated data and performing opinion mining on it can provide some beneficial results. "Politics using social media" [2] is now becoming a popular research area. Opinion mining is a kind of Natural Language Processing (NLP) used for identifying the mood of the public about a particular person, product, or service. In this research, opinion mining is performed on the dataset which is downloaded from one of the popular microblogging website Twitter [3] to compare the accuracy of classifiers in determining the popularity of various political parties by finding the solution to the following mentioned problems:

1. Classification of the downloaded tweets based on which political party or leader they belong to.
2. Determining the polarity of the tweets as positive, neutral, or negative.
3. Conclusion based on the overall polarities of the tweets for individual parties or leaders.

For example, "So proud of Corbyn, thanks for representing us. What a man-I#election#myvote4labour". Here, "Corbyn" is the leader of the Labour party in the United Kingdom General Election 2017, "proud of Corbyn" and "myvote4labour" make this tweet's polarity as positive.

The opinionated words (usually adjectives) and important hashtags like (#myvote4labour) are first extracted and classified as positive, negative, and neutral using the appropriate library and packages. Based on that, the polarity of the whole tweet is determined and finally summarization of all the tweets can be used for decision-making of which party or leader is more likely to perform better in the upcoming election. Opinion mining in this research is performed using R Programming Language [4].

The sections in this paper are categorized in the following manner. Section 1 describes the need and advantages of the opinion mining or sentiment analysis. Section 2 describes the related work performed by distinguished researchers in the field of opinion mining and sentiment analysis. Section 3 describes the proposed methodology and workflow. Section 4 describes the obtained results. The last section describes the comparison and conclusion of the obtained results along with the future work.

2 Related Work

Shengyi et al. [5] have introduced an improved version of k-NN algorithm named INNTC for text categorization. They have used one-pass clustering algorithm with approximately linear time complexity which is an incremental clustering algorithm capable of handling large and high-dimensional data. After forming the clusters, they have used the k-NN classifier for the text categorization. And finally, they compared the accuracy and showed that their algorithm performs better than SVM in many of the datasets.

Karim et al. [6] have worked in two very popular data mining techniques: the Naïve Bayes and the C4.5 decision tree algorithms. They have used publically available dataset UCI data for training and testing the models in order to predict the chances of whether a client will be subscribing to a term deposit or not. They got the accuracy of 93.96% for C4.5 decision tree and 84.91% for Naïve Bayes classifiers. Later in their work, they also used decision tree for extraction of interesting and important decision in business area.

Hsu et al. [7] have worked on the SVM classifier and showed when to use linear kernel and when to use RBF kernel. They proved that if the number of features is small, then data can be mapped to higher dimensional spaces by using nonlinear kernel. But in case of large number of features, one should not map data to a higher dimensional space as nonlinear mapping will not enhance the performance. So in case of large number of features, the linear kernel can perform good enough. They also proved that RBF kernel is at least as good as linear kernel (which holds true only after searching the (C, γ) space, where C and γ are two parameters for an RBF kernel.)

Kousar Nikhath et al. [8] have implemented an email classification application based on text categorization, using k-NN classification algorithm and achieved the accuracy of 74.44%. They implemented the application using two processes: training process and classification process, where the training process uses a previously categorized set of documents (training set) to train the system to understand what each category looks like. And then the classifier uses the training model to classify new incoming documents. They also used Euclidean distance as a similarity function for measuring the difference or similarity between two instances.

In [9], Alexandra Balahur and Marco Turchi have first implemented a simple sentiment analysis system for English tweets and later they extended it to multilingual platform. With the help of Google machine translation system, they translated the training data to four different languages—French, German, Italian, and Spanish. And finally they proved how overall accuracy of sentiment analysis can be improved from 64.75% (in English) to 69.09% (in all 5 languages) by the jointly using the training data from different languages.

Dilara et al. [10] have performed sentiment classification on Twitter Sentiment 140 datasets and proved that using their method Naïve Bayes can outperform the accuracy of SVM classifier also. Since Naïve Bayes has a disadvantage, whenever it is applied to high-dimensional data, i.e., for text classification it suffers from sparsity, so they deduced the smoothing as the solution of this problem. He used Laplace Smoothing in his work.

In [11], Minqing Hu and Bing Liu have proposed a technique which is based on Natural Language Processing (NLP) and Data Mining method for mining and summarizing the product reviews. They have first identified the features of the product mentioned by the users and then identified the opinionated sentence in each of the reviews present and classified it as positive or negative opinions and finally produced a conclusion from the determined information.

Chien-Liang et al. [12] have designed and developed a movie rating and review summarization system in a mobile environment, where movie rating is based on the sentiment classification results. They used feature-based summarization for the generation of condensed descriptions of the movie review. They proposed a novel approach to identify the product features which is based on latent semantic analysis (LSA). In addition to the classification's accuracy, they also considered system response time in their system design. Their work also has an added advantage where users, based on their interest, can choose the features; this module employees an LSA-based filtering mechanism, which could efficiently minimize the size of summary.

3   Methodology

Figure 1 represents the flow of data through various modules used in the proposed methodology.

3.1   Collection of Corpus

The datasets are directly downloaded from the microblogging website Twitter using the Twitter API and twitteR [13] package in R. Since we are considering the

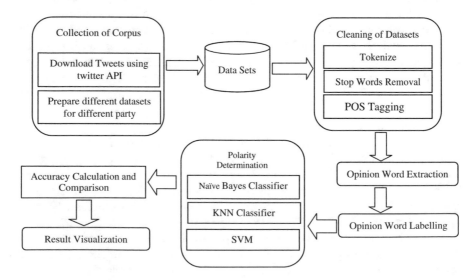

Fig. 1 Flow chart for the proposed methodology

"General Election 2017" and tweets about this election generally contains hashtags as "#GE2017" or "GeneralElection17", we have first downloaded the tweets having these hashtags. After this, the tweets are divided into different datasets based on which political party belongs to.

For example, the tweets about "Labour" party whose leader is "Jeremy Corbyn" contain words like "labour" or "corbyn", so tweets with such keywords are placed in Labour's dataset.

Then, from each of the datasets, 240 tweets are randomly selected which consists of all positive, negatives, and neutral tweets upon which opinion mining is to be performed. This method of corpus collection can be used for collecting any kind of datasets in many different languages because twitter API allows to select the languages of the tweets to be collected.

3.2 Cleaning of Datasets

The collected dataset contains a lot of ambiguous words, so as to remove those words some cleaning processes need to be done on the collected datasets. All the blank spaces and punctuation marks are removed first, then tokenization is performed in order to convert all the words into individual tokens. The next step is to remove stop words; these are most commonly used words in a language and cannot help in polarity determination. Then, the final step is to apply sentence token annotations and word token annotations to the datasets.

3.3 Collection of Opinionated Words

Extraction of opinionated words from the cleaned datasets is then performed. This step involves two process: first is to do POS tagging and second is to extract the hashtags present. POS tagging refers to assigning each of the tokens present in a corpus with its corresponding parts of speech (nouns, pronouns, adjectives, etc.). Polarity of the sentences is mostly determined by the adjective words; therefore, after performing POS tagging we can directly get the opinionated words [14]. But in the case of election, hashtags play more important role than adjective words; here public usually uses hashtags like "#myvote4conservative" to express positive feelings toward "Conservative" party, and "#toriesout" for expressing negative feeling for the same party. So along with adjective words, hashtags are also extracted to be used as opinionated words.

3.4 Labeling of Opinionated Words

The collected words are then checked upon the dictionary comprising of two files positive words and negative words, and based on that a polarity is assigned to the tweet. If there is a positive word or hashtag in the tweet, it will be assigned as +1 and for each of the negative words, −1 is assigned.

For example, "Corbyn will win#myvote4corbyn" in this tweet "win" and "myvote4corbyn" are found in positive words and hence 1 + 1 − 2, so +2 will be the score of this, and then based on the score is positive or negative, the polarity will be assigned as +1 or −1, respectively. For the words not found in dictionary, the score of 0 is assigned and tweets with score 0 will be considered as neutral with polarity as 0.

The dictionary used can be modified as per the user requirements, as in different elections and for different political parties, people use different hashtags. So accordingly user can add the new hashtags with the polarity in the dictionary.

3.5 Classification of the Tweets

Classification algorithm will be applied to the datasets so that the model can be trained on the training datasets and can be used for predicting the future trends of the data. There are many classification algorithms available; out of them, we are considering three algorithms in this paper, namely, Support Vector Machine (SVM) [15], Naïve Bayes Classifiers, and k-Nearest Neighbors (k-NN) classifiers.

4 Experiment Results

There are six political parties participating in the UK General Election 2017, out which there are only two major parties which play significant role in the election. So in this work only those two political parties, "Labour" and "Conservative", are considered. Datasets of 240 tweets comprising all positive, neutral, and negative tweets for both of the parties are used for Opinion Mining.

4.1 Experimental Steps

1. Download the datasets for the required elections.
2. Two datasets are prepared, one for the "Labour" party and the another for the "Conservative" party.

Table 1 Classifiers with their obtained accuracy for both political parties' datasets

Classifiers	Political parties		Average accuracy (%)
	Conservative party (%)	Labour party (%)	
Naïve Bayes	64.58	61.46	63.02
k-NN	92.71	91.67	92.19
SVM	72.92	75	73.96

3. Read the datasets separately.
4. Perform the cleaning of the datasets.
5. Polarity assignment is done.
6. 60% of the datasets are considered as training sets and remaining 40% as test sets are prepared.
7. Three different classifiers are used for classification of the tweets, i.e., Naïve Bayes Algorithm [16], SVM (Support Vector Machine) [17], and k-NN (k-Nearest Neighbors) [18].
8. Calculation of accuracy for each of the classifiers.
9. Comparison of the performance of the above-stated algorithms and visualization.
10. Comparison of the performance of two parties and visualization for the prediction.

By analyzing the experimental results as shown in Table 1, it is observed that performance of k-NN classifier is more accurate than the other two classifiers. Here, k-NN is giving more accuracy as it is used similarly to the INNTC [5], where clustering is performed by collecting just the adjective and hashtags of the tweets from the training sets, which reduces the test similarity computation to a great extent and therefore reduces the impact on performance which is affected by single training sample (noisy sample) and then the k-NN algorithm is applied. Graphical representation of the same is shown in Figs. 2 and 3.

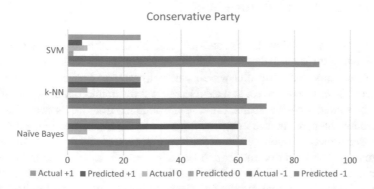

Fig. 2 Comparison of classifiers on Conservative party datasets

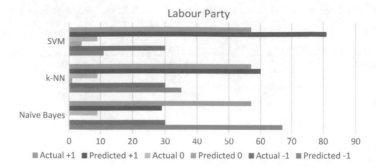

Fig. 3 Comparison of classifiers on Labour party datasets

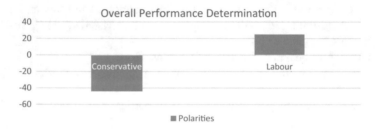

Fig. 4 Overall performance determination

The graph below is drawn for determining the future performance of both of the political parties in the upcoming elections using k-NN classifiers.

Considering the above-drawn graphs, in Fig. 4, it can be easily predicted that Conservative party's performance is likely to be decreased in the next election, whereas Labour party is predicted to perform better in the next election.

5 Conclusion and Future Work

This proposed work shows how datasets can be directly downloaded from the microblogging websites and opinion mining or sentiment analysis can be performed on it to get some useful results which can help later in decision-making. Three classifiers SVM, Naïve Bayes, and k-NN are used for opinion mining out of which k-NN gives most accurate results. Later result obtained from k-NN classifier is used to determine the performance of political parties in the future election. Future work can be performed to improve the accuracy of the predictions based on previous performance. The proposed model can be used for both election campaigning and performance prediction of the political parties. Also, it can be used by common people to know the present reputation of any political party or any politician.

References

1. Yadav, S.K.: Sentiment analysis and classification: a survey. Int. J. Advanc. Res. Comput. Sci. Manag. Studies, **3**(3) (2015)
2. Taimur, I., Ataur, R.B., Tanzila, R., Mohammad, S.U.: Filtering political sentiment in social media from textual information. In: 5th International Conference on Informatics, Electronics and Vision (2016)
3. Alexander, P., Patrick, P.: Twitter as a corpus for sentiment analysis and opinion mining. In: International Conference on Language Resources and Evaluation (2010)
4. Akhil Kumar, K.V., Manikanth Sai, G.V., Shetty, N.P., Chetana, P., Aishwarya, B.: Aspect based sentiment analysis using R programming. In: Proceedings of Fourth International Conference on Emerging Research in Computing, Information, Communication and Applications (ERCICA-2016)
5. Shengyi, J., Guansong, P., Meiling, W., Limin, K.: An improved K-nearest-neighbour algorithm for text categorization. Exp. Syst. Appl. **391**(3) (2012)
6. Karim, M., Rahman, R.M.: Decision tree and Naïve Bayes algorithm for classification and generation of actionable knowledge for direct marketing. J. Softw. Eng. Appl. **6**, 196–206 (2013)
7. Hsu, C.W., Chang, C.C., Lin, C.J.: A Practical Guide to Support Vector Classification. www.csie.ntu.edu.tw/~cjlin/papers/guide/guide.pdf
8. Kousar Nikhath, A., Subrahmanyam, K., Vasavi, R.: Building a K-nearest neighbour classifier for text categorization. Int. J. Comput. Sci. Informat. Technol. **7**(1), 254–256 (2016)
9. Alexandra, B., Marco, T.: Improving sentiment analysis in twitter using multilingual machine translated data. In: Recent Advances in Natural Language Processing (2013)
10. Dilara, T., Gurkan, T., Ozgun Sagturk, Ganiz, M.C.: Wikipedia based semantic smoothing for Twitter sentiment classification. IEEE (2013)
11. Minqing, H., Bing, L.: Mining and summarizing customer reviews. In: Proceedings of the Tenth ACM SIGKDD International Conference on Knowledge Discovery and Data Mining, KDD'04 pp. 168–177, ACM New York (2004)
12. Chien-Liang, L., Wen-Hoar, H., Chia-Hoang, L., Gen-Chi, L., Emery, J.: Movie rating and review summarization in mobile environment. IEEE Trans. Syst. Man Cybernet. Part C Appl. Rev. **42** (2012)
13. twitteR Package. https://cran.r-project.org/web/packages/twitterR/twitteR.pdf
14. Pujari, C., Aiswarya, Shetty, N.P.: Comparison of classification techniques for feature oriented sentiment analysis of product review data. In: Data Engineering and Intelligent Computing, pp 149–158 (2017)
15. Vladimir, N.V.: The Nature of Statistical Learning Theory. Springer, New York (1995)
16. Naïve Bayes. https://cran.r-project.org/web/packages/naivebayes/naivebayes.pdf
17. David, M.: Support Vector Machines: The Interface to libsvm in package e1071 (2017)
18. Package Class. https://cran.r-project.org/web/packages/class/class.pdf

Analysis of Passenger Flow Prediction of Transit Buses Along a Route Based on Time Series

Reshma Gummadi and Sreenivasa Reddy Edara

Abstract India's transport sector has a prominent role in transportation of passengers. Andhra Pradesh State Road Transport Corporation (APSRTC) is a major transportation in the state. State has many routes, and there are so many towns on a particular route. Most of the population mainly depends on transportation system; hence, it is necessary to predict the occupancy percentage of the transit buses in a given particular period for the convenience of passengers. There is a need of advancement in transportation services for effective maintainability. Identifying passenger occupancies on a different number of buses is found to be a major problem. A promising approach is the technique of forecasting the data from previous history and the better predictive mining technology must be applied to analyze the passenger to predict the passenger flow. In this work, ARIMA-based method is analyzed for studying the APSRTC transit bus occupancy rate.

Keywords Transit buses · Passenger flow · ARIMA

1 Introduction

1.1 Overview

Andhra Pradesh State Road Transport Corporation (APSRTC) recently took an initiative for implementation of information technology in the state of Andhra Pradesh, and the effective use of IT helps in many ways for APSRTC. It helps in providing better services to passengers and also leads to effective maintenance

R. Gummadi (✉)
Acharya Nagarjuna University, Guntur 522510
Andhra Pradesh, India
e-mail: reshma.gorripati@gmail.com

S. R. Edara
Department of Computer Science & Engineering,
Acharya Nagarjuna University, Guntur 522510, Andhra Pradesh, India

© Springer Nature Singapore Pte Ltd. 2018
S. C. Satapathy et al. (eds.), *Information and Decision Sciences*,
Advances in Intelligent Systems and Computing 701,
https://doi.org/10.1007/978-981-10-7563-6_4

31

management of vehicles. The role of Information Technology plays a key role in better inventory control and also better managerial controls. In view of that, Ticket Issuing Machines (TIMs) were introduced in APSRTC in May 2000. The main objective is to issue tickets even though ground booking is completed and also to pick up number of passengers in a route. These are introduced in a view that management can derive information like punctuality analysis and travel patterns of the public, generating MIS reports from the database. At present, there are 14,500 TIMs being utilized.

The data collected using TIMs from passengers includes starting stage, ending stage, ticket time, ticket percentage, number of adults, and number of children. The digitization of information helps in so many number of ways to increase the bus services. As the data is already collected, and this data is helpful for further investigations and forecasting the information can be done. By analyzing the data, we can predict the passenger occupancies which are helpful for estimating the flow of passengers at critical period of time also. The predictive analysis must be done by considering all the cases like festivals, weekdays, and normal days. As ticket time is also collected, the estimation of the peak times and proportion of passengers can be done accordingly.

1.2   Objectives

The objective of this paper is to analyze various ARIMA models to implement the estimation of the passenger flow at APSRTC. Based on the data which is collected from APSRTC Head Office, the passenger flow was analyzed with findings and suggestions. In this paper, the modeling methods for passenger occupancies were discussed and the analysis of modeling techniques on real time data is presented. There are so many model-based prediction algorithms like Kalman filtering, regression models, and artificial neural networks. As time plays a key role in this application, an ARIMA analysis model can be used to predict the passenger flow over a period of time. SARIMA model is analyzed as there is seasonality in monthly data for which the high values may tend to occur in few months, and low values may tend to occur in some other months. In this scenario, we consider $S = 12$ is the span of the periodic seasonal behavior. The model can be selected in such a way that total number of passengers to be predicted for a particular month in a year is based on previous year data.

2   Data Source

This paper examined novel methods to estimate the passenger occupancies over a period of time. The model dataset used in this paper based on the data which was the primary source is provided by APSRTC on transit buses information.

2.1　Transit Data

APSRTC provides transport services within AP. Among those services, the transit information of Macherla to Chilakaluripet route ticket-wise data of 24 services extracted from three depot databases for the period of April 1, 2016 to December 31, 2016 was given and the dataset snapshot is displayed below.

The dataset consists of various measures and dimensions but for analysis purpose, we choose ETMDate, FromStagecode, ToStageCode as dimensions, and NoOfAdults considered to be a measure.

From Macherla to Chilakaluripet, there are 20 intermediate stations and each station is identified by bus number which is a unique id as shown in Fig. 1. The data given in excel table consists of Date, Trip No, Ticket percentage, Start Stage, End Stage, No of adults, Number of Childs, and corresponding Ticket time given. Among those fields, Date, Start Stage, End stage, and total number of passengers are to be considered for analysis.

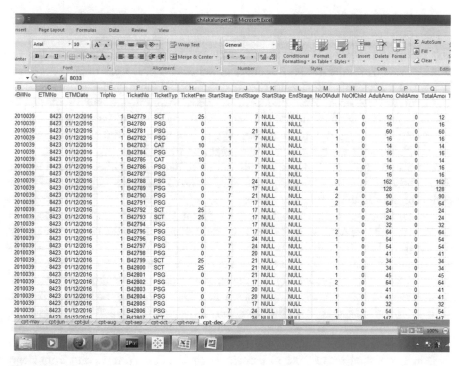

Fig. 1 Snapshot of data given by APSRTC

2.2 Model Evaluation

To analyze the characteristics of historical data, time series plays a prominent role. In the dataset provided by APSRTC, they have given data with minute duration. Time series is chosen because data is available on daily basis. Here, data is in probabilistic in nature. It is practically impossible for 100% accuracy. Let us define $d^n(t)$ the demand of the passengers at time interval $(t - 1)$ for day n time series $T^n_m(t)$ consists of data n_m time intervals before $d^n(t)$ on the same day. As data is available in minutes, that should be grouped into day-wise and finally to monthly data. In this case, the previous time intervals to be considered with differences measured at each interval.

$$T^n_d(t) = \{d^n(t-1),\ d^n(t-2),\ \ldots d^n(t-n_m)\}$$

Seasonality can be defined in this case as number of passenger's data in months.

2.3 Analyzing Bus Schedule

The model is complex as there is difficulty in planning route definition. Let us consider a service operates with number of scheduled buses per minute duration, i.e., varies from 1 min duration to 15 min duration and these type can be treated as low-frequency services. High-frequency services are more number of buses per hour. For the performance measures to be accurate, reliability of bus location data is the essential starting point. Kalman filtering can be applied for the number of inputs from different sources to establish the connection of the bus.

The starting time of all buses that are in one route segment checks the previous time schedules and then all buses that have covered the same lag over the same period of time in the previous intervals.

For example, prediction is being calculated on 1 day in a month service. The figures for the previous minutes on that route segment would be compared by removing outliers from the dataset, and forecasting the passengers is derived and added to the current number of passengers to provide total number of passengers for that segment. This is potentially calculated with each bus location for each bus stop. Trip-wise pattern analysis checks whether the current trip has similar pattern as that of previous patterns; on the same day, Z-test can be applied in this case and by identifying positive skewness and negative skewness, it would give results for pattern analysis.

Graph is plotted for number of passengers that represents rows and from stage represents columns as shown in Fig. 2. The data over a particular route from each stage is plotted. Number of passengers observed is increasing over particular stages and decreasing over some stages. Similarly, after analyzing 42000 records,

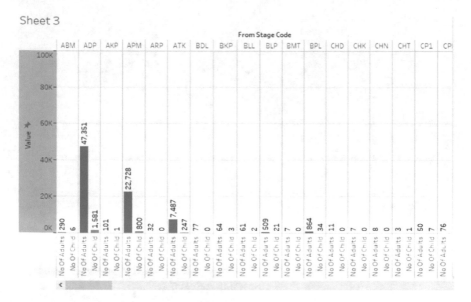

Fig. 2 Number of passengers on a particular stage

predicting the number of passengers along route from start stage to end stage for every month is to be done.

The data can be extracted month-wise, and for each month the amount can be calculated. In a similar way, the number of passengers can be calculated month-wise which is shown in Fig. 3. Once the data is plotted in monthly manner, ARIMA modeling can be applied to predict the passenger flow for each month in the next year. Hence, from next year number of buses to be increased or decreased can be known by forecasting the number of passengers.

2.4 ARIMA Modeling

ARIMA modeling can be applied to the available data. At the initial step, identify whether the variable which is to be predicted is stationary in time series or not; if the variable is not stationary, it should be made stationary using the following process. Let x_t denote the number of passengers at time t; the proposed general ARIMA formulation is as below.

Let $\{x_t | t \hat{I} T\}$ denote a time series such that $\{w_t | t \hat{I} T\}$ is an **ARIMA (p, q)** time series where $w_t = D^d x_t = (I - B)^d x_t = $ the *dth* order differences of the series x_t. Then, $\{x_t | t \hat{I} T\}$ is called an **ARIMA (p, d, q)** time series (an **integrated autoregressive moving average** time series); here, B denotes the backshift operator which is used

Sheet 4

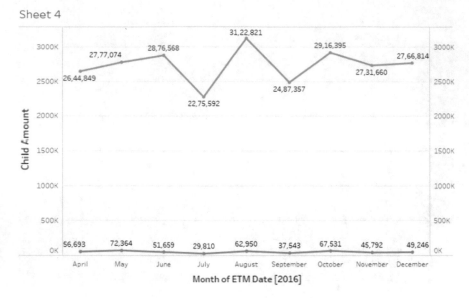

Fig. 3 Monthly report of passengers

to forecast the values based on previous values; p denotes the autoregressive parameters, d is the number of differencing passes, whereas q represents the moving average which was computed for the series after it was difference once. Input series of ARIMA must have a constant mean, variance, and autocorrelation through time.

The equation for the time series $\{w_t | t \hat{I} T\}$ is $b(B)w_t = d + a(B)u_t$. Suppose that d roots of the polynomial $f(x)$ are equal to unity, then f(x) can be written as $f(B) = (1 - b_1 x - b_2 x^2 - \cdots - b_p x^p)(1 - x)^d$, and f(B) could be written as $f(B) = (I - b_1 B - b_2 B^2 - \cdots - b_p B^p)(I - B)^d = b(B)D^d$. In this case, the equation for the time series becomes $f(B)x_t = d + a(B)u_t$. In **ARIMA (1,1,1)** $x_t = (1 + b_1)x_{t-1} - b_1 x_{t-2} + d + u_t + a_1 u_{t-1}$. If a time series is $\{x_t: t \hat{I} T\}$, that is, seasonal we would expect observations in the same season in adjacent days to have a higher autocorrelation than observations that are close in time (but in different seasons in this case seasonality can be defined as month estimated for 1 year). This model satisfies the equation incorporating seasonality.

$$(I - B^k)^{ds}(I - B)^d \beta^{(s)}(B)\beta(B)x_t = \alpha^{(s)}(B)\alpha(B)u_t + \delta$$

Forecasting an ARIMA (p, d, q), time series can be done by the following process:

$$x_T(l) = E(x_{T+l}|P_T)$$

Let P_T denote $\{\ldots, x_{T-2}, x_{T-1}, x_T\}$ = the "past" until time T. Then, the optimal forecast of x_{T+l} given P_T is denoted by this forecast that minimizes the mean square error.

For the given data, there is a need of calculating the first difference, log first difference, seasonal, log seasonal, and seasonal first difference which apply SARIMA modeling with order SARIMA (0,1,0) * (0,1,1,12) and SARIMA (0,1,0) * (2,1,1,12) from which state-space model results can be obtained. The goodness of fitted value can be known by observing AIC value. The modeling is selected for lower AIC value. At the next step, the predicted values can be obtained by applying the above-selected model and calculates the new values of the series.

3 Conclusion

This paper has mainly focused the analysis of passenger flow prediction of transit buses based on time series. APSRTC dataset was used for analysis. ARIMA modeling is analyzed. Based on the investigated simulation and analysis report, comprehensive study was done. This information helps in improving the service efficiency and reduction of waiting time of passengers due to lack of buses when there is more passenger flow. The analysis carried out in this work will benefit the public transport authority in helping them to plan accordingly to place number of buses over a particular time span to reduce the overcrowding or under crowding. As a future research, other forecasting methods available in the literatures can be experimented.

References

1. Al-Deek, H., D'Angelo, M., Wang, M.: Travel time prediction with non-linear time series. In: Proceedings of the ASCE 1998 5th International Conference on Applications of Advanced Technologies in Transportation, Newport Beach, CA, pp. 317–324 (1998)
2. Chien, S.I.J., Kuchipudi, C.M.: Dynamic travel time prediction with real-time and historic data. J. Transport. Eng. **129**(6), 608–616 (2003)
3. Bin, Y., Zhongzhen, Y., Baozhen, Y.: Bus arrival time prediction using support vector machines. J. Intell. Transport. Syst. **10**(4), 151–158 (2006)
4. Chen, S. et al.: The Time Series Forecasting: From the Aspect of Network. arXiv:1403.1713 (2014)
5. Vagropoulos, S.I., Chouliaras, G.I., Kardakos, E.G., Simoglou, C.K., Bakirtzis, A.G.: Comparison of SARIMAX, SARIMA, Modified SARIMA and ANN-based Models for short-term PV generation forecasting. In: IEEE International Energy Conference, Leuven pp. 1–6 (2016)
6. Takaomi, H., Takashi, K., Masanao, O., Shingo, M.: Time series prediction using DBN and ARIMA. In: International Conference on Computer Application Technologies, IEEE (2015)

Smart Fire Safety: Serious Game for Fire Safety Awareness

George Jacob, R. Jayakrishnan and Kamal Bijlani

Abstract Serious game is used as an educational tool to teach about some complex actions in the real world. Serious game helps to visualize different situations which the learner has not experienced in real life. Fire accidents are occurring generally due to careless handling of flammable substances. From the analysis of experiments, we understood that most of the people did not get any training on fire and safety. The objective of this study is to give proper guidance and training to handle fire. This can be implemented through a serious game with some alert messages. This paper proposes to create awareness about fire risks that can happen in different scenarios through a serious game. This game focuses on common people. This game will provide an experience to the player.

Keywords Serious game · Game-based learning · Fire risks

1 Introduction

There are three elements that lead to a fire which are fuel source, presence of oxygen, and required heat to ignite. These elements are together in a triangle called fire triangle [1]. Fire risk assessment should be done to identify fire hazards and to know precautions to be taken against fire.

Game-based learning is a pedagogical method that can be used as an effective tool to teach difficult concepts through virtually developed environment. Educational games are mostly used for training purposes because games can increase

G. Jacob (✉) · R. Jayakrishnan · K. Bijlani
Amrita e-Learning Research Lab (AERL), Amrita School of Engineering,
Amritapuri, Amrita Vishwa Vidyapeetham, Amrita University, Kollam, India
e-mail: georgekunchattil@gmail.com

R. Jayakrishnan
e-mail: jayakrishnanr88@gmail.com

K. Bijlani
e-mail: kamalb@am.amrita.edu

© Springer Nature Singapore Pte Ltd. 2018
S. C. Satapathy et al. (eds.), *Information and Decision Sciences*,
Advances in Intelligent Systems and Computing 701,
https://doi.org/10.1007/978-981-10-7563-6_5

motivation and engagement of the learner [2]. So it becomes an alternate method to traditional learning [3]. Well-designed serious games [4] provide the user with higher levels of immersion and experiences. According to NFPA report [5] and Wildfire Activity Statistics Report [6], fires cause huge damage to buildings and loss of many lives of people every year. This is occurring due to improper training about handling fire [7]. This game helps to give proper training to understand fire risks and prevention of fire accidents. Educational games [8, 9] are used to design concept which helps to visualize the topics very easily. We introduce a serious game based on fire and safety [10, 11] which is developed in Unity 3D.

2 Related Works

An important research has been done in the area of various learning domains. Soler et al. [12] assessed the impact of video game-based learning on the basis of academic performance. The main goal of this paper is to evaluate the performance of students through game-based learning. Jose et al. did research on students' performance. Senderek et al. [13] evaluated the impact of game-based learning on vocational and corporate training. Kuo [14] assessed motivation of learner by means of surveys and tests. After analysis of learners, author understood that the level of user engagement has increased after playing the game. Based on the study, the authors recommended the use of the serious game. The use of game-based learning methods helps to teach vocational subjects which help to increase efficiency and reduce costs.

Blaze [15] is a serious game which is to improve household fire safety awareness. This game is developed using Unity Pro game development tool. Many of the fire safety methods are not handled by people due to lack of "hands-on" training. So serious games allow users to experience fire risks and situations that are difficult to achieve in reality. This game is more specific on how to perform actions in the event of a kitchen or cooking fire. Score is rewarded at the end of each quiz.

EVA [16] is a serious game which is designed as a training tool for healthcare professionals. This game is used to train and evaluate healthcare personnel's behavior in hazardous situations such as fire. This game is developed using Unity3D. EVA is a game focusing mainly on evacuation training. Models of the game are developed by Autodesk. The game uses an FPS support. The models in the game were developed by Autodesk. It contains map to guide the player. If player moves to wrong place, an alert message is displayed. Player's actions are saved in a log data file.

The game [17] is a 2D game which is developed using NeoAxis game engine. The objective of the game is to acquire personal fire safety skills. Code of the game is developed by using .NET code scripts. Model for the game is developed using 3D Studio Max. The game is organized into different levels based on difficulty. By playing through each level, the player can acquire knowledge about building and what actions to be performed in different situations. A score is awarded based on task time completion and player's actions in the game.

The game [18] is a 3D game which is developed by using Unity. The first objective is to gain situational awareness through the completion of a Rapid Damage Assessment. Through this process the players learn the elements of an R.D.A. The second objective is to teach the players how to use an incident management system framework in a post disaster environment. The third objective is to highlight the importance and need for unified interagency communication and collaboration. Code of the game is developed by using C#.NET, Node.js and WebRTC API.

Fire Escape [19] game is 2D game which is designed for kids. This game helps to teach about fire risks in a kitchen [20], how to extinguish a fire, and what actions to be performed when the fire happens. When kitchen fire happens, a player needs to shout to alert people about fire. Game also teaches about fire escape plan. Feedback is given based on player's actions. Help Mikey [21] is a 2D game designed for kids. The objective of the game is to teach about the importance of smoke alarms, what actions to be performed when a fire happens, and how to escape from fire. Player should get feedback based on task selected by the player.

3 Smart Fire Safety Serious Game

3.1 Game Description

The name of the game is "Smart Fire Safety" which is a 3D game developed using Unity3D. The game is played by a single player. It includes some options namely Instructions, Play, and Quit. Instructions are the first option in home screen of the game. Instructions include the overview and description of the game. It also includes information about player controls. Arrow keys in the keyboard are used for movements, and mouse is used to rotate the screen to view the environment of the game. When the player clicks Play button in home screen of the game, the player can see two situations such as Kitchen and Petrol Pump. The first situation mainly deals with fire risks in "kitchen" which create awareness about substances that can lead to fire accidents. This scenario teaches by giving an alert message when a player comes near substances that are likely to catch fire. The first object is "Lighted Candle". When player moves toward the candle, we can see an alert message that says that the candle should not be placed near the carton box. If ignited candle falls on the carton box, it can catch fire. Later on, fire from the cardboard can spread to the wooden table.

The second object is towel. If the player comes near the towel, the player can see an alert message "There is a towel which is near a stove. If you ignite stove and place a towel nearby it, it may leads to the fire. This fire can also occur due to a gas leakage in the LPG gas cylinder". Player gets an idea that towel should not be placed on the stove when the stove is ignited. The third object is a gas cylinder. When the player comes near gas cylinder, the player can see an alert message "Check the proper connection of a cable to gas cylinder and stove. Check if there is any leakage in gas cylinder". The player can get awareness that cable should be

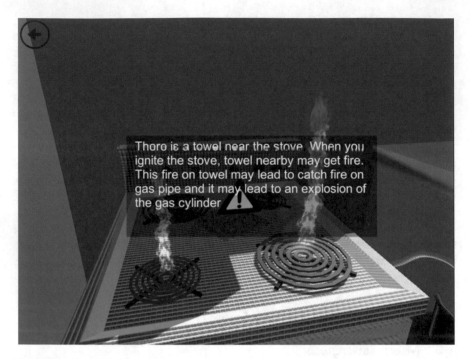

Fig. 1 Towel alert in kitchen

properly connected between stove and gas cylinder. If the cable is not properly connected, there may be a chance of gas leakage which can lead to fire explosion in the kitchen. When the player comes near the gas cylinder, an alert message is shown. The alert message is newspaper should not be placed near the stove. If stove is ignited and newspaper is placed near it, the newspaper can catch fire from the stove (Fig. 1).

The second situation mainly deals with fire risks in "Petrol Pump" which gives an awareness about fire-causing substances. This situation teaches by giving some alert messages when the player comes near fire-causing substances. The first object is petrol pump cable. When the player comes near the petrol pump cable, the player can see an alert message "Check for petrol leakage. Check that the tube is properly connected while transferring petrol from petrol truck to petrol tank". The player can get awareness that cable should be properly connected between petrol truck and petrol tank while transferring petrol. If the cable is not properly connected, there may be an occurrence of petrol leakage. This petrol leakage may lead to fire explosion. The second object is a mobile phone. In petrol pump, the player can see a person who is using a mobile phone. When the player moves near a person who is using a mobile phone, the player can see an alert message "Don't use the mobile phone in petrol pump. Defective battery in mobile phone creates a spark on mobile due to overheating. This leads to fire on petrol pump and mobile phone may explode." (Figs. 2 and 3).

Fig. 2 Gas leakage alert in petrol pump

3.2 Technical Details

The game is developed in Unity3D game engine. The 3D models used in the game were obtained from the internet. Some textures and graphics were developed in graphic editors like Adobe Photoshop. C Sharp and Javascript were scripts used to program the game. Microsoft Visual Studio using a plugin which is integrated with Unity was used as editor for creating coding for the game.

3.3 Game Design

The idea is basically to create a serious game in which player can understand about precautions and safety methods to be followed. This idea is demonstrated in kitchen situation and petrol pump situation. Learner as a player will not be having much stress in learning when observing and understanding the different situations in the game. The merits of being a game are plenty (Fig. 4).

Fig. 3 Mobile alert in petrol pump

4 Results and Analysis

An experimental study was conducted on different age groups for exploring the effectiveness of game-based learning over traditional learning. This experiment is conducted among 46 people.

Figure 5 explains that 6.5% have only got experience in fire and safety training, and 93.5% have not get any experience in fire and safety. 63% have not get any fire and safety training through the serious game. So, the game is an effective tool for people to get training about fire and safety. A quiz is also conducted among the people. There are two assessments conducted for the game such as pre-assessment and post-assessment (Figs. 6 and 7).

After analysis, performance of people is increased after playing the game. So, we can say that game-based learning is an effective tool than traditional-based learning.

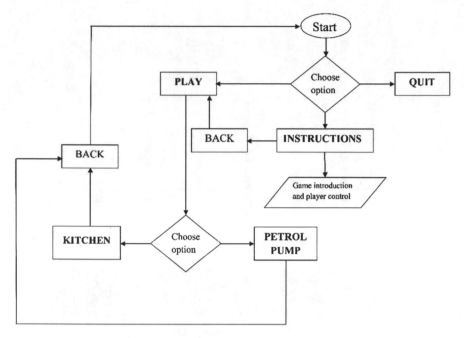

Fig. 4 Flowchart of smart fire safety game

Fig. 5 Experience in fire and safety training, experience in serious game

5 Conclusion and Future Enhancement

The key role of this serious game is to give an awareness about fire hazards and safety measure. The game gives an introduction to fire risks in kitchen and petrol pump. By playing this game, the player gets an awareness of handling fire-causing substances in order to prevent the occurrence of fire. In future, the game will include

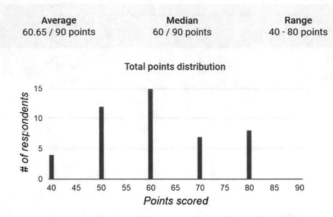

Average	Median	Range
60.65 / 90 points	60 / 90 points	40 - 80 points

Fig. 6 Score distribution after pre-assessment of the game (traditional learning)

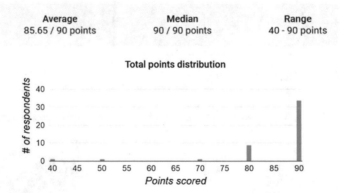

Average	Median	Range
85.65 / 90 points	90 / 90 points	40 - 90 points

Fig. 7 Score distribution after post-assessment of the game (game-based learning)

fire hazards in chemistry laboratory. It will be redesigned by including categories of fire, training for using a fire extinguisher, and fire escape plan. It will contain assessment at the end of each task. A score will be awarded based on the player's actions.

Acknowledgements We are grateful for the support provided by Amrita E-Learning Research Lab and Amrita Vishwa Vidyapeetham University.

Declaration Authors hereby declare the due permissions are taken from the authorities of Amrita Vishwa Vidyapeetham University, India for conducting this survey and carry out simulations in their research lab.

References

1. Yeturu, S.K., Annapurani, R., Janakiram, C., Joseph, J., Pentapati, K.C.: Assessment of knowledge and attitudes of fire safety—an institution based study. J. Pharm. Sci. Res. (2016)
2. Hendrix, M., Arnab, S., Dunwell, I.: Integrating serious games in adaptive hypermedia applications for personalised learning experiences. In: The Fifth International Conference on Mobile, Hybrid and On-line Learning (2013)
3. Wendek, V., Gutjahr, M.: Designing collaborative multiplayer serious games for collaborative learning. In: Proceedings of the CSEDU, vol. 2, pp. 199–210 (2012)
4. Sreelakshmi, R., McLain, M.L., Jayakrishnan, R., Rajeshwaran, A., Rao, B., Bijlani, K.: Gamification to enhance learning using Gagne's learning model. In: 6th International Conference on Computing, Communication and Networking Technology (ICCCNT) (2015)
5. Fire Statistics and Report from NPFA. http://www.nfpa.org/news-and-research/fire-statistics-and-reports/fire-statistics
6. Fire Statistics. http://cdfdata.fire.ca.gov/incidents/incidents_stats
7. Petrol Stations Safety. http://www.hsa.ie/eng/Your_Industry/Petrol_Stations/
8. Veeramanickam, M.R.M., Radhika, N.: A study of educational games application model in E-learning cloud system. In: International Conference on Information Communication and Embedded Systems (ICICES) (2014)
9. Game Based Learning. http://edtechreview.in/298-what-is-game-based-learning
10. Measures to Assess Fire Safety. http://www.livemint.com/Leisure/Mtxy3Pg73KpOCeINtSSrfI/10-measures-to-assess-fire-safety.html
11. Fire Safety. https://en.wikipedia.org/wiki/Fire_safety
12. Soler, J.L., Mendoza, P., Vasquez, S.P.: Impact of Video Game-Based Learning on Academic Performance and Motivation: A Case Study
13. Senderek, R., et al.: The implementation of game based learning as part of the corporate competence development. In: International Conference on Interactive Collaborative and Blended Learning (ICBL), pp. 9–11 (2015)
14. Kuo, M.-J.: How does an online game based learning environment promotes students intrinsic motivation for learning natural science and how does it affect their learning outcomes? In: First IEEE International Workshop on Digital Game and Intelligent Toy Enhanced Learning (DIGITEL07) (2007)
15. DeChamplain, A., Rosendale, E., McCabe, I., Stephan, M., Cole, V., Kapralaos, B.: Blaze: a serious game for improving household fire safety awareness. In: IEEE International Games Innovation Conference, pp. 2–3 (2012)
16. Silva, J.F., Almeida, J.E., Rossetti, R.J.F., Coelho, A.L.: A Serious Game for EVAcuation Training. IEEE (2013)
17. Chittaro, L., Ranon, R.: Serious games for training occupants of a building in personal fire safety skills (2009)
18. Babu, S.K., McLain, M.I., Bijlani, K.: Collaborative game based learning of post disaster management. In: 8th International Conference, T4E (2016)
19. Safety for Kid—Fire Escape. https://play.google.com/store/apps/details?id=vn.commage.safetykid.lesson4_1
20. Home Fires. https://www.ready.gov/home-fires
21. Fire Safety: Help Mikey. https://play.google.com/store/apps/details?id=air.animatusapps.helpmikey

An OpenMP-Based Algorithmic Optimization for Congestion Control of Network Traffic

Monika Jain, Rahul Saxena, Vipul Agarwal and Alok Srivastava

Abstract The last decade being a web revolution in the field of electronic media, data and information exchange in various forms has significantly increased. With the advancement in the technological aspects of the communication mechanism, the textual form of data has taken the shape of audiovisual format, and more and more content over internet is being shared in this form. Data sharing in this form calls for the need of high bandwidth consumption which may slow down the network resulting in performance degradation of content delivery networks due to congestion. Several attempts have been made by the researchers to propose various techniques and algorithms to achieve optimal performance of the network resources under high-usage circumstances. But due to high-dense network architectures, the performance implementations of suggested algorithms for congestion may not be able to produce the desired results in real time. In this paper, we have presented an optimized multi-core architecture-based parallel version of two congestion control algorithms—*leaky bucket* and *choke packet*. The experimental results over a dense network show that optimized parallel implementation using OpenMP programming specification gets the network rebalancing in a very short span of time as compared to its serial counterpart. The proposed approach runs 60% faster than the serial implementation. The graphical map for the speed up continues to increase with the size of the network and routers. The paper throws the light on the implementation aspects as well as result analysis in detail along with some existing algorithms for the problem.

Keywords Congestion control algorithm · Leaky bucket · Choke packet
Multi-core · High-performance computing

M. Jain (✉) · R. Saxena · V. Agarwal · A. Srivastava
Department of Information Technology, Manipal University Jaipur, Jaipur, India
e-mail: Monikalnct@gmail.com

R. Saxena
e-mail: rahulsaxena0812@gmail.com

© Springer Nature Singapore Pte Ltd. 2018
S. C. Satapathy et al. (eds.), *Information and Decision Sciences*,
Advances in Intelligent Systems and Computing 701,
https://doi.org/10.1007/978-981-10-7563-6_6

1   Introduction

In the current era of digital communication, web interface has become a vital platform for information exchange among people. Today, almost every person uses multimedia utilities like video conferencing, online meeting, voice call, online games, and much more. To accomplish this multimedia-based data transactions over the network, efficient networking mechanism and techniques are required in order to facilitate the communication efficiently. Drastically increasing users along with heavy usage of data in this form puts a challenge in front of the network managers to deduce techniques in order to shed down the load on the network routers and devices properly. For this, various kinds of routing protocols are being used to transfer the data among multiple nodes around the globe. But due to excessive data traffic over the network, some packets are lost and do not reach the intended destination as shown in Fig. 1.

Possible reasons for this can be one among the following:

1. over usage of bandwidth,
2. selfish node that consumes whole network bandwidth,
3. any attack over the network,
4. due to multiple users using the network at the same time,
5. router less storage capacity,
6. processors are sluggish, and
7. network links are slow.

In this paper, we have targeted the problem of over usage of bandwidth. Due to excessive amount of data packets transferred at the same time, some packets do not reach the destination, or queuing delay is introduced in the network which deteriorates network quality. Along with the degradation in network quality, congestion also results in

- queuing delay,
- frame or data packet loss, and
- blocking of new connections [1].

Fig. 1 Data packets lost due to congestion

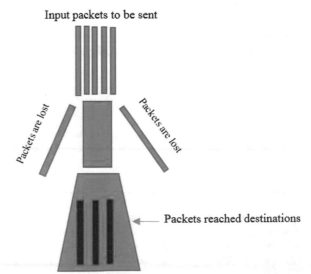

Input packets to be sent

Packets are lost

Packets are lost

Packets reached destinations

To handle the problem of congestion in networks, various congestion control algorithms have been proposed. Some congestion control algorithms are as follows.

- Leaky bucket algorithm: Swarna et al. [2] have presented an efficient version of leaky bucket algorithm for congestion control in which the authors controlled the data transmission rate by setting the bucket size and output flow rate.
- Token bucket algorithm: Valenzuela et al. [3] have presented a hierarchical token bucket traffic shaper for IEEE 802.11 wireless local area networks. The proposed scheduler was implemented over IEEE 802.11b testbed to perform analysis of the implementation results for improved QoS.
- Choke packet method: Chandra et al. [4] proposed an adaptive virtual queue-based congestion control mechanism using choke packets. The basic idea behind queue management is to identify congestion and control the transmission rates before queues in the network overflow and packets are dropped.
- Back-Pressure Technique: Ahmed and Paulus [5] proposed a congestion avoidance and mitigation technique in which routes are selected based on the distance between sender and receiver, relative success rate (RSR) value of node, and buffer occupancy of a node. Based on these parameters, the most appropriate neighborhood node, i.e., non-congested node for transmitting the packet, is chosen.
- Weighted fair queuing: Aramaki et al. [6] have presented a queue-based scheduling method for low-cost M2M devices for wireless networks. In the method, priority-based packet scheduling is done where each packet is assigned a priority and it keeps changing dynamically at intermediate devices. The method takes up into consideration the requested quality level for each service.
- Load Shedding: Choumas et al. [7] described an energy-efficient distributed load shedding mechanism where a multi-objective function-based decision is made so as to forward the data packets while some data packets are shed off depending on the energy consumption and resources available.

AIMD: Zhou et al. [8] discussed a generic congestion control of additive increase and multiplicative decrease to propose two AIMD-based multipath congestion control solutions.

ECN: Ramakrishna et al. [9] for the first time proposed the concept of ECN (Explicit Congestion Notification) for TCP networks where the network routers.

The above-discussed algorithmic measures show a continuous attempt to resolve the problem of congestion which is an obvious problem for the modern-day complex networks which hampers the end-user experience badly. Figure 2 shows the map of multicast backbone network where the data travels from multiple links to reach the destination.

In this paper, we come up with a proposal to enhance the efficiency of leaky bucket and choke packet methods to handle the problem of congestion at a node in the network. The paper is divided into major four sections. In Sect. 1, we have discussed the overview of congestion control algorithms, i.e., what are the reasons for congestion and techniques for congestion control algorithms. Further, leaky bucket and choke packet algorithms are discussed in detail. Section 2 discusses in

Fig. 2 Map of multicast
backbone network [11]

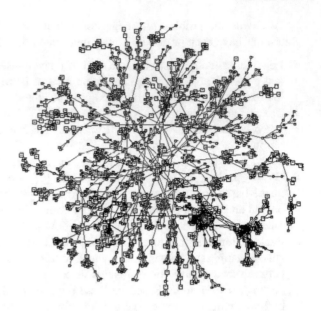

brief the need of high-performance computing measures in the current era. Further, it throws a light in brief on the OpenMP programming standards and specification that has been used in the proposed implementation for optimized results in terms of execution time. Section 3 presents the serial and parallel implementations of the algorithms mentioned in Sect. 2. Section 4 discusses in detail the performance analysis of leaky bucket and choke packet algorithm for serial and parallel versions with graphical map plot. Finally, the paper sums up with a conclusion stating parallel implementation to be 60% faster than the traditional implementation.

1.1 Leaky Bucket Algorithm

Leaky bucket algorithm is a popular congestion control technique in which packet sent does not depend on the incoming packets. It is based on single-server queue. It is like bucket with a small hole at the bottom. No matter how much water is filled in the bucket; the water out will be through that hole only at a constant speed. If the bucket is filled, then the extra water will be discharged from the bucket. Same mechanism is followed in data communication network where instead of water, data packets are used [2]. Figure 3 describes the mechanism of leaky bucket with water (a) as well as with packets (b).

Choke Packet algorithm
Choke packet algorithm is a closed-loop technique. In a data communication network, each router monitors and maintains the resource utilization for the network. Whenever a router starts spilling off the data packets due to congestion, the router

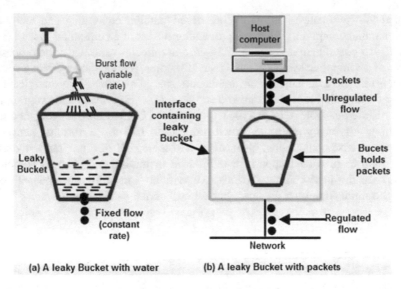

(a) A leaky Bucket with water   (b) A leaky Bucket with packets

Fig. 3 **a** Mechanism of leaky bucket with water **b** mechanism of leaky bucket with packets [12]

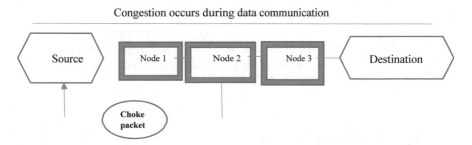

Fig. 4 Choke packet method

sends the routing information to the source notifying to slow down the data packet discharge rate. Figure 4 shows the implementation mechanism of choke packet algorithm.

2 Overview of Parallel Computing Measures and OpenMP

Processing of large amount of data and complex computations takes number of days to solve various problems over single-core architecture. It demands high computational power to solve the problems quickly and efficiently. Based on the processing power capacity of the machine, parallel processing-based computing can be broadly divided into two categories: multi-core based parallel computing and many-core based parallel computing When the processing power of a computing system is less than 10–15 cores, then it is called multi-core architecture and when the processing power of a computing system is more than 100 cores, (Graphical processing system) then it is referred to as a many-core architecture-based machine. To accomplish these kinds of computing modifications to a serial code, various APIs and programming standards are available. CUDA [10, 15] (Compute Unified Device Architecture) is one among the many platforms which provide an API for performing many-core architecture-based computing. OpenMP is a similar kind of a programming speci-fication which provides support for multi-core or multiple CPU-based computing support. In the next coming sections, we will be presenting an OpenMP-based parallel optimization over the leaky bucket and choke packet algorithms.

2.1   OpenMP (Open Multi-processing)

OpenMP is one of the most suitable platforms to write a parallel code based on the shared memory parallelism of the modern-day machines [14, 16]. It consists of three main components, namely, directives, clauses, and environmental variables. Based upon the concept of multithreading, where work is divided among the threads, the programmer has to explicitly specify which code to run in parallel using compiler directives. OpenMP supports C and Fortran language. In the next section, we will be discussing the traditional algorithmic code for congestion control fol-lowed by its parallel version using OpenMP.

3   Algorithm for Congestion Control

ALGORITHM 1: Congestion Control

Procedure congestion_serial (incoming, limit [], store [], t_hold [], number)
        Begin
for data transfer to destination node do
While (data transfer to node i)
if (incoming packet size < t_hold[i] − store [i] and limit [i])
        Accept the incoming packet in store[i].
        else {
        limit [i]= t_hold[i]-store[i] //Report it as a choke packet.
        x=limit[i] // Acceptable size of new incoming packet }
        end if
        store[i] = store[i] - outgoing[i];
if (store [i] has no incoming space)
        store[i]=0
        end of while end of for loop End

In the above mentioned algorithmic procedure, knowing the router capacity and bandwidth limit, incoming data packet rates are adjusted [13]. The procedure continues until the desired data packet is not transferred to the destination node. At every node, if the incoming packet size is acceptable, the packet is stored in array *store [i]* and the buffer storage for the router is adjusted. If the packet size is greater than the node capacity or the bandwidth of the channel, it is reported as a *choke packet* and the new acceptable size of the packet is identified as *limit [i]*.

A parallel version of the pseudocode with C++ algorithmic implementation is mentioned here below:

ALGORITHM 2: Congestion Control

```
Procedure congestion_control_parallel (incoming, limit [], store [], t_hold [], number)
            Begin
                        omp_set_num_threads (20);
                        #pragma omp parallel for
                                    for (i=0; i<dest; i++)
                                    { store[i]=0;
                        limit[i] = t_hold[i];        }
                                    limit[i] = '\0';
                        #pragma omp parallel for schedule(dynamic)
                                    for (i=0; i<dest; i++)
                                    { for (number=num1; number! =0; number--)
            {           if (incoming <= (t_hold[i] - store[i]) && incoming <= i_limit[i])
                        { store[i] += incoming;
                                    flag=0; }
                        else {
                                                limit[i]=t_hold[i]- store[i];
flag++;
x=i_limit[i]; } }
            store[i] = store[i] - outgoing[i];
            store[i]= (store[i]< 0? 0: store[i]);
            } End
```

The code is spawned over 20 threads which runs the loop parallely for the nodes to make a check for the conditions. *Schedule* [13, 14] clause lets the loop iteration to be scheduled dynamically. Due to independent calculations at each node, dynamic scheduling performs well as no iteration has to wait for the results of other. Increasing the number of threads for a large-sized network will further increase the efficiency of the algorithm in terms of execution time taken. However, the performance improvement is upper bounded by the system resources, capacity, and thread workload balancing.

Fig. 5 **a** Output execution results for Algorithm 1, **b** output execution results for Algorithm 2

(a)
```
Incoming packet size 391 for 9 destination:
Accepted
Buffer size 4603 out of 5231
After outgoing, 4309 packets left out of 5231 in buffer

Incoming packet size 2811 for 9 destination:
Choke packet send for packet 2811
Now incoming packet size should be less than 922
Dropped 2811 no of packets
Buffer size 4309 out of 5231
After outgoing, 4015 packets left out of 5231 in buffer

Incoming packet size 110 for 9 destination:
Accepted
Buffer size 4125 out of 5231
After outgoing, 3831 packets left out of 5231 in buffer

Incoming packet size 2847 for 9 destination:
Choke packet send for packet 2847
Now incoming packet size should be less than 1400
Dropped 2847 no of packets
Buffer size 3831 out of 5231
After outgoing, 3537 packets left out of 5231 in buffer


Time taken by code is 0.811000
```

(b)
```
After outgoing, 368 packets left out of 5338 in buffer

Incoming packet size 41 for 7 destination:
Accepted
Buffer size 614 out of 8290
After outgoing, 402 packets left out of 8290 in buffer

Incoming packet size 41 for 5 destination:
Accepted
Buffer size 614 out of 6235
After outgoing, 393 packets left out of 6235 in buffer

Incoming packet size 41 for 9 destination:
Accepted
Buffer size 614 out of 5184
After outgoing, 352 packets left out of 5184 in buffer

Incoming packet size 41 for 1 destination:
Accepted
Buffer size 614 out of 8860
After outgoing, 331 packets left out of 8860 in buffer

Incoming packet size 3610 for 0 destination:
Accepted
Buffer size 614 out of 7130
After outgoing, 339 packets left out of 7130 in buffer


Time taken by code is 0.147000
```

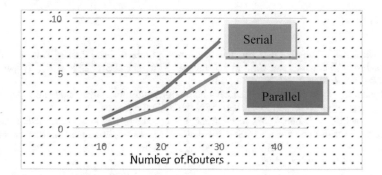

Fig. 6 Graphical plot serial versus parallel; Y-axis: execution time of the algorithm in seconds; X-axis: number of routers in the network

4 Experimental Results and Performance Analysis

The program has been tested over for both the implementation procedures as discussed above by varying the number of routers and network size. For all the given cases, results have been reported in the form of a graphical plot. Parallel implementation of the algorithm over OpenMP is found to be approximately 67% faster than the serial implementation as shown in below-mentioned figures. Figure 5a shows the snapshot of running sequential code, and Fig. 5b shows the snapshot of running parallel Open MP code (Fig. 6).

The graphical plot clearly indicates that as the network intensity increases and more and more data packets are to be transmitted, serial implementation slows down. On the other hand, parallel version of the algorithm also grows almost in a linear passion with the increase in the number of routers but the performance efficiency is still high in comparison to the serial code.

5 Conclusion

With the massive demand of internet users, congestion is a normal problem today. Traditional algorithm takes too much time to run over single-core architecture which does not adapt well to the real-time scenarios of networking domain. This paper presents an optimized algorithm which can run faster over multi-core architecture. The serial code implementation is mapped to parallel architecture of the machine utilizing the parallel processing power of the multiple cores available which makes the congestion control algorithm to run 60% faster approximately than the serial code. In future, the algorithm proposed can be further reinvestigated and modified to work over cluster of computers for a severely large number of network nodes in order to gain the results comparable to what has been achieved over a single machine for a comparatively smaller size of network of nodes.

References

1. Akiene, P.T., Kabari, L.G.: Simulation of an Optimized Data Packet Transmission in a Congested Network: Network and Complex Systems. ISSN 2224-610X (Paper) ISSN 2225-0603 (Online), vol. 5, no. 8, 2015
2. Swarna, M., Ravi, S., Anand, M.: Leaky bucket algorithm for congestion control. Int. J. Appl. Eng. Res. **11**(5), 3155–3159 (2016). ISSN 0973-4562
3. Valenzuela, J.L., Monleon, A., San Esteban, I., Portoles, M., Sallent, O.: A hierarchical token bucket algorithm to enhance QoS in IEEE 802.11: proposal, implementation and evaluation. In: 2004 IEEE 60th Vehicular Technology Conference, 2004. VTC2004-Fall, vol. 4, pp. 2659–2662. IEEE (2004)
4. Chandra, M.A., Kavitha, M.T.: Adaptive virtual queue with choke packets for congestion control in MANETs. Int. J. Comput. Netw. Wirel. Commun. (IJCNWC) ISSN: 2250-3501 (2014)
5. Ahmed, A.M., Paulus, R.: Congestion detection technique for multipath routing and load balancing in WSN. Wirel. Netw. **23**(3), 881–888 (2017)
6. Aramaki, T., Kinoshita, K., Tanigawa, Y., Tode, H., Watanabe, T.: A congestion control method for multiple services on shared M2M network. In: 2016 5th IIAI International Congress on Advanced Applied Informatics (IIAI-AAI), pp. 873–877. IEEE (2016)
7. Choumas, K., Paschos, G.S., Korakis, T., Tassiulas, L.: Distributed load shedding with minimum energy. In: The 35th Annual IEEE International Conference on Computer Communications, IEEE INFOCOM 2016, pp. 1–9. IEEE (2016)
8. Zhou, D., Song, W., Cheng, Y.: A study of fair bandwidth sharing with AIMD-based multipath congestion control. IEEE Wirel. Commun. Lett. **2**(3), 299–302 (2013)
9. Ramakrishna, K., Floyd, S.: A proposal to add explicit congestion notification (ECN) to IP (No. RFC 2481) (1998)
10. Saxena, R., Jain, M., Sharma, D.P., Mundra, A.: A review of load flow and network reconfiguration techniques with their enhancement for radial distribution network. In: 2016 Fourth International Conference on Parallel, Distributed and Grid Computing (PDGC), pp. 569–574. Waknaghat, (2016). http://www.doi.org/10.1109/PDGC.2016.7913188
11. Amir, E.: A map of the MBone: August 5th, 1996. http://www.cs.berkeley.edu/~elan/mbone.html
12. Web Page: http://ecomputernotes.com/computernetworkingnotes/communication-networks/what-is-congestion-control-describe-the-congestion-control-algorithm-commonly-used
13. Dike, D., Obiora, V., Eze, C.: Improving congestion control in data communication network using queuing theory model. IOSR J. Electr. Electron. Eng. **11**(2), 49–53 (2016). ISSN: 2278-1676, p-ISSN: 2320-3331
14. Saxena, R., Jain, M., Bhadri, S., Khemka, S.: Parallelizing GA based heuristic approach for TSP over CUDA and OPENMP. In: 2017 International Conference on Advances in Computing, Communications and Informatics (ICACCI), pp. 1934–1940, Udupi, (2017). http://www.doi.org/10.1109/ICACCI.2017.8126128
15. Saxena R., Jain M., Sharma D.P.: GPU-based parallelization of topological sorting. In: Somani A., Srivastava S., Mundra A., Rawat S. (eds) Proceedings of First International Conference on Smart System, Innovations and Computing. Smart Innovation, Systems and Technologies, vol 79. Springer, Singapore (2018)
16. Saxena, R., Jain, M., Singh, D., & Kushwah, A.: An enhanced parallel version of RSA public key crypto based algorithm using openMP. In: Proceedings of the 10th International Conference on Security of Information and Networks (pp. 37–42). ACM (2017, October)

Comparative Analysis of Frequent Pattern Mining for Large Data Using FP-Tree and CP-Tree Methods

V. Annapoorna, M. Rama Krishna Murty, J. S. V. S. Hari Priyanka and Suresh Chittineni

Abstract Association rule mining plays a crucial role in many of the business organizations like retail, telecommunications, manufacturing, insurance, banking, etc., to identify association among different objects in the dataset. In the process of rule mining, identify frequent patterns, which can help to improve the business decisions. FP-growth and CP-tree are the well-known algorithms to find the frequent patterns. This work performs comparative analysis of FP-growth and CP (compact pattern)-tree based on time and space complexity parameters. The comparative analysis also focuses on scalability parameter with various benchmark dataset sizes. Outcomes of this work help others to choose the algorithm to implement in their application.

Keywords Knowledge discovery · Frequent pattern · Compact pattern tree

1 Introduction

In today's world, finding frequent patterns (or item sets) becomes a challenging task for most of the business entrepreneurs, in data mining and knowledge discovery process. Frequent pattern is an item set or subset that occurs more number of times than the user-specified threshold, in a transactional database. Association rule mining is one of the widespread concepts for generating frequent patterns, and this

V. Annapoorna (✉) · M. Rama Krishna Murty · J. S. V. S. Hari Priyanka · S. Chittineni
ANITS, Sangivalasa, Bheemunipatnam, Visakhapatnam, India
e-mail: annapoorna.vaddadi@gmail.com

M. Rama Krishna Murty
e-mail: ramakrishna.malla@gmail.com

J. S. V. S. Hari Priyanka
e-mail: priyapatnaik.hari@gmail.com

S. Chittineni
e-mail: sureshchittineni@gmail.com

© Springer Nature Singapore Pte Ltd. 2018
S. C. Satapathy et al. (eds.), *Information and Decision Sciences*,
Advances in Intelligent Systems and Computing 701,
https://doi.org/10.1007/978-981-10-7563-6_7

method enables the user to identify correlations, sequential patterns, episodes, multidimensional and maximal patterns, etc. among variables in large databases, through various data mining techniques [1]. Basically, association rule mining process neither depends on the arrangement of products in each record of a transaction nor the arrangement of records in a transactional database to discover interesting patterns.

Frequent pattern [1] generation process is defined as follows: consider a transactional dataset DB, and let {R1, R2 ... Rn} be a set of n transactions in DB. Let P = {P1, P2 ... Pm} be a set of m items or products. In DB, every transaction or record is assigned with an ID, i.e., TID, which includes a few set of products from P.

$$\text{Let } M \Rightarrow N \tag{1}$$

Equation (1) represents an association rule such that M, N <= P, and MN. In this rule, M and N are item sets called as antecedent and consequent of an association rule, and the pattern M, N is called as a frequent pattern when the occurrence pattern M, N is above the threshold value specified by the user. The interestingness of an association rule is determined using two measures like support and confidence. These two refers to the usefulness and certainty of the rule.

A widespread concept for frequent pattern generation in transactional dataset is Apriori algorithm. As the candidate generation step of Apriori method requires multiple database scans, we analyzed that frequent pattern mining is a time-consuming process using Apriori algorithm. We observed that FP-growth and CP-tree algorithms provide efficient frequent pattern mining with less number of database scans. Frequent pattern growth method needs two database scans construct frequent pattern tree, from which frequent patterns can be identified [2–5]. CP-tree method uses prefix tree-based frequent pattern mining with one scan of the database. CP-tree is periodically restructured using branch sorting method and provides effective pattern mining [2–4, 6, 7]. This work compares the performance of FP-growth and CP-tree methods for finding frequent patterns and concluded that CP-tree method is more efficient and scalable when compared with FP-growth method [8].

The remaining portions of this paper are structured as follows: Sect. 2 discusses about few of the well-known methods to generate frequent patterns. Section 3 explores proposed methods for frequent pattern generation and it also provides the dataset details employed for the implementation of this work. Section 4 provides the outcomes of this experiment and summarizes comparative analysis of the proposed methods of identifying the frequent patterns and also discusses about some of the performance issues related to these algorithms. Section 5 concluded and mentions the future work of this paper. Finally, Sect. 6 lists the references used in this paper.

2   Related Work

Many algorithms are already available for efficient discovery of frequent patterns from the large databases but those algorithms differ with their performance aspects such as time and space efficiency.

The first and a well-known algorithm for finding the frequent item sets is the Apriori [9, 10] algorithm introduced by Agrawal in 1993. Agrawal defined that Apriori uses a hash tree (Horizontal) data structure and follows BFS fashion to identify the frequent patterns. In Apriori algorithm, the candidate generation step increases the count of database scans, which affects the performance of Apriori. VIPER [9] algorithm is proposed by Shenoy in 2000, frequent pattern mining uses vertical data structure, and it also follows the same BFS process in VIPER. Zaki proposed ECLAT [9, 10] method in 2000 and used a vertical data structure and DFS fashion in mining frequent patterns.

Han introduced FP-growth [11] algorithm in 2000 which mines the frequent patterns with only two database scans by constructing a prefix tree (horizontal) data structure and follows DFS method. Burdick defined the MAFIA algorithm in 2001 with vertical data structure and BFS procedure for frequent pattern mining. Yabu Xu proposed PP-mine method with prefix tree data structure and DFS-based mining in 2002. El-Haji and Zaki defined COFI and DIFFSET algorithms, respectively, in 2003. COFI constructs the same prefix tree (horizontal) data structure and follows DFS method as like FP-growth method, whereas DIFFSET algorithm follows a method similar to VIPER. Song defined TM algorithm and used vertical data structure with DFS method of traversal to find frequent item sets in 2006. Show-jane yen defined a prefix tree-based TFP algorithm in 2009. Yen introduced SSR algorithm which uses horizontal data structure and DFS method to recognize frequent item sets in 2012.

Though many methods are evolved to find frequent item sets, those are not much popularized due to their inefficient performance. Finally, we analyzed that cp-tree method uses only one database scan and provides efficient frequent mining compared to FP-growth algorithm.

3   Implementation

In this work, implemented familiar association rule methodologies namely FP-tree and CP-tree mining using retails dataset in Java IDE. The working procedure of these two algorithms is given below.

3.1 Generation of Frequent Patterns Using Frequent Pattern Tree Method

The frequent pattern tree method requires two scans of the dataset in which initial scan of the database is to rearrange the order of items according to their frequency. The subsequent scan of the dataset is used to construct a frequent pattern tree that reduces the time complexity [12, 13–15]. In the frequent pattern growth method, there are few challenges. One among them is two number of scans required, which leads more time complexity against size of the dataset. Another challenge is to generate conditional pattern base for finding frequent items.

To overcome the above challenges, compact pattern tree method is suitable, which can reduce number of scans of the database to one scan and thereby reduces the time complexity. The advantages of the CP-tree method and its working procedure are discussed in Sect. 3.2.

3.2 Generation of Frequent Patterns Using Compact Pattern Tree Method

Compact pattern tree method builds a frequency-descending tree structure to generate frequent patterns with one scan of the transactional database.

Compact pattern tree construction is mainly a two-step process:

(i) Inclusion step: The items from each transaction are included into the CP-tree as per the present item order of item list (I), and the occurrence of corresponding items is updated in the I-list.

(ii) Reorganization step: This step reorganizes the I-list as per the frequency-decreasing order of items and rearranges the CP-tree nodes according to the order of items in newly reorganized I-list.

These two steps are iteratively executed in alternate manner, beginning with the inclusion step by including the foremost portion of dataset, and ending with the reorganization step at the end of dataset. The main purpose of the intervallic tree reorganization method is to build a frequency-descending compact tree with abridged overall reorganization price.

3.2.1 Working of CP-Tree Method

Initially, tree construction starts with an empty root node, named as null. Then, each item from each transaction is added to the tree according to the predefined product ordering of the transaction. I-list is used to maintain the products order of compact tree, and it also retains frequency or count of each product of CP-tree. After inserting some of the transactions in the process of constructing the CP-tree, tree

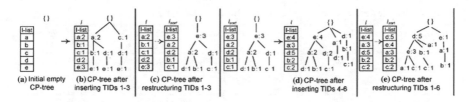

Fig. 1 Various stages required for construction of compact pattern tree

has to be restructured dynamically when the product order of item list (I) deviates, current decreasing count of products to a specified level, and I-list needs to be updated with the item order with the current list.

Compact pattern tree construction is illustrated in Fig. 1, for a sample dataset of transactions D as shown in Table 1. As discussed earlier, the CP-tree construction starts with an empty or null node as tree root as shown in Fig. 1a, b shows the structure of compact tree after inserting few transactions of D and this process is called as inclusion step of CP-tree and here the tree looks like frequency independent tree. Like FP-tree, path of nodes represents an item set in CP-tree. For simplicity, node pointers are ignored in the CP-tree but item count is maintained as I-list. The reorganization step of CP-tree method starts, after each inclusion step in order to simplify the frequent pattern generation. CP-tree restructuring process requires to sort the item frequency of I-list and it should be maintained in another list called as I-sort. I-sort holds items according to their decreasing frequency, and CP-tree is restructured according to list I-sort as shown in Fig. 1c, and thereby it is called as frequency-decreasing tree. Figure 1d shows the structure of CP-tree by including remaining transactions of D and the status of the lists I and I-sort. Now, the tree is again subjected to a second reorganization step as shown in Fig. 1e, to obtain its frequency-descending tree. This process is iteratively applied using CP-tree method, to insert all the transactions of D into CP-tree. The tree at the end of this iterative process is a compact tree with higher prefix sharing among patterns in the tree nodes, and these patterns are said to be frequent if their frequency is above the user-specified threshold. In this way, the CP-tree method generates the frequent patterns using single scan of the database.

Table 1 Transactional dataset	TID	Items purchased
	1	b, a, e
	2	a, d, e
	3	c, e, d
	4	a, d
	5	b, c, d
	6	e, d

Table 2 Sample dataset of online retails used for implementation

Order#	Stock-code	Product	Product-quantity	Order-date	Price per unit	Cust-id	Country-name
536366	22633	Hand Warmer Union Jack	6	1-12-2010 8:28	1.85	17850	UK
536366	22632	Hand Warmer Red Polka Dot	6	1-12-2010 8:28	1.85	17850	UK
536367	84879	Assorted ColourBird Ornament	32	1-12-2010 8:34	1.69	13047	UK
536367	22745	Poppy's Playhouse Bedroom	6	1-12-2010 8:34	2.1	13047	UK
536367	22748	Poppy's Playhouse Kitchen	6	1-12-2010 8:34	2.1	13047	UK
536367	22749	Felt craft Princess Charlotte Doll	8	1-12-2010 8:34	3.75	13047	UK
536367	22310	Ivory Knitted Mug cosy	6	1-12-2010 8:34	1.65	13047	UK

3.3   Dataset Information

Dataset

Transnational dataset of UK-based and registered non-store online retail is used for implementation of this work. Dataset contains 541909 numbers of instances of the academic year 2010–2011, out of which 2016 number of samples are considered for experimentation. The dataset has been collected from http://archive.ics.uci.edu/ml/machine-learning-databases/00352/ (Table 2).

4   Experimental Results

As earlier mentioned, the implementation of two frequent pattern algorithms in the Java IDE and experimented with the benchmark dataset called online retails from the UCI machine repository. Frequent pattern and compact pattern methods are experimented and got efficient results in both algorithms, but CP-tree performs better than FP-tree in terms of time complexity parameter due to single scan of

Table 3 Time complexity comparison between FP- and CP-tree methods

S. no	Algorithm	No. of samples	Threshold value	Time consumed (in milliseconds)
1.	FP-growth	2016	1000	4364
2.	CP-tree		1000	204

dataset enough in its procedure. The comparative analysis of both the algorithms in terms of time complexity is shown in Table 3.

Scalability: Most of the frequent pattern mining algorithms work well on small dataset having few data objects probably below 200. However, large databases may contain millions of data objects. The FP- and CP-tree methods are used to handle with such a huge datasets. Though these two methods are scalable to find frequent patterns, the CP-tree performance slightly dominates the FP-tree performance.

Another interesting point observed in this experimentation is that the CP-tree method supports enhancing tree construction dynamically for the data streams or growing real-time datasets. At this juncture, the FP-growth algorithm requires

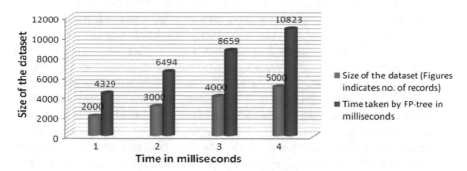

Fig. 2 Processing time against dataset size to generate frequent item set using FP-tree

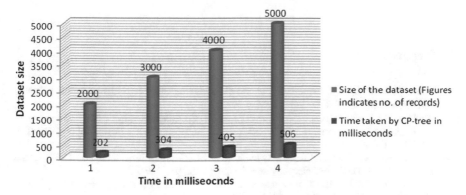

Fig. 3 Processing time against dataset size to generate frequent item set using CP-tree

Table 4 Time taken for frequent item generation using FP-tree and CP-tree for large dataset

Size of the dataset (Figures indicates no. of records)	Time taken by FP-tree in milliseconds	Time taken by CP-tree in milliseconds
2000	4329	202
3000	6494	304
4000	8659	405
5000	10823	506

repeating the complete procedure of finding the frequent patterns from the beginning when the new transactions are added in the dataset (Figs. 2, 3, and Table 4).

5 Conclusion and Future Work

In today's world, many of the business organizations data has been growing rapidly. Because of the huge availability of business data, frequent pattern mining had become a challenging task for a business decision-maker to make appropriate business decisions and to improve the organizations profits. In this process, we have studied so many algorithms like Apriori, VIPER, ECLAT, FP-growth, etc. and we identified that FP-growth method is efficient with respect to time complexity parameter in mining frequent patterns from a transactional database and it requires two scans of the database. In this work, we used a dynamic tree restructuring method called as CP (compact pattern)-tree mining algorithm to generate frequent patterns with only one scan of the database. Our experimentation focuses on a comparative analysis of FP-tree and CP-tree methods, in mining frequent patterns. Our results show that performance of CP-tree slightly dominates FP-tree performance in time and space complexity parameters and through our analysis, we also conclude that CP-tree method is better scalable than FP-tree method. Our future work aims at finding frequent patterns for stream data and big data using CP-tree method.

References

1. Patro, S.N., Mishra, S., Khuntia, P., et. al.: Construction of FP-tree using Huffman Coding. Int. J. Comput. Sci. (IJCSI) **9**(3), 2 (May 2012)
2. Tanbeer, S.K., Ahmed, C.F., Jeong, B.-S., Lee, Y.-K.: Efficient single-pass frequent pattern mining using a prefix-tree. Inf. Sci. (Elsevier) **179**, 559–583 (2008). https://doi.org/10.1016/j. ins.2008.10.027
3. Shrivastava, N., Khanna, R.: FP-Growth tree based algorithms analysis: CP-Tree and K Map. Bin. J. Data Min. Netw. **5**, 26–29 (2015). ISSN: 2229-7170
4. Pandya, M., Trikha, P.: A new tree structure to extract frequent pattern. Int. J. Emerg. Technol. Adv. Eng **3**(3) (March 2013). ISSN: 2250-2459

5. Ghatage, R. A.: Frequent pattern mining over data stream using compact sliding window tree & sliding window model. Int. Res. J. Eng. Technol. (IRJET) **02**(04) (July 2015). p-ISSN: 2395-0072, e-ISSN: 2395-0056
6. Pandya, M., Trikha, P.: An efficient prefix tree structure to extract frequent pattern. Int. J. Adv. Eng. Technol. **6**(3), 1220–1227 (July 2013)
7. Zhang, S., Zhang, J., Zhang, C.: EDUA: an efficient algorithm for dynamic database mining. Inf. Sci. (Elsevier) **177**, 2756–2767 (2007)
8. Srimania, P.K., Patilb, M.M.: Frequent item set mining using INC_MINE. In: Massive Online Analysis Frame work, Science Direct, Procedia Computer Science, vol. 45, pp. 133–142. (Elsevier) (2015)
9. Meenakshi, A.: Survey of frequent pattern mining algorithms in horizontal and vertical data layouts. Int. J. Adv. Comput. Sci. Technol. ISSN **4**(4), 2320–2602 (April 2015)
10. Nasreen, S., Azam, M.A., Shehzad, K., et.al.: Frequent pattern mining algorithms for finding associated frequent patterns for data streams. In: A Survey, International Conference on Emerging Ubiquitous Systems and Pervasive Networks (EUSPN) Science Direct, Pro: Computer Science, vol. 37, pp. 109–116 (2014)
11. Koh, Y.S., Dobbie, G.: SPO-Tree: Efficient Single Pass Ordered Incremental Pattern Mining. Springer, Berlin, vol. 6862, pp. 265–276 (2011). https://doi.org/10.1007/978-3-642-23544-3-20
12. Lodhi, N.S., Dangra, J., Rawat, M.K.: A compact table based time efficient technique for mining frequent items from a transactional data base. Int. J. Adv. Res. Comput. Sci. Softw. Eng. **5**(1) (January 2015). ISSN: 2277 128X
13. Fole, M.D., Choudhary, C.: Finding an efficient approach for generating frequent patterns in large database. Int. J. Adv. Res. Comput. Eng. Technol. (IJARCET) **4**(2) (Februray 2015)
14. Narvekar, M., Syed, S.F.: An optimized algorithm for association rule mining using FP Tree. Procedia Comput. Sci. (Elsevier) **45**, 101–110 (2015)
15. Shashikumar, G., Totad, R.B., Geeta, P.V.G.D., Reddy, P.: Batch incremental processing for FP-tree construction using FP-Growth algorithm. In: Knowledge and Information Systems. Springer (2012). https://doi.org/10.1007/s10115-012-0514-9

Wireless Seatbelt Latch Status Indicator

S. P. Adithya and O. V. Gnana Swathika

Abstract Road transportation is an integral part of our daily life. Millions of deaths occur due to road accidents. Thus, safety becomes crucial when it comes to opting for road transport. This paper proposes a wireless seatbelt latch system that is cost-effective and requires a set of electronic components and a sensor which detects the status of the seatbelt and transmits to the Electronic Control Unit (ECU) of the car. The proposed system is tested and validated using Proteus software.

1 Introduction

Nowadays, seatbelts in cars are equipped with Hall Effect sensors. If the seatbelt is latched, the Hall Effect sensor experiences magnetic field from a neodymium magnet attached in the movable part of the buckle. This activates the transmitter in the buckle to transmit the information to the receiver in the ECU wirelessly that the seatbelt is latched [1]. In [2], Light-Emitting Diodes (LEDs) and Complementary Metal–Oxide–Semiconductor (CMOS) cameras are used as transmitters and receivers, respectively. But the suggested seatbelts prove to be costly and large in size.

Data may be transmitted through Radio Frequency (RF) waves with shorter range and reliability. It works exactly like Dedicated Short-Range Communication (DSRC) except the fact that it is a one-way communication system [3]. Alternative communication systems in an automobile system are discussed in [4, 5].

S. P. Adithya · O. V. Gnana Swathika (✉)
VIT University, Chennai 600127, Tamil Nadu, India
e-mail: gnanaswathika.ov@vit.ac.in

© Springer Nature Singapore Pte Ltd. 2018
S. C. Satapathy et al. (eds.), *Information and Decision Sciences*,
Advances in Intelligent Systems and Computing 701,
https://doi.org/10.1007/978-981-10-7563-6_8

2   Proposed Wireless Seatbelt Latch Status Indicator

The system consists of transmitter units and one receiver unit. RF waves are used to transmit data. ATmega328p µController is used to compute data to be transmitted and processes the data that is received. ATmega328p TQFP is chosen because of its compact size and also because of it being automotive graded.

2.1   Construction and Working

The system consists of two units, the transmitting unit and the receiving unit as shown in Fig. 1. Each transmitter is assigned with a unique address, and all the transmitter addresses are saved in the receiver to track which transmitter has currently sent the corresponding data.

Transmitter Circuit: It consists of µController ATmega328p, which is the heart of the circuit. The clock frequency is set as 8 MHz by connecting 8 MHz oscillator grounded with two 22 pF capacitors. The transmitter circuit is powered by ATmega328p only while data is transmitted, thus conserves energy. The Hall Effect sensor is connected in the input terminal of ATmega328P.

Hall Sensor gives positive output voltage in the presence of a magnetic field and gives 0 V output in the absence of magnetic field. This is given as input to the µController. ATmega328p will power up the transmitter only when there is a change in the input, i.e., positive to zero voltage or zero to positive voltage. Once there is a change in the input, the transmitter will read the voltage. If the voltage is positive, the µController interprets that the seatbelt is latched up in the buckle. Latched up data and the address data are sent through the transmitter. Once the data transition is done, transmitter turns OFF. And if it is zero voltage, the µController interprets that seatbelt is not latched up in the buckle. Unlatched up data and the address data are sent through the transmitter. Once the data transition is done,

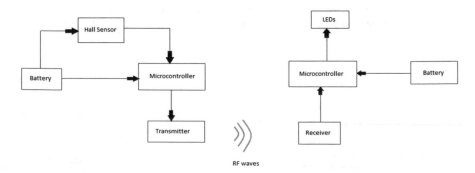

Fig. 1 Block diagram of the complete system, **a** transmitter unit, **b** receiver unit

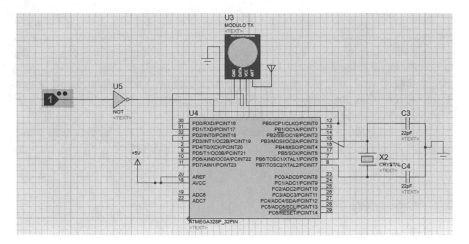

Fig. 2 Transmitter circuit diagram—digital logic state switch is used instead of hall effect sensor

transmitter turns OFF. Power supply for the transmitter is provided by the μController as shown in Fig. 2. This helps to overcome the signal clashing problem with too many transmitters.

Receiver Circuit: Circuit construction is similar to transmitter circuit with few modifications. Clock frequency is set as 8 MHz by connecting 8 MHz oscillator grounded with two 22 pF capacitors as shown in Fig. 3. The LEDs represent the status of seatbelt. If not latched LED turns ON, and if latched LED turns OFF. More LEDs may be incorporated with each LED representing a seat's latch status.

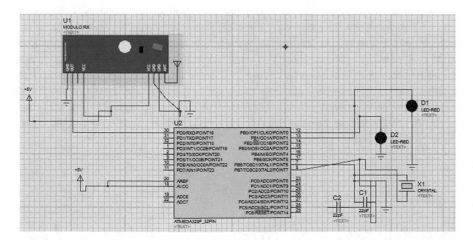

Fig. 3 Receiver circuit diagram—the LEDs, D1 corresponds to transmitter 1 and D2 corresponds to transmitter 2

The receiver always remains in ON state and waits to receive a signal. Signal received is converted to data and sent to ATmega328p. μController compares the address with the preloaded addresses and executes the corresponding function that turns ON or OFF the specific LED related to the transmitter's address. The transmitted data is read, and the corresponding LEDs glow. The whole circuit is connected to the ECM board supply unit for power. If the address match does not exist, it erases the data and waits for the signal again.

Coding for both the circuits is done using embedded C++ language in Atmel Studio 7. Simulations are done using Proteus 8 Professional.

2.2 Difficulties Faced

A basic RF transmitter pairs with RF receiver when switched ON. Thus, when another transmitter tries to transmit data to the same receiver, the receiver does not receive the data as it is paired with a transmitter already. In this case not more than one combination of transmitter and receiver is used in a system. This limitation is broken by switching on the transmitter only when there is a data to transmit and remains switched OFF rest of the time. Transmitter requires less than 10 ms to transmit four bytes of data. The first byte is synchronizing byte, second and third bytes are the address bytes, and the fourth byte tells the status of the seatbelt latch, i.e., latched (switched *off*) or unlatched (switched *on*). Multiple transmitters are incorporated in the system now with a single receiver and the chance of signal collision practically is impossible as each transmitter will be switched ON only when there is a change in the status of the seatbelt latch in the buckle.

This RF system falls under short message broadcast category (for *safety* applications) as discussed in [3]. The cost of production can be reduced to a great extent when compared with [2] which uses Optical Wireless Communication (OWC) with LED TX and camera RX. Not only the cost, it is easy to install and eliminate the need of laying wires from the buckle to the ECU in the dashboard of the car, thus making installation process neat and quick. Each car has a unique address and synchronizing byte, this leads to having 16,777,216 cars with unique address. This number can be extended by allotting more address bytes. The speed of the transmission can be improved using 16 MHz crystal oscillator instead of 8 MHz but practically it is said to be less reliable as it can distort more and travels lesser distance than 8 MHz.

3 Simulations and Results

Simulations are done in Proteus 8 Professional. Since a magnet cannot be moved during the runtime of the simulation, Hall Effect sensor and magnet are replaced with a digital logic switch that gives 5 V when switched *ON* and 0 V when

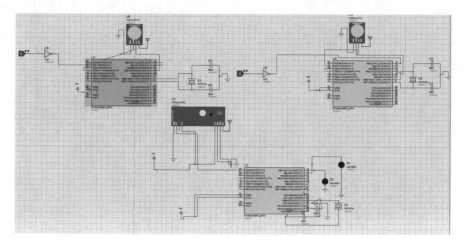

Fig. 4 Wireless seatbelt latch system with one receiving unit and two transmitting units

Table 1 Inputs and outputs of simulation. Inputs are T1 and T2 (T1—*switch state of transmitter 1* and T2—*switch state of transmitter 2*). Outputs are D1 and D2 (D1 and D2 are *LEDs in the receiving unit*)

T1	T2	D1	D2
Low	Low	On	On
High	Low	Off	On
Low	High	On	Off
High	High	Off	Off

switched *OFF*. A car may be incorporated with five transmitting units each in different buckles; in simulation, two transmitting units and a receiving unit are shown for simplicity. Each transmitting unit corresponds to a specific LED in the receiver side. The complete simulation setup is shown in Fig. 4.

As there are two transmitters used, there are four different combinations of input that may be given from transmitters as shown in Table 1. Also, it is evident that the LEDs respond inversely to the transmitting unit. This is because when seatbelt is not latched in the buckle, an LED will indicate the driver to fasten the seatbelt, i.e., if T1 is Low, D1 will be switched ON, and vice versa.

Simulation Outputs: The output for the four different combinations of inputs is shown in Proteus 8 Professional as shown in Table 1. The results for all the test cases are discussed. Figure 5 shows that receiver's LEDs are switched ON (Test case 1) when inputs T1 and T2 are Low. Fig. 6 shows D1 turned OFF (Test case 2) as T1 goes high and Fig. 7 represents the vice versa (Test case 3). In Fig. 8, both D1 and D2 are switched OFF (Test case 4) as T1 and T2 are High.

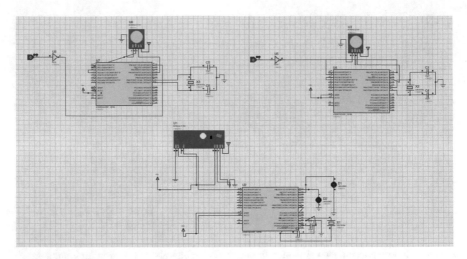

Fig. 5 Test case 1—receiver LEDs D1 and D2 are turned *ON* as T1 and T2 are *Low*

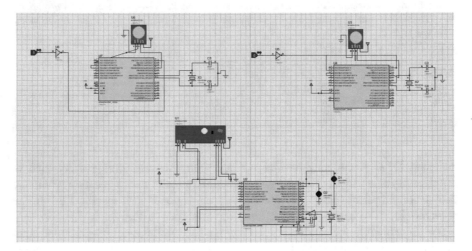

Fig. 6 Test case 2—receiver LED D1 turned *ON* as T1 alone is *Low*

4 Conclusion

The seatbelt latch status is read wirelessly through RF transmitters. Transmitters are powered by the μController that eliminates the signal clashing problem. Addressing transmitters individually make the receivers identify the transmitters and their seatbelt buckle conveniently. The ability to extend the address enables the car

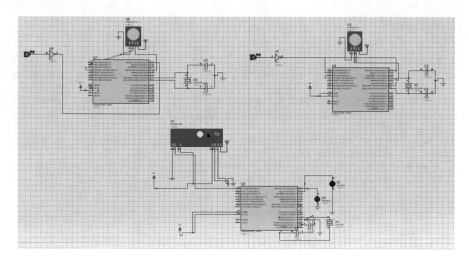

Fig. 7 Test case 3—receiver LED D2 turned *ON* as T2 alone is *Low*

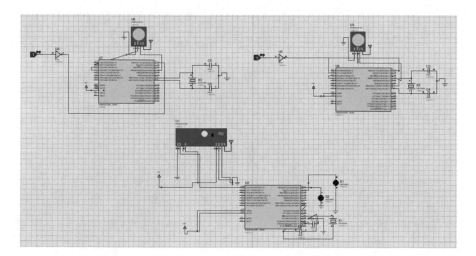

Fig. 8 Test case 4—receiver LEDs turned *OFF* because T1 and T2 are *Low*

manufactures to produce more unique addresses. The system is simple to construct making this idea significant. The mentioned points are tested and validated through simulations done in Proteus 8 Professional.

References

1. Leen, G., Heffernan, D.: Expanding Automotive Electronic Systems (2002)
2. LED and CMOS Image Sensor Based Optical Wireless Communication System for Automotive Applications (2013)
3. Bai, F., Krishnan, H., Sadekar, V., Holland, G., ElBatt, T.: Towards characterizing and classifying communication-based automotive applications from a wireless networking perspective (2006)
4. Bai, F., Krishnan, H.: Reliability Analysis of DSRC Wireless Communication for Vehicle Safety Applications (2006)
5. Kisic, M.G., Blaz, N.V., Babkovic, K.B., Zivanov, L.D., Damnjanovic, M.S.: Detection of Seat Occupancy Using a Wireless Inductive Sensor (2017)

A Study of Various Varieties of Distributed Data Mining Architectures

Sukriti Paul, Nisha P. Shetty and Balachandra

Abstract Owing to the explosion of data in today's world, datasets are enormous, geographically distributed and heterogeneous. Data mining aims extracting useful information from voluminous repositories where data is stored. Predictive analysis of hidden patterns in massive datasets poses to be a challenge. The problems faced while using the data warehousing model for such datasets were privacy, centralization of the data present at multiple independent sites, bandwidth limitation, complexity of integration, and analysis of the data at a global level. Distributed algorithms have been designed to address the same. Distributed data mining (DDM) techniques regard the distributed datasets as one virtual table and assume the existence of a global model which could be designed if the data were combined centrally. This paper presents distributed data mining systems and frameworks for analyzing data and mining the required knowledge from it. Emphasis has been laid on the architectures of such models. Factors like computation resources, communication, hardware, and usage of distributed resources of data have been considered while analyzing or designing distributed algorithms. Such algorithms primarily aim at memory expense and average distribution of working load. Distributed data finds its application in e-commerce, e-business, intrusion detection systems, and sensor networks.

Keywords Data mining · Distributed computing · Grid computing
P2P · Similarity model

1 Introduction

This section contains an introduction to the data warehousing model and the concept of distributed data mining.

S. Paul (✉) · N. P. Shetty · Balachandra
Manipal Institute of Technology, Manipal University, Manipal 576104, India
e-mail: sukritipaul05@gmail.com

© Springer Nature Singapore Pte Ltd. 2018
S. C. Satapathy et al. (eds.), *Information and Decision Sciences*,
Advances in Intelligent Systems and Computing 701,
https://doi.org/10.1007/978-981-10-7563-6_9

1.1 Data Warehousing Model (Old, Traditional Data Mining)

Local data is distributed at independent sites, in the form of tables. These tables are linked to a central repository from which integrated knowledge is mined at a global level, as shown in Fig. 1. Data is stored in dimension tables and fact tables. Maintaining privacy of the distributed sites, while drawing a global inference, can be challenging for massive and heterogeneous datasets. The warehouse knows the information associated with each site.

1.2 Distributed Data Mining—The Concept

Due to the arising need of integrating knowledge from distributed data sources comprising massive datasets, in the global markets, the concept of distributed data mining emerged. Traditional data models proved to be inefficient and the centralized "knowledge discovery and data mining" (KDD) architectures were under scrutiny. "Distributed data mining" comprises analysis of data in a distributed manner, taking into consideration both, the centralized collection and distributed analysis of data [1]. The Internet, networks of "Virtual organizations", sensor networks, corporate private networks (intranets), LANs, and ADHOC wireless networks are some examples of distributed environments. In order to mine data in such environments, effective distributed algorithms are being devised which takes into account communication overhead, computation time, storage limitation, requirements, and hardware.

DDM frameworks comprise user, data, hardware, and software for computation and cost-efficient communication. These frameworks make provisions for scalability, hence dividing datasets with high dimensionalities into smaller subsets which

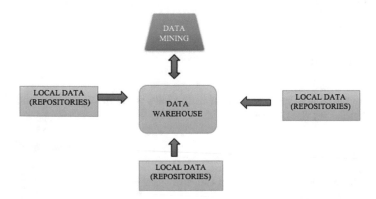

Fig. 1 The traditional data mining warehouse model

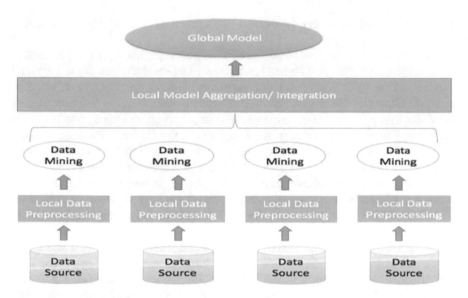

Fig. 2 Data mining architecture

can be computationally analyzed individually, as shown in Fig. 2. The data tables present at the individual sites are parts of a single virtual table belonging to a global model. The global table maybe partitioned horizontally or vertically based on whether the datasets are homogeneous (various sites storing data with the exact set of features) or heterogeneous (sites storing data with a different collection of features, some of which may or may not be common) [2].

2 Distributed Data Mining in P2P Networks

Although most DDM techniques follow the centralized approach, this fails for privacy-sensitive data mining. There may be restrictions on local datasets due to privacy at the individual sites, where the data cannot be collected to a central repository and further analyzed. For example, the various chains of an insurance company may want to analyze patterns, regularities, and trends in case of a fraudulency, without disclosing the customers' transactions and raw data to the central branch. Connectivity poses to be another problem as it may be infeasible to have a massive database perform "non-distributed data mining". The cost of data transmission may be unreasonably high and the databases may witness frequent changes. Data integration from heterogeneous sources is challenging due to the mixing of many structures of complex data and mixing both static and dynamic data. P2P networks emphasize on flexibility, self-organization, scalability, and performance, and have been researched upon, to overcome the above-mentioned shortcomings.

According to [3], the peer-to-peer class of applications makes use of "resources, storage cycles, content, and human presence available at the edges of the Internet." Gaining access to such decentralized resources involves operating in an environment of "unpredictable IP addresses and unsteady connectivity." Thus, peer-to-peer nodes have significant or total autonomy of central servers and must operate outside the DNS. P2P networks are distributed application architectures which comprise a network of nodes. Each peer is independent and equipotent with respect to participation in the application, i.e., each node acts as a server and client without any centralized control. Computer resources are shared among the nodes. Message passing is adopted over an asynchronous network for local nodes to communicate with their neighbouring nodes, in order to perform tasks [4]. A huge variety of data is gathered from various sources and is stored by the peers, collectively. *BitTorrent* is one of the most commonly used peer-to-peer file sharing protocols. File sharing applications like *Napster* and *Kazaa* follow the P2P architecture. Certain searching and indexing applications, and P2P network threat analysis implement P2P networks. Figure 3 describes some desirable characteristics for P2P algorithms in [4].

DDM approaches can be categorized into centrally coordinated and P2P data mining. Centrally coordinated approaches comprise distributed clustering, distributed classifier learning, and distributed association rule mining (ARM) techniques. "Peer-to-peer data mining" comprises primitive operations and complicated data mining. Distributed averaging, majority voting, and gossiping fall under primitive operations. They are used in complicated data mining. Complicated data mining consists of P2P clustering, P2P classification, and P2P association rule mining (ARM) [3]. Limited bandwidth range, constraints with regard to memory, limited battery power, and limited CPU capacity reduce the efficiency of mining in P2P networks. Non-scalability and exorbitant communication costs pose to be the issues in P2P mining.

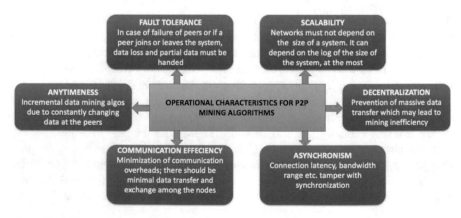

Fig. 3 Operational characteristics for P2P data mining algorithms

2.1 Primitive Operations in DDM in P2P Networks

Calculating aggregate values over the components of a network in a decentralized manner are possible via a gossip-based protocol proposed in [5]. Counts, averages, sums, products, extreme values, max, and min are some of the primitive aggregates that are obtained using algorithms described in [5]. This is suited for dynamic systems where changes in the system can be monitored constantly as all the nodes accept aggregate values continuously. The value at each node is calculated only once. Averages, sums, and counts can also be obtained using iterative procedures over an asynchronous network [6] via probabilistic frameworks.

Gossip algorithms like the Push-Sum algorithm involve communicating with immediate neighbours which are chosen at random, at each round. Also being known as epidemic protocols, they do not require an error recovery mechanism and are highly fault tolerant [7]. A number of aggregated statistics can be computed from the statistics sent by each peer to another peer in an ADHOC manner. Research work has been done in finding the distributed average based on local algorithms [8] and majority voting over P2P networks [9]. Sequential ARM, executed locally at each peer, is combined with a "majority voting protocol" to discover all of the association rules present in the consolidated database, according to the latter.

2.2 Complicated Data Mining in P2P Networks

Complicated data mining scenarios include the domains as shown in Fig. 4.

- P2P Association Rule Mining (ARM)

Using this algorithm, the peers participating or involved in the network determine whether each item set is a frequent item set or not. Suppose one wants to discover patterns and relations like "people who are interested in movies of genre A are also interested in movies of genre B" or "students who purchase physics text books in July also buy lab manuals." P2P ARM algorithms enable this over distributed datasets. Research is being carried out regarding the "LSD Majority Algorithm" [10] for large-scale distributed systems [9]. Sequential ARM is executed locally and coupled with a majority voting protocol at each node, which discovers the association rules present in the collaborated database at each peer [4].

Fig. 4 Complicated data mining problem scenarios

- P2P Classification

Classification models are learnt from the locally distributed datasets, taking into consideration the computational power and time, storage capacity, bandwidth limit, and privacy. The dynamism of datasets, scalability, and fault tolerance are few shortcomings with regard to P2P classification. Web retrieval and web exploration applications like focused crawlers require topic-based cataloging of web pages under document organization. Users across a P2P network may have common topics of interest which could be helpful in constructing machine learning models. An effective methodology is implemented in [11] to build reliable machine learning applications in P2P networks by building meta-methods which are ensemble based. An advanced decision model is derived and constructed from multiple peers, where the generalization performance of various local predictive models is taken into consideration.

- P2P Clustering

The probe and echo mechanism for modified k-means clustering described in [12] is one such algorithm used by information retrieval systems to classify documents. Such grouping provides a description of the constituents of the document collection in respect of the classes that the documents fall under. Each partial document collection is described in a convenient short form by document classes, which is further interchanged with other peers in the network. A set of cluster centroids are decided and guessed initially via an initiator. These centroids, in company with a probe, are sent to its adjacent nodes. On receiving a probe, each peer re-sends the probe to all its neighbouring peers except for its predecessor. Following this, "k-means clustering" is performed on the obtainable and necessary documents [12]. A "Hierarchically-distributed Peer-to-Peer (HP2PC) architecture and clustering algorithm" has been implemented for document clustering in [13]. The clustering problem is divided across neighbourhoods in a modular fashion, and a distributed K-means variant is used to resolve each part. Supernodes are recursively assembled to form higher level neighbours. We arrive at global solutions by combining the clusters up the multilayer overlay network hierarchy. The algorithm used in [13] aims at reducing communication overheads and is a frequency term-based text clustering algorithm for peer-to-peer networks.

3   Distributed Data Mining (DDM) via Grid

Starting from gene and protein databases to climatic databases, there is an explosion of complex data in today's world. Datasets maybe heterogeneous, distributed (geographically), and huge due to which large-scale resource sharing may pose to be a challenge. They are being collected and stored in local databases at very high speeds, especially in e-commerce applications. A grid is a geographically distributed computation infrastructure or framework consisting of a group of heterogeneous machines that the users can access via a sole interface. Apart from resource access and resource sharing technology, grids provide operational services over broadly distributed dynamic virtual organizations comprising institutions or individuals [14]. Parallel and distributed computing paradigms are being carried forward via grids and they are used in both computer-intensive and data-intensive applications. Grid computing primarily focuses on large-scale coordinated resource sharing and high performance in a distributed environment, where systems are tightly coupled with software to work on related problems.

The grid services in Fig. 5 consist of information grid, computation grid, and data grid. Data grids enable massive datasets to be collected and stored in local repositories and to be moved easily. Data grid middleware is essential for the replication and transfer of data over grids [14]. Massive datasets are managed without repeated authentication, in data-intensive applications. "Knowledge Grid" [15] is an enhancement to data grids and provides advanced tools, models, and resources for the DDM and KDD on local data repositories distributed across the grid.

3.1   Distributed Data Mining—Grid Services

3.1.1   Service-Oriented Architecture (SOA)

A web service is a software component that is written via a service description language. The services offered are registered with a service registry [16], which are discovered by querying the registry appropriately. It is platform independent. The published services can be invoked using the necessary binding protocols. The above-mentioned processes facilitate the client to invoke nonlocal services. SOA is a programming model which enables interoperability across heterogeneous platforms and exposes some or all of its applications on different domains and machines [17], irrespective of the deployment location and implementation features of applications. It helps in developing modular software applications.

Fig. 5 Constructing
Knowledge Grid services

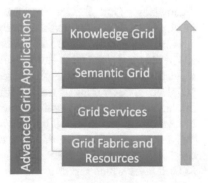

3.1.2   Open-Grid Services Architecture (OGSA)

It provides a distinct and extensible collection of elementary interfaces and services for the development of applications pertaining to grid. A uniform exposed-service semantics, called "*Grid Service*", is defined by OGSA. Each and every resource is represented as a grid service [18]. The uses of the OGSA have been highlighted below [18]:

- Provides location transparency for service instances.
- Standard conventions for naming, creating, and discovering grid services are listed.
- Service instances are provided with numerous protocol bindings.
- Integrates with native platform facilities.
- Complex distribution systems like change, notification, and life management systems are created.

3.1.3   Web Service Resource Framework (WSRF)

"WSRF" bridges the gap between the Web and the grid. It is a refactoring and enrichment of grid services and primarily focuses on exploiting new Web Service standards and on progressing Open-Grid Services Infrastructure (OGSI). Basic services for supporting DDM operations in grids are defined via WSRF. These include processes dealing with data selection, collection, analysis, transport, knowledge models' representation, and data visualization [18]. In order to carry out the above, services have to be defined based on the criteria mentioned in [18].

3.1.4   The Knowledge Grid Framework

The "Knowledge Grid framework" is a reference software architecture that supports the establishment of DDM and "distributed KDD" processes over a grid. "Parallel and distributed knowledge discovery" (PDKD) systems are implemented on top of grid systems like the Globus Toolkit with the help of such an infrastructure. The "Knowledge Grid Service" can be implemented using WSRF conventions and mechanisms. It is exhibited as a web service that exports one or many operations.

The "Knowledge Grid" is structured into two hierarchical parts—"the Core K-Grid layer and the High-level K-Grid layer". The "Core K-Grid layer" consists of services that are directly implemented on generic and basic grid services while the "High-level K-Grid layer" deals with services that are used to define, develop, and implement PDKD computations over the knowledge grid. User-level applications invoke operations exported by "High-level K-Grid services", as depicted in Fig. 6. "High-level and Core K-Grid services invoke operations provided by Core K-Grid services". The KMR, KBR, and KEPR are repositories managed by the Core K-Grid layer which serves as a knowledge repository layer. Each of the services of

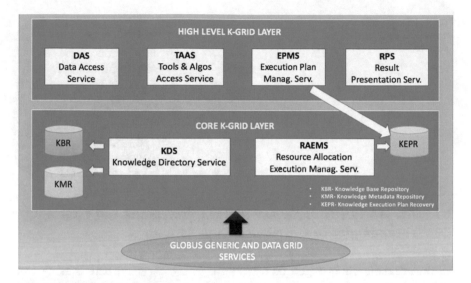

Fig. 6 Constructing Knowledge Grid services

the knowledge grid in Fig. 6 has been discussed in elaborate, in [15]. Knowledge Grid functionalities can be accessed via a client interface on a heterogeneous machine, and the interaction between a client interface and High-level K-Grid services are possibly remote interactions [18]. Some of the basic concepts behind the architecture design of the "Knowledge Grid" are scalability, security, privacy, algorithm integration and independence, data heterogeneity, and compatibility with the grid infrastructure [18].

3.2 Globus Toolkit Services

The most popular middleware used in grid applications is the Globus Toolkit. It manages resource management, data management, security, information discovery, communication, fault detection, and issues regarding portability [14]. The "Globus Toolkit" consists of both web service and non-web service components and makes use of different third-party software services. The important services offered by the Globus Toolkit are described in [14, 17]. Some of them are "grid security infrastructure (GSI), heartbeat monitors (HBM), Globus resource allocation manager (GRAM), monitoring and discovery service (MDS), Grid FTP and dynamically-updated resource online co-allocator (DUROC)." Figure 7 shows how Knowledge Grid services are built.

Fig. 7 Globus generic
service and data grid
hierarchy

3.3   The Apriori Grid Service and Implementation of Grid Technologies in Distributed Data Mining

An organization within a grid framework may have a central branch along with multiple local branches as shown in Fig. 8. Numerous grid nodes (GN) are organized in forming a branch. The aim of the data mining task in such a context is to discover association rules within the databases of the various local branches. The association rule discovery service interacts with grid services like service registry, notification, authorization, and manageability. A comparison of the two categories of grid services, i.e., Predictive Apriori and Apriori, has been made in [14] via a case study. The association rules discover the task which is to be performed and launched, from the central branch, after which data is mined from the central and local branches.

The above-mentioned infrastructure makes use of data mining methodologies and associated technologies in a distributed environment. A discovery service contained in the service data access records data regarding registration of grid service instances with registry services. The "service data access" also describes a standard interface and details required for the creation of services of the "Apriori Grid Service", dynamically.

findServiceData() is a method called by the client-side application to obtain service-related data from the various Apriori grid service instances. On identifying and creating the Apriori grid service, the user sends parameters pertaining to connection and location of remote transaction databases, minimum support and

Fig. 8 Virtual organization framework using grid technologies

confidence with regard to the Apriori algorithm, invoking preprocessing functions, etc. Once the client sends commands along with parameters to the various system components, the configuration parameters of both, the infrastructure and of the Apriori algorithm, call the data mining methods. The association rules discovery task is performed independently on remote transaction databases present at the central or local branches of the virtual organization [14].

4 Conclusion

Distributed data mining aids in operations on large amount of data situated in diverse geographical locations. This paper showcases various architectures of distributed data mining in different environments. Further focus should be in combining data provided by Google and Yahoo into the above-described models to eliminate some of the concerns associated with distributed data mining such as privacy preservation, efficient processing of humongous data, and so on.

References

1. Kargupta, H., Kamath, C., Chan, P.: Distributed and parallel data mining: emergence, growth, and future directions. In: Advances in Distributed and Parallel Knowledge Discovery, pp. 409–416. AAAI/MIT Press (2000)
2. Li, T., Zhu, S., Ogihara, M.: A New Distributed Data Mining Model Based on Similarity. ACM (2003)
3. Datta, S., Giannella, C., Kargupta, H.: K-means clustering over a large dynamic network. In: SDM, pp. 153–164 (2006)
4. Rekha Sunny, T., Thampi, S.M.: Survey on distributed data mining in P2P networks. CoRR (2012)
5. Jelasity, M., Montresor, A., Babaoglu, O.: Gossip-based aggregation in large dynamic networks. ACM Trans. Comput. Syst. 23(3), 219–252 (2012)
6. Mehyar, M., Spanos, D., Pongsajapan, J., Low, S.H., Murray, R.M.: Asynchronous distributed averaging on communication networks. IEEE/ACM Trans. Netw. 15(3) (2007)
7. Kempe, D., Dobra, A., Gehrke, J.: Gossip-based computation of aggregate information. In: Proceedings of 44th Annual IEEE Symposium on Foundations of Computer Science (2003)
8. Kowalczyk, W., Jelasity, M., Eiben, A.E: Towards data mining in large and fully distributed peer-to-peer overlay networks. In: Proceedings of BNAIC'03, pp. 203–210 (2003)
9. Wolff, R., Schuster, A.: Association rule mining in peer-to-peer systems. IEEE Trans. Syst. Man Cybern. 34(6) (2004)
10. Schuster, A., Wolff, R.: Association rule mining in peer-to-peer systems. In: Proceedings of the 3rd International Conference on Data Mining (2003)
11. Siersdorfer, S., Sizov, S.: Automatic document organization in a P2P environment. LNCS, vol. 3936, pp. 265–276 (2006)
12. Eisenhardt, M., Müller, W., Henrich, A.: Classifying Documents by Distributed P2P Clustering (2003)
13. Hammouda, K.M., Kamel, M.S.: Hierarchically distributed peer-to-peer document clustering and cluster summarization. IEEE Trans. Knowl. Data Eng. 21(5), 681–698 (2009)

14. Ahamed, B.B., Hariharan, S.: A survey on distributed data mining process via grid. Int. J. Database Theory Appl. **4**(3) (2011)
15. Cannataro, M., Talia, D.: Knowledge grid an architecture for distributed knowledge discovery. ACM **46**(1), 89–93 (2003)
16. von Laszewski, G., Ruscic, B., Amin, K., Wagstrom, P., Krishnan, S., Nijsure, S.: A framework for building a scientific knowledge grid applied to thermochemical tables. Int. J. High Perform. Comput. Appl. **17**(4) (2003)
17. Wankar, R.: Grid computing with globus: an overview and research challenges. Int. J. Comput. Sci. Appl. **5**(3), 56– 69
18. Talia, D.: Grid-based Distributed Data Mining Systems, Algorithms and Services

Graph Representation of Multiple Misconceptions

Sylvia Encheva

Abstract Subject-related misconceptions take place at nearly all levels of education where the process of identifying and overcoming them is most of the time quite laborious. The majority of related research is addressing single misconceptions in particular subjects. In this paper, we propose a graphical representation of interrelations among multiple misconceptions indicated by test results. Appropriate actions based on tests outcomes implying misconceptions are also suggested.

Keywords Misconceptions · Test outcomes · Graphical representation

1 Introduction

Students knowledge and skills are regularly evaluated throughout their studies where most of the time conclusions refer to degrees to which they have learned a subject. While the amount of research on testing students learning is really enormous, it is easy to notice that approaches to identify and overcome misconceptions have received much less attention. One possible explanation to that might be that some students have to take large-enrollment courses where opportunities for the active pursuit of the unknown during a lecture appear to be somewhat difficult even while taking advantages of up to date information technologies. Another one is related to the fact that it is not always that simple to distinguish between lack of knowledge and misconceptions. In this paper, we are discussing possibilities for what can be done in case of multiple misconceptions being detected.

Bipartite graphs [3] are employed to facilitate a search for possible patterns among misconceptions. Ordered sets [2] are to be applied for ranking already identified misconceptions. The outcome is very helpful while choosing appropriate actions to enhance students learning.

S. Encheva (✉)
Western Norway University of Applied Sciences, Bjørnsonsg. 45,
5528 Haugesund, Norway
e-mail: sbe@hvl.no

© Springer Nature Singapore Pte Ltd. 2018   89
S. C. Satapathy et al. (eds.), *Information and Decision Sciences*,
Advances in Intelligent Systems and Computing 701,
https://doi.org/10.1007/978-981-10-7563-6_10

2   Orderings and Graphs

The problem of determining a consensus from a group of orderings and the problem of making statistically significant statements about orderings are addressed in [2].

A partial ordering whose indifference relation is transitive is called a *weak ordering*. All partial orderings of a set with three elements can be seen in [2].

A bipartite graph is a graph whose vertices can be divided into two disjoint sets where every edge connects a vertex in one set to a vertex in the other set, [5].

Different approaches for remediation of misconceptions in physics are presented in [4, 6]. Fostering conceptual change in different domains is discussed in [7].

3   Misconceptions

The main idea in this section is to find patterns among possible students misconceptions. Students are suggested to take a test. Their understandings of four concepts are to be evaluated by analyzing tests' outcomes. Such tests are usually expected to address problems in a unit within a subject, but the below-presented structure allows various applications. This means that if a test is addressing understanding of concepts initially represented in different subjects one can detect possible transfer of misconceptions from one subject to another. These types of transfers may easily go unnoticed when assessment of knowledge and skills is essential for grading.

In a case submitted answers indicate misconceptions, appropriate assistance will be provided in the form of personalized theoretical explanations and examples, after which the student may retake the test. Can misconceptions be related in a sense that if there is evidence for one or several misconceptions then there is a strong possibility for other particular misconceptions to be detected? In this work, we present a graphical illustration of misconceptions possibly appearing together in students responses.

To the rest of this work letters "i, j, m, n" stand for correct answers to the first, second, third, and fourth test questions where "k, o, v, w" stand for incorrect answers to these questions. Omitted answers are treated as wrong answers.

All types of responses are assigned to the vertices of the graph in Fig. 1 previously shown in [1]. Needless to say, a couple of tests responses can differ in one, two, three, or four answers (each test contains four questions). This can become very handy when a lecture is focusing on either individual or groups progress. For single students, it will assist in the process of providing personalized guidance where clustering of group responses simplifies the search for teaching materials that need modifications.

Vertices whose labels differ in one position only are connected by a single edge, Fig. 2.

Examples of response types which differ in two positions are listed below.

Any of the four groups of vertices

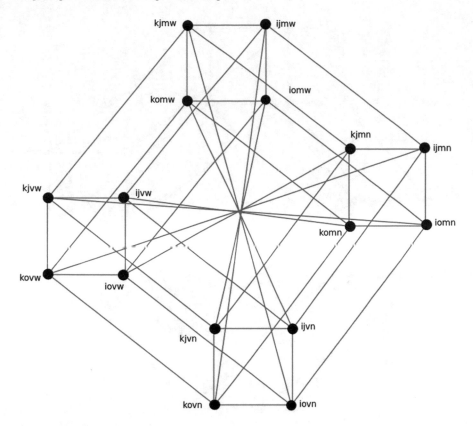

Fig. 1 Response types

- $\{k,j,m,w\}$, $\{k,j,v,w\}$, $\{k,j,v,n\}$, $\{k,j,m,n\}$ (highlighted in red, Fig. 3)
- $\{i,j,m,w\}$, $\{i,j,m,n\}$, $\{i,j,v,n\}$, $\{i,j,v,w\}$
- $\{k,o,m,w\}$, $\{k,o,m,n\}$, $\{k,o,v,n\}$, $\{k,o,v,w\}$
- $\{i,o,m,w\}$, $\{i,o,m,n\}$, $\{i,o,v,n\}$, $\{i,o,v,w\}$

contains the same response types in the first two positions.

Any of the four groups of vertices

- $\{k,j,m,w\}$, $\{k,o,m,w\}$, $\{i,o,m,w\}$, $\{i,j,m,w\}$ (highlighted in blue, Fig. 3)
- $\{k,j,v,w\}$, $\{k,o,v,w\}$, $\{i,o,v,w\}$, $\{i,j,v,w\}$
- $\{k,j,v,n\}$, $\{k,o,v,n\}$, $\{i,o,v,n\}$, $\{i,j,v,n\}$
- $\{k,j,m,n\}$, $\{k,o,m,n\}$, $\{i,o,m,n\}$, $\{i,j,m,n\}$

contains the same response types in the last two positions.

Any of the four groups of vertices

- $\{i,j,v,w\}$, $\{i,o,v,w\}$, $\{i,o,v,n\}$, $\{i,j,v,n\}$ (highlighted in green, Fig. 3)
- $\{k,j,v,w\}$, $\{k,o,v,w\}$, $\{k,o,v,n\}$, $\{k,j,v,n\}$

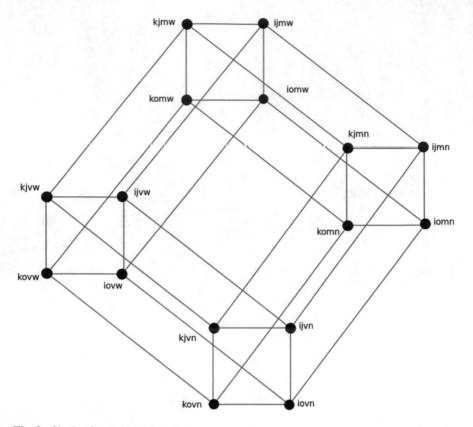

Fig. 2 Single edge connected vertices

- $\{k,j,m,w\}$, $\{k,o,m,w\}$, $\{k,o,m,n\}$, $\{k,j,m,n\}$
- $\{i,j,m,w\}$, $\{i,o,m,w\}$, $\{i,o,m,n\}$, $\{i,j,m,n\}$

contains the same response types in the first and third positions. These are only three of many other existing groups of vertices with similar properties.

The four vertices $\{i,j,v,w\}$, $\{k,j,v,w\}$, $\{k,o,m,w\}$, and $\{k,o,v,n\}$, (highlighted in read) whose labels differ in exactly three positions from vertex $\{i,j,m,n\}$ (highlighted in green), are shown in Fig. 4. The rest of such related vertices are placed in a similar configuration.

Any two vertices that do not share any response type are connected by an edge highlighted in blue in Fig. 4.

If several students answers appear to be in one of the groups described above it is beneficial to look at the particular pattern, provide appropriate assistance and follow the development. Further adjustments of teaching materials, questions' formulations, and help functions might be needed.

Let us now look into more details at how a vertex is connected to other vertices and keep in mind that all vertices are connected in the same way due to this graph

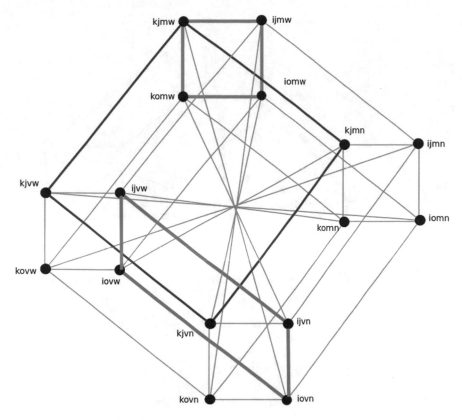

Fig. 3 Connected vertices implying similar responses

properties. A vertex $\{k, o, v, w\}$ in Fig. 5 is connected to five other vertices where: two of them (highlighted in green) are its neighbor vertices in the same square ($\{k, o, v, w\}$, $\{j, o, v, w\}$, $\{i, j, v, w\}$, $\{k, j, v, w\}$), two other vertices (highlighted in red) belong to the two neighbor squares ($\{k, o, m, w\}$, $\{i, o, m, w\}$, $\{i, j, m, w\}$, $\{k, j, m, w\}$) and ($\{k, o, v, n\}$, $\{i, o, v, n\}$, $\{i, j, v, n\}$, $\{k, j, v, n\}$), and a vertex belonging to the opposite square ($\{k, o, m, n\}$, $\{i, o, m, n\}$, $\{i, j, m, n\}$, $\{k, j, m, n\}$) where they do not share any response type (highlighted in blue).

Suppose a student is taking a test for the first time and her answers are "$\{k, o, m, w\}$". She is then suggested to go through selected learning materials and then take the test second time. All tests appear seemingly different, i.e., they are addressing the same concepts with different questions. If the answer to the second test is, Fig. 5:

- "$\{k, j, v, w\}$" or "$\{i, o, v, w\}$" (highlighted in green) then either the original presentation of the last two concepts might need adjustment and/or the related learning materials should be revised,

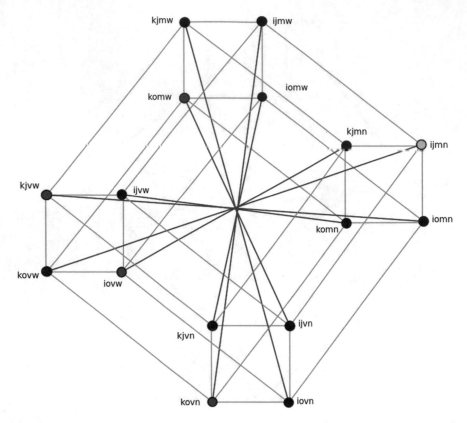

Fig. 4 Vertices whose labels differ in exactly three positions

- "$\{k, o, m, w\}$" or "$\{k, o, v, n\}$" (highlighted in red) then the original presentation of the first two concepts might need adjustment and/or the related learning materials should be revised,
- "$\{i, j, m, n\}$" (highlighted in blue) then it is of interest to find out whether the same help is appropriate for more students.

Suppose two consecutive responses from a student appear in vertices not connected by a single edge. In such cases, the provided help should follow recommendations related to single concepts being misunderstood.

Thus presented graphical illustration of students responses gives a good indication of the clarity of tests contents and usefulness of provided help.

While offered to take a test, students will also be asked to rank the concepts with respect to difficulties. Based on their responses the four concepts will be ordered and appear in one of the forms in Fig. 6. Concepts are denoted by a, b, c, d in order to be independent of local notations. They are further on placed in sets IV, III, II according to the number of concepts being compared by a student. Thus, the set IV contains all cases where all four concepts in a test are compared, in the set III only

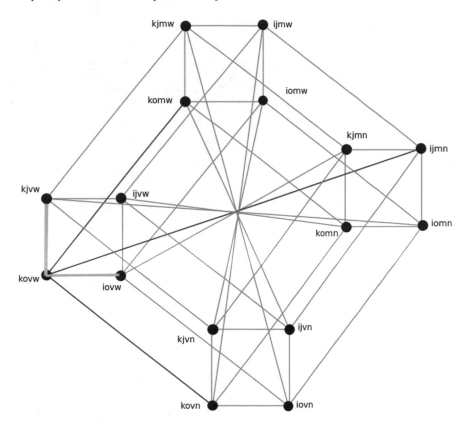

Fig. 5 Connected vertices

three of the four are compared, in the set *II* only two of the four are compared, and the set containing cases where students do not provide any opinions is trivial and therefore omitted. The top concept in any of the sets *IV, III, II* is supposed to be the most difficult one to understand and when concepts appear unrelated, it means they have not been compared by test takers.

Note that ranking concepts with respect to difficulties is not always related to correct responses. The purpose of ranking is to receive feedback from students and take some actions afterward to facilitate better understanding.

Remark: Distances in graphs can be also applied to facilitate automated evaluations of tests responses. Suppose a group of students takes a test only two times. This data can be used to compare distances among responses from another group of students that take the test only two times. As a result, modifications of concepts presentations might be introduced. When students take the test three times or more, both individual and group data can be used to improve knowledge transfer.

Fig. 6 Concepts ranking

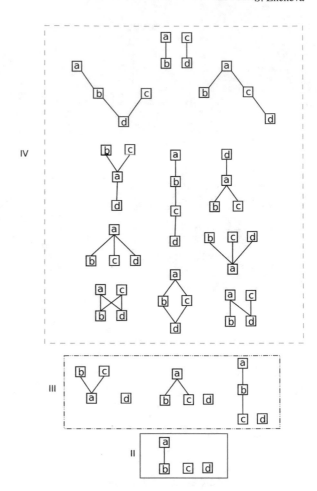

4 Conclusion

Conceptual understanding can often be very demanding and energy consuming. To
prevent the latter, we propose a test of only four questions addressing four concepts.
Thus presented graphical illustration of students responses gives a clear indication
about the quality of test's contents and usefulness of provided help. Furthermore
detailed consideration of tests outcomes and concept ranking can contribute to the
adjustment of teaching materials.

References

1. Abay-Asmerom, G., Hammack, R.H., Larson, C.E., Taylor, D.T.: Direct product factorization of bipartite graphs with bipartition-reversing involutions. SIAM J. Discrete Math. **23**(4), 2042–2052 (2010)
2. Bogart, K.P.: Some social sciences applications of ordered sets. In: Rival, I. (ed.) Ordered Sets, pp. 759–787. Reidel, Dordrecht (1982)
3. Bollobas, B.: Extremal Graph Theory. Dover (2004)
4. Brown, D.E.: Using examples and analogies to remediate misconceptions in physics: factors influencing conceptual change. J. Res. Sci. Teach. **29**(1), 17–34 (1992)
5. Hammack, R.: On uniqueness of prime bipartite factors of graphs. Discrete Math. **313**(9), 1018–1027 (2013)
6. Kikas, E.: University students' conceptions of different physical phenomena. J. Adult Develop. **10**(3), 139–150 (2003)
7. Mason, L.: Developing epistemological thinking to foster conceptual change in different domains. In: Limon, M., Mason, L. (eds.) Reconsidering Conceptual Change: Issues in Theory and Practice, pp. 301–335. Kluwer Academic Publishers, Netherlands (2002)

Classification Through Discriminant Analysis Over Educational Dataset

Parag Bhalchandra, Aniket Muley, Mahesh Joshi,
Santosh Khamitkar, Hanumant Fadewar and Pawan Wasnik

Abstract This work highlights the effective strategy for classification values related to categorical dependent using discriminant function analysis. A personal dataset of students was created with social, economical, and academic variables for the actual empirical analysis. The discriminant function analysis was implemented to study the behavior patterns against the performance of students. It was observed that some of the variables are playing an important role in discrimination and based on our data, as we found 67.4% is the only correct classification.

Keywords Discriminant function analysis · Data mining · Analysis of patterns Statistical analysis

P. Bhalchandra (✉) · S. Khamitkar · H. Fadewar · P. Wasnik
School of Computational Sciences, S.R.T.M. University,
Nanded 431606, Maharashtra, India
e-mail: srtmun.parag@gmail.com

S. Khamitkar
e-mail: s.khamitkar@gmail.com

H. Fadewar
e-mail: fadewar_hsf@rediffmail.com

P. Wasnik
e-mail: pawan_wasnik@yahoo.com

A. Muley (✉)
School of Mathematical Sciences, S.R.T.M. University,
Nanded 431606, Maharashtra, India
e-mail: aniket.muley@gmail.com

M. Joshi
School of Educational Sciences, S.R.T.M. University,
Nanded 431606, Maharashtra, India
e-mail: maheshmj25@gmail.com

© Springer Nature Singapore Pte Ltd. 2018
S. C. Satapathy et al. (eds.), *Information and Decision Sciences*,
Advances in Intelligent Systems and Computing 701,
https://doi.org/10.1007/978-981-10-7563-6_11

99

1   Introduction

In general, appropriate classification is an important part of the analysis for prediction of certain tasks. The discriminate analysis between two and more naturally
occurring groups with respect to a variable is often taken into consideration via
discriminate functional analysis as per Grimm and Yarnold [7]. It is basically a
classification which acts for distributing variables into groups of the same type. It
ultimately aids for the decision-making process. Mathematically, it is expressed as
an investigation for average group differences. The outcome, that is gained
knowledge, is then used to understand group membership of new variables. It is alike
to the analysis of variance (ANOVA), if worked out computationally. Discriminant
analysis has a base of matrix theory. The ANN, soft computing, and machine
learning can be alternatives to discriminate analysis as studied by Huberty and
Olejnik [9]. The discriminant analysis is useful when groups are known already [3].
This is how it is different than clustering, as highlighted by Sasirekha and Baby [14].
However, the prerequisite in doing so is the knowledge of quantitative predictor
measures, and group measures as seen in Stevens [16], and Tatsuoka and
Lohnes [17].

   We have an operational student's dataset with us. It has been tailored by us from
actual student's data. For time being, this dataset has been processed through
several data mining studies, in order to explore deep insights. In the underlined
study, we have made one more attempt to deploy the discriminant function to
establish split between the groups. We also strive to determine which the variable(s)
of the dataset are the best predictors of student's satisfaction? The study is further
extended to determine the best of discrimination between the types of students as
well as to interpret group differences and to classify new objects. The performance
of students and the satisfaction of students related to education is the chief spotlight
of our study. This work is carried out collaboratively within three Schools of our
University. The primary source of data was involved for the creation of own
dataset. The data was collected in the form questionnaire being circulated to the
students during their stay at School of Computational Sciences.

2   Methodology

Multivariate statistical analysis typically evaluates to diminish the original dimension of data. Discriminant analysis has been widely used in sympathetic classification problems dealt by some researchers, including Shin and Fong [15],
Durán et al. [4], Wilson [19], Toldo and Zouain [18], Caeiro et al. [2], Khan
mohammadi et al. [11], Roshani et al. [13], and Geranian et al. [5]. The general
discriminant equation, in unstandardized units, is expressed in the following form
[7, 16]:

$$f = B_0 + B_1X_1 + B_2X_2 + \ldots + B_nX_n, \tag{1}$$

where f is the discriminant function,

X_1, X_2, ... X_n are the independent variables, and

Canonical coefficients, i.e., B_1, B_2, ... B_n for each variable.

Equation (1) represents the probability of group function to be calculated for all cases. This method is used to determine cases which have been the accurate classification of a data set. The Statistical Package for the Social Sciences, i.e., SPSS 22.0v license copy is used for analyzing the dataset.

In order to do proper empirical analysis using discriminate analysis, a student's data set was devised out using a questionnaire which has questions with predefined options as per Shoukat Ali et al. [1], Gratez [6], and mainly related to social, economical, and physical aspects of the students. This questionnaire was defined as per Minaei-Bidgoli et al. [12], Shoukat Ali et al. [1]. Only the attributes related to satisfaction of students were considered. The dataset was having 360 student records. Every record in dataset consists of 46 fields. There were 43 questions related to student's aspects. Remaining 03 questions were related to personal details. The dataset was processed using SPSS platform as per the remarks of Han and Kamber [4, 8] and SPSS [10]. The research hypothesis was set to classification through discriminant analysis over the educational dataset. The selected variables for this hypothesis includes the following:

1. SCHOLERSHIP—Do you get a scholarship?
2. SELF-LIB—Do you have an own book library?
3. SELF-PC—Do you have an own personal computer at home?
4. PLACELVING—Where do you live? (Own house, Shared Room, Hostel?)
5. INTERNET—Do you use Internet?
6. F_T_STUDY—Free time available for self-study
7. F_T_FRIEND—Free time to spare with Friends
8. FAILURES—Number of failures in past examinations
9. F-INCOM—Father's Income
10. M-INCOM—Mother's Income.

3 Results and Discussions

In this section, the discovery of discriminating variables is illustrated. After completion of data collection, it was processed through the cycle of data processing where answers to questions were coded into numerical figures. The collected data was investigated as a whole and trash. The accuracy and completeness were checked. Further, for cross tabulation, comparison and inferential analysis have been performed through SPSS software. We have observed that some of the

Table 1 Group statistics

PER_SATISF	Mean	Std. deviation	Valid N (listwise)	
			Unweighted	Weighted
SCHOLERSHIP	0.54	0.501	95	95.000
SELF-LIB	0.39	0.490	95	95.000
SELF-PC	0.59	0.495	95	95.000
PLACELVING	1.60	0.777	95	95.000
INTERNET	0.94	0.245	95	95.000
F_T_STUDY	2.47	0.742	95	95.000
F_T_FRIEND	2.12	0.666	95	95.000
FAILURES	0.37	1.072	95	95.000
F-INCOM	1.60	0.927	95	95.000
M-INCOM	1.06	0.285	95	95.000

variables are playing the important role in discrimination. Further, based on our data, we have found that 67.4% is the only correct classification. Following facts remain as the base of our results.

Table 1 represents the ultimate group statistics. The distribution of observations was recorded via group statistics. There were two categories or groups like 1 (means Yes) and 2 (means No). We have recorded 359 numbers of total observations (as 100% observations). The SPSS platform computes one function only.

Once group statistics is over, we strive for eigenvalues to understand the discrimination degree of the underlined function. As only one function participated, it accorded 100% of the variance. Then cumulative percentage (%) of the variance is calculated. Again it accorded for 100% as only one function participates in the discriminate analysis. Table 2 highlights the canonical correlations of our variables. It is to be noted that the one discriminant function is used to compute for analysis.

The reviewed literature highlights that higher correlations values always prettify the function. When there are two groups, the canonical correlation is the most useful measure (Table 2) and is the same as Pearson's correlation between discriminant scores and groups. A function having discriminate value 1 is always considered as a perfect function. Our analysis in SPSS gave a correlation of 0.251. It is positive in nature but comparatively not high. Hence, we need to test the hypothesis regarding discriminating power of the variables. We propose the following two hypotheses:

1. Null Hypothesis H_0: This highlights zero or nil power of significant discrimination among the variables.

Table 2 Discriminant function summary: Eigen values

Function	Eigen values	% Variance	Cum. (%)	Canonical correlation
1	0.067	100.0	100.0	0.251

2. Alternate Hypothesis H_1: A significant discriminating power in the variables may be there.

After the execution of Chi-square test with a confidence set to 95%, we obtained P-value = 0.000. The SPSS platform has a provision for multivariate statistics, often implemented as Wilk's lambda. If its value is less, the Wilk's lambda is considered as better. This means that it signifies greater discriminating ability of the function. We got this value as 0.937. If it is one, the group means are equal. Further, the Chi-square is 23.246 with 2 degrees of freedom. Our zero value of Wilk's lambda means group means appear to differ. Since x = 0.05, we reject H_0 and accept H_1. This forces us to go ahead with the discriminant analysis. This is illustrated in Table 3.

Discriminant analysis is carried out by looking for standardized coefficients and their comparison. This comparison is given in Table 4. It is witnessed that the free time to study has the highest discriminating power due to the highest discriminant coefficient of 0.762 followed by mother's income.

Here, we have two predictor factors that are free time to study and mother's income. Since the predictive equation is being constructed, the nonstandardized canonical coefficient will be used to construct the discriminant function as follows:

$$Z = -6.509 + 1.182(\text{F\_T\_STUDY}) + 3.778(\text{M} - \text{INCOM}). \qquad (2)$$

Thus, we got the nonstandardized scores of the independent variables through the Canonical Discriminant Function coefficients. This is illustrated in Tables 5 and 6.

Huberty and Olejink [9], Dunham [3] initiated Group Centroid method to have an average discriminant score of the participating groups. There are two groups in our study. Hence, there are two scores. These two groups, viz., the *Yes group* and the *No group,* are not equal (95 vs. 264). Their absolute value scores are same but the discriminating scores are opposite. In order to deal with these facts, a divide

Table 3 Wilk's λ

Test of function(s)	Wilk's λ	Chi-square	df	sig
1	0.937	23.246	2	0.000

Table 4 Standardized canonical discriminant function coefficient

Details	Function 1
F_T_STUDY	0.762
M-INCOM	0.587

Table 5 Coefficients of the function

	Function 1
F_T_STUDY	1.182
M-INCOM	3.778
(Constant)	−6.509

Table 6 Group centroid analysis

PER_SATISF	Function 1
No	0.432
Yes	0.155

point strategy on the centroid is used. The weights are assigned on centroids. The decision rule classification will be as follows:

1. Foresee and sort as Yes, if $-0.155 < Z < 0.0003343$.
2. Foresee and sort as No, if $0.0003343 < Z < 0.432$.

Our empirical analysis was successful in the sense that all 359 observations have been processed successfully. The classification summary is illustrated in Table 7 and prior probabilities are illustrated in Table 8.

Now we need to analyze the confusion matrix. This can be done by verifying discriminant function's ability to predict, that is, how many of these systems were NO or YES systems? In order to do so, we have compared actual data with the discrimination function. There will be no error if these two values are same. Otherwise, if these two values are different, there are errors in our model. This consideration is mentioned in Table 9.

Table 7 SPSS output: classification processing summary

	Processed	359
Excluded	Missing group codes	0
	Minimum one missing	0
	Actually used in output	359

Table 8 SPSS output prior probabilities for groups

PER_SATISF	Prior	Cases used in analysis	
		Unweighted	Weighted
No	0.500	95	95.000
Yes	0.500	264	264.000
Total	1.000	359	359.000

Table 9 Classification accuracy results

		PER-SATISF	Predicted group membership		Total
			No	Yes	
Original	Count	No	46	49	95
		Yes	68	196	264
	%	No	48.4	51.6	100.0
		Yes	25.8	74.2	100.0

Our discriminant function has correctly classified 67.4% of data in terms of *Yes and No systems*. Thus, out of the 359 systems, 196 systems have been correctly classified as *Yes systems*. Similarly, out of the 95 systems, 46 systems have been correctly classified as *No systems*. Further, 49 schemes have been wrongly classified as *Yes systems*. The exactness of the model is thus satisfactory. This highlights a considerable predictive capacity of the discriminant function. Table 2 reveals the significance of Wilk's λ for all the independent variables (less than 0.05) and its least value shows the importance of variable for the discrimination. The group means case score for each group (centroid) is shown in Table 6. Table 4 explores that the variable, i.e., free time to study represents positive relationship and it explains classification of the first group of samples. The results of the data predicition show correct classification about 67.4% and there is reclassification from group one two is only 3 samples, and second groups 128 samples were reclassified in (Table 9). The outcome of this study reveals that free time to study and mother's income are the major factors inappropriately included in the discriminant function.

4 Conclusion

In this study, we have focused on student's satisfactory performance based on the collected information about 359 student's dataset. Discriminant analysis is performed to test certain parameters affecting performance. It has been observed that 67.4% of data was correctly classified as Yes and No of the system by the discriminant function. Mother's income and free time to study are the parameters which play important role in discrimination or classification of student's satisfactory performance. This study focuses on various personal, educational, and socioeconomical information parameters. Limitation of the study comes from the reality that this research was carried out only on the students of computer science school and thus there may be other data sets which will show different characteristics.

Declaration Authors hereby declare that the student data has been collected from college with informed consent from the college management as well as higher authorities.

References

1. Ali, S., Haider, Z., Munir, F., Khan, H., Ahmed, A.: Factors contributing to the students academic performance: a case study of Islamia University Sub Campus. Am. J. Edu. Res. **1** (8), 283–289 (2013)
2. Caeiro, S., Costa, M.H., Goovaerts, P., Martins, F.: Benthic biotope index for classifying habitats in the Sado estuary: Portugal. Marine Environ. Res. **60**(5), 570–593 (2005)
3. Dunham, M.H.: Data mining: introductory and advanced topics. Pearson Education India (2006)

4. Durán, R., García-Gil, S., Vilas, F.: Side Scan Sonar application to seabed mapping of the Ría de Pontevedra (Galicia, NW Spain). A case of study. J. Iber. Geol. **26**, 45–66 (2000)
5. Geranian, H., Tabatabaei, S.H., Asadi, H.H., Carranza, E.J.M.: Application of discriminant analysis and support vector machine in mapping gold potential areas for further drilling in the Sari-Gunay gold deposit. NW Iran. Natural Resour. Res. **25**(2), 145–159 (2016)
6. Graetz, B.: Socio-economic status in education research and policy in John Ainley et al., socio-economic status and school education DEET/ACER Canberra. J. Pediatr. Psychol. **20** (2), 205–216 (1995)
7. Grimm, L.G., Yarnold, P.R.: Reading and understanding multivariate statistics. Am. Psychol. Aεεoc. (1995)
8. Han, J., Pei, J., Kamber, M.: Data mining: concepts and techniques. Elsevier (2011)
9. Huberty, C.J., Olejnik, S.: Applied MANOVA and discriminant analysis, vol. 498. Wiley
10. IBM SPSS Statistics 22 Documentation on internet. https://www.ibm.com/support/docview. wss?uid=swg27038407
11. Khanmohammadi, M., Garmarudi, A.B., Samani, S., Ghasemi, K., Ashuri, A.: Application of linear discriminant analysis and attenuated total reflectance fourier transform Infrared microspectroscopy for diagnosis of colon cancer. Pathol. Oncol. Res. **17**(2), 435–441 (2011)
12. Minaei-Bidgoli, B., Kashy, D.A., Kortemeyer, G., Punch, W.F.: Predicting student performance: an application of data mining methods with an educational web-based system. In: Frontiers in Education, vol. 1, p. T2A-13. FIE 2003 33rd annual, IEEE 2003 Nov
13. Roshani, P., Mokhtari, A.R., Tabatabaei, S.H.: Objective based geochemical anomaly detection—application of discriminant function analysis in anomaly delineation in the Kuh Panj porphyry Cu mineralization (Iran). J. Geochem. Explor. **130**, 65–73 (2013)
14. Sasirekha, K., Baby, P.: Agglomerative hierarchical clustering algorithm-A. Int. J. Sci. Res. Publ. 83
15. Shin, P.K.S., Fong, K.Y.S.: Multiple discriminant analysis of marine sediment data. Mar. Pollut. Bull. **39**(1), 285–294 (1999)
16. Stevens, J.P.: Applied multivariate statistics for the social sciences, pp. 510–511. Lawrence Erlbaum. Mahwah, NJ (2002)
17. Tatsuoka, M.M., Lohnes, P.R.: Multivariate analysis: techniques for educational and psychological research. Macmillan Publishing Co, Inc (1988)
18. Toldo, E.E., Zouain, R.N.A.: Environmental monitoring of offshore drilling for petroleum exploration (MAPEM): a brief overview. Deep Sea Res. Part II **56**(1), 1–3 (2009)
19. Wilson, D.I.: Derivation of the chalk superficial deposits of the North Downs, England: an application of discriminant analysis. Geomorphology **42**(3), 343–364 (2002)

Dynamic and Secure Authentication Using IKC for Distributed Cloud Environment

M. Ranjeeth Kumar, N. Srinivasu and Lokanatha C. Reddy

Abstract Distributed computing is one of the most significant recent paradigms facing IT organizations. Since this new handling technology requires clients to believe in their information being useful to providers, there have been problems regarding the enhancement of security and details were explored. A few strategies using quality based security have been prescribed for the administration of access of abbreviated subtle elements in cloud computing. However, be that as it may, the greater part of them are resolute in actualizing complex the rules of accessibility administration. To provide effective secure authentication for multi-user data sharing in the cloud computing environment, Transmitted Team Key Management (TTKM) has traditionally been used for the sharing of distributed data between multiple users in cloud computing settings. This allows the users to share their data securely using Shamir secret key sharing. One major limitation of TTKM is in the provision of inner side security in data sharing for multiple files with single security considerations in the distributed cloud environment. To access this limitation properly in the distributed environment, in this paper we propose the Integrated Key Cryptosystem (IKC) for multiple file sharing with a single aggregate key for single user data sharing. It is the combination of different security systems in attributes-based encryption. Our experimental results shows effective data utilization in the real time distributed cloud environment with different file sharing in a cloud computing environment.

Keywords Distributed computing · Transmitted team key management
Integrated key cryptosystem · Aggregate key · Cloud environment

M. R. Kumar (✉) · N. Srinivasu
Department of CSE, KL University, Guntur, Andhra Pradesh, India
e-mail: maduri.ranjith@gmail.com

N. Srinivasu
e-mail: srinivasu28@kluniversity.in

L. C. Reddy
Department of CS, Dravidian University, Kuppam, Andhra Pradesh, India
e-mail: lokanathar@yahoo.com

© Springer Nature Singapore Pte Ltd. 2018   107
S. C. Satapathy et al. (eds.), *Information and Decision Sciences*,
Advances in Intelligent Systems and Computing 701,
https://doi.org/10.1007/978-981-10-7563-6_12

1   Introduction

Cloud computing is a model for efficient authentication of extensive system access in order to share configurable PC assets. Multiple data sharing and accumulated choices furnish customers and organizations with the ability to load and systematize their data in visitor data offices [1]. It defines different sources to explore services related frameworks of extent, like an application (like the force network) above a framework. At the base of cloud preparation is the more extensive idea of con-solidated offices and disseminated administrations.

As shown in Fig. 1 above, distributed computing defines and provides three services regarding cloud setup and the processing of different services in the dis-tributed environment. SAAS (Software As A Service), and PAAS (Platform As A Service) are the two different services used for cloud handling of storage space data and which preserve privacy of information, including both of users and file infor-mation which are the motivation for the provision of privacy [2]. Some companies currently have ACPs controlling which clients can access which information. These ACPs are frequently demonstrated as far as the characteristics of the clients, and for the most part are known as distinguishing proof components, utilizing available administration dialects such as XACML. Such a methodology, usually referred to as property-based availability and controllability, encourages a fine-grained open-ness administration which is pivotal for high-affirmation data security and peace of mind.

ABE (Attribute Based Encryption) with TTKM allows different encrypted fea-tures to generate cipher text data. Normally, for multi-user data sharing with dif-ferent privacy with user authentication in the distributed environment as shown in

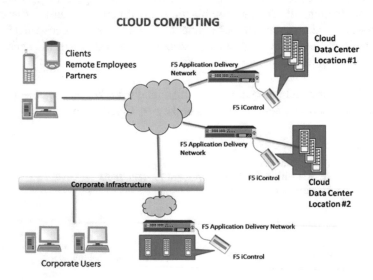

Fig. 1 Distributed computing with different resource provisioning [3]

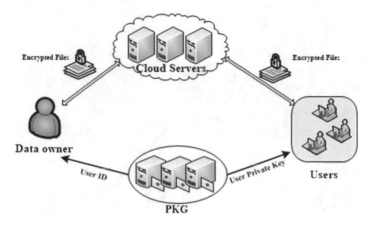

Fig. 2 Privacy based encryption over multi user access control in cloud computing [2]

Fig. 2, TTKM performs multi-user authentication with normal encryption and decryption using different attribute-based encryption methods. The ABE strategy identifies different features (consisting of a set of protected and decrypted options in data outsourcing). One of the main limitations of using TTKM with ABE is the extensive resource utilization for multi-file data sharing with different expressive resources with combined features to decrypt and provide privacy-oriented authentication with reduced complexity for different files with single key procedures. The TTKM strategy can result in the issue of owners needing to use every accepted client's group key to protected details. In order to provide a solution to this drawback in the cloud computing environment, in this paper we propose IKC (Integrated Key Cryptosystem) for multiple file sharing with an integrated key service policy in the distributed cloud environment. The following procedure is an example of data sharing in a distributed environment.

In terms of the procedure for implementation of IKC with different user authentication in different file sharing in the cloud environment, we are using DropBox to store different pictures for secure storage purposes. The example uses the scenario of two different people named Alice and Bob. Based on our concept of data sharing in the cloud, Alice "friends" Bob regarding sharing different pictures, using outside environment of DropBox. Alice manages all the keys with different progressions to allow Bob's pictures into her data representations. It is Alice's decision to convey the vital variables to share one person's pictures with other people that have accessible files in secure cloud sharing. At present, there are two representative systems for her under the conventional view of security.

a. Alice gets overall information from other users present in the distributed cloud environment with their own records, in order to forward different formations.
b. Alice gets data with an unwanted imperative for individual components for key alliance with critical variables.

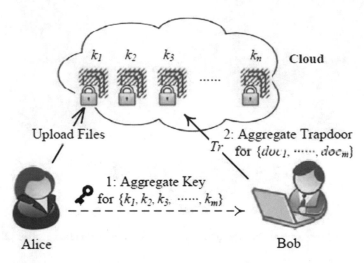

Fig. 3 Based on Integrated key specification, Alice shares secure keys for uploaded files

As shown in Fig. 3, Alice checks individual file sharing with different files in the communication process. In this way, the optimum solution is that Alice climbs information data with novel undone clues, yet only conveys skip stand out (steady size) decoding clue [4]. Since the scrambling clue is sent by means of a protected arm and retain key, minimal clue measurement is continually suitable. For example, it is not possible to suspect gigantic stockpiling for unscrambling imperative elements such as different things such as approach cells, astute charge cards or Wi-Fi pointer axis.

These key essential variables are typically spared in the carefully designed capacity, which is generally spending. The adjacent investigation activities chiefly focus on minimizing the cooperation particulars such as information exchange utilization, and data with different operations.

Organization of the paper: Background work on multi-user data sharing is discussed in Sect. 2; Aggregated key cryptosystem procedure is defined in Sect. 3; the implementation procedure of IKC is discussed in Sect. 4; the experimental evaluation of proposed approach is formalized in Sect. 5 and Sect. 6 provides and overall conclusion for integrated key management in the distributed environment.

2 Background Work

This section describes and discusses the background of Transmitted Team Key Management (TTKM) approach to secure data storage and the transmission of data between different users. The TTKM procedure is shown below:

TTKM consists of two entities for data maintenance i.e. a key server (svr) and group members (user's), and is based on server configuration to send or distribute

Fig. 4 Distributed key
process between different
users in TTKM procedure

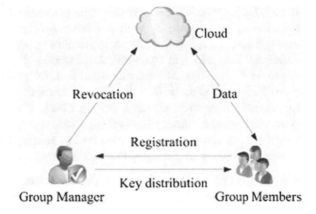

different files for different users into sequential data forwarding between different files to different users. The TTKM procedure is as follows:

ParamGen: Server requires security parameters i.e. k,s,r for individual users into the distributed environment in order to setup communication based on possible key operations and principles on the usr side.

TkDeliv: Server (Svr) delivers files into each usr based on personal information present in the cloud for the accessing of data with a secure format.

KeyGen: Server (svr) selects a distributed group key based on personal information category data representation in the distributed environment. Subsequently, svr deploys all the user specifications based on transmitted channel representation.

KeyDer: Based availability of user data with respect to key management with different parameters. Server produces public information to multiple users with a distributed key via transmitted channel representation.

The procedure of TTKM with different users and a distributed key specification is shown in Fig. 4.

It is based on distributed key management for different users, multiple file accessing with key distribution and also performs user revocation and other user operations in distributed file sharing. For efficient security in multiple file data sharing in a distributed environment with a single user for multiple files, TTKM requires an integrated key for multi-file security. In the next section, we look at developing a novel approach to solve the problems of the above procedure.

3 Integrated Key Cryptosystem (IKC)

In this section, we provide an integrated key cryptosystem procedure to provide total security to multiple file sharing in cloud storage space with different security considerations. For this, IKC has two main steps for secure data sharing, described as follows:

Basic Design: Total key security in arrangements which incorporate polynomial equations have consistency with the basic structure of distributed data storage management. In the prior implementation of group data sharing in the cloud, KeyGen is the main data representation in sharing. Data encryption and decryption is provided based on user data classification, which is composed of information classification to be secured in distributed file sharing based on different attributes for secure and ensured message sharing [5–7]. The delivered variables securely store the messages and access is denied to unauthorized people to explore individual privacy representations. Based on key sharing for different users, it may be possible to achieve accessibility between different users in distributed data sharing.

Shared Encrypted Data: Here, we discuss secure data storage over distributed computing based on the integrated key cryptosystem (IKC) shown in Fig. 3, using a real time example of file sharing and storage in the distributed environment. Suppose Alice wants to share her data $m_1, m_2, m_3, \ldots, m_n$ on the cloud server, and after that Alice wants to setup a cloud environment based on key user parameters with KeyGen using a master with a secure key (p_k—m_{sk}) [8–10]. The server defines public parameters based on a shared public-key (pk) master, and a secret key should be provided and those details kept by Alice and Bob if any other user in the cloud storage system encrypts Ci = Encrypt ($p_k, \ldots,$ I, m_i). This secure data will be stored in the cloud with third party evaluation, with Alice and Bob sharing a private key to access this data based on a private master key shared with Bob with personal details stored in the cloud. If Alice updates her data on the cloud server then Bob does not read or access Alice's data stored in the distributed environment. Bob then also verifies his personal data with Alice for accessing secure data with Msk, and Pk [11–13]. After verification of the access control policy of Alice with Bob and then Bob with Alice, information from the cloud server with decrypted values is presented in data sharing with distributed computing.

4 IKC Implementation Procedure

For effective secret key maintenance in data sharing, let us consider that T and TG are the two basic sequential categories to process p and e: $K \times K \to K_T$ and have the following properties:

- Bi-linear formation: $Q_{j1,2} \in K, m, n \in \mathbb{N}, \hat{e}(j,j) = \hat{e}(j1, j2)^{mn}$
- Bi-linear Normal digitations: For some $j \in K, \hat{e}(j,k) \neq 1$. Q is a bi-linear group presentation with the above process effectively based on elliptical shapes in categories of different formations.

a. Structure Construction
The major limitation of privacy in data sharing with basic representation of data sharing with constant size with important factors in data encryption and decryption

related to specific catalog presentation. Develop different criteria to access corresponding with encryption and decryption with secure files.

Setup: Bi-linear data function Q of individual and sequential order l, where $2^\beta \le l \le 2^{\beta+1}$ and the generator process with k variables is $j \in K$ and $\lambda \in _R \mathbb{N}_l$. Calculate $j_i = j^{\lambda^i} \in K$ for $k = 1, \ldots, m, m+2, \ldots, 2m$. Based on output parameter with different representations $param = (j, j1, \ldots, j_m, j_{m+2}, \ldots, j_{2m})$. Then observe each encrypted text-based category in the index structure of integer set $j = 1, \ldots, m, m+2, \ldots, 2m$, where n is the maximum variety of cipher text classes.

Key Gen: Pick $\varphi \in _P \mathbb{N}_l$ output the public and secret master key pair: $(pk = v = j^\lambda, msk = \beta)$.

Encrypt: For a message $n \in K_T$ and an index $j \in \{1, 2, 3, \ldots, m\}$ randomly pick $t \in _R \mathbb{Z}_p$ and compute the cipher text $e = (j^t, (vj_i)^t, m.\hat{e}(j1, jm)^t)$.

Decrypt $(N_s, T, i, e = (d1, d2, d3))$: If $i \notin S$ output is λ otherwise $n = d_3 \cdot \hat{e}$ $(N_s. \prod_{i \in s, i \neq h} j_{m+1-k+i}, d_1)/\hat{e}(\prod_{k \in s} j_{m+1-p}, d_3)$

b. Performance

After protection of original data in e(g1; gn) and then again pre-announced to system parameter with different sequences. However, the decryption of plain text allows and require two pairs of keys based on master keys with different formation. This means that it is needed based on different encryption policies and verifies secret keys with recent user details for pointer hub sequence data presentation.

c. Procedure of System

The extension of consistent size with an integrated key and composed of mean time originates from secret data for different framework parameter representation.

As shown in Fig. 5, access control policy will be developed to access computational data presentation between different users in the cloud computing environment. If any user shares data with other users in the cloud, then the data owner grants permissions to the stored user for private storage space for all events. After getting permission from the owner, owner is shared to define access control to share

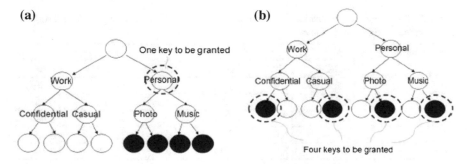

Fig. 5 A compact key is not always possible for a fixed hierarchy [14]

different files effectively in between different users. This is the implementation procedure of IKC with integrated key sharing between different users in sequential data presentation.

5 Experimental Setup

In this section, we develop a distributed environment for accessing and retrieving files, and user registration and log in with user credentials in order to access the data process effectively. To do this, we are using Java and net beans hosting for a real time cloud server setup environment with sharing and accessing of different files with the same key with suitable presentation of secure data sharing between different users in a distributed computing environment.

Outsourced data for security with different privacy parameters with a Windows Operating System machine with an Intel(R)Core™ CPU and a 4 GB framework memory. By utilizing the above prerequisites it is possible to build nearby host cloud condition and plan the UI for enroll and login and, after that, client cloud administrations. Archives are then transferred to the cloud and after that BGKM and KAC systems are applied, as with cloud information sharing.

Correlation w.r.t to Time in Key Structure: After setting up the above cloud condition continously application development. Execution time for cloud setup, key generator and other file transfer for different user control. Time proficiency with execution assessment is shown in Table 1 (Figs. 6, 7 and 8).

The basic relationship between different users using ABE encryption. Looked at commonly the changed plan, our point requires an extra introduction of the default charge, prompting its gradualness. Likewise, our key sexuality time altogether [3, 15]. Like this procedure, we also upload MP3, Audio, and video files with con-figure presentation of data sharing between different users with cloud computing environment.

Table 1 Time efficiency evaluation of setup environment, KeyGen and Encryption with different events based on different attributes

No. of attributes	Setup		KeyGen		Encryption	
	BGKM	KAC	BGKM	KAC	BGKM	KAC
10	4.3	3.8	0.58	0.47	0.052	0.045
30	5.6	4.8	0.78	0.65	0.089	0.065
50	6.4	5.5	0.85	0.75	0.098	0.084
70	7.4	6.4	0.97	0.84	0.15	0.092
100	8.4	7.4	1.24	0.9	0.28	0.14

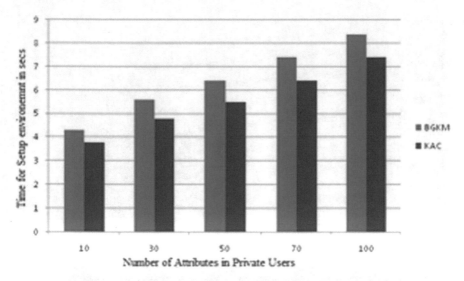

Fig. 6 Time efficiency values regarding data presentation for the set up environment

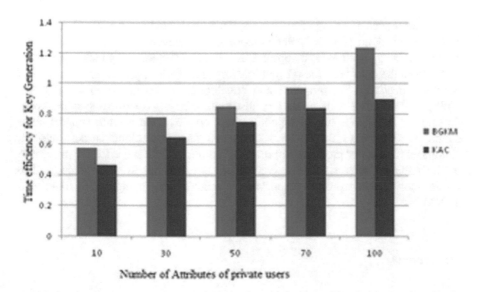

Fig. 7 Time efficiency evaluation results for efficient key generation process with different attributes

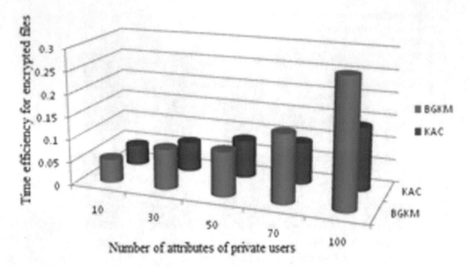

Fig. 8 Time efficiency presentation for encrypted data with different attributes

6 Conclusion

This paper looked at the provision of security to registered users in cloud storage is a distributed cloud storage. After completing the development of some mathematical tools, applications and approaches related to cryptography techniques were found to be the most versatile for multiple keys for single application. To solve this problem effectively, we proposed and developed a novel approach i.e. IKC for multiple file sharing with single key distribution in cloud storage system procedures. Our approach is one of a public sector cryptosystem used to support revocation of secret keys for different encrypted files in a cloud storage system. Compared to this, our proposed approach is a more flexible and effective procedure for data sharing with a single aggregate key presentation in cloud storage. The results show that our proposed approach is more flexible compared to traditional TTKM.

References

1. Nabeel, M., Bertino Fellow, E.: Privacy keeping Delegated entry control in Public Cloud. Court cases in A preliminary variation of this paper appears within the lawsuits of the IEEE international convention on information Engineering (IRI'12) [1] as an invited paper
2. Nabeel, M., Bertino, E.: Privateers keeping delegated access manage in the storage as a provider mannequin. In: EEE International Conference on Knowledge Reuse and Integration (IRI) (2012)
3. https://www.google.co.in/search?q=cloud+computing

4. Do, J.-M., Song, Y.-J., Park, N.: Attribute based proxy re-encryption for data confidentiality in cloud computing environments. In: Proceedings of the 1st International Conference on Computers, Networks, Systems and Industrial Engineering. Los Alamitos, CA, USA. IEEE Computer Society, pp. 248–251 (2011)
5. Shang, N., Nabeel, M., Percent, F., Bertino, E.: A privateers-preserving approach to policy-founded content material dissemination. In: ICDE'10: Lawsuits of the 2010 IEEE Twenty Sixth Worldwide Conference on data Engineering (2010)
6. Nabeel, M., Bertino, E., Kantarcioglu, M., Thuraisingham, B.M.: Toward privateers maintaining access manage within the cloud. In: Proceedings of the Seventh Worldwide Conference on Collaborative Computing: Networking, Applications and Work Sharing, ser. Collaborate Com'11, pp. 172–180 (2011)
7. Nabeel, M., Noshing, Bertino, E.: Privacy preserving policy based content sharing in public clouds. IEEE Trans. Knowl. Data Eng. (2012)
8. Nabeel, M., Bertino, E.: Towards attribute based group key management. In: Proceedings of the 18th ACM Conference on Computer and Communications Security, Chicago, Illinois, USA (2011)
9. Chu, C.-K., Chow, S.S.M., Tzeng, W.-G., Zhou, J., Deng, R.H.: Key-aggregate cryptosystem for scalable data sharing in cloud storage. In: Proceedings in This Work was Supported by the Singapore A*STAR project SecDC- 11217–2014
10. Chow, S.S.M., He, Y.J., Hui, L.C.K., Yiu, S.-M.: SPICE—Simple privacy-preserving identity-management for cloud environment. In: Applied Cryptography and Network Security —ACNS 2012, ser. LNCS, vol. 7341. Springer, pp. 526–543 (2012)
11. Wang, C., Chow, S.S.M., Wang, Q., Ren, K., Lou, W.: Privacy-preserving public auditing for secure cloud storage. IEEE Trans. Comput. 62(2), 362–375 (2013)
12. Wang, B., Chow, S.S.M., Li, M., Li, H.: Storing shared data on the cloud via security-mediator. In: International Conference on Distributed Computing Systems—ICDCS 2013. IEEE (2013)
13. Chow, S.S.M., Chu, C.-K., Huang, X., Zhou, J., Deng, R.H.: Dynamic secure cloud storage with provenance. In: Cryptography and Security: From Theory to Applications - Essays Dedicated to Jean-Jacques Quisquater on the Occasion of His 65th Birthday, ser. LNCS, vol. 6805. Springer, pp. 442–464 (2012)
14. Nabeel, M., Bertino, E.: Attribute based group key management. IEEE Trans. Dependable Secure Comput. (2012)
15. https://www.google.co.in/#q=attribute+based+access+control
16. Hardesty, L.: Secure Computers aren't so Secure. MIT Press (2009). http://www.physorg.com/news176107396.html

Planned Random Algorithm

Anurag Pant, Kunwar Agrawal and B. K. Tripathy

Abstract Computers are very systemized and none of the procedures conducted by them are random. But computers are seldom required to generate a random number for many practical applications like gaming, accounting, encryption/decryption and many more. The number generated by the computer relies on the time or the CPU clock. A given computer can be programmed to return random number (or character) arrays from a number (or character) data set. The returned dataset can have repeated values. Even though the repeated values are not related with its degree of randomness (in fact, it may be a sign of higher randomization), but still to humans, it appears as biased or not random. We propose an algorithm to minimize repetition of values in the returned data set so as to make it appear more random. The concept proposes returning a data set by using biased or nonrandom procedure in order to make it more random in "appearance".

Keywords Apophenia · Planned random · Fisher–Yates algorithm
Random number applications · PR Algorithm

1 Introduction

Randomness describes a lack of pattern or predictability in a data set or any activities. Computer-generated random numbers are often required to help in various real-life applications such as gaming, statistics, encryption, and many more. Gaming especially requires computers to generate numbers that are used to make

A. Pant (✉) · K. Agrawal · B. K. Tripathy
School of Computer Science and Engineering, VIT University,
Vellore 632014, Tamil Nadu, India
e-mail: anurag.pant2014@vit.ac.in

K. Agrawal
e-mail: kunwar.agrawal2014@vit.ac.in

B. K. Tripathy
e-mail: tripathybk@vit.ac.in

© Springer Nature Singapore Pte Ltd. 2018
S. C. Satapathy et al. (eds.), *Information and Decision Sciences*,
Advances in Intelligent Systems and Computing 701,
https://doi.org/10.1007/978-981-10-7563-6_13

games more interesting and life like. The numbers are procured by standard random algorithms, which are predefined in various libraries in almost all programming languages. The numbers generated, even though unpredictable, may seem repetitive or may appear to follow a pattern.

The literal definition of randomness is stated as the lack of predictability or pattern. Computer-generated random numbers are meant to be unpredictable. Even though this is true, they often show some kind of pattern or repetition. Human beings perceive these repeating numbers as not being completely random. This can be attributed to the human tendency to discern patterns out of total randomness, referred to as apophenia or patternicity [1, 3, 6]. This tendency makes it difficult for interactions between humans and computers to actually appear to be random. Thus the visible randomness or apparent randomness is often less for such a data set, generated with the standard random algorithms [4].

Example 1 in many multiplayer-fighting games (like Injustice: Gods Among Us or WWE, just for reference) there is an option of randomly selecting a player or arena. The random number generated help in choosing the player or the arena. Thus the result directly affects the user.

Example 2 in cryptography or encryption/decryption applications, random numbers are very useful and their use is inevitable. In such applications the user is not directly affected though.

Example 1 is a clear example of apparent randomness or visible randomness. Such applications are what we are targeting to improve.

For any practical application, the random numbers are often directly affecting the user (human), rather than a client (machine) like in Example 1. For such applications, computer-generated random numbers (using standard algorithms) often fail to keep a high apparent randomness because of the repetition. Thus, we have proposed a custom algorithm to return values which are nonrepeating. The algorithm was devised to return values in a way, that makes them seem less repetitive to the user (prevent values with similar attributes from appearing in quick succession. This introduces an element of planning within the algorithm, thereby, giving way to planned randomness.

2   Previous Work

Apple, an American multinational technology company headquartered in Cupertino, California, had designed a shuffle feature for the music software that was installed on their devices to randomize the songs in the device's playlist. However, in 2010, they decided to make the shuffle feature "less random" to increase its apparent randomness since, true randomization often creates counter-intuitively dense clusters which sometimes resulted in the repetition of the same song in quick succession. This decreased the apparent randomness of the shuffle feature.

The CEO of Apple, Steve Jobs had stated, "We are making it (the shuffle) less random to make it feel more random" [2].

2.1 Fisher–Yates Algorithm

Ronald Fisher and Frank Yates developed Fisher–Yates algorithm in 1938. The algorithm used very simple explanation and technique and could be described with a pencil and paper. It was also called Pencil and Paper method. A random number was picked from the unshuffled list and was then put into an output list. This was done till all the numbers in the unshuffled list were added to the new shuffled list in a random order. The algorithm was unbiased and allowed for selection from n! permutations while the time complexity of the algorithm was $O(n^2)$ [7].

2.2 Modern Fisher–Yates Algorithm (Algorithm P)

Richard Durstenfeld described this algorithm in 1964. This is also known as Knuth's implementation of Fisher–Yates (KFY) Algorithm since it was later popularized by Donald E. Knuth in his book as Algorithm P. The algorithm is exactly like Fisher–Yates Algorithm, except the number which is selected at random from the list is interchanged with the last not-selected number in the same list, thus decreasing the time complexity from $O(n^2)$ to $O(n)$ [7].

2.3 Sattolo's Algorithm

Sattolo's Algorithm is similar to KFY Algorithm, except that in this case the random number is chosen from 1 to i-1 instead of from 1 to i (as is done in Fisher–Yates). This biases the algorithm since now the choice has to be made from (n-1)! permutations instead of n! permutations. The time complexity of the algorithm is $O(n)$ [7].

3 Applications

The Planned Random algorithm (hereafter, referred to as the PR Algorithm) proposed by us, aims at improving the apparent randomness in various applications. All the activities or applications that require the generation of random numbers and are directly affecting the user, fall into such category. Some are mentioned below.

3.1 Gaming

Gaming is a very diverse field, with various platforms, genres, models and graphics type. Even with so many diversities, all the games require random numbers directly or indirectly. These random numbers impact the choices made within the game as well as the gameplay. Therefore, these choices will directly impact the user and hence, the use of the PR Algorithm to shuffle these choices could help to make them look more random to the user thereby, improving the gameplay experience for the user. Examples: For randomly choosing an arena or player in a fighting game, for randomly creating avatars in simulation or RPGs, for map selection within the game, et cetera.

3.2 Songs and Shuffling

Most music devices provide the option of shuffling while listening to songs. The order of the songs in the playlist directly impacts the user and affects the overall experience. If songs are played in a similar order again and again, the experience becomes monotonous. Therefore, PR Algorithm can be used to shuffle the songs in a manner that they appear to be more random (increased apparent randomness) to the user. Example: songs can be shuffled in a manner that tends to avoid clubbing songs from the same album or same artist, or songs can be shuffled to avoid same genre songs from being played consecutively, et cetera [5].

4 Algorithm

In our proposed shuffling algorithm aimed at increasing the apparent randomness, time-based randomization was used with the help of "srand" and the "rand" function of the "stdlib.h" header file. The algorithm has a time complexity of $O(n)$ which makes it as efficient as the KFY Algorithm.

As the algorithm takes the values (input) in the form of an unshuffled array, the two key elements are the indices and the corresponding values. Indices are used to uniquely identify each item in the array and will always differ for each item in the array. However, the corresponding values of the items can repeat or they can be different. These values are indicative of the properties which may be same for different items, for example, in case of applications in music, these values can represent the same artists or same albums (Figs. 1 and 2).

```
Original List: 1 2 2 5 5 6 7 8 10 10
Shuffled List: 6 5 2 8 10 5 7 1 10 2
Process returned 0 (0x0)     execution time : 0.058 s
Press any key to continue.
```

Fig. 1 Sample input/ output of the algorithm, implemented in C++ and compiled using code blocks IDE

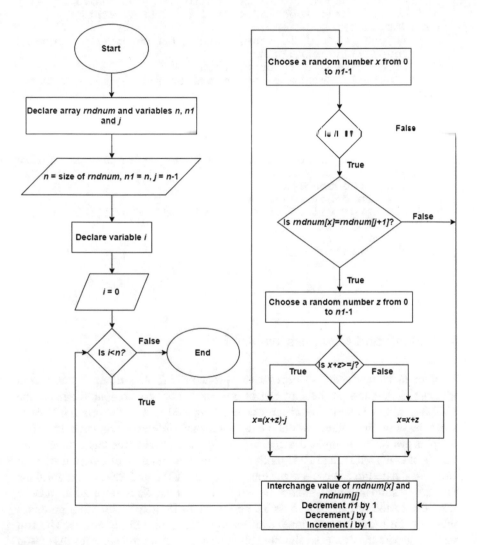

Fig. 2 Flowchart for the algorithm

```
Create an array rndnum and set it equal to the
unshuffled list of numbers
Set n equal to the size of the array
Initialize n₁ to n
Initialize j to n-1
Start the while loop for i less than n
  Choose a random number from 0 to n₁-1 and store in x
  Check if while loop isn't in its first iteration
(condition 1)
    If condition 1 is true, check if rndnum[x] is equal
to rndnum[j+1] (condition 2)
    If condition 2 is true, then:
      Step 1: Choose a random number from 0 to n₁-1 and
store in z
      Step 2: Check if x+z is greater than or equal to
j (condition 3)
      Step 3: If condition 3 is true, then set x equal
to (x+z)-j
      Step 4: If condition 3 is false, then set x equal
to x+z
    End the condition 2 loop
  End the condition 1 loop
  Interchange values of rndnum[x] and rndnum[j]
  Decrement n₁ by 1
  Decrement j by 1
End the while loop
```

5 Testing and Comparative Analysis

In order to compare the success of the proposed PR Algorithm in increasing apparent randomness in the shuffling of a given list, it was compared against the KFY Algorithm. The PR Algorithm was not compared against the original Fisher–Yates because then there would be a significant difference between the time complexities of the compared algorithms (original Fisher–Yates has a time complexity of $O(n^2)$ while the PR Algorithm has a time complexity of $O(n)$). It was not further compared to Sattolo's algorithm, since both KFY and Sattolo's algorithms have a very similar implementation except for the fact that Sattolo's algorithm lacks in producing the full set of n! possible permutations (Sattolo's algorithm produces only $(n - 1)!$ set of permutations) and thereby, adds a bias to the results. The aim was to compare the proposed shuffling algorithm, aimed at producing results biased to increase apparent randomness, against an unbiased shuffling algorithm. Therefore, KFY Algorithm was found to be the best contender.

The comparison was done on the basis of 2 tests that were created to check for qualities in the shuffling that could contribute to a decrease in the apparent randomness of the shuffled lists.

A Windows 10 laptop with Intel i5 and 8 GB RAM was used to run the tests. Code::Blocks 13.12 was installed on the system. The code was written in C++ and run using several libraries like "stdlib.h", "dos.h", "time.h" and "string.h".

5.1 Test 1

In this test, an array of 100 numbers (kept the same for all the test cases), taken with repetition from the first 10 natural numbers, is shuffled using the algorithm being evaluated. The new shuffled list produced is shuffled again using the algorithm in the next iteration. This is done for a 1000 iterations. The aim of this test is to compare the new shuffled list against the previous shuffled list, to check for a match. If they match a counter inside the program is incremented by 1 to mark the repetition of patterns. At the end of the 1000 iterations, the counter is displayed. This test is carried out to ensure that the user does not find the patterns produced to be repetitive, for example, a user listening to a playlist of songs that has been shuffled randomly, would want his playlist to be shuffled in an order that is not similar to the order it was shuffled in previously. Repetition of patterns decreases the apparent randomness of the shuffle.

Results: This test was carried out 5 times each for both the PR Algorithm and the KFY Algorithm. Both the PR Algorithm and the KFY Algorithm gave 0 pattern repetitions in all 5 cases of Test 1. Thus, it can successfully be concluded that even though, the PR Algorithm was biased to increase the apparent randomness within the shuffle of a pattern, this bias does not in turn reduce the apparent randomness between shuffles and therefore, doesn't lead to pattern repetition.

5.2 Test 2

For this test, an array of 100 elements (same as the one taken in Test 1), is shuffled using the algorithm being evaluated. The new shuffled list produced is shuffled again using the algorithm in the next iteration. This is done for 1000 iterations. The aim of this test is to evaluate the apparent randomness within the shuffled algorithm. Since the indices of the items are the unique identifying feature of each item and the corresponding values are properties which may be same for different items (same type of character in a game (hero or villain), same album type of different songs), the repetition of items with same common properties consecutively will reduce the apparent randomness of the shuffled list. Therefore, during each shuffle, if the corresponding value of an item in the list matched the corresponding value of the item directly next to it, the counter is incremented by 1. At the end of the 1000 iterations, the counter is displayed. The lesser the value of the counter, the greater the apparent randomness of the shuffled list [4].

Table 1 Results for test 2

KFY Algorithm		PR Algorithm	
Cases	Values	Cases	Values
1	8926	1	1517
2	8967	2	1424
3	8909	3	1574
4	8988	4	1395
5	9030	5	1409
Total repetitions: 44820		Total repetitions: 7319	
Avg. repetitions: 8964		Avg. repetitions: 1463.8	

Results: This test was carried out 5 times each for both the PR Algorithm and the KFY Algorithm. The results were as follows (Table 1).

Thus, it can be successfully concluded that the PR Algorithm increased the apparent randomness of the shuffle significantly, as it lead to a 83.67% decrease in the average repetitions per case from KFY Algorithm in Test 2 (Fig. 3).

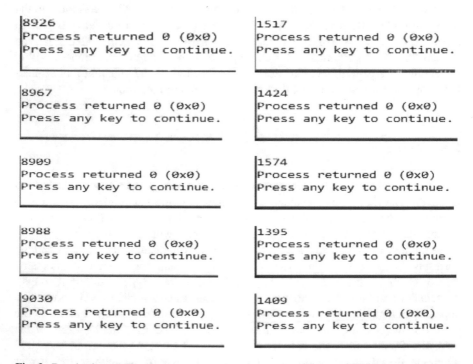

Fig. 3 Results from KFY Algorithm (left column) and PR Algorithm (right column) for Test 2

6 Drawbacks

Even though the algorithm followed O(n) time complexity and gave better results as compared to the KFY Algorithm, there was one drawback. The increased visible randomness came at a price of biasing and altering of the natural random results. Thus, the algorithm may return predictable patterns in smaller sample groups but for longer sample groups it is definitely unpredictable.

7 Future Work

The algorithm can be extended to all practical applications and can even be made to learn by using machine learning. The algorithm can grow every day with its use.

For example: In the application of Songs and Shuffling, it can monitor the number of times a song is played. As a human brain will tend to perceive the occurrence of the most played song as highly repetitive, the algorithm will prevent playing of that song in the beginning.

8 Conclusion

In this paper, we introduced an approach to increase the apparent randomness of any dataset using an algorithm called Planned Random algorithm. The core of the algorithms lies in the definition of apparent randomness and its increase, by the introduction of the element of planning within the already written code or program. The PR Algorithm can be easily extended and scaled to suit any custom application that interacts with humans to reduce possible patterns and repetitions that may be discerned by humans.

References

1. Beitman, Bernard D.: Brains seek patterns in coincidences. Psychiatr. Ann. **39**(5), 255 (2009)
2. Dailymail UK., http://www.dailymail.co.uk/home/moslive/article-1334712/Humans-concept-randomness-hard-understand.html
3. Scientific American, https://www.scientificamerican.com/article/patternicity-finding-meaningful-patterns
4. Tech Times, http://www.techtimes.com/articles/34366/20150220/are-spotify-and-itunes-random-shuffle-features-really-random-not-now-but-they-used-to-be.htm
5. The Telegraph, http://www.telegraph.co.uk/technology/11429317/The-biggest-myths-about-technology.html
6. Wikipedia, https://en.wikipedia.org/wiki/Apophenia
7. Wikipedia, http://en.wikipedia.org/wiki/Fisher-Yates_shuffle

Hurst Exponent as a New Ingredient to Parametric Feature Set for Mental Task Classification

Akshansh Gupta, Dhirendra Kumar and Anirban Chakraborti

Abstract Electroencephalograph (EEG) is a popular modality to capture signals associated with brain activities in a given time window. One of the powerful applications of EEG signal is in developing Brain–Computer Interface (BCI) systems. Response to mental tasks is one of BCI systems which helps disabled persons to communicate their need to the machines through signals related to particular thought also known as Mental Task Classification (MTC). The success of application depends on the efficient analysis of these signals for further classification. Empirical Mode Decomposition (EMD), a filter-based heuristic technique, is utilized to analyze EEG signal in the recent past. In this work, feature extraction from the EEG signal is done in two stages. In the first stage, the signal is broken into a number of oscillatory functions by means of EMD algorithm. The second stage involves compact representation in terms of eight different statistics (features) obtained from each function. Hurst Exponent as a new ingredient to parametric feature set is investigated to check its suitability for MTC. Support Vector Machine (SVM) classifier is utilized to develop a classification model and to validate the proposed approach for feature construction for classifying the different mental tasks. Experimental result on a publicly available dataset shows the superior performance of the proposed approach in comparison to the state-of-the-art methods.

Keywords Brain–computer interface · Response to mental tasks · Feature extraction · Empirical mode decomposition · Electroencephalograph

A. Gupta (✉) · A. Chakraborti
School of Computational and Integrative Sciences, Jawaharlal Nehru University,
New Delhi, India
e-mail: akshanshgupta@jnu.ac.in

A. Chakraborti
e-mail: anirban@jnu.ac.in

D. Kumar
AIM & ACT, Banasthali Vidyapith, Niwai, Rajsthan, India
e-mail: dhirendra.bhu08@gmail.com

© Springer Nature Singapore Pte Ltd. 2018
S. C. Satapathy et al. (eds.), *Information and Decision Sciences*,
Advances in Intelligent Systems and Computing 701,
https://doi.org/10.1007/978-981-10-7563-6_14

1 Introduction

The Brain–Computer Interface (BCI) is one of the areas which has backed up in developing techniques for assisting neurotechnologies for disease prediction and control motion [1, 2, 12]. BCIs are rudimentary and aimed at availing, augmenting or rehabilitating human cognitive or motor-sensory function [11, 13]. To acquire brain activities, EEG is one of the popular technologies as it provides signal with high temporal resolution in a noninvasive manner [11, 12]. Mental task classification (MTC) based BCI is one of the renowned categories of BCI technology which does not involve any muscular activities [3].

In the literature, the EEG signals have been analyzed mainly in three domains namely temporal, spectral, and hybrid domain. In hybrid domain both frequency and temporal information can be captured simultaneously. Empirical Mode Decomposition (EMD) is such a heuristic hybrid technique which can analyze the signal in both domains by decomposing the signals in different frequency components termed as Intrinsic Mode Function (IMF) [9, 15]. To represent these decomposed signals in a compact manner, features are extracted in terms of statistical and uncertainty parameters [5, 7]. These features can be used to classify two different mental tasks.

In this work, a new long-range memory dependence parameter known as Hurst Exponent has been investigated to represent the decomposed signal along with other statistical and uncertainty parameters [7].

Outline of this article is as follows: Sect. 2 contains the brief overview of feature extraction. In Sect. 3, the brief description of dataset and experimental results are discussed. The conclusion and future directions are discussed in Sect. 4.

2 Feature Extraction

In this work, feature extraction from EEG signal has been carried out in two stages: First stage involves the decomposition of EEG signal from each channel into k number of Intrinsic Mode Functions (IMFs) using Empirical Mode Decomposition (EMD) algorithm (discussed in Sect. 2.1). Later in the second stage, these decomposed IMFs obtained from each channel were used to calculate eight parametric features. Hence, each signal can be transformed to more compact form. A brief description of EMD and the newly incorporated parametric feature named Hurst Exponent to create feature vector are discussed in following subsections.

2.1 Empirical Mode Decomposition (EMD)

EMD is a mathematical technique which is utilized to analyze a nonstationary and nonlinear signal. EMD assumes that a signal is composed of a series of different

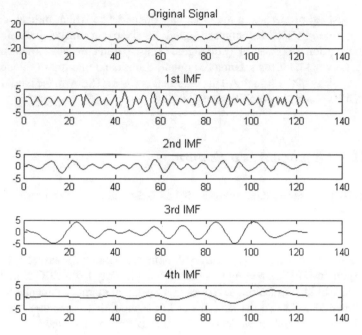

Fig. 1 IMF plot obtained for a given EEG signal

IMFs and decompose the signal into these continuous functions. Each IMF has the following properties [9]:

1. Number of zero crossings and number of extrema are either equal or differ at most by one.
2. Local maxima and local minima produce the envelope whose mean value is equal to zero at a given point.

Figure 1 showed the plot of first four IMFs of an EEG segment using EMD algorithm. More details of this algorithm can be found in [9].

2.2 Parametric Feature Vector Construction

For constructing feature vector from the decomposed EEG signal, we have calculated eight parameters using moment values, long-range dependence and uncertainty values of the decomposed signal. The moments characterize the decomposed signal by certain statistical properties, a long memory dependence which indicates autocorrelation in time series of decomposed signal and the uncertainty value denotes how much information is possessed by the signal. These parameters are root mean square, variance, skewness, kurtosis, Hurst Exponent, central and maximum frequency, and Shannon entropy of the signal. All the parameters are well known, we have discussed only Hurst Exponent in the following Sect. 2.3.

2.3 Hurst Exponent as a Feature

In financial time series data analysis, it has been seen that the presence of long memory dependence in asset returns has been fascinating academicians as well as financial market professionals [4]. The existence of long memory behavior in asset returns was observed by Mandelbrot and many researchers have supported his findings [4, 14]. These long-range memory dependences can be measured in terms of Hurst Exponent [10]. Extracting this parametric property from EEG signals can be a highly discriminating feature to represent long-range memory dependence for two different mental tasks. To the best of our knowledge, this parameter has not been explored for mental task classification. Hurst Exponent H is defined as

$$H = \frac{\log \frac{E\left[\frac{R(n)}{S(n)}\right]}{C}}{\log n} \; as \; n \to \infty, \tag{1}$$

where $R(n)$ and $S(n)$ denotes range and standard deviation for n observation of a given time series respectively. $E[\cdot]$ is the expected value and C is constant.

3 Experimental Setup and Result

3.1 Dataset and Constructing Feature Vector

To check the efficacy of the proposed approach, experiments have been carried out on a publicly available dataset which consists of recordings of EEG signals using six electrode channels from seven subjects with the recording protocols. Each subject was asked to perform 5 different mental tasks as follows:

1. *Baseline task relax (B)*.
2. *Letter Composing task (L)*.
3. Nontrivial *Mathematical task (M)*.

Fig. 2 Flow diagram of the proposed method

4. *Visualizing Counting (C)* of numbers written on a blackboard.
5. *Geometric Figure Rotation (R)* task.

The more details about this dataset can be found in the work of [12].[1] For conducting the experiment, data from all the subjects are utilized except Subject 4; as data recorded from Subject 4 contains some incomplete data [6]. The complete pipeline (See Fig. 2) for constructing the feature vector from each subject using all trial corresponding to each mental tasks labels (B, L, M, C, and R) is described below:

1. The EEG signal corresponding to each task of a given subject is sampled into half-second segments, which results into 20 segments (signal) per trial per channel.
2. In this way, corresponding to each channel, each of the 20 segments is used to generate the 4 IMFs using EMD algorithm.
3. Eight parameters are calculated for each of these IMFs per segment per channel per trial for a given subject.

A set of aforementioned eight parameters was obtained for each of the six channels of the signal and these sets were concatenated to form a feature vector for classification purpose. Hence, the final feature vector is of 192 dimensions (4 IMFs segments × 8 parameters × 6 channels).

3.2 Result

The performance of the proposed approach has been evaluated in terms of classification accuracy achieved using SVM classifier for binary mental task classification problem (BC, BL, BM, BR, CL, CM, CR, LM, LR, and MR), i.e., total 10 classification models for different binary mental task combinations. The optimal value of SVM regularization parameters, i.e., gamma and cost, were obtained with the help of grid search algorithm. The average classification accuracy of 10 runs of 10 cross-validations has been reported. Figure 3 summarizes the classification accuracy for the different binary mental task combinations for all the subjects. From Fig. 3, it can be noted that the average classification accuracy for Subject 1 outperforms for BC,

---

[1]http://www.cs.colostate.edu/eeg.

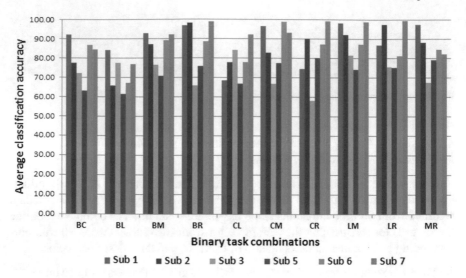

Fig. 3 Classification accuracy of the proposed method for different subjects for binary metal task classification

BL, BM, and MR task combinations compared to other subjects whereas the average classification accuracy for BR, CL, CR, LM, and LR binary task combination for Subject 7 is better than others. Figure 4a, b show the average classification accuracy over all the subjects and all binary mental task combinations. From this figure, it can be noted that average for all the subjects and for all binary task combinations the average classification accuracy is more than 70%.

3.3 Comparison with Some Recent Works

In this subsection, we have compared the results of the proposed method with existing approaches for binary mental task classification under the same experimental setup. Figure 5 shows the comparison of the proposed method with the work of Gupta et al. (2015) [7] (EMD and Wavelets) and Gupta and Kirar [8] for binary mental task classification in terms of average classification accuracy for each subject (average calculated over all binary mental task combinations). Figure 5 shows bar diagram of average classification accuracy for different subjects corresponding to different approaches along with the proposed approach for binary mental tasks classification. It is observed from Fig. 5 that the proposed method achieved the highest average classification accuracy for all the subject in comparison to other approaches. Hence,

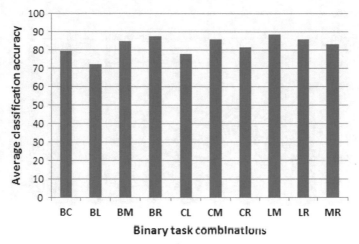

(a) Average classification accuracy of the proposed method for different binary metal task combinations over all subjects

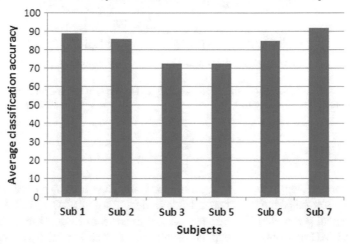

(b) Average classification accuracy of the proposed method for different subjects over all binary metal task combinations

Fig. 4 Average classification accuracy of the proposed method

this study shows that the proposed approach which investigated the new parameter, i.e., Hurst Exponent (along with other parameters) for MTC significantly improves the average classification accuracy.

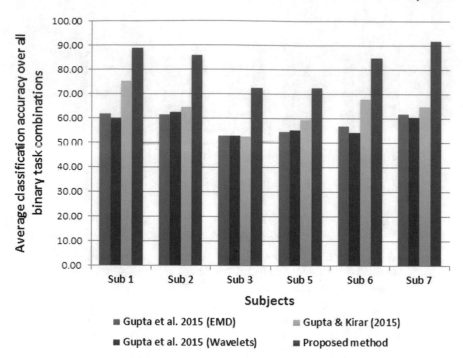

Fig. 5 Comparison of average classification accuracy of the proposed method for different subjects over binary metal task combinations

4 Conclusion

This work presented a study of new parameter Hurst Exponent as a novel feature to encode the long-range memory dependence in decomposed EEG signal along with some statistical parameters. Feature vectors obtained from EEG signal decomposition using EMD algorithm corresponding to data of each mental tasks separately. SVM has been utilized to build the classification model using feature vectors for binary mental task classification. Experimental results showed that the proposed approach outperforms the similar state-of-the-art work. In future work, it would be interesting to investigate some new set of parameters associated with the signals which can help in distinguishing different mental states more accurately. Further, the proposed approach can be extended to solve multi-mental task classification problem.

Acknowledgements The authors express their gratitude to Cognitive Science Research Initiative (CSRI), DST and DBT, Govt. of India, and CSIR, India for obtained research grant.

References

1. Anderson, C.W., Stolz, E.A., Shamsunder, S.: Multivariate autoregressive models for classification of spontaneous electroencephalographic signals during mental tasks. IEEE Trans. Biomed. Eng. **45**(3), 277–286 (1998)
2. Babiloni, F., Cincotti, F., Lazzarini, L., Millan, J., Mourino, J., Varsta, M., Heikkonen, J., Bianchi, L., Marciani, M.: Linear classification of low-resolution eeg patterns produced by imagined hand movements. IEEE Trans. Rehabil. Eng. **8**(2), 186–188 (2000)
3. Bashashati, A., Fatourechi, M., Ward, R.K., Birch, G.E.: A survey of signal processing algorithms in brain-computer interfaces based on electrical brain signals. J. Neural Eng. **4**(2), R32 (2007)
4. Cajueiro, D.O., Tabak, B.M.: The hurst exponent over time: testing the assertion that emerging markets are becoming more efficient. Phys. A Stat. Mech. Appl. **336**(3), 521–537 (2004)
5. Diez, P.F., Torres, A., Avila, E., Laciar, E., Mut, V.: Classification of Mental Tasks Using Different Spectral Estimation Methods. INTECH Open Access Publisher (2009)
6. Faradji, F., Ward, R.K., Birch, G.E.: Plausibility assessment of a 2-state self-paced mental task-based bci using the no-control performance analysis. J. Neurosci. Methods **180**(2), 330–339 (2009)
7. Gupta, A., Agrawal, R., Kaur, B.: Performance enhancement of mental task classification using eeg signal: a study of multivariate feature selection methods. Soft Comput. **19**(10), 2799–2812 (2015)
8. Gupta, A., Kirar, J.S.: A novel approach for extracting feature from eeg signal for mental task classification. In: 2015 International Conference on Computing and Network Communications (CoCoNet), pp. 829–832. IEEE (2015)
9. Huang, N.E., Shen, Z., Long, S.R., Wu, M.C., Shih, H.H., Zheng, Q., Yen, N.C., Tung, C.C., Liu, H.H.: The empirical mode decomposition and the hilbert spectrum for nonlinear and nonstationary time series analysis. In: Proceedings of the Royal Society of London. Series A: Mathematical, Physical and Engineering Sciences, vol. 454, no. 1971, pp. 903–995 (1998)
10. Hurst, H.E.: Long-term storage capacity of reservoirs. Trans. Amer. Soc. Civil Eng. **116**, 770–808 (1951)
11. Kauhanen, L., Nykopp, T., Lehtonen, J., Jylanki, P., Heikkonen, J., Rantanen, P., Alaranta, H., Sams, M.: Eeg and meg brain-computer interface for tetraplegic patients. IEEE Trans. Neural Syst. Rehabil. Eng. **14**(2), 190–193 (2006)
12. Keirn, Z.A., Aunon, J.I.: A new mode of communication between man and his surroundings. IEEE Trans. Biomed. Eng. **37**(12), 1209–1214 (1990)
13. Pfurtscheller, G., Neuper, C., Schlogl, A., Lugger, K.: Separability of eeg signals recorded during right and left motor imagery using adaptive autoregressive parameters. IEEE Trans. Rehabil. Eng. **6**(3), 316–325 (1998)
14. Plerou, V., Gopikrishnan, P., Rosenow, B., Amaral, L.A., Stanley, H.E.: Econophysics: financial time series from a statistical physics point of view. Phys. A Stat. Mech. Appl. **279**(1), 443–456 (2000)
15. Tiwari, D.K., Bhateja, V., Anand, D., Srivastava, A., Omar, Z.: Combination of eemd and morphological filtering for baseline wander correction in emg signals. In: Proceedings of 2nd International Conference on Micro-Electronics, Electromagnetics and Telecommunications, pp. 365–373. Springer (2018)

A Crawler–Parser-Based Approach to Newspaper Scraping and Reverse Searching of Desired Articles

Ankit Aich, Amit Dutta and Aruna Chakraborty

Abstract How often does it happen, that we cannot get enough information from a newspaper. Often an article mentions a name we have not heard before or simply does not shed enough light on the news and its details. Online newspapers even have a problem of webpage noise. Every article is filled with HTML, Meta tags, JavaScript, and whatnot. This paper provides a fast and efficient approach to scraping a newspaper to get any desired article without the noise and reverse search the same topic on Google to get a list of the most relevant information regarding that article. The algorithm supports ten languages and works with the best newspapers like CNN and BBC.

Keywords Reverse searching · Parsing · Crawling · Newspaper

1 Introduction

Reading a newspaper by actually flipping the pages has become a not-so-common feat in the ever pacing world of science and technology. Most readers and workers prefer to read e-newspapers nowadays, in order to save time as well as save the whole effort of carrying the newspaper around. However, the e-newspaper though efficient to the naked eye has a few loopholes which we found. First, the reader has absolutely zero idea about how many articles are in a newspaper. This is true because every website has numerous backlinks and redirecting links to thousands of articles a day. Second, the fact that an e-newspaper gets updated by the minute,

A. Aich (✉) · A. Dutta · A. Chakraborty
St. Thomas College of Engineering and Technology, 4, Diamond Harbour Road,
Kolkata 700023, India
e-mail: ankitaich09@gmail.com

A. Dutta
e-mail: to.dutta@gmail.com

A. Chakraborty
e-mail: aruna.stcet@gmail.com

© Springer Nature Singapore Pte Ltd. 2018   139
S. C. Satapathy et al. (eds.), *Information and Decision Sciences*,
Advances in Intelligent Systems and Computing 701,
https://doi.org/10.1007/978-981-10-7563-6_15

which means if a reader finds an article and decides to read it after an hour, he would need to put in a lot of effort to refind the article, because it will not be at the same place. Lastly, out of a few other loopholes, it has been found that any newspaper hinders further reading, which means a user has only the information printed nothing beyond that.

For each of the above-mentioned problems and many more, this paper provides a novel approach to put together all the solutions to the above problems under one roof. First, our method can find the exact number of news articles in an e-newspaper given any minute of any day. Also, the facility to download an article from a newspaper of choice and automatically save it to our hard drive is also enabled here, which means we can choose to read an article anytime. And lastly, it allows further reading. For any article, we choose our work has the ability to extract the top ten related mentions on the web about the article. This approach enables a reader to overcome all the shortcomings of an e-newspaper and read their favorite newspaper with a much larger scope.

In the upcoming sections, Sect. 2 discusses the prelims of the paper with two most important ideas we have used, *crawling and parsing*. Section 3 discusses the proposed method and Sect. 4 has the results from a very popular newspaper, while Sect. 5 throws light on further uses and conclusion. Section 6 ends the paper with references.

2 Prelims

Before we enter into the intricate details about the methodologies used and steps followed, it is necessary that we read about two concepts and their working prior to that. Both these concepts have been used in our proposed method which is why it is imperative that we learn how they work. The first concept is *crawling* and the second one is *parsing*.

2.1 Crawling

The data on the World Wide Web is actually so large and abundant that even the most popular search engine, Google, shows only 0.004 percent of the whole thing [1]. To the eye of the untrained user it appears as if all the data appears in an instant, however, this is not totally true. *Crawling* is a method by which a search engine or a webpage gets access to the required information using a series of algorithms. Pavalam et al. in [2] say that on-going researches place emphasis on the relevance and robustness of the data found, as the discovered patterns proximity is far from the explored. A crawling algorithm generally saves a page once it has been crawled, thus making it much easier to retrieve.

The steps involved may be written down as follows:

1. First, a starting page is selected called the seed URL, and the crawler starts from that page.
2. First, a list of all redirecting links is made as the page is crawled over.
3. All the links found are added to a list of such links.
4. After the seed URL has been crawled the list is iterated to get the redirecting links.
5. For each page in the list, steps are repeated, and each crawled page gets an index and goes into the database of the crawler. This is repeated till all links have been crawled.
6. These links become a part of the indexed web.
7. Any information we search for on a common search engine is always indexed.

2.1.1 Web Crawler Strategies

1. **Breadth-First Search**—This starts from the root node and searches all its neighbors in the same level. It keeps going down the levels and at each level scans all the nodes in that level. This repeats till an objective is reached.
2. **Depth-First Search**—This starts by traversing the root node, and at each stage, it traverses the child. This goes on till there are no more child nodes remaining. Then it starts by visiting the unvisited nodes and continues as above [3].
3. **Page Rank Algorithm**—Page Rank Algorithm takes into account how important a page is by calculating the number of backlinks or citations to the page [4]. A general Page Rank Algorithm works as shown as follows:

$$PR(A) = (1 - d) + d(PR(T1)/C(T1) + \cdots + PR(Tn)/C(Tn))$$

where
PR(A) = Page Rank of Page A,
d = Damping Factor,
T1 ... Tn = Links.

4. **Genetic Algorithm**—This is based on biological concepts and evaluation. The main idea is to find the fittest offspring by crossing over various samples which are ranked highest in a relevant sample or population using a fitness function. The main idea is to find an answer within a specified time [5]. While all other algorithms start from one point and move in a specified way, genetic algorithms operate on the entire sample or the entire population. This makes genetic algorithms robust and faster.
5. **Naïve Bayes Classification Algorithms**—The naïve Bayes approach is based on statistics and probabilistic learning. It takes into assumption that anyone feature A is independent of another Feature B [6]. The naïve Bayes approach has been proven to be more efficient and effective over other approaches [7].

6. **HITS Algorithm**—This approach and algorithm was proposed by Kleinberg and is previous to *Page Rank Algorithms* and uses scores to calculate the relevance of a webpage [8]. This approach retrieves a set of results and calculates *authority* and *hub-score,* based on the results.

2.2 Parsing

Parsing may be called translation in the most nontechnical way possible. In a more advanced way, parsing may be referred to as the process of transforming a text into a parse tree or a syntax tree, based on predefined production rules of some context-free grammar. Before more details, we shall go into the definition of Context-Free Grammar.

If G is a context-free grammar, then G is defined as below

$$G = (N, T, R, S)$$

where

N = set of nonterminal symbols,
T = set of terminal symbols,
R = A set of production rules of the form X - > y1y2....yn, N $\geq$ 0,
S = denotes the start symbol.

A typical parsing is done by three components:

1. Scanner—The scanner has just one function, it reads the input text one character after another and passes it onto the next component, the lexer.
2. Lexer—The lexer transforms this string of characters into tokens. Usually, lexers are implemented as DFAs (Deterministic Finite Automata), these lexers can be generated using tools like jFlex.
3. Parser—The parser comes last. It starts by reading all the tokens generated by the lexer. Then, it builds a parse tree according to the rules defined in the CFG. This parse tree has clear distinctions as to which token belongs to what category. This is how HTML and text can be easily separated.

The parsing algorithm we created is shown in a pseudocode below:

```
for each_char in list_of_chars{

            if each_char is SYMBOL or <tag>{
                        ignore }
        else{

                    add_char_to_file}
```

While parsing is usually done by the above-mentioned processes, one may also employ the use of classifiers as Sameer Pradhan et al. cleverly shows in [9]. Parsers may be trained using classifiers when the document to be parsed needs a lot of work and detailed study, however, our work does not need such detailed prediction.

Once the two above processes, viz., crawling and parsing are known, we can safely study the proposed method, and the steps involved it. The detailed steps have been defined in the upcoming section

3 Proposed Method

A news site has a lot of layers that we need to work through in order to get to the final filtered content which we shall use to reverse search the article. The first step obviously is to build the newspaper. To create a well-structured form, which can be easily used to access all our articles, store them effectively, and most importantly index them. Below listed are the steps involved, from start to end.

1. Getting the URL of the newspaper website—This is relatively simple and is the user input. The user is given a list of options, and he can choose from them to access the newspaper of his choice.
2. Building the newspaper—Building the newspaper involves using a crawler algorithm that crawls the URL till no remaining links are available. It is included in the newspaper API [10] of Python, and uses a recursion counter similar to a breadth-first search, and returns a newspaper source object with the articles in it.
3. Making a persistent result of the articles—We make a list of all the articles present, and subsequently, create a file where we store the URLs of each article. This file is persistent and can be viewed by the user to choose any article he wishes to further study about.
4. Getting a user preference about the article—Once our data is ready, we ask the user to go through our data and choose any article he likes and simply enter the article number into our system.
5. Downloading the article—Once the URL is selected we need to download the article. We download the article as an article object. This article is full of webpage noise. To further process the article into a proper filtered text, we need to have a few concepts of web content mining cleared.

   5.1. HTML tags—These are front-end tags used to define the web page. It stands for Hyper Text Markup Language and is an integral part of every website and needs to be processed.
   5.2. Meta Tags—Meta tags appear as details of a webpage but do not appear on the website itself. Every webpage when searched for in a search engine shows some details which are not available on the website. These tags contribute to website noise as well.
   5.3. Other formatting and designing—From advertisements, to images, to JavaScript and CSS, the noise in a website is both myriad and many and needs to be dealt with.

5.4. Parsing—Parsing is basically dividing the text or content of a file into distinct categories so that they can be extracted separately. In our case, we aim to extract the text from the file.

6. Parsing the article—We created a parser based on the logic and steps we defined in the prelims of this paper. To do so, we followed the following steps.

   6.1 Tokenize and break the entire text into characters.
   6.2 Remove all symbols
   6.3 Remove all punctuations
   6.4 Remove meaningless words and useless spaces
   6.5 Concatenate the article text to get the exact desired news article we wanted.

7. Extracting the heading—It is a news article we are working with and every news article has a heading. We extract the heading by reading the first five words of the extracted text. Usually, a Google search query has an average of 7–10 words, so when the main content of our search query is ready we can build the query.
8. Building the search query—Based on if the user wants more text-based articles or image-based we build a search query, by concatenating a relevant string to our extracted heading. Example: "Articles about" + *search query*.
9. Searching Google—We pass this query through our code into Google specifying the language and results in requirement. Then we copy the URLs of all the top search results into our final reverse searched file.
10. Getting more articles and extracting keywords—Once we have a lot of articles from a single domain, we may consider extracting keywords from them using a tf–idf approach.

   10.1. Term Frequency [11]—The number of times each term occurs in a single document. Higher the tf score, more the chance of the word being a keyword.
   10.2. Inverse Document Frequency [12]—This is the inverse of the frequency of each word, across multiple documents. Idf helps us filter our stop words and another verbiage. A high idf score indicates more chances of a word being a keyword.
   10.3. Tf-idf—The product of the two terms. Since tf and idf scores are directly proportional to the chances of a certain word being a keyword, we multiply the two and get the tf-idf score. This score is more or less a determining factor for a word to be a keyword. Higher this score, more the chances of this being a keyword.

The above method for parsing after removal of symbols and punctuations are left with fragments of code, owing to the fact that a webpage at the end of the day is formatted using front-end coding technologies. For instance, the HTML tags once removed leaves behind the text that was enclosed in the angular brackets. The example below has been used to explain the same more clearly.

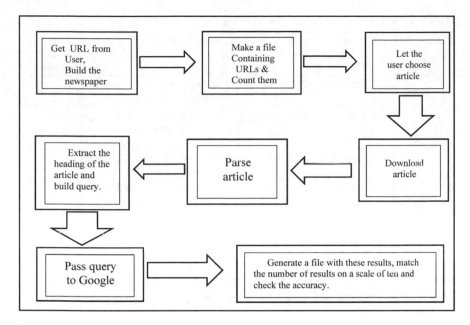

Fig. 1 Showing the steps of the proposed method

Initially, if we have a code segment <h1> this is text </h1>

Once we remove the symbols and punctuations we are left with

h1 this is text h1

Now, the intended article should not have the words h1 in them, because those are tags and need not be added. These words are classified as meaningless words in the parsing stage.

If these words are not removed then they will be included in the final text. Since such tags occur a lot in a document, the chances of them having a high-term frequency increases. This may decrease the accuracy of any future NLP operations performed on the article text. A diagram below has been used to illustrate the above operations in details (Fig. 1).

4 Experimental Results

Every newspaper website gets updated every minute if not second. This phenomenon causes the nature of the output to vary with time. Which means at any two timestamps no URL will be at the same index. Our experimental results were taken from articles from a span of many days to show robustness and effectiveness of our method. The articles were all generated from https://cnn.com (Table 1).

As is visible and evident from the table above, we can find an average of 2434 articles per news website whereas blank opening of the newspaper hardly shows

Table 1 Results of scraping and reverse searching

SL no.	Date	News URL	URL nos.	Article no.	Headline of article	Reverse search matches
1	07/05/17	cnn.com	2320	3	America's NAFTA nemesis: Canada not Mexico	10
2	07/13/17	cnn.com	2464	1	Separated twins move to rehab after emotional farewell	6
3	07/13/17	cnn.com	2460	5	Canada Geese removed from Virginia Beach	9
4	07/13/17	cnn.com	2460	7	Nadal upset by Gilles Muller in five-set marathon	10
5	07/15/17	cnn.com	2460	2	JP Morgan tees off on American Gridlock	10

us a 100. Our reverse search matches show an accuracy of 90% (on 10). The method takes minimal time and took an average of 3–5 s to build an entire newspaper.

5 Conclusion and Future Scope

The future scope of our paper and work can be myriad. First and foremost it helps the reader with all the three major advantages as stated in our introduction. Our primary aim was to bring together all the demands of a reader under one umbrella and allow hassle-free and broadened prospects for something as simple as newspaper reading.

The future scope of this paper may be in fields of NLP with domain-specific keyword extraction. We hope to have completed our primary objective and have helped as many readers as possible with our approach.

References

1. Dominiguez, T.: How Much of The Internet is Hidden. Seeker, 2 Sept 2015
2. Pavalam, S.M., Kashmir Raja, S.V., Akorli, F.K., Jawahar, M.: A survey of web crawler algorithms. Int. J. Comput. Sci. Issues **8**(6), no 1, 309–313 (2011)
3. Shen, A.: Algorithms and Programming: Problems and Solutions, 2nd edn. p. 135. Springer (2010)
4. Brin, S., Page, L.: Anatomy of a large scale hypertextual web search engine. In: Proceedings of the WWW Conference (2004)
5. Sivanandam, S.N., Deepa, S.N.: Introduction to Genetic Algorithms, p. 20. Springer (2008)

6. Zhang, H.: The Optimality of Naive Bayes. American Association for Artificial Intelligence (2004)
7. Caruana, R., Niculescu-Mizil, A.: An empirical comparison of supervised learning algorithms. In: Proceedings of the 23 rd International Conference on Machine Learning, Pittsburgh, PA (2006)
8. Kleinberg, John "Hubs, Authorities, and Communities" ACM computing survey (1998)
9. Pradhan, S., Ward, W., Hacioglu, K., Martin, J.H., Jurafsky, D.: Shallow Semantic Parsing using Support Vector Machines. University of Colorado, Stanford University (2004)
10. Lucas, O.-Y.: Newspaper. github.com/codelucas/newspaper
11. Luhn, H.P.: A statistical approach to mechanized encoding and searching of literary information. IBM J. Res. Dev. IBM **1**(4), 315 (1957)
12. Spark, Jones K.: A statistical interpretation of term specificity and its application in retrieval. J. Doc. **28**, 11–21 (1972)

Impact of Cloud Accountability on Clinical Architecture and Acceptance of Healthcare System

Biswajit Nayak, Sanjaya Kumar Padhi and Prasant Kumar Pattnaik

Abstract Cloud computing is a technology that provides facilities to the client to access shared resources, software, and information through Internet from servers on the cloud. It is very interesting to see the way cloud computing will put importance on the health care, since it is a very diverged application as well as complex and unique and presents several challenges such as protecting member health records in addition to the guidelines set based on federal compliance rules and regulations. It is also very important to see how cloud computing will address and contribute toward the issues in the healthcare industry and improve clinical and quality outcomes for patients. The advent of cloud computing technology provides effective and dependable results to support healthcare services. There is a discussion to customize architecture and for a prospective rule of the recent expertise in healthcare information system. Along with this, it represents the global confronts and technical difficulties for the recent technology. The purpose of this paper is to explore the current state and trends of cloud computing in health care.

Keywords Cloud computing · Clinical architecture · Cloud healthcare system
Healthcare cloud architecture

B. Nayak (✉)
Information Technology, Sri Sri University (SSU), Cuttack, Odisha, India
e-mail: biswajit.nayak.mail@gmail.com

S. K. Padhi
Computer Science & Engineering, Biju Patnaik Technical University (BPUT),
Rourkela, Odisha, India
e-mail: sanjaya2004@yahoo.com

P. K. Pattnaik
School of Computer Science & Engineering, Kalinga Institute of Industrial Technology
(KIIT) University, Bhubaneswar, Odisha, India
e-mail: patnaikprasant@gmail.com

B. Nayak
Biju Patnaik Technical University (BPUT), Rourkela, Odisha, India

© Springer Nature Singapore Pte Ltd. 2018   149
S. C. Satapathy et al. (eds.), *Information and Decision Sciences*,
Advances in Intelligent Systems and Computing 701,
https://doi.org/10.1007/978-981-10-7563-6_16

1 Introduction

Most of the organizations in the healthcare industry always fail to adopt new technology compared to the other industries, and the cloud computing technologies are tools that have the potential to justify facilities and clinical benefit expenditure. The cloud computing technology provides facility to access applications that were previously developed but not unattainable which is one of the most desired benefits. Digitizing the vast medical record puts impact as these are managed through cloud services. The organization would have a huge cost to access and store vast medical records but it is different now as you will pay as you will use.

Cloud services can be used to improve the patient care more efficiently. It is possible to review from anywhere to determine various decisions instantly within the very short period. As the patient information is centrally located and accessible to authorized users different group of health professionals can take allied decision. The experts can spend the time and effort to implement the best practices for each component, which ultimately delivers added benefit to the clinical users and their patients. The cloud is such a technology that changes the pattern for the consistent delivery of healthcare information technology services and the hardware and software that are scalable on a pay-per-use model ensuring effective delivery of healthcare information services [1, 2].

2 Background History

In the year 2005, Peter G Goldsmith realizes the widespread adaption of Health Information Technology (HIT) due to the increase in health expenditure and desire to improve in health care [3].

In the year 2006, Richard Lenz and Manfred Reichert describe the current IT solution supports organizational pattern to some extent to support the integration of heterogeneous component and expecting that the system must support distributed healthcare network [4].

In the year 2007, Sing et al. focused on the framework that can inform the design and testing of IT-based interventions to improve the effectiveness of communication [5].

In the year 2009, Sahay analyzes how to resolve the centralized planning to accommodate decentralized healthcare management information system implementation [6].

In the year 2011, Alexander Kaletsch and Ali Sunyaev focused on the framework to protect personal health records to increase the privacy and build trustworthy personal health information system [7].

In the year 2012, Eman AbuKhousa highlighted the different facets that contribute to building e-health cloud under four categories as cloud-based storage solution, platform solution, implementation models, and security solution [8].

In the year 2013, Daniel E. Rivera and Holly B. Jimision illustrated the way systems and modeling approaches that will impact the change and optimizing the behavioral outcomes of health [9].

In the year 2014, Abdul Manan and Imran Ashraf focus on data security, privacy, data ownership, trust, and legitimate access to patients most secure data [10].

In the year 2015, Pavel et al. highlighted on the multiclass computational models ranging from the sensor to behavioral decision for improving better healthcare behaviors [11].

In the year 2016, Hassan A Aziz focused on methods to deal with the increase in patient data. Electronic health record paired with the cloud to bring the solution for the huge data challenge [12].

3 The Role of Cloud In Improving Health Care

Like other industries, health care also concerns with high availability which is a must and also some regulatory compliance issues like security as well as privacy. It also focuses on important data movement across borders and ownership. Most of the organizations in the area of health care are implementing cloud-based solutions or operating them. These technologies are limited but due to the availability of some tool the clinical healthcare system growing rapidly.

Use of cloud computing probably is not a solution for the entire problem. However, it is great amending which will enhance the efforts and results in an improved healthcare system. Cloud technology reduces and even removes the burden of infrastructure management by providing access to all type of resources and services. This provides an environment that minimizes the expenditure and provides an easy way to adapt required technology. It is not the thing that all the provider taking time for adopting new technologies. All the health organizations are looking for digitization because to increase the quality of the patient care or health information system [12, 13] (Fig. 1).

Besides academic researchers, many world-class software companies (Table 1) have heavily invested in the cloud, extending their new offerings for medical records services, promising an explosion in the storage of personal health information online. The demand for health cloud computing is accounted worldwide.

This is only due to the several factors like greater accessibility means information can be accessed easily, also from anywhere but there should be an Internet connection.

Storage is another major concern. Reduced cost because nothing to buy permanently that means we can use as per requirement and also for the required time period. It does not require building infrastructure.

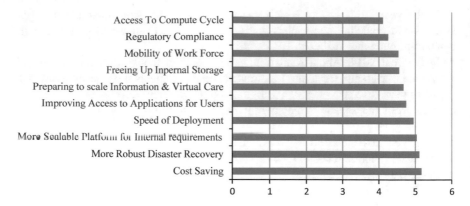

Fig. 1 The given parameters are surveyed and analyzed by taking the value "0" for no-motivating vector and "7" for highly motivating factor (Research and analysis conducted by HIMMS Analytics)

Table 1 Different cloud services provided by different organizations

Organization	Cloud service
Microsoft	HealthVault
Oracle	Exalogic elastic cloud
Amazon	Amazon web services

4 Current State of Health Care

This industry has always underutilized as organizations depend upon on hand-written notes and records to make decisions and notify. Healthcare information in the digital form between departments and applications to access the patient's information is difficult, as it is not impossible. This access to healthcare information wastes millions of dollars every year. Distribution of healthcare information among required stack holders is very complex. The countries with diagnostic imaging seem to have had more success with the components of the patient record.

Most of the information technology departments are familiarized to the traditional technologies as they look for traditional licensed s/w platforms, and complex infrastructures hold up by a large workforce. The organization requires skilled employee or staff in almost all areas of information technology.

As the time progress new expertise are bringing in orders the information technology transportation and due to the demand of time, it starts to expose its limit towards its efficiencies.

Due to the technological advancements in the recent scenario, patients are very cautious regarding their own health information; they are aware of different types of information related to different diseases as well as more alert to their healthcare

Fig. 2 Use of cloud technology by different healthcare organizations in the year 2014 (Research and analysis conducted by HIMMS Analytics)

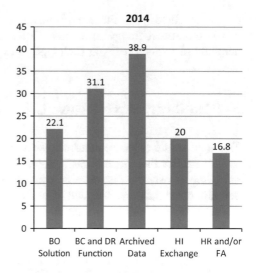

problems and rising request to avail the newest technologies. Patient seeks the best health care at a reasonable cost and accesses their health information.

It can be termed as, if everyone is able to go through their personal information in the bank from anywhere, then why not their personal security information regarding health information. To provide such types of facility many developed countries developed data centers for healthcare information, however, still some of them have challenges. According to the Figs. 2 and 3 (Research and analysis conducted by HIMMS Analytics), it shows the approach to use cloud health care increasing exponentially. In the year 2014, the demand for use cloud technology in

Fig. 3 Use of cloud health care by different healthcare organizations in the year 2016 (Research and analysis conducted by HIMMS Analytics)

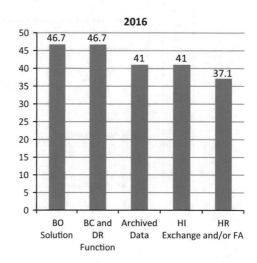

the various fields of healthcare organization is very less as compared to the demand of use cloud technology in the year 2016. The diagram shows the growth exponentially in various fields in such type of organization [11, 13].

5   Healthcare Architecture Using Cloud

The architecture of the healthcare system in the diagram shows the collaboration of several entities like cloud infrastructure, cloud services, security, management, and most importantly the organization where it will be implemented. As it is a public domain, more than one user will have the facility to use the same resource also for the different computing as well as storage operation. The data provider and the technology providers interact with each other and these activities are recorded and encrypted before uploading to the space at the cloud. It is uploaded as a jar file. The data owner is provided with the log data through emails in the push mode and they can view log data using pull mode. The communication system holds the picture archive and communication systems as well as digital image and communication. The log data is provided to the data owner by mail [14, 15] (Fig. 4).

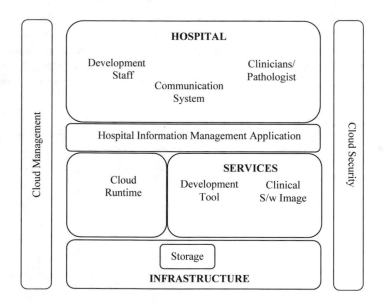

Fig. 4 Architectural building block of healthcare organization using cloud

6 Cloud Healthcare Accountability Criteria

It is always necessary to evaluate the requirement that helps to accelerate the cloud healthcare performance. Several attributes are considered for accountability assess.

Security: It is the major concern when it is associated with the health information system. So it is required to surmount associated risks when dealing with the health information through the cloud, cloud providers must demonstrate security measures need to be deployed so that unauthorized access to health information must be prevented [15–17].

High Availability: Cloud providers should have information of high availability of data or required available requirements and this must ensure the guaranteed delivery of information.

Standards-Based Data Management: Standards-Based Data Management is driving the development of standards throughout many different areas. The use of standards in managing data will future proof the data to ensure that access and migration of data will always be possible.

Table 2 Different cloud services and their dimensions provided by different organization

Attributes	Dimensions
Security	• Make safe access to the facility • Network protection • Data protection • Staff training and regulatory compliance awareness
Availability	• Downtime for maintenance • Sensitivity to data growth • Latency and idleness of network • Redundancy of hardware
Standards-based data management	• DICOM/HL7 • XML metadata • NIST • IHE framework
Scalability	• Provisioning • Plug-and-play growth • Dynamic scaling
Remote access	• Users capacity • Peak access times performance • Mobile device's flexibility
Contractual assurance	• Insurance for breach of privacy • Service-level agreements • Migration assistance • Scalability

Scalability: When the system is online and volume of data grows exponentially, the cloud provider should be able to scale up the requirement so that there should not be any impact on performance and if there is any impact then that should be negligible.

Remote Access: As the data stored remotely, there must be flexibility to access is a major concern for the healthcare organizations as they access through the cloud. So it is required to provide adequate services to the users, for which various aspects are taken into account.

Contractual Assurance: There should be strong conformities like in the agreement which will make sure uninterrupted assured delivery of services (Table 2).

7 Conclusion

The research discovers the various facets of technology like cloud environment and the way providers of health care can go further based on a solution like a cloud. It leads to the emerging, broad-based clinical decision support system. It also will provide broad operational departmental systems with EMR integrations well as emerging data warehousing and analytics solutions. The framework presented here represents the architectural framework which will accompany healthcare system in cloud computing for smooth work and consider accountability from a systematic perspective, with the intent to respond to call for attention to how best to regulate, supervise, and monitor cloud providers.

References

1. Brinkerhoff, D.W.: Accountability and health systems: toward conceptual clarity and policy relevance. Health Policy Plann. **19**(6), 371–379 (2004) © Oxford University Press, all rights reserved. https://doi.org/10.1093/heapol/czh052
2. Ahuja1, S.P., Mani1, S., Zambrano1, J.: A survey of the state of cloud computing in healthcare. Netw. Commun. Technol. **1**(2) (2012). ISSN 1927-064X E-ISSN 1927-0658
3. Goldschmidt, P.G.: HIT and MIS: implications of health information. Technology and Medical Information Systems Communications of the ACM Oct 2005, vol. 48(10)
4. Lenz, R., Reichert, M.: IT support for healthcare processes premises, challenges, perspectives. Elsevier, Data Knowl. Eng. **61**, 39–58 (2007)
5. Singh, H., Naik, A.D., Rao, R., Petersen, L.A.: Reducing diagnostic errors through effective communication: harnessing the power of information technology. Gen. Intern. Med. **23**(4), 489–94 (2007) © Society of General Internal Medicine. https://doi.org/10.1007/s11606-007-0393-z
6. Anifalaje, A.A.: Decentralisation and health systems performance in developing countries. Int. J. Healthc. Deliv. Reform Initiat. **1**(1), 25–47(2009)
7. Kaletsch, A., Sunyaev, A.: Privacy engineering: personal health records in cloud computing environments. In: Thirty Second International Conference on Information Systems, Shanghai (2011)

8. AbuKhousa, E., Mohamed, N., Al-Jaroodi, J.: e-Health cloud: opportunities and challenges. Future Internet **4**, 621–645 (2012). https://doi.org/10.3390/fi4030621

9. Rivera, D.E., Jimison, H.B.: Systems modeling of behavior change: two illustrations from optimized interventions for improved health outcomes. IEEE Pulse **4**(6), 41–47 (2013). [6656980]. https://doi.org/10.1109/MPUL.2013.2279621

10. Manan, A.A., Ashraf, I.I.: Opportunities and threats of cloud computing in HealthCare. Int. J. Comput. Appl. **101**(2), 0975–8887 (2014)

11. Pavel, M., Jimison, H.B., Korhonen, I., Gordon, C.M., Saranummi, N.: Behavioral informatics and computational modeling in support of proactive health management and care. IEEE Trans. Biomed. Eng. **62**(12), 2763–2775 (2015). https://doi.org/10.1109/tbme.2015.2484286

12. Aziz, H.A., Guled, A.: Cloud computing and healthcare services. J. Biosens. Bioelectron. **7**, 220 (2016). https://doi.org/10.4172/2155-6210.1000220

13. Hu, H., Bai, G.: A systematic literature review of cloud computing in eHealth. Health Inform. Int. J. (HIIJ) **3**(4) (2014)

14. Jasim, O.K., Abbas, S., El-Horbaty, E.M., Salem, A.M.: Advent of cloud computing technologies in health information. Int. J. Inf. Theor. Appl. **21**(1) (2014)

15. Hanen, J., Kechaou, Z., Ayed, M.B.: An enhanced healthcare system in mobile cloud computing environment. Vietnam J. Comput. Sci. **3**(4), 267–277 (2016)

16. Nayak, B., Padhi, S.K., Patnaik, P.K.: Understanding the mass storage and bringing accountability. In: National Conference on Recent Trends in Soft Computing & It's Applications (RTSCA) (2017)

17. Zhejiang, G., Lingsong, H., Hang, T., Cong, L.: A cloud computing based mobile healthcare service system. In: Smart Instrumentation, Measurement and Applications (ICSIMA), 2015 IEEE 3rd International Conference, 05 Sept 2016. https://doi.org/10.1109/icsima.2015.7559009

A Stability Analysis of Inverted Pendulum System Using Fractional-Order MIT Rule of MARC Controller

Deep Mukherjee, Palash Kundu and Apurba Ghosh

Abstract In this paper, modification of MIT rule of MARC (Model Adaptive Reference Controller) using fractional derivative concept has been proposed for an integer-order-inverted pendulum system which is highly unstable. Here, the G-L fractional derivative method has been proposed to design fractional-order MIT rule of MARC controller. This controller has been tuned by adaptive gain and an additional degree of freedom to the stable angular displacement of the pendulum and to track the reference model better with respect to time domain specifications. Next, this stability of inverted pendulum using fractional-order MIT rule has been analyzed with normal integer-order MIT rule.

Keywords Inverted pendulum · MARC · MIT rule · Fractional-order MIT rule

1 Introduction

A case study has been shown to implement model adaptive reference control of inverted pendulum system which is one of the most challenging tasks of control engineering, using normal MIT rule followed by fractional-order MIT rule and to compare their performances on stabilization of angular movement of the inverted pendulum. The MIT rule was designed in Massachusetts Institute of technology.

D. Mukherjee (✉)
School of Electronics Engineering, Kalinga Institute of Industrial Technology,
KIIT University, Bhubaneswar, India
e-mail: deepmukherjee@rediffmail.com

P. Kundu
Department of Electrical Engineering, Jadavpur University, Kolkata, India
e-mail: palashm.kushi@gmail.com

A. Ghosh
Department of Instrumentation Engineering, Burdwan University, Burdwan, India
e-mail: apurbaghosh123@yahoo.com

© Springer Nature Singapore Pte Ltd. 2018
S. C. Satapathy et al. (eds.), *Information and Decision Sciences*,
Advances in Intelligent Systems and Computing 701,
https://doi.org/10.1007/978-981-10-7563-6_17

159

Several researchers proposed various schemes to analyze the performance of inverted pendulum using different control techniques. Mohammad Ali [1] proposed the control of inverted pendulum cart system by use of PID controller. Reza [2] proposed the design of PID controller for inverted pendulum using a genetic algorithm. Adrian-Duka [3] proposed MARC using lyapunov theory and fuzzy model reference control for an inverted pendulum. The adaptive controller has been chosen as it is more effective than fixed gain PID controller to handle difficult situations.

In our work, the design of model adaptive reference control for inverted pendulum has been suggested by normal MIT rule [4] based on adaptation gain but this rule by itself does not improve stability and adaptive controller designed using MIT rule is very sensitive to the amplitudes of the signals. So the adaptive gain is generally kept small to make stable. Recently nowadays, fractional order [5] provides an effective means of capturing the approximate nature of the real world. So, MIT rule has been modified as fractional-order MIT rule with an extra degree of freedom which has been varied with fixed adaptation gain to test the characteristics of error between the reference model and plant and to analyze the nature of performance of pendulum using fractional-order rule over integer-order MIT rule.

1.1 Inverted Pendulum

The inverted pendulum is among one of the difficult systems to control in the field of control system [6] and has been taken as a benchmark for analyzing control strategies. An inverted pendulum is a nonlinear dynamic system including a stable equilibrium point, when the pendulum is at pending position and a stable equilibrium point when pendulum is at upright position. In our work, the inverted pendulum has been assumed which has been shown in in Fig. 1.

The pendulum is a weight suspended from a pivot, so that it can swing freely. When a pendulum is displaced sideways from its resting, equilibrium position, it is

Fig. 1 Rotational inverted pendulum system that shows a free body diagram

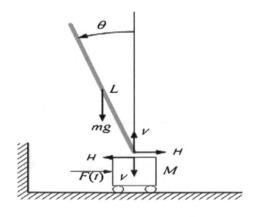

subject to a restoring force due to gravity that will accelerate it back toward the equilibrium position at $\theta = 0$. When released, the restoring force combined with the pendulum's mass causes it to oscillate about the equilibrium position, swinging back and forth. In the above figure, let d_1 = length of pendulum, c = frictional constant, m = mass of pendulum, g = acceleration, d_0 = half length at centre, and T = tension.

The equation of motion for a pendulum is

$$J\frac{d^2\theta}{dt^2} + c\frac{d\theta}{dt} - mgd_0\cos\theta = (d_1)t. \tag{1}$$

Take the Laplace transform as

$$\frac{\theta(s)}{T(s)} = \frac{d_1}{Js^2 + cs - mgd_0}. \tag{2}$$

The parameters are given as

a. J (Inertia) = 0.2453 N/S
b. C (Frictional Constant) = 0
c. m (Mass) = 900 g
d. g (Acceleration) = 9.81
e. d_1 (Length of Pendulum) = 0.102 m
f. d_0 (Half length at center) = 0.945 cm.

Substituting the values of parameters for a real-time process, the overall system transfer function is

$$\frac{\theta(s)}{T(s)} = \frac{0.102}{0.2453s^2 - 0.049}. \tag{3}$$

1.2 MARC (Model Adaptive Reference Controller)

Model adaptive reference control [7] has been chosen to design the adaptive controller, which is dependent on the adaptive gain by altering its output of real plant tracks the output of a reference model having same reference input and the adaptation law uses the error between the process and the model output. The basic block diagram of model adaptive reference control is shown in Fig. 2.

The transfer function of the reference model has been taken as follows:

$$T.F = \frac{9}{s^2 + 6s + 9}. \tag{4}$$

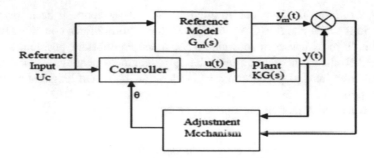

Fig. 2 Block diagram of model adaptive reference controller (MARC)

The inverted pendulum system has been taken as a plant to track the reference model.

1.3 MIT Rule

Adaptive gain plays a vital role to design MIT rule [8] and this adaptive gain is incrementally adjusted to minimize the error between the plant and model. The gradient process has been considered to design MIT rule. The cost function is defined as as follows:

$$J(\theta) = \frac{e^2}{2},$$ (5)

where θ is the adjustable parameter and e is the error between the plant and model. θ is kept in the direction of the negative gradient of J. The normal MIT rule expression has been shown as follows:

$$\frac{d\theta}{dt} = -\gamma e \frac{de}{d\theta}$$ (6)

$$e = y(t) - y_m(t)$$ (7)

$$e = G_p u - G_m u_c$$ (8)

$$u = \theta_1 u_c - \theta_2 y_{plant}$$ (9)

γ is called as an adaptive gain and its range lies between 0.3 and 5 to track the model better. The reference model has been considered very close to plant.

$$\frac{de}{d\theta_1} = \frac{a_{1m}s + a_{0m}}{s^2 + a_{1m}s + a_{0m}} u_c \tag{10}$$

$$\frac{de}{d\theta_2} = -\frac{a_{1m}s + a_{0m}}{s^2 + a_{1m}s + a_{0m}} y_{plant} \tag{11}$$

1.4 Fractional-Order MIT Rule

The controller is very sensitive to the changes in the amplitude of the reference input using MIT rule and to overcome this problem an advanced method [9] of MIT rule has been introduced for parameter adjustment by to develop the control law. In our work, G-L fractional derivative [10] has been approached on error signal following the normal MIT rule expression and the new equation becomes as shown in below.

$$\frac{d\theta}{dt} = -\gamma e \frac{d^\alpha e}{d\theta^\alpha} \tag{12}$$

where the rate of change of parameter θ depends on both adaptation gain and the derivative order alpha. So, G-L fractional derivative [11, 12] has been defined as

$$D^\alpha f(t) = \lim_{h \to 0} \frac{1}{h^\alpha} \sum_{k=0}^{n} (-1)^k \binom{n}{k} f(t - kh), \tag{13}$$

where h is defined as step size.

Now assuming, $D^\alpha f(t) \approx D_h^\alpha f(t)$ it has been obtained as

$$D_h^\alpha f(t) = h^{-\alpha} \sum_{j=0}^{k} (-1)^j \binom{\alpha}{j} f(kh - jh). \tag{14}$$

Now $\binom{\alpha}{j}$ can be approximated as

$$\frac{\alpha!}{j!(\alpha - j)!} = \frac{\Gamma(\alpha + 1)}{\Gamma(j + 1)\Gamma(\alpha - j + 1)}, \tag{15}$$

where Γ is defined as gamma function. Now, this gamma function [13] has been applied on the error signal.

$$\frac{d\theta}{dt} = -\gamma \left(\frac{d^{\alpha}}{d\theta^{\alpha}} e \right) y_m \tag{16}$$

where γ is adaptive gain, e is error between model and plant, y_m is reference output, and α is an additional degree of freedom. Using this mathematical expression fractional-order rule has been designed in Simulink.

2   Result and Analysis

Fractional-order MIT rule has been approached to study the nature of performance with one extra degree of freedom alpha [14], which has been varied with fractional order less than one keeping fixed adaptive gain value. The Simulink model is shown in Fig. 3.

The nature of the response of inverted pendulum using fractional-order MIT rule is shown in Fig. 4.

Now from Table 1, it has been analyzed that keeping the gain fixed as 0.4 and varying alpha between 0.5 and 0.75, the reference model has been tracked better using only the value of alpha with 0.5 with respect to rise time followed by settling time and peak overshoot as performance metrics (Table 2).

Now to compare with fractional-order technique design of normal MIT rule of MARC controller has been shown in Fig. 5.

The nature of the response of inverted pendulum with integer-order MIT rule is shown in Fig. 6 (Table 3).

So, from the above table, it has been studied that using the gain value 0.4 reference model has been tracked better with respect to performance metrics rise

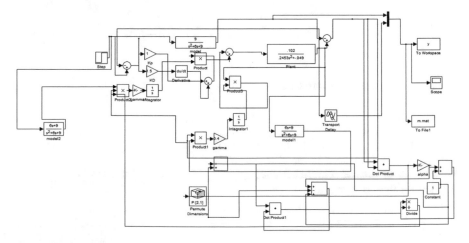

Fig. 3 Simulink model of fractional-order MIT rule

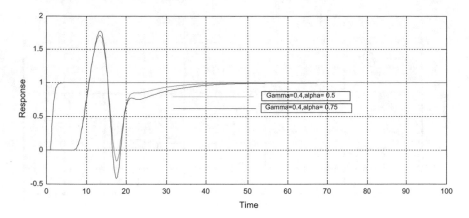

Fig. 4 Blue response shows reference output, red and green response shows plant output using fractional-order MIT rule

Table 1 Performance indices for plant using fractional-order MIT rule with fixed gain

Alpha	Gamma	T_r (s)	T_s (s)	$\%M_p$
0.5	0.4	5.40	32.60	1.71
0.75	0.4	5.40	40.00	1.77

Table 2 Performance indices for reference model

T_r (s)	T_s (s)	$\%M_p$
2.40	2.80	1.00

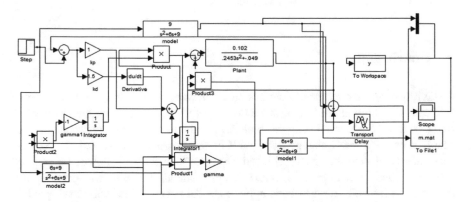

Fig. 5 Simulink model of integer-order MIT rule

time, settling time, and overshoot. But, keeping the fixed adaptive gain fractional-order rule is applied with one extra degree of freedom with value 0.5 reference model has been tracked with desired performance criteria where peak

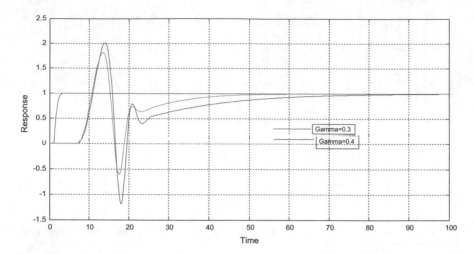

Fig. 6 Blue response shown model output, red and green response shows plant output with gain 0.3, 0.4 using integer-order MIT rule

Table 3 Performance indices for plant using normal MIT rule

Gamma	T_r (s)	T_s (s)	%M_p
0.3	5.80	75.00	2.01
0.4	5.60	45.00	1.81

overshoot followed by rise time and settling time are less than the value of performance metrics using normal MIT rule.

3 Conclusion

A stability of an inverted pendulum has been analyzed approaching fractional-order MIT rule and normal MIT rule on model adaptive reference controller to track a stable reference model. Using normal MIT rule to track the reference model adaptive gain has been changed and with better adaptive gain as 0.4 the desired performance of plant has been achieved with rise time 5.60 s followed by settling time 45 s and overshoot 1.81. Next keeping fixed the adaptive gain as 0.4 fractional-order MIT rule has been applied on the current plant and varying additional degrees of freedom as 0.5 followed by 0.75, it has been studied that performance of plant has been achieved better with less rise time, settling time and overshoot than normal MIT rule. Using fractional-order rule the reference model has been tracked better than normal MIT rule with better stability.

Acknowledgements I take this opportunity to express my sincere gratitude to all the authors following their research articles through which I figured out an innovative idea to present this work.

References

1. Ali, M.: Control of an Inverted Pendulum System by use of PID Controller. ISSN 1013-5316, CODEN: SINTE 8
2. Reza: Design of optimal PID control for an inverted pendulum using genetic algorithm. Int. J. Innov. Manage. Technol. **3** (2012)
3. Duka, A.: Model reference adaptive control and fuzzy model reference control for inverted pendulum comparative analysis. In: WSEAS International Conference on Dynamical Systems and Control, ISBN 960-8457-37-8, pp. 168–173. Italy (2005)
4. Nikranjbar, A.: Adaptive PID controller design guidelines for a class of nonlinear systems Int. J. Eng. Syst. Model. Simul. **6** (2014)
5. Podlubny, I.: Fractional order system and $PI^{\lambda}D^{\mu}$ controllers. IEEE Trans. Autom. Control **44** (1), 208–214 (2004)
6. Kurdekar, V.: Inverted pendulum control: a brief overview. Int. J. Mod. Eng. Res. **3** (2013)
7. Annaswamy, A.M.: Model reference adaptive control. Control Syst. Robot. Autom. **X**
8. Adrian, C., Corneliu, A.: The simulation of adaptive system using MIT Rule. In: 10th WSEAS International Conference on Mathematical Methods and Computational Technique in Electrical Engineering, pp. 301–305. Bulgaria (2008)
9. Jain, P., Nigam, M.J.: Design of model reference adaptive controller using modified MIT rule for a second order system. In: Advance in Electronic and Electrical Engineering, ISSN 2231-1297, vol. 3, pp. 477–484. Research India Publications (2013)
10. Charef, A.: Model Reference Adaptive Control with Fractional Derivative
11. Abedini, M.: Model Reference Adaptive Control in Fractional Order Systems using Discrete time Approximation Methods, vol. 25, Issue 1–3. Elsevier (2015)
12. Loverro, A.: Fractional Calculas: History, Definitions and Applications for the Engineers. Department of Aerospace and Mechanical Engineering, USA (2004)
13. Podlubny, I.: Using Fractional Order Adjustment Rules and Fractional Order Reference Models in Model Reference Adaptive Control, vol. 3, pp. 269–279. Springer, Netherlands (2002)
14. Coman, S.: Fractional Adaptive Control for a Fractional Order Insuline-Glucose Dynamic Model. IEEE, Romania (2017)

KMDT: A Hybrid Cluster Approach for Anomaly Detection Using Big Data

Santosh Thakur and Ramesh Dharavath

Abstract In the current digital era, huge data are being generated in a voluminous state from different sources. This lead towards a processing repository called Big Data. Managing and processing such data in parallel clusters is a big challenge. To capture this problem, in this paper, we propose a hybrid algorithm for cluster analysis using the Spark framework for analyzing the Big Data instances. The proposed algorithm is the combination of two machine learning techniques namely, K-Means (KM) and C5.0 Decision Tree (DT). As per the factor of cluster, euclidean distance is used to find the nearest cluster and the related DT is built for each cluster using C5.0 DT algorithm. The inferences of the DT are used to classify each anomaly and the normal instances of the large datasets. Experimental results show that the proposed hybrid algorithm outperforms with other existing algorithms and produces better classification accuracy for anomaly detection.

Keywords Hadoop · Spark · K-means · Decision tree · Big Data

1 Introduction

The importance of anomaly detection is due to important and actionable information for different application domains. Anomaly detection has been initiated in the early nineteenth century. Another variant of anomaly detection is the detection of

S. Thakur · R. Dharavath (✉)
Department of Computer Science and Engineering, Indian Institute of Technology (ISM), Dhanbad 826004, India
e-mail: ramesh.d.in@ieee.org

S. Thakur
e-mail: santosh.t68@gmail.com

© Springer Nature Singapore Pte Ltd. 2018
S. C. Satapathy et al. (eds.), *Information and Decision Sciences*,
Advances in Intelligent Systems and Computing 701,
https://doi.org/10.1007/978-981-10-7563-6_18

novelty, which aims to detect unobserved previous data. Anomaly detection is the identification of patterns which are abnormal in nature and does not conform to the expected pattern or event [1]. An anomaly is something that deviates from what is standard, normal, and or expected. Anomaly detection is important for several application domains such as public health, decision making, and climate studies. The exact patterns of anomalies are different for various domains. Anomalies are outlined in three different forms; Point anomalies, Contextual anomalies, and Collective anomalies [2].

Several anomaly detection techniques have been specially developed for certain application domains. The applicability of these methods depends on various data dynamics. Conceptually, some detecting anomalies are relatively composed of common properties for the static data [3]. Some variants of methods have been proposed and analyzed to detect anomalies, which fall into three different categories: unsupervised, supervised, and semi-supervised. These techniques range from training the detection algorithm using totally unlabeled data and classify the normal or abnormal instances [4]. To preserve this instance, in our proposed model, we use K-means with C5.0 decision tree algorithm. As a clustering commodity, initially, k-means clustering is used to partition the dataset into K neighboring clusters using Euclidian distance. After the partitioning phase, the C5.0 decision tree methodology along with its heuristics are applied to each neighboring cluster to detect the anomalies.

1.1 Preliminaries

K-means algorithm: K-means is one of the most common and partition-based clustering algorithm. At the time of assigning data points, k-means works in an iterative manner. The data points are randomly chosen to update the cluster center. This process will continue until no changes take place in a cluster for a fixed number of iterations. The data points within the cluster are similar and dissimilar which are outside the cluster. In k-means, we define two measures in the form of distance between two clusters and data points respectively. Euclidean distance is the most popular method to measure the distance [5], where the points $p = (p_1 + p_2 + p_3)$ and $q = (q_1 + q_2 + q_3)$ are used in its space $\sqrt{(p_1 + q_1)^2 + (p_2 + q_2)^2 + (p_3 + q_3)^2}$.

Algorithm₁: K means
Input : *Numerical, There must be a distance metric over the variable space i.e. Euclidian distance* **Output** : *The centres of each discovered cluster, and the assignment of each input data to a cluster i.e. centroid*
1. let X=$\left(x_1, x_2, ..x_n \right)$ and $\left(c_1, c_2, ..c_n \right)$ **2.** Randomly select C cluster centres **3.** *while m>itr* do **4.** *for* each x_i Distance with all centres c_j is calculated Assign x_i to nearest c_j calculate new centroid **5.** *end while* **6.** *end*

C5.0 Decision tree technique: C5.0 is a machine learning algorithm developed by Quinlan based on decision tree [6]. It is an extension of C4.5, and better than the C4.5 in terms of efficiency and memory. In C5.0 model, samples are devided on the basis of information gain field [7]. C5.0 generates classifiers expressed as decision trees, but it also constructs the classifiers in a comprehensible rule set form.

Improvement in C5.0 from C4.5 algorithm: C5.0 decision tree is smaller in comparison with C4.5. In some unseen cases, C5.0 rule set has lower error rates [8]. It is easy to understand the C5.0 model in comparison to some other type of models, where it requires less training time to estimate. The scalability of both the decision tree and rule set is greatly improved [6, 9].

Algorithm₂: Decision Tree
1. Generate a rule set **2.** Pick the most informative attribute **3.** Find the partition with the highest information gain **4.** at each result node, repeat **1** and **2** **5.** End

The rest of the paper is organized in the following manner. Section 2 presents some suitable methodologies applied and proposed in the literature. Section 3 states the working strategy of the proposed methodology with suitable datasets. Section 4 represents the result formation based on the proposed approach followed by the conclusion Sect. 5.

2   Related Literature

Extensive research has been done in the field of anomaly detection. Various techniques have been proposed and applied for detecting the anomalies in different domain applications. Different clustering algorithms such as k-means, fuzzy c-means, hierarchical-based clustering, etc., conclude that clustering is an efficient approach for anomaly detection [2]. Simple k-means is the time efficient among all four clustering algorithms. At the same time, k-means have some limitations. Where, k-means works only for shaping clusters in fixed numbers. This property made it difficult to predict the suitable K value. This problem is further enhanced and solved by a modified version of k-means algorithm [10].

Modified k-means overcomes the problem of finding the optimal number of clusters. But, this approach suffers from a major drawback in the form of time complexity, where its time complexity is more than k-means for larger datasets [11]. To achieve the better accuracy, K-means is integrated with other techniques such as apriori algorithm, ID3, and Decision Tree [9]. Rao et al. [12] proposed a technique by combining k-means with different classification techniques such as k-means +ID3, where it overcomes the disadvantage of both ID3 and k-means. But, it is a time-consuming process. Muniyandi et al. [13] proposed an anomaly detection method by combining k-means and C4.5 decision tree method. These methods achieve better performance in comparison to k-means, C4.5, and KNN.

Ghanem et al. [14] suggested a hybrid approach for anomaly detection for large-scale dataset using a metaheuristic method. This approach shows a better accuracy in generating a suitable number of detectors compared to other algorithms like naive Bayes, decision tree, multilayer feedback neural network [14]. Pandya et al. [8] compared ID3, C4.5, and C5.0 decision tree classifiers with each other and conclude that C5.0 records better accuracy and also generate efficient result among all the classifiers. Shilton et al. [15] presented a new form of SVM approach for the multiclass classification and anomaly detection in a single step, which improves the performance if the data contain vectors, which are not represented in the training dataset. At the same time, this approach faces difficulty in setting one of the algorithm's parameters [6].

By considering the above constraints and to solve the related accuracy issues, in this paper, we use the combination of k-means and DT C5.0 as a hybrid approach to detect unexpected patterns in the large datasets. The reason for choosing the k-means is due to its time complexity $O(nkm)$ and space complexity $O(n + k)$, where n is the number of clusters, k is the number of patterns, and m is the number of iterations. The reason for choosing C5.0 is, it is more efficient and its decision tree is smaller in comparison with C4.5. On the other hand, the strategy of C5.0 eliminates unnecessary attributes.

3   Proposed Approach

In this section, we present the methodology of our proposed model. The proposed hybrid approach is computed on Spark with Hadoop framework. The objective of the model is to work on clusters to process the high dimensional data i.e. big data. Initially, data are distributed on different nodes of the cluster and stored on Hadoop Distributed File System (HDFS). In HDFS, each data chunk maintains three replicas by default to avoid the loss of data. On the other hand, this replica factor can be selected by the programmer according to the requirement. Resilient Distributed Datasets (RDD) are used to fetch the data from HDFS to Spark. With the advantage of adding spark configuration, the proposed methodology handles a large amount of data clusters in less time. Initially, in k-means, euclidean distance was used to divide the dataset into K closest clusters then C5.0 method is applied on each neighboring cluster to construct the decision tree for all clusters and classify each occurrence into a normal instance or anomaly using the results of decision trees. The proposed approach consists of two phases which are named as selection phase and classification phase. The proposed algorithm named KMDT is presented in Algorithm 3.

- **Selection phase**: The closest cluster is selected for each test instance. In the selection cluster, the decision tree corresponding to the cluster is generated.
- **Classification phase**: The test instance is classified into normal and anomaly using the result of C5.0 decision tree and the cluster label as normal or anomaly.

Algorithm₃: KMDT Algorithm

1. Test the instances z(i) *where i=1...n.*
2. Read the dataset.
3. Select the K initial centroid of the cluster randomly.
4. *for* each instance *z(i)* in the dataset, find the closest cluster using Euclidian distance

$$D((z(i),c(j)),j=1...k,\ d((z(i),c(j))=\sqrt{\sum_{a=1}^{m}(z_{ia}-c_{ja})^{2}}$$

5. Compute Decision Tree algorithm for the closest cluster using highest information gain

$$\text{gain (A)}=I\left(T_{1}+T_{2}...T_{n}\right)-\sum\left(\left(T_{1}+T_{2}...T_{n}\right)/T\right)*I\left(T_{1}+T_{2}...T_{n}\right)$$

6. Apply the test instance *z(i)* using the decision tree include it into the cluster
7. Classify the test instance *z(i)* using the decision tree and include it into the cluster.
8: Update the cluster center
9: *end*

3.1 Datasets

To evaluate the proposed hybrid algorithm (KMDT) on large-scale cluster sequences, we use synthetic datasets related to credit card transactions. The related datasets are taken from the repository of ccFraud [16] to evaluate the efficiency of KMDT. To perform the experiments and to validate the working efficiency of the model, we use synthetic datasets [17]. These datasets contain ten million samples with seven attributes. As per the file formation strategy of HDFS, data blocks are divided into 64 MB size by default which are stored across various clusters.

4 Result Analysis

This section presents the performance of the proposed algorithm. Experiments are performed on a single node cluster with Ubuntu operating system 14:04 LTS. The Spark version 1.6.0 is installed on top of the Hadoop and credit card datasets are used for anomaly detection. We apply k-means on the datasets to form K clusters (i.e. K=10). If any overlapping anomaly type of data occurs, it cannot be eliminated by k-means. C5.0 technique is used to classify each cluster. We create a new data frame as final variables which contains clustered variable as a target value and partition the dataset into two parts; train and test set. For testing, we take 80% values for the training and 20% values to compute C5.0 of the closest clusters.

Experiments were conducted by using the above experimental setup to evaluate the efficiency of the KMDT algorithm. The performance measure of the proposed algorithm is compared with other different algorithms as shown in Table 1. Figure 1 illustrates the performance of k-means, k-means+C4.5, ANN, and proposed hybrid approach with an average over 4 trails on credit card datasets. The result of the proposed algorithm is more accurate in comparing with other classification algorithms.

Table 1 Performance evaluation of different algorithms

Classification algorithms	Performance measures	
	Exec. time (s)	Accuracy (%)
K-means	61	87.3
K-means + C4.5	58	92.7
ANN	59	93.4
Proposed (KMDT)	41	96.1

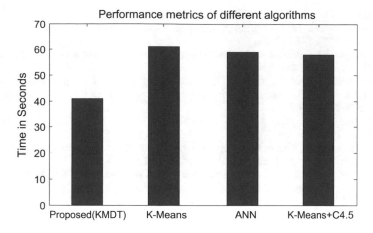

Fig. 1 Performance metrics of different algorithms with KMDT

5 Conclusion

In this paper, we have proposed a Spark based algorithm by combining two different techniques named k-means and C5.0 decision tree for detecting the anomalies. Firstly, K-means is applied to partition the dataset into K clusters and then C5.0 decision tree is built for each cluster for better classification. C5.0 decision tree and cluster labels are used to classify the instances. The proposed KMDT approach can handle high-performance systems in a distributed environment for complex data analysis problems. Based on the predicted results, the proposed algorithm performs better with other algorithms presented in the literature. This experiment is performed on Hadoop with Spark using MLlib. KMDT algorithm can be applied in different fields for anomaly detection such as credit card fraud, monitoring electric fluctuations and mobility detection. As a future work, we further extend the scope of KMDT to improve the accuracy in a more suitable fashion.

Acknowledgements This work is partially supported by Indian Institute of Technology (ISM), Govt. of India. The authors wish to express their gratitude and thanks to the Department of Computer Science and Engineering, Indian Institute of Technology (ISM), Dhanbad, India for providing their support in arranging necessary computing facilities.

References

1. Hayes, M.A., Capretz, M.A.: Contextual anomaly detection in big sensor data. In: 2014 IEEE International Congress on Big Data (BigData Congress), June, pp. 64–71. IEEE (2014)
2. Chandola, V., Banerjee, A., Kumar, V.: Anomaly detection: a survey. ACM Comput. Surv. (CSUR) **41**(3), 15 (2009)

3. Rettig, L., Khayati, M., Cudré-Mauroux, P., Piórkowski, M.: Online anomaly detection over Big Data streams. In: 2015 IEEE International Conference on Big Data (Big Data), October, pp. 1113–1122. IEEE (2015)
4. Hayes, M.A., Capretz, M.A.: Contextual anomaly detection framework for big sensor data. J. Big Data 2(1), 2 (2015)
5. Bottesch, T., Bühler, T., Kächele, M.: Speeding up k-means by approximating Euclidean distances via block vectors. In: International Conference on Machine Learning, June, pp. 2578–2586 (2016)
6. Rani, M.S., Xavier, S.B.: A hybrid intrusion detection system based on C5. 0 decision tree and one class SVM. Int. J. Current Eng. Technol. 5(3) (2015)
7. Patil, N., Lathi, R., Chitre, V.: Comparison of C5.0 & CART classification algorithms using pruning technique. Int. J. Eng. Res. Technol. 1(4) (2012)
8. Pandya, R., Pandya, J.: C5. 0 algorithm to improved decision tree with feature selection and reduced error pruning. Int. J. Comput. Appl. 117(16) (2015)
9. Wu, X., Kumar, V., Quinlan, J.R., Ghosh, J., Yang, Q., Motoda, H., McLachlan, G.J., Ng, A., Liu, B., Philip, S.Y., Zhou, Z. H.: Top 10 algorithms in data mining. Knowl. Inf. Syst. 14(1), 1–37 (2008)
10. Maulik, U., Bandyopadhyay, S.: Performance evaluation of some clustering algorithms and validity indices. IEEE Trans. Pattern Anal. Mach. Intell. 24(12), 1650–1654 (2002)
11. Shafeeq, A., Hareesha, K.S.: Dynamic clustering of data with modified k-means algorithm. In: Proceedings of the 2012 Conference on Information and Computer Networks, pp. 221–225 (2012)
12. Rao, K.H., Srinivas, G., Damodhar, A., Krishna, M.V.: Implementation of anomaly detection technique using machine learning algorithms. Int. J. Comput. Sci. Telecommun. 2(3), 25–31 (2011)
13. Muniyandi, A.P., Rajeswari, R., Rajaram, R.: Network anomaly detection by cascading k-Means clustering and C4. 5 decision tree algorithm. Proc. Eng. 30, 174–182 (2012)
14. Ghanem, T.F., Elkilani, W.S., Abdul-Kader, H.M.: A hybrid approach for efficient anomaly detection using metaheuristic methods. J. Adv. Res. 6(4), 609–619 (2015)
15. Shilton, A., Rajasegarar, S., Palaniswami, M.: Combined multiclass classification and anomaly detection for large-scale wireless sensor networks. In: 2013 IEEE Eighth International Conference on Intelligent Sensors, Sensor Networks and Information Processing, April, pp. 491–496. IEEE (2013)
16. ccFraud Dataset: Apr. 2017. http://packages.revolutionanalytics.com/datasets/. Accessed 01 June 2017
17. Kamaruddin, S., Ravi, V.: Credit card fraud detection using Big Data analytics: use of PSOAANN based one-class classification. In: Proceedings of the International Conference on Informatics and Analytics, August, p. 33. ACM (2016)

Extraction and Sequencing of Keywords from Twitter

Harkirat Singh, Mukesh Kumar and Preeti Aggarwal

Abstract Social media has been the game changer of this generation much like telephony was for the previous. The amount of information available on this platform is huge. This information if extracted and analyzed, can be an immensely helpful source of news and latest developments around the world. As a source and sink of information, it is much faster than traditional news channels and media platforms. This paper uses Twitter data to extract keywords and then sequence them to give useful information. Keywords are extracted from graph constructed from users' posts by heaviest k-subgraph problem. We then proposed a method to sequence extracted keywords in a particular order to get some meaningful information by using Edmonds' algorithm.

1 Introduction

There are two categories of approaches for keyword extraction (1) supervised or (2) unsupervised. The annotated data source is mandated for supervised approaches, while there is no such requirement for unsupervised approaches. Supervised approaches have two main concerns which need to be considered (1) requirement of manually annotated keywords for preparing training data (2) and that the trained data is biased towards the particular domain. Due to these issues in supervised approaches, unsupervised methods have gained more attention, specifically

H. Singh (✉) · M. Kumar · P. Aggarwal
Computer Science and Engineering, University Institute of Engineering
and Technology, Panjab University, Chandigarh 160014, India
e-mail: harkiratsingh.tu@gmail.com

M. Kumar
e-mail: mukesh_rai9@yahoo.com

P. Aggarwal
e-mail: pree_agg@pu.ac.in

© Springer Nature Singapore Pte Ltd. 2018   177
S. C. Satapathy et al. (eds.), *Information and Decision Sciences*,
Advances in Intelligent Systems and Computing 701,
https://doi.org/10.1007/978-981-10-7563-6_19

graph-based methods which have been developed by statistics of the source which gets reflected in the structure of the graph.

In this paper, we also have used graph-based approach, i.e., branch and bound implementation [1] of the heaviest k-subgraph problem to extract keywords from Twitter dataset and then propose a method to sequence these keywords to get meaningful information related to news, stories, and events. Algorithms for finding dense subgraphs (i.e., subgraphs with a relatively large number of edges) have proved to be an effective tool for event detection in social media [2]. For finding important related keywords, the weighted graph is constructed from twitter dataset. In this graph, nodes represent each keyword whereas weighted edge between nodes represent a number of times they co-occur in tweets. When an event occurs, the keywords related to it will co-occur with greater frequency in the dataset. Thus, the subgraph of those keywords will have larger density. In the proposed method to extract sequences, we used Edmonds' algorithm [1] to find spanning arborescence of maximum weight which is the directed analog of the maximum spanning tree. From spanning arborescence, directed paths are extracted to get various sequences. In summary, we first find heaviest k-subgraph using the branch and bound implementation [1] of heaviest k-subgraph to extract important keywords and then we proposed a method to find a sequence from these keywords by using Edmonds' algorithm [3].

2 Related Work

Research using various methods has been done for keywords extraction using graph-based approaches. Ohsawa et al. in [4] presented an algorithm in which graph is segmented into clusters each representing the terms that co-occur. Top-ranked terms are selected as keywords based on each term's relationship to these clusters.

Palshikar in [5] proposes a hybrid structural and statistical approach in which the word in terms of its centrality in graph qualifies as a keyword.

Tsatsaronis et al. in [6] present SemanticRank, a ranking algorithm for keyword and sentence extraction from text. The algorithm constructs a semantic graph using implicit links, which are based on semantic relatedness between text nodes and consequently ranks nodes using different ranking algorithms.

Abilhoa et al. [7] propose a keyword extraction method for tweet collections that represent texts as graphs and applies centrality measures for finding the relevant vertices (keywords).

Litvak et al. in [8] introduce DegExt, a graph-based language-independent keyphrase extractor, which surpasses GenEx [9] and TextRank [10] in terms of precision, implementation simplicity, and computational complexity.

Zhou et al. in [11] investigate weighed complex network-based keyword extraction incorporating the exploration of the network structure and linguistics knowledge.

Lahiri et al. in [12] extract keywords and key phrases from co-occurrence networks of words and from noun phrases collections' network. Eleven measures (degree, strength, neighborhood size, coreness, clustering coefficient, structural diversity index, page rank, HITS hub authority score, betweenness, closeness, and eigenvector centrality) are used for keyword extraction from directed/undirected and weighted networks.

Xu et al. in [13] propose a method LET (LDA&Entropy&TextRank) to extract topic keywords from Sina Weibo topics text sets. LET considers both topic influence of keywords and topic discrimination of keyword that combines the merits of LDA, Entropy and TextRank. In [14], Tugba identifies keywords using latent semantic analysis (LSA). In [1], heaviest k-subgraph is used for event detection which plays an important role in this paper. After analyzing various works, we concluded that there not much work has been done on sequencing of extracted keywords to extract meaningful information.

3 Proposed Technique

In this section, we presented a proposed technique for sequencing the extracted keywords from twitter dataset. Flow diagram of various steps is shown in Fig. 1.

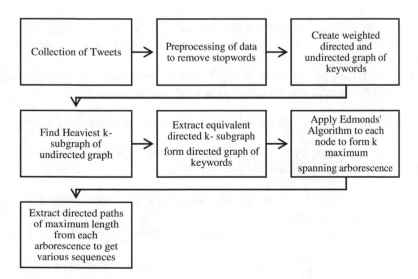

Fig. 1 Flow diagram of the proposed technique

3.1 Preprocessing of Data

In this step, preprocessing of Twitter data is done. Each tweet is first tokenized and then Stanford POS Tagger is used to remove the stop words. Remaining words in the particular tweet are stored in the same sequence as in an original tweet.

Let each tweet be represented by $T_i = \{st_1x_1, st_2, x_2, x_3, st_3 \ldots st_n.x_m\}$, which contains stop words st_n and other words x_n. After preprocessing, stop words are removed and tweet is represented by $T_i = \{x_1, x_2, x_3, \ldots x_m\}$.

3.2 Graph Formation

Let $T = \{T_1, T_2, T_3, \ldots T_n\}$ be the set of tweets and $T_i = \{x_1, x_2, x_3, \ldots x_m\}$ be one of the tweets where a set of x_m is the sequence of words in a tweet T_i.

A graph $G = (V, E)$ is the structure specifying relationships between the words of a collection, where V corresponds to the set of words, called vertices or nodes and E is the set of relations among words. Each word x in T_i represents a node in G and an edge between any two nodes corresponds to the simultaneous occurrence of the two words in a tweet. The weight denotes the number of such instances.

Let $Edge(x, y)$ be the boolean function which tells whether their exists edge between node x and y.

$$Edge(x, y) = \begin{cases} true, & x \: and \: y \: co - occur \: in \: atleast \: one \: tweet \\ false, & otherwise \end{cases}$$

We construct an undirected and directed graph. An undirected graph is required in heaviest k-subgraph to extract keywords and a directed graph is required for forming sequences.

For undirected graph, let it be G, (x, y) is an unordered pair and edge between x and y is bidirectional, thus $Weight(x, y)$ is same as $Weight(y, x)$

$$Weight(x, y) = \{ w, \quad where \: w \: is \: the \: number \: of \: co - occurences \: of \: x \: and \: y$$

For directed graph let it be DG, $Edge(x, y)$ and $Edge(y, x)$ are two different edges, so $Weight(x, y)$ and $Weight(y, x)$ are different

$$Weight(x, y) = \left\{ w_{xy} \left| \begin{array}{c} where \: w_{xy} \: is \: the \: number \: of \: co - occurences \: of \\ x \: and \: y \\ such \: that \: x \: occurs \: before \: y \: in \: tweets \end{array} \right. \right\}$$

$$Weight(y, x) = \left\{ w_{yx} \left| \begin{array}{c} where \: w_{yx} \: is \: the \: number \: of \: co - occurences \: of \\ x \: and \: y \\ such \: that \: y \: occurs \: before \: x \: in \: tweets \end{array} \right. \right\}$$

Fig. 2 Example of **a** Undirected. **b** Directed graph

For example, particular words like "Earth" and "day" come in five tweets together. Then the weight of edge becomes five in an undirected graph. But for directed graph, order of keywords matter, for example, if three times "Earth" comes earlier and two times "day" comes earlier in total five tweets then one edge from "earth" to "day" will have weight three and another edge from "day" to "earth" will have weight two as shown in Fig. 2.

3.3 Heaviest k-Subgraph

In this section, we extract important keywords which are related to each other. To extract important keywords we find the heaviest subgraph of a particular size (called heaviest k-subgraph problem.) from the graph formed after preprocessing of tweets.

Problem definition (*Heaviest k-Subgraph problem*). Given an undirected weighted graph G and an integer $k > 1$, we wish to find a subgraph containing k nodes such that the sum of the weights of its edges is maximum.

We use branch and bound algorithm described in [1] on G for finding the solution of heaviest k-subgraph problem. We take the value of k = 7, 10, 15. Then equivalent directed subgraph *EDG* is extracted from the directed graph *DG*.

3.4 Edmonds' Algorithm for Finding Sequences

In this step, we proposed a technique for finding sequences. We use Edmonds' algorithm [3] to find spanning arborescence of maximum weight which is directed analog of the maximum spanning tree. An arborescence is a directed graph in which, for a vertex u called the root and any other vertex v, there is exactly one directed path from u to v. Spanning arborescence of maximum weight is also called optimum branching as discussed below.

A branching in a directed graph $G = \langle V, E \rangle$ is defined as a set of edges $B \subseteq E$ such that

1. B does not contain a cycle, and
2. No two edges of B enter the same vertex, i.e., if (u, v), and (u', v') are distinct edges of B, then $v \neq v'$.

An optimum branching is then defined as a branching with optimum weight among all possible branchings for *G*. Here the weight of the branching is the total sum of the weights of the edges of the branching. Depending on context, optimum can mean either minimum or maximum. We used here maximum as we are using spanning arborescence of maximum weight.

We find optimum branching from each vertex of equivalent directed subgraph *EDG* and then from each vertex, in its corresponding optimum branching we extract directed paths of maximum length to form sequences. Many of these sequences give meaningful information. Section 4.1 shows the various results of this proposed technique.

3.5 *Proposed Algorithm*

```
Input: undirected graph G ,directed graph DG and integer k
Output: Set of sequences of keywords
  //branchAndBound(Graph G, int k) returns nodes of
heaviest k-subgraph
setOfExtractedKeywords = branchAndBound(G,k);

directedSubGraph = extractDirectedSubGraph(DG,
setOfExtractedKeywords)

for each node u in setOfExtractedKeywords do
      optimalBranching = edmondsAlgorithm(u,
      directedSubGraph);

      sequences=getSimplePathsFromNode(optimalBranching ,
      u);

      SetOfSequences = SetOfSequences + sequences;
Return SetOfSequences;
```

4 Experiments and Results

We used Edinburgh Twitter FSD corpus which contains a collection of tweets related to particular topics. These tweets are used for extracting keywords and sequencing them to give meaningful information related to particular topic. The algorithms were implemented in Java and were run on a machine with 2.53 GHz clock.

4.1 Results

Below are the tables which show various keywords extracted related to various topics and some of the sequences formed from these keywords

TOPIC: earthquake in Virginia	
K = 7 York Felt Carolina Earthquake Virginia Washington 5.8	**Extracted sequences** 1. York 5.8 earthquake Virginia felt Carolina 2. Felt 5.8 earthquake Virginia Washington 3. Carolina felt 5.8 earthquake Virginia Washington 4. Earthquake Virginia felt Carolina 5. Virginia 5.8 earthquake 6. Washington York 5.8 earthquake Virginia felt Carolina 7. 5.8 earthquake Virginia felt Carolina
K = 10 New, York Felt, Carolina Earthquake North, City Virginia Washington 5.8	1. Felt 5.8 earthquake Virginia New York 2. Carolina felt 5.8 earthquake Virginia New York 3. Earthquake Virginia felt Carolina 4. North felt 5.8 earthquake Virginia New York 5. City 5.8 earthquake Virginia felt Carolina 6. Virginia 5.8 earthquake 7. Washington City 5.8 earthquake Virginia felt Carolina 8. 5.8 earthquake Virginia felt Carolina
K = 15 New BREAKING EF York Felt Was Earthquake City Magnitude Washington @AP Magnitude #earthquake Virginia 5.8	1. New @AP BREAKING 5.8 earthquake Virginia felt City 2. BREAKING 5.8 earthquake Virginia felt City 3. York @AP BREAKING 5.8 earthquake Virginia felt City 4. Felt @AP BREAKING 5.8 earthquake Virginia New York 5. Was 5.8 earthquake Virginia felt @AP BREAKING 6. Earthquake Virginia felt @AP BREAKING 5.8 #earthquake 7. Earthquake Virginia felt @AP BREAKING 5.8 magnitude 8. City @AP BREAKING 5.8 earthquake Virginia New York 9. Magnitude 5.8 earthquake Virginia felt @AP BREAKING 10. Washington @AP BREAKING 5.8 earthquake Virginia felt City 11. @AP BREAKING 5.8 earthquake Virginia felt City 12. Magnitude 5.8 earthquake Virginia felt @AP BREAKING 13. #earthquake 5.8 earthquake Virginia felt @AP BREAKING 14. Virginia felt @AP BREAKING 5.8 earthquake Magnitude 15. 5.8 earthquake Virginia felt @AP BREAKING

Topic: plane carrying Russian hockey team Lokomotiv crashes, 44 dead	
K = 7 Plane Hockey Lokomotiv Team KHL Crash Russia	**Extracted Sequences** 1. Plane crash Russia 2. Plane hockey team 3. Hockey plane crash Russia 4. Lokomotiv plane crash Russia 5. Lokomotiv plane hockey team 6. Team plane crash Russia 7. KHL Lokomotiv plane crash Russia 8. KHL Lokomotiv plane hockey team 9. Crash Russia plane hockey team

(continued)

(continued)

Topic: plane carrying Russian hockey team Lokomotiv crashes, 44 dead	
K = 10	1. Plane carrying Russian hockey team
Plane	2. Hockey team dead Russian plane crash Russia
Hockey	3. Hockey team dead Russian plane crash Russia
Russian	4. Russian plane crash Russia
Lokomotiv	5. Lokomotiv dead Russian plane carrying KHL
Dead	6. Lokomotiv dead Russian plane crash Russia
Team	7. Dead Russian plane carrying KHL
KHL	8. Team dead Russian plane carrying KHL
Crash	9. Team dead Russian plane crash Russia
Russia	10. KHL Russian plane crash Russia
Carrying	11. Crash Russian plane carrying KHL
	12. Russia Russian plane carrying KHL
	13. Carrying Russian plane crash Russia
K = 15	1. Plane carrying Russian hockey team
Plane	2. Hockey killed Russian plane carrying KHL
Hockey	3. Russian plane carrying KHL
Russian	4. 36 dead Russian plane carrying KHL
36	5. Plane Russian plane carrying KHL
Plane	6. Dead Russian plane carrying KHL
Dead	7. Ice killed Russian plane carrying KHL
Ice	8. Ice killed Russian plane crash Russia
Team	9. Team killed Russian plane carrying KHL
Killed	10. Killed Russian plane carrying KHL
KHL	11. KHL killed Russian plane crash Russia
Crash	12. Crash killed Russian plane carrying KHL
Russia	13. Russia killed Russian plane carrying KHL
Lokomotiv	14. Lokomotiv killed Russian plane carrying KHL
Crashed	15. Crashed killed Russian plane carrying KHL
Carrying	

5   Conclusion

A lot of research has been done on keyword extraction methods. In this paper, we explored a method to sequence extracted keywords, so that some useful information can be extracted. We proposed a method for extracting sequences from these keywords. Looking at the results we concluded that not all sequences give meaningful information, but there exist some sequences which provide useful information according to the value of k in heaviest k-subgraph. Future work includes improvement in results with more meaningful sequences and dealing with noisy results of keyword extraction methods.

References

1. Letsios, M., Balalau, O.D., Danisch, M., Orsini, E., Sozio, M.: Finding heaviest k- subgraphs and events in social media. In: 2016 IEEE 16th International Conference on Data Mining Workshops (ICDMW), Barcelona, pp. 113–120 (2016)
2. Angel, A., Koudas, N., Sarkas, N., Srivastava, D., Svendsen, M., Tirthapura, S.: Dense subgraph maintenance under streaming edge weight updates for real-time story identification. VLDB J. **23**(2), 175–199 (2014)
3. Edmonds, J.: Optimum branchings. J. Res. Nat. Bur. Stand. **71B**, 233–240 (1967)
4. Ohsawa, Y., Benson, N.E., Yachida, M.: KeyGraph: automatic indexing by co-occurrence graph based on building construction metaphor. In: Proceedings of the Advances in Digital Libraries Conference, ADL '98, p. 12, Washington, DC, USA (1998)
5. Palshikar, G.K.: Keyword extraction from a single document using centrality measures. Pattern Recogn. Mach. Intell. LNCS **4815**, 503–510 (2007)
6. Tsatsaronis, G., Varlamis, I., Nørvag, K.: SemanticRank: ranking keywords and sentences using semantic graphs. In: Proceedings of the 23rd International Conference on Computational Linguistics, pp. 1074–1082, Beijing, China (2010)
7. Abilhoa, W.D., de Castro, L.N.: A keyword extraction method from twitter messages represented as graphs. Appl. Math. Comput. **240**, 308–325 (2014)
8. Litvak, M., Last, M., Aizenman, H., Gobits, I., Kandel, A.: DegExt—a language-independent graph-based keyphrase extractor. In: Advances in Intelligent Web Mastering—3, vol. 86, pp. 121–130. AISC (2011)
9. Turney, P.D.: Learning to extract keyphrases from text. Technical Report ERB-1057, National Research Council of Canada, Institute for Information Technology, 1999
10. Mihalcea, R., Tarau, P.: TextRank: bringing order into texts. In: Proceedings of the Conference on Empirical Methods in Natural Language Processing (EMNLP 2004), pp. 104–411, Barcelona, Spain, July 2004
11. Zhou, Z., Zou, X., Lv, X., Hu, J.: Research on weighted complex network based keywords extraction. Chin. Lex. Semant. LNCS **8229**, 442–452 (2013)
12. Lahiri, S., Choudhury, S.R., Caragea, C.: Keyword and keyphrase extraction using centrality measures on collocation networks, Cornell University Library (2014). arXiv:1401.6571
13. Xu, S., Guo, J., Chen, X.: Extracting topic keywords from Sina Weibo text sets. In: 2016 International Conference on Audio, Language and Image Processing (ICALIP), Shanghai, pp. 668–673 (2016)
14. Suzck, Tugba: Using latent semantic analysis for automated keyword extraction from large document corpora. Turkish J. Electr. Eng. Comput. Sci. **25**, 1784–1794 (2017). https://doi.org/10.3906/elk-1511-203

A Modular Approach for Social Media Text Normalization

Palak Rehan, Mukesh Kumar and Sarbjeet Singh

Abstract The normalized data is the backbone of various Natural Language Processing (NLP), Information Retrieval (IR), data mining, and Machine Translation (MT) applications. Thus, we propose an approach to normalize the colloquial and breviate text being posted on the social media like Twitter, Facebook, etc. The proposed approach for text normalization is based upon Levenshtein distance, demetaphone algorithm, and dictionary mappings. The standard dataset named lexnorm 1.2, containing English tweets is used to validate the proposed modular approach. Experimental results are compared with existing unsupervised approaches. It has been found that modular approach outperforms other exploited normalization techniques by achieving 83.6% of precision, recall, and F-scores. Also 91.1% of BLUE scores have been achieved.

1 Introduction

Social media networks like Twitter, Facebook, WhatsApp, etc., are most commonly used medium for sharing news, opinions, and to stay in touch with peers. Messages on Twitter are limited to 140 characters. This led users to create their own novel syntax in tweets to express more in lesser words. Free writing style, use of URLs, markup syntax, inappropriate punctuations, ungrammatical structures, abbreviations, etc., make it harder to mine useful information from them. There is no standard way of posting tweets. This lack of standardization hampers NLP and MT tasks and renders huge volume of social media data useless. Therefore, there is a

---

P. Rehan (✉) · M. Kumar · S. Singh
Computer Science & Engineering Department, University Institute of Engineering
and Technology, Panjab University, Chandigarh, India
e-mail: palakrehan@gmail.com

M. Kumar
e-mail: mukesh_rai9@yahoo.com

S. Singh
e-mail: sarbjeet@pu.ac.in

© Springer Nature Singapore Pte Ltd. 2018   187
S. C. Satapathy et al. (eds.), *Information and Decision Sciences*,
Advances in Intelligent Systems and Computing 701,
https://doi.org/10.1007/978-981-10-7563-6_20

need to reform such text forms into standard forms. This can be achieved by normalization which is a preprocessing step for any application that handles social media text. Process of converting ill-formed words into their canonical form is known as normalization. In this paper, we propose a modular approach for the lexical text normalization which is applied on English tweets. First, it is necessary to categorize text into two classes: Out-Of-Vocabulary (OOV) and In-Vocabulary (IV). Words that are not in its standard form are considered as OOV while words which are part of standard orthography fall under IV category. For example, "talkin" is an OOV word (nonstandard word) having "talking" as its IV form. Some words, although may be correct, coincide with other IV words, like "wit" is an IV word which may indicate "with" as its correct form. Such words are neglected for normalization task.

Due to the presence of unsurprisingly long tail of OOV words, a method that does not require annotated training data is preferred. Thus, an unsupervised approach has been proposed with different modular stages for preprocessing, candidate generation, and candidate selection steps. The proposed cascaded method achieves state-of-the-art results on English twitter dataset and can be applied to any other social media dataset. In addition, a pilot study is conducted on peer methodologies employed inside the proposed approach.

2 Related Work

Previous work attempted noisy channel model as one of the text normalization techniques. Brill and Moore characterized the noisy channel model based on string edits for handling the spelling errors. Toutanova and Moore [1] improved the above model by embedding information regarding pronunciation. Choudhury et al. [2] proposed a supervised approach based on Hidden Markov Model (HMM) for SMS text by considering graphemic/phonetic abbreviations and unintentional typos. Cook and Stevenson [3] expanded error model by introducing probabilistic models for different erroneous forms according to sampled error distribution. This work tackled three common types: stylistic variation, prefix clipping, and subsequence abbreviations. Yi yang et al. [4] presented a unified log-linear unsupervised statistical model for text normalization using maximum likelihood framework and novel sequential Monte Carlo training algorithm.

Some of the previous work was based on Statistical Machine Translation (SMT) approach for normalization. SMT deals with context-sensitive text by treating noisy forms as the source language and the standard form as the target language. Aw et al. proposed an approach for Short Messaging Service (SMS) text normalization using phrase-level SMT and bootstrapped phrase alignment techniques. The main drawback of SMT approach is that it needs a lot of training data and it cannot accurately represent error types without contextual information.

Gouws et al. [5] developed an approach based on string and distributional similarity along with dictionary lookup method to deal with ill-formed words. Han et al. introduced similar technique based on distributional similarity and string similarity. Selection of correct forms was performed on pairwise basis. Mohammad Arshi et al. [6] proposed a tweet normalization approach. First, candidates were generated by targeting lexical, phonemic, and morphophonemic similarities. More recent approaches handle the text normalization using CRFs and neural networks. Min et al. [7] proposed a system where Long Short-Term Memory (LSTM) recurrent neural networks using character sequences and Part-Of-Speech (POS) tags had been used for predicting word-level edits. Yang and Kim used an CRF-based approach. CRF using both brown clusters and word embeddings that were trained using canonical correlation analysis as features.

Abiodun Modupe [8] developed a semi-supervised probabilistic approach for normalizing informal short text messages. Language model probability had been used to enhance the relationships between formal and informal word. Then, string similarity was employed with a linear model to include features for both word-level transformations and local context similarity.

Proposed approach in this paper also adopts lexical based text normalization using unsupervised methods in order to handle wide variety of ill-formed words. This approach can be applied to different social media text messages.

3 Proposed Work

Inspired by earlier work done on text normalization, a modular approach for lexical text normalization is proposed, which has been applied to English tweets. Preprocessing, Candidate Generation, and Candidate Selection are the three main stages of the proposed system.

Text refining is applied on extracted strings and thereafter categorization into IV and OOV lexicons is performed. Candidate Generation stage generates list of possible correct words for an input OOV word. In the end, Candidate Selection stage selects a best possible candidate from all generated candidates. Raw tweets are fed into preprocessor whose output will act as input for the token qualifier which will segregate input into two heaps: OOV and IV words. Out of total OOV tokens, filtration is performed with the help of available python packages as punctuation symbols, hashtags, URLS, @mentions, and proper nouns. *(Name of persons, locations, brands, movies, etc., comes under proper noun).

OOV tokens detected by token qualifier will be processed by the candidate generator which will generate possible normalized candidates via three different techniques: Levenshtein distance, demetaphone algorithm, and dictionary mappings. Candidate selector module will work on candidate list generated by the candidate generator and will generate best possible candidate for each OOV.

Preprocessing module takes raw tweets as input. Tokenization in the form of strings is performed and then unwanted strings containing @, hashtags, URLS, punctuations, emoticons, and any foreign language are filtered out. Output strings generated after filtration are considered as lexicons. This is initial step required to carry any NLP task. Output generated by preprocessing module is fed as input to the Token Qualifier. It performs classification of lexicon into two categories: OOV and IV. To predict whether a given lexicon is OOV, many Standard English spell checker like GNU Aspell, Hanspell, and dictionaries (Pyenchant corpus) are available.

According to research, correct formed English words having repeating characters are found to have maximum of two character repetitions. Thus, repetition of more than two characters in a string is trimmed off to two characters (helloooo → hello, gooood → good). Regular expressions are applied to OOV strings with alphanumeric text. Some of the transformations with examples are given as: 4 → fore (B4 → bfore), 2 → to (2night → tonight), 9 → ine (F9 → fine), etc. After applying trimming and regular expressions, OOV words that are going to be processed further are obtained. First technique to generate candidates for OOV word is through Levenshtein distance (also known as edit distance). Edit distance is number of applied insertions, deletions, and alterations in order to transform one string into another. It is used to handle spelling errors. Edit distance >2 results in generation of large number of candidates most of which are inappropriate and at same time are complex to process. So, we prefer edit distance with <=2. Algorithm 4 takes input of Algorithm 3 and generates strings having edit distance <=2 with respect to input OOV. In order to have precise and limited generated candidate list, string similarity measures are applied on candidate list generated via edit distance (<=2).

Algorithm 1: Levenshtein_candidates (Modified_OOV)
{

1. *Levenshtein_set = []*
   near_by_editcandidates = []
2. *Input the Modified_OOV*
3. *Generate strings having edit distance ≤ 2 from input OOV.*
4. *If generated string (from step 3) english vocabulary:*

   Add generated strings to Levenshtein_set only

5. *Apply string similarity measures (fuzzy_ratio) between input and each of corresponding strings in Levenshtein_set:*

   5.1 *Select those pairs having maximum similarity ratio.*
   5.2 *Add above pairs in near_by_editcandidates*

6. *Return near_by_editcandidates*

   }

Nowadays, Internet slangs like lol → laughing out loud and abbreviations (Cuz → because) are common in social media text. So that we generate candidates using dictionary mapping by applying algorithm 2.

Algorithm 2: Slang_candidates (Modified_OOV)
{

1. *Slang_output = []*
2. *Input the Modified_OOV*
3. *Check input in Slang dictionay*
   // Slang dictionary is prepared by collecting internet abbreviations
   // from www.noslang.com *on 4 nov, 2016*

   3.1 *If input is found in dictionary:*

      output corresponding mapping to Slang_output

4. *Return Slang_output* }

In order to handle errors due to phonemes (words that sound same), demeta-phone algorithm is used. Words like nitc and night are phonemes of each other. In order to have limited, precise candidate is set and to reduce processing complexity, string similarity measures are applied on phonemes generated by the demetaphone Algorithm.

Algorithm 3: Demetaphone_candidates (Modified_OOV)
{

1. *Demetaphone_set = []*
   Relevant_demetaphone = []
2. *Input the Modified$_{OOV}$*
3. *Generate demetaphone code for each input and english vocabulary word pair.*
4. *Add pairs having same code to Demetaphone_set.*
5. *In order to have only relevant and limited pairs, apply string similarity measures (fuzzy_ratio) to each pair in Demetaphone_set.*

   5.1 *Select only those pairs that have maximum similarity ratio.*
   5.2 *Add above pairs to Relevant_demetaphone set.*

6. *Return Relevant_demetaphone.*
   }

Candidate list generated by all three techniques (output of Algorithm 1–3) acts as input to candidate scorer (Algorithm 4). Equal probability to each candidate in list corresponding to a OOV lexicon is assigned. Aggregate probabilities of all those candidates which are present in more than one list are calculated by performing summation on their probabilities. This will act as score. Prepare an aggregate list by combining candidate lists of all three candidate generation techniques.

Algorithm 4: Candidate_Scorer (near_by_editcandidates, Relevant_demetaphone, Slang_output)

{

1. $Edit_{score} = []$, $demetaphone_{score} = []$, $slang_{score} = []$, $combine_{score} = []$, $aggregate_{candidates} = []$

2. *Assign equal probability to each candidates in near_by_editcandidates and store probabilities in Edit_score set.*

3. *Assign equal probability to each candidates in Relevant_demetaphone and in Slang_output and store them in demetaphone_score and slang_score respectively.*

4. *Aggregate probabilities for common candidates that are present in all above three sets. Combine score corresponding to a Modified_OOV token is computed as:*

$$Edit_{score[Modified_{OOV}]} + demetaphone_{score[Modified_{OOV}]} + slang_{score[Modified_{OOV}]}$$

5. *Prepare aggregate$_{candidates}$ set by combining candidates from near_by_edit-candidates, Relevant_demetaphone, Slang_output*

6. *Return combine$_{score}$, aggregate$_{candidates}$* }

Aggregate candidate list and score list prepared by Algorithm 4 will act as input to Algorithm 5. Select that candidate from aggregate candidate list (for an OOV lexicon) corresponding to which maximum scores are present in score list. In case there are more than one candidate with same scores, then apply POS tagging. During POS tagging, assign scores according to the importance of context like nouns will be given highest weight followed by the verb and then the adjective. This will return a single best candidate for each incorrect word.

Proposed modular approach works on raw tweets. Preprocessing is done by removing unwanted strings (punctuations, hashtags, etc.). Token qualifier is then called to detect OOV and IV words. Rules are applied to the output of the token qualifier to generate OOV tokens which will be used for further processing. Candidates are generated via Levenshtein distance (Algorithm 1), demetaphone algorithm (Algorithm 2), and dictionary approach (Algorithm 3). In order to select best possible normalized word corresponding to an OOV word, candidate scorer (Algorithm 4) and candidate selector (Algorithm 5) are employed.

Algorithm 5: Candidate_selector (aggregate$_{candidates}$, combine$_{score}$)

{

1. *Best_candidate = []*

2. *Read candidates from aggregate_candidates.*

3. *Select those candidates from aggregate$_{candidates}$ corresponding to which maximum probability is present in combine_score set.*

4. *If (only one candidate is outputed from step 3): Store it in best_candidate set.*

5. *Else:*

   5.1 *Apply POS (Part of speech) tag.*

5.2 *Assign maximum score (say 1) to noun, followed by verb (0.5) and then adjective (0.25)*

5.3 *Select candidate with maximum score // obtained after scoring of* 5.2

5.4 *Store result in Best_candidate*

6. *Return Best_candidate*

  }

4 Experimental Setup and Results

Proposed modular approach has been implemented on LexNorm 1.2 dataset which was an updated version of dataset for lexical normalization described in [9]. This dataset contains English messages sampled from Twitter API (from August to October, 2010). Results are evaluated on the basis of precision, recall, F-score, and BLEU score. The proposed work is performed on Python 2.7 version for windows and natural language processing inbuilt python packages are utilized to execute modules. Let $T_{dataset}$ be all tokens from dataset and let OOV_t be the list of all detected OOV in dataset $\in T_{dataset}$ • gen_{oov}^t be the generated candidates for an oov $\in OOV_t$ · sel_{oov}^t be the best normalized candidate selected by system for an oov token, oov $\in OOV_t$ · cor_{oov}^t be the tagged correction for an oov $\in OOV_t$ · $norm_{oov}^t$ be the set of normalized oov tokens $\in OOV_t$ normalized by system.

$$Precision(P) = \frac{\sum_{t \in T_{dataset}} \left| \left\{ sel_{oov}^t : sel_{oov}^t = cor_{oov}^t, sel_{oov}^t \in gen_{oov}^t, oov \in OOV_t \right\} \right|}{\sum_{t \in T_{dataset}} \left| \left\{ norm_{oov}^t : norm_{oov}^t, oov \in OOV_t \right\} \right|} \tag{1}$$

$$Recall(R) = \frac{\sum_{t \in T_{dataset}} \left| \left\{ sel_{oov}^t : sel_{oov}^t = cor_{oov}^t, sel_{oov}^t \in gen_{oov}^t, oov \in OOV_t \right\} \right|}{\sum_{t \in T_{dataset}} \left| \left\{ oov : oov \in OOV_t \right\} \right|} \tag{2}$$

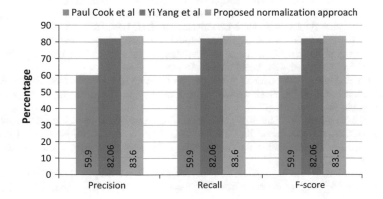

Fig. 1 Comparative results with unsupervised methods

$$F - score = \frac{2 * Precision * Recall}{Precision + Recall} \tag{3}$$

Figure 1 shows that the proposed modular approach yields better accuracy as compared to existing unsupervised methods. Modular approach has 1.54% better results than log linear model for unsupervised text normalization [4]. Moreover, an unsupervised model for text normalization proposed by Paul cook et al. [3] also has low performance (57.9% accuracy) than the proposed approach (having 83.6% performance).

5   Conclusion

Text normalization addresses all forms of OOV words and aimed at standardization of these words. Proposed approach is based on three methods: Levenshtein distance, demetaphone algorithm, and dictionary approach. Experimental results are calculated for Lexnorm 1.2, standard dataset for twitter messages. The proposed system is compared with existing unsupervised text normalization methods. It has been found that modular approach outperforms other exploited normalization techniques by achieving 83.6% of precision, recall, and F-scores. Also 91.1% of BLUE scores have been achieved.

References

1. Toutanova, K., Moore, R.C.: Pronunciation modeling for improved spelling correction. In: Proceedings of the 40th Annual Meeting on Association for Computational Linguistics, ACL 02, pp. 144–151, Philadelphia, USA (2002)
2. Choudhury, M., Saraf, R., Jain, V., Mukherjee, A., Sarkar, S., Basu, A.: Investigation and modeling of the structure of texting language. Int. J. Doc. Anal. Recogn. **10**, 157–174 (2007)
3. Cook, P., Stevenson, S.: An unsupervised model for text message normalization. In: Proceedings of the Workshop on Computational Approaches to Linguistic Creativity, pp. 71–78. Association for Computational Linguistics, Boulder, USA, June (2009)
4. Yang, Y., Eisenstein, J.: A log-linear model for unsupervised text normalization. In: Proceedings of the 2013 Conference on Empirical Methods in Natural Language Processing (EMNLP 2013), pp. 61–72, Seattle, USA, Oct 2013
5. Gouws, S., Hovy, D., Metzler, D.: Unsupervised mining of lexical variants from noisy text. In: Proceedings of the First workshop on Unsupervised Learning in NLP, pp. 82–90, Edinburgh, Scotland (2011)
6. Saloot, M.A., Idris, N., Shuib, L., Raj, R.G., Aw, A.: Toward tweets normalization using maximum entropy. In Proceedings of the ACL 2015 Workshop on Noisy User-generated Text, pp. 19–27. Association for Computational Linguistics, Beijing, China, 31 July 2015 (2015)
7. Min, W., Mott, B., Lester, J., Cox, J.: Ncsu_sas_wookhee: a deep contextual long-short term memory model for text normalization. In: proceedings of WNUT, Beijing, China (2015)

8. Modupe, A., Celik, T., Marivate, V., Diale, M.: Semi-supervised probabilistics approach for normalising informal short text messages. In: Conference on Information Communication Technology and Society (ICTAS). IEEE (2017)
9. Han, B., Baldwin, T.: Lexical normalisation of short text messages: makn sens a# twitter. In: Proceedings of the 49th Annual Meeting of the Association for Computational Linguistics: Human Language Technologies, vol. 1, pp. 368–378. Association for Computational Linguistics, Portland, Oregon, June (2011)

Author Verification Using Rich Set of Linguistic Features

A. Bhanu Prasad, S. Rajeswari, A. Venkannababu
and T. Raghunadha Reddy

Abstract Author Verification is a type of author identification task, which deals with identification of whether two documents were written by the same author or not. Mainly, the detection performance depends on the used feature set for clustering the documents. Linguistic features have been utilized for author identification according to the writing style of a particular author. Disclosing the shallow changes of the author's writing style is the major problem which should be addressed in the domain of authorship verification. It motivates the computer science researchers to do research on authorship verification in the field of computer forensics. In this work, three types of linguistic features such as stylistic, syntactic, and semantic features are used to improve the accuracy of author verification. The Naïve Bayes multinomial classifier is used to build the classification model and good accuracy is achieved for Author Verification.

Keywords Author verification · Stylistic features · Syntactic features
Semantic features · Naïve Bayes multinomial

A. Bhanu Prasad (✉)
CSE Department, Vardhaman College of Engineering, Hyderabad, India
e-mail: andrajub4u@gmail.com

S. Rajeswari
CSE Department, VR Siddhartha Engineering College, Vijayawada, India
e-mail: rajeswari.setti@gmail.com

A. Venkannababu
CSE Department, Sri Vasavi Engineering College, Tadepalligudem, AP, India
e-mail: venkannababu.alamuru@gmail.com

T. Raghunadha Reddy
IT Department, Vardhaman College of Engineering, Hyderabad, India
e-mail: raghu.sas@gmail.com

© Springer Nature Singapore Pte Ltd. 2018
S. C. Satapathy et al. (eds.), *Information and Decision Sciences*,
Advances in Intelligent Systems and Computing 701,
https://doi.org/10.1007/978-981-10-7563-6_21

197

1   Introduction

There is a vast amount of data on the Internet and it is growing rapidly every day. Such a high rate of growth also brings some problems with it. Fraudulent, stolen, or unidentified data are encountered online on a daily basis. These problems can be dangerous and serious problems in places like the public websites, government, forensics, and schools. Because of these threats, and in detection of truth, it is important to know the author of a text.

Authorship Analysis is divided into three categories including Authorship Attribution, Authorship Verification, and Authorship Profiling. Authorship Attribution studies a text in dispute and finds the corresponding author in a set of candidate authors. Authorship Verification compares multiple pieces of written text and determines whether they are written by the same author or not without identifying the author. Authorship Profiling detects unique characteristics like gender, age, location, nativity language, and educational background of an author's written texts and creates an author profile. In this work, the Author Verification task is concentrated. Author Verification techniques are important in several information processing applications.

In the context of cyberspace, a digital document found can be used as an evidence to prove that a suspect is a criminal if he/she is the author of the document. If the suspect authors are unknown, i.e., there is no suspect, thus this is commonly known as an authorship identification problem. However, there are also some cases when the identification of the author is not necessary, i.e., it is enough just to know if the document in dispute was written by the author of the documents that are given. This is a problem faced by many forensic linguistic experts which are called as authorship verification problem.

This paper is organized as follows. Section 2 demonstrates the existing approaches already implemented and tested in authorship verification. Section 3 introduces the set of linguistic features used for document representation in authorship verification. The classification procedure and our approach for finding accuracy of author verification are explained in Sect. 4. In Sect. 5, the experimental results obtained will be discussed and Sect. 6 presents the conclusion.

2   Literature Review

Authorship Verification is the process of verifying an author by checking whether the document is written by the suspected author or not [1]. Victoria Bobicev proposed [2] a method to automatically detecting the author of a given text when the corpus contains small training sets with known authors. They used the prediction by partial matching (PPM) method based on statistical n-gram model. Without feature engineering, PPM obtains total information from the original corpus. They experimented with a corpus of 30 authors, 100 posts of each author and approximately each post length is 150–200 words. It was observed that their system accuracy measure F-measure is not increased when the document length was increased.

Vanessa Wei Feng et al. adopted [3] an unmasking approach, which is used to enhance the quality of features used in building weak classifiers. They experimented with 538 features for English, 568 for Greek, and 399 for Spanish language. The features include coherence features and stylometric features. They observed that their work achieved best accuracy for English and Spanish texts, but less accuracy for Greek texts.

Darnes Vilariño et al. used [4] syntactic, lexical, and graph based features to represent the document vectors. Subdue data mining tool is used to extract the graph-based features. A support vector machine is used to prepare the classification model. Lexical-syntactic features include phrase level features such as word suffixes, stopwords, punctuation marks and trigrams of POS, and character level features such as vowel combination and vowel permutation. It was observed that their system run time is greater than most of the other submissions.

Cor J. Veenman et al. used [5] the compression dissimilarity measure to compute the compression distance between the documents. They proposed three approaches such as nearest neighbor with compression distances, two class classifications in compression prototype space and bootstrapped document samples for author verification task. It was observed that they obtained best accuracy among the submissions in PAN 2013 competition.

Michiel van Dam used [6] the profile-based approach and they applied common N-gram (CNG) method which utilized the normalized distance measure between short and unbalance text. In CNG method, each document is represented with character n-grams. It was observed that their approach obtained good accuracy for English and Spanish languages, but fails for Greek language.

Shachar Seidman proposed [7] a general impostors method which is based on comparing the similarity between given documents and number of external documents. It was observed that their approach achieved overall first rank in the competition. Timo Petmanson extracted [8] frequent significant features such as nouns, punctuations, verbs, and first words of sentences or lines, they used principal component analysis to compute the Matthews correlation coefficient for all pairs of extracted features.

Alberto Bartoli et al. proposed [9] a machine learning approach by using a set of linguistic features. They extracted various features such as word n-grams, character n-grams, POS tag n-grams, word lengths, sentence lengths, sentence lengths n-grams, word richness features, punctuation n-grams, and text shape n-grams. Their approach obtained first rank in author verification for Spanish language in PAN 2015 competition.

3   Linguistic Features

A feature is an attribute of an object that can characterize the document. Most objects and entities have more than one feature. In machine learning, such objects are represented as a vector of features. Features help us to differentiate the objects from one another and also help to describe them. It is essential to select useful and distinctive features in order to achieve high classification scores. In this work, the experimentation carried out with numeric and semantic features and also experiment on each type of feature in isolation as well as experimenting by merging them together gradually.

A numeric feature is a measurement. Numeric features represent a feature of a document with numbers. Two types of numerical features such as stylistic features and syntactic features are used in our experiment. For example, the word count in a document is a numeric feature which contains numeric values. The following set of numerical features was used for the experiments covering almost all the aspects of the previously defined stylistic features in the literature. Typically, these stylistic features include total number of characters, average length per word, number of sentences, words per sentences, words longer than six characters, total number of short words, number of syllables, syllables per word, number of complex words (more than 3 syllables), number of capital letters, number of small letters, ratio of capital letters to small letters, capital letters words, number of words, contraction words, the number of words with hyphens, words followed by digits, unique terms, ratio of number of words which contain more than 3 syllables to total number of words, number of acronyms, number of foreign words, number of words that occur twice (hapax dis legomena), and number of specific words.

Syntactic features include part of speech based features such as number of nouns, number of passive verbs, number of base verbs, number of adjectives, number of clauses and number of phrases, number of articles, number of prepositions, number of coordinate conjunctions, and number of auxiliary verbs. In this work, another syntactic measure such as punctuation measures which is not in the literature as important and those includes number of commas, number of colons (:), number of semicolons (;), number of single quotes ("), number of double quotes ("), number of exclamation marks (!), number of question marks (?) and the number of "etc.". Syntactic features have been extracted by using the parse trees of the sentences. These parse trees are obtained by using the Stanford Parser.

A numeric feature was representing features with numbers. Semantic features represent features with sets of meanings. Synonym sets are used as semantic features. The semantic features are used to directly tie the features to the meaning of word. The meaning of the words is used as semantic features. A WORDNET of synonym set is created for each author as a model which can represent an author's writing topic. To use synonyms for semantic features, WORDNET is needed.

For our work, the experimentation is carried out on PAN 2014 competition author verification dataset. Table 1 shows the characteristics of the corpus used in our work.

Table 1 Dataset characteristics of PAN 2014 competition for English language

Features	Testing data	Training data
Number of authors	100	100
Number of documents	100	500
Vocabulary size	12764	41583
Number of documents per author	1	5
Average words per sentence	21	25
Average words per document	1121	1135

4 Our Approach

The procedure of author verification process is represented in Fig. 1. In this procedure, first, the preprocessing techniques such as stopwords removal and stemming are performed on the collected corpus. Then, the features that differentiate the writing style of the author from the updated corpus are extracted. The document vectors are generated by using extracted features from the corpus. The document vectors are given to classification algorithm to generate the classification model. Finally, the classification model is used to analyze the unknown document and predicts whether the document is written by the particular author or not.

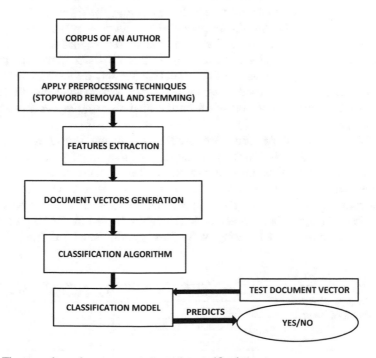

Fig. 1 The procedure of our approach for author verification

Classification is a problem of identifying which category a new input belongs in. An algorithm that implements a classification is called a classifier. There are many classification algorithms like Naïve Bayes multinomial, random forest, decision trees, bagging, support vector machines, and neural networks used for classification. In this work, Naïve Bayes multinomial classifier is used to predict the accuracy of author verification.

K-fold cross-validation testing method is used to test our classifier. K-fold cross-validation is widely used for classifiers. K-fold cross-validation has K iterations. On each iteration, one random unit is selected for testing and the remaining K-1 is used for training. This process is repeated K times while each randomly selected unit is used exactly once. With this method, we ensure that all data is used for both training and testing.

5   Empirical Evaluations

In this work, experimentation carried out with machine learning practices to solves this problem. Naïve Bayes multinomial classifier is identified to generate a efficient classification model because it has high scalability due to number of features/ predictors it can have. At the beginning of each classification, the documents are transformed into feature vectors. This transformation/extraction process is performed only once. After extracting the features from the documents, add these features to a feature vector. As discussed above, in this classification process, 10-fold cross-validation is used. For each fold of the validation, construct the training vectors and test vectors. The training vectors are passed to the classifier and the classifier will create a classification model by iterating all the training vectors. This classification model is used to test the test vector and calculate the efficiency of classifier. The average of all folds will be the final accuracy of our classification methodology.

Precision and recall measures are used as evaluation measures to find the accuracy of our approach. Precision is the ratio of number of problems that correctly answered to total number of problems. Recall is the ratio of number of problems that correctly answered to total number of answers. The accuracies of various combinations of features are represented in Table 2. The combination of stylistic features, syntactic, and semantic features achieved the good precision of 97.8% and recall of 96.7% by using Naïve Bayes multinomial classifier.

Table 2 The accuracies of precision and recall measures for various set of features

Features used	Precision (%)	Recall (%)
Stylistic features	86.8	84.1
Syntactic and POS features	88.2	90.7
Semantic features	91.6	92.9
Combination of all features	97.8	96.7

6 Conclusion

The selections of features vary by the nature of the document. In this work, experimentation carried out with three types of features such as stylistic features, syntactic features, and semantic features. Our work obtained good precision of 97.8% for Author Verification by using Naïve Bayes multinomial classifier.

References

1. Koppel, M., Schler, J., Argamon, S.: Computational methods in authorship attribution. J. Am. Soc. Inform. Sci. Technol. **60**(1), 9–26 (2009)
2. Bobicev, V.: Authorship detection with PPM. In: Proceedings of CLEF 2013 Evaluation Labs (2013)
3. Feng, V.W., Hirst, G.: Authorship verification with entity coherence and other rich linguistic features. In: Proceedings of CLEF 2013 Evaluation Labs (2013)
4. Vilariño, D., Pinto, D., Gómez, H., León, S., Castillo, E.: Lexical-syntactic and Graph-based features for authorship verification. In: Proceedings of CLEF 2013 Evaluation Labs (2013)
5. Veenman, C.J., Li, Z.: Authorship verification with compression features. In: Proceedings of CLEF 2013 Evaluation Labs (2013)
6. van Dam, M.: A basic character n-gram approach to authorship verification. In: Proceedings of CLEF 2013 Evaluation Labs (2013)
7. Seidman, S.: Authorship verification using the impostors method. In: Proceedings of CLEF 2013 Evaluation Labs (2013)
8. Petmanson, T.: Authorship identification using correlations of frequent features. In: Proceedings of CLEF 2013 Evaluation Labs (2013)
9. Bartoli, A., Dagri, A., Lorenzo, A.D., Medvet, E., Tarlao, F.: An author verification approach based on differential features. In: Proceedings of CLEF 2013 Evaluation Labs (2015)

Deterministic and Randomized Heuristic Algorithms for Uncapacitated Facility Location Problem

Soumen Atta, Priya Ranjan Sinha Mahapatra and Anirban Mukhopadhyay

Abstract A well-known combinatorial optimization problem, known as the Uncapacitated Facility Location Problem (UFLP) is considered in this paper. Given a set of customers and a set of potential facilities, the objective of UFLP is to open a subset of the potential facilities such that sum of the opening cost for opened facilities and the service cost of customers is minimized. In this paper, deterministic and randomized heuristic algorithms are presented to solve UFLP. The effectivenesses of the proposed algorithms are tested on UFLP instances taken from the OR-Library. Although the proposed deterministic algorithm gives optimal results for most of the instances, the randomized algorithm achieves optimal results for all the instances of UFLP considered in this paper including those for which the deterministic algorithm fails to achieve the optimal solutions.

Keywords Uncapacitated facility location problem (UFLP) · Simple plant location problem (SPLP) · Warehouse location problem (WLP) · Heuristics Randomization

1 Introduction

The *uncapacitated facility location problem* (*UFLP*) is the problem of finding the optimal placement of facilities of unrestricted capacities among n potential facility locations such that the cost of satisfying demands of all the customers is minimized [1–5]. Here, the cost is of two types: the *service* or *connection cost* to

S. Atta (✉) · P. R. S. Mahapatra · A. Mukhopadhyay
Department of Computer Science and Engineering, University of Kalyani,
Nadia, Kalyani 741235, West Bengal, India
e-mail: soumen.atta@klyuniv.ac.in

P. R. S. Mahapatra
e-mail: priya@klyuniv.ac.in

A. Mukhopadhyay
e-mail: anirban@klyuniv.ac.in

© Springer Nature Singapore Pte Ltd. 2018   205
S. C. Satapathy et al. (eds.), *Information and Decision Sciences*,
Advances in Intelligent Systems and Computing 701,
https://doi.org/10.1007/978-981-10-7563-6_22

provide service to a customer by a facility and the *opening cost* to open a facility. UFLP is also known as the *Simple Plant Location Problem* (SPLP) [1, 6] and the *Warehouse Location Problem* (*WLP*) [7]. UFLP is known to be an NP-hard problem [8, 9]. So, different heuristic approaches are used to solve this problem to obtain near-optimal solution. Some of the approaches are branch-and-bound algorithm [10, 11], tabu search [4, 5], constant factor approximation algorithm [12], greedy heuristic [13], neighborhood search [14], hybrid multi-start heuristic [15], semi-Lagrangian relaxation [16], message-passing [17], surrogate semi-Lagrangian dual [18], discrete unconscious search [19], etc.

In this paper, two heuristic algorithms are proposed. We call these two algorithms as the *deterministic BFR* and the *randomized BFR*. Here, *BFR* is the acronym for *backward–forward–replacement phase*. As the name suggests the first algorithm is deterministic in nature, i.e., it always produces same output for a particular input data set instance. The output of the second algorithm depends on random behavior of some steps. Both the algorithms consist of three phases. These phases are *forward phase*, *backward phase*, and *replacement phase*. The detailed description of these phases is given in Sect. 3. The effectiveness of these two algorithms is tested on UFLP instances of different sizes taken from the literature.

The organization of the rest of the paper is as follows: Sect. 2 formally defines the problem. The proposed algorithms are described in Sect. 3. Computational results are reported and compared in Sect. 4. Finally, Sect. 5 concludes the paper.

2   Problem Definition

The uncapacitated facility location problem (UFLP) [1, 3–5] can be stated as follows: A set $J = \{j_1, j_2, \ldots, j_M\}$ of M customers or cities and a set $I = \{i_1, i_2, \ldots, i_N\}$ of N potential facility locations (sites) are given. A nonnegative opening cost f_i associated with each facility location and a nonnegative service or connection cost c_{ij} between facility i and each customer or city j are also given. The objective of UFLP is to connect each customer or city to the nearest opened facility such that the *total cost*, i.e., the sum of service or connection cost and opening cost of opened facilities is minimized. It is worthy to note that the demand of any customer is fulfilled by only one facility and hence the capacity of each facility is assumed to be infinite.

Using the above descriptions of variables, the mathematical formulation of UFLP [4] is as follows:

$$minimize \sum_{i=1}^{N} \sum_{j=1}^{M} c_{ij} x_{ij} + \sum_{i=1}^{N} f_i y_i$$

subject to

$$\sum_{i=1}^{N} x_{ij} = 1, \quad j = 1, \ldots M,$$

$$x_{ij} \leq y_j, \quad i = 1, \ldots N, \; j = 1, \ldots M,$$

$$x_{ij}, y_i = \{0, 1\}, \quad i = 1, \ldots n, \; j = 1, \ldots M,$$

where

$$x_{ij} = \begin{cases} 1, & \text{if customer } j \in J \text{ is served from site } i \in I \\ 0, & \text{otherwise;} \end{cases}$$

$$y_l = \begin{cases} 1, & \text{if a facility is established at location } i \in I \\ 0, & \text{otherwise.} \end{cases}$$

3 Proposed Heuristic Algorithms

Each of the proposed algorithms, viz., deterministic BFR and randomized BFR consists of forward phase, backward phase, and replacement phase. So, these phases are described in details in the following Sects. 3.1, 3.2, and 3.3, respectively, before the two proposed heuristic algorithms. Each of these phases takes two data structures, viz., cost matrix and a set of opened facilities as its inputs. Throughout the paper, the cost matrix, the set of opened facilities, and the set of non-opened facilities are denoted by C, $\mathcal{F}$ and $\overline{\mathcal{F}}$, respectively. The cost matrix C is a matrix of order $N \times (M + 1)$ where (i) $C(i, j)$ denotes the service cost between the ith facility location to the jth customer, $1 \leq i \leq N$, $1 \leq j \leq M$ and (ii) $C(i, M + 1)$, $1 \leq i \leq N$ denotes the opening cost of the ith facility.

In the proposed algorithms, a function, named as *TotalCost*, is used to compute the total cost. This function takes C and $\mathcal{F}$ as its inputs and gives the corresponding total cost. The time complexity of this function is $\mathcal{O}(N|\mathcal{F}|)$.

3.1 Forward Phase

In this phase, new facilities are opened if and only if the total cost is reduced. The *if* block from line 4 to 7 of Algorithm 1 is executed only when the existing set of opened facilities, $\mathcal{F}_e$ is empty. The sum of each row of C is computed and then these sums are sorted in line 5 to find the *index* of each facility. The first facility according to this sorted index is assigned to $\mathcal{F}_e$ in line 6. So, after the execution of the *if* block from lines 4 to 7, the cardinality of $\mathcal{F}_e$ must be at least one. The *while* loop in line 8 is executed at least once to add a new facility, if possible, in $\mathcal{F}_e$ and the execution of this loop stops when there is no improvement in terms of total cost by adding a new facility in $\mathcal{F}_e$. So, this *while* loop in line 8 may be executed at most $(N - |\mathcal{F}_e|)$-times. For each $i \in \overline{\mathcal{F}_e}$, the total cost corresponding to the set $\mathcal{F}_e \cup \{i\}$ is computed and the ith facility for which the total cost is minimum is opened if the total cost is reduced. The *for* loop in line 11 is executed $(N - |F_e|)$-times. At the end of the *while* loop in line 27, F_e is assigned to the new set of opened facilities, F_n in line 28. The pseudocode of forward phase is given in Algorithm 1 and the time required for the execution of each statement is mentioned there.

Algorithm 1: Forward Phase

Input: Cost matrix C, existing set of opened facilities F_e
Output: New set of opened facilities F_n

1   $N \leftarrow$ no. of facilities; $//$ $\mathcal{O}(1)$
2   $F_n \leftarrow [\ ]$; $//$ $\mathcal{O}(1)$
3   *improve* $\leftarrow$ *True*; $//$ $\mathcal{O}(1)$
4   **if** F_e *is empty* **then**
5   Compute the sum of each row of C and sort these sums in ascending order
       to find sorted *index* of each facility; $//$ $\mathcal{O}(N \log N)$
6      $F_e \leftarrow index(1)$; $//$ $\mathcal{O}(1)$
7   **end**
8   **while** *improve* = *True* **do**
     $//$ This `while loop` may iterate at most
       $(N - |F_e|)$-times.
9      $old\_tcost \leftarrow TotalCost(C, F_e)$; $//$ $\mathcal{O}(N|F_e|)$
10     $flag \leftarrow 0$; $//$ $\mathcal{O}(1)$
11   **for** *each* $i \in \overline{F}_e$ **do**
     $//$ This `for loop` is iterated $(N - |F_e|)$-times.
12        $F_{temp} \leftarrow [\ ]$; $//$ $\mathcal{O}(1)$
13        $F_{temp} \leftarrow F_e \cup \{i\}$; $//$ $\mathcal{O}(1)$
14        $temp\_tcost \leftarrow TotalCost(C, F_{temp})$; $//$ $\mathcal{O}(N|F_{temp}|)$
15   **if** $temp\_tcost < old\_tcost$ **then**
16           $min\_cost\_f \leftarrow i$; $//$ $\mathcal{O}(1)$
17           $old\_tcost \leftarrow temp\_tcost$; $//$ $\mathcal{O}(1)$
18           $flag \leftarrow 1$; $//$ $\mathcal{O}(1)$
19   **end**
20   **end**
21   **if** $flag = 1$ **then**
22        $F_e \leftarrow F_e \cup \{min\_cost\_f\}$; $//$ $\mathcal{O}(1)$
23   **else**
24   *improve* $\leftarrow$ *False*; $//$ $\mathcal{O}(1)$
25   **end**
26      $new\_tcost \leftarrow TotalCost(C, F_e)$; $//$ $\mathcal{O}(N|F_e|)$
27   **end**
28   $F_n \leftarrow F_e$; $//$ $\mathcal{O}(|F_e|)$
29   **return** F_n

3.2 Backward Phase

In this phase, the facilities are closed from the already opened facilities if and only if the total cost is reduced. The *if* condition in line 4 of Algorithm 2 checks the cardinality of $\mathcal{F}_e$. If the set $\mathcal{F}_e$ is empty or its cardinality is 1 then $\mathcal{F}_n$ is assigned as $\mathcal{F}_e$ in line 5 and the algorithm is terminated. If the set $\mathcal{F}_e$ contains more than one facility then the algorithm performs the following steps to close one facility at a time. The *while* loop in line 7 may iterate at most $(|\mathcal{F}_e| - 1)$-times and the execution of this loop stops when there is no improvement in terms of total cost by deleting a facility from $\mathcal{F}_e$. For each $i \in \mathcal{F}_e$, the total cost corresponding to the set $\mathcal{F}_e \backslash \{i\}$ is computed in line 13 and the ith facility for which the total cost is minimum is closed provided that it reduces the total cost. These steps are repeated for closing one facility at a

---

Algorithm 2: Backward Phase

Input: Cost matrix C, existing set of opened facilities $\mathcal{F}_e$
Output: New set of opened facilities $\mathcal{F}_n$

1   $N \leftarrow$ no. of facilities; // $\mathcal{O}(1)$
2   $\mathcal{F}_n \leftarrow [\ ]$; // $\mathcal{O}(1)$
3   *improve* $\leftarrow$ *True*; // $\mathcal{O}(1)$
4   **if** $\mathcal{F}_e$ *is empty or* $|\mathcal{F}_e| = 1$ **then**
5      $\mathcal{F}_n \leftarrow \mathcal{F}_e$; // $\mathcal{O}(|\mathcal{F}_e|)$
6   **else**
7   **while** *improve* $=$ *True* **do**
         // This while loop may iterate at most $(|\mathcal{F}_e| - 1)$-times.
8          $old\_tcost \leftarrow TotalCost(C, \mathcal{F}_e)$; // $\mathcal{O}(N|\mathcal{F}_e|)$
9          $flag \leftarrow 0$; // $\mathcal{O}(1)$
10   **for** *each* $i \in \mathcal{F}_e$ **do**
            // This for loop is iterated $|\mathcal{F}_e|$-times.
11            $\mathcal{F}_{temp} \leftarrow [\ ]$; // $\mathcal{O}(1)$
12            $\mathcal{F}_{temp} \leftarrow \mathcal{F}_e \backslash \{i\}$; // $\mathcal{O}(1)$
13            $temp\_tcost \leftarrow TotalCost(C, \mathcal{F}_{temp})$; // $\mathcal{O}(N|\mathcal{F}_{temp}|)$
14   **if** $temp\_tcost < old\_tcost$ **then**
15               $min\_cost\_f \leftarrow i$; // $\mathcal{O}(1)$
16               $old\_tcost \leftarrow temp\_tcost$; // $\mathcal{O}(1)$
17               $flag \leftarrow 1$; // $\mathcal{O}(1)$
18   **end**
19   **end**
20   **if** $flag = 1$ **then**
21            $\mathcal{F}_e \leftarrow \mathcal{F}_e \backslash \{min\_cost\_f\}$; // $\mathcal{O}(1)$
22   **else**
23   *improve* $\leftarrow$ *False*; // $\mathcal{O}(1)$
24   **end**
25         $new\_tcost \leftarrow TotalCost(C, \mathcal{F}_e)$; // $\mathcal{O}(N|\mathcal{F}_e|)$
26   **end**
27      $\mathcal{F}_n \leftarrow \mathcal{F}_e$; // $\mathcal{O}(|\mathcal{F}_e|)$
28   **end**
29   **return** $\mathcal{F}_n$

time as long as the total cost is reduced. The *for* loop in line 10 is iterated $|\mathcal{F}_e|$-times. The pseudocode of backward phase is given in Algorithm 2 and the time required for the execution of each statement is mentioned there.

3.3 Replacement Phase

The objective of replacement phase is to check whether it is possible to replace already opened facilities in $\mathcal{F}_e$ with non-opened facilities in $\overline{\mathcal{F}}_e$ to reduce the total cost without changing the number of opened facilities. The algorithm performs the following steps to replace the opened facility j in $\mathcal{F}_e$ with a non-opened facility i in $\overline{\mathcal{F}}_e$. For each $i \in \overline{\mathcal{F}}_e$, the total cost for each of the sets $(\mathcal{F}_e\backslash\{j\}) \cup \{i\}$ is computed and the ith facility for which the total cost is minimum is opened and the facility j is closed if it improves the total cost. These steps are repeated as long as improvement occurs in terms of the total cost. The pseudocode of replacement phase is given in Algorithm 3 and the time required for the execution of each statement is mentioned there.

Algorithm 3: Replacement Phase

Input: Cost matrix C, existing set of opened facilities $\mathcal{F}_e$
Output: New set of opened facilities $\mathcal{F}_n$

1 **for** *each $j \in \mathcal{F}_e$* **do**
   // This *for loop* is iterated $|\mathcal{F}_e|$-times.
2   *old_tcost* $\leftarrow$ *TotalCost*$(C, \mathcal{F}_e)$; // $\mathcal{O}(N|\mathcal{F}_e|)$
3   *flag* $\leftarrow$ 0; // $\mathcal{O}(1)$
4   **for** *each $i \in \overline{\mathcal{F}}_e$* **do**
   // This *for loop* is iterated $(N - |\mathcal{F}_e|)$-times.
5      $\mathcal{F}_{temp} \leftarrow [\,]$; // $\mathcal{O}(1)$
6      $\mathcal{F}_{temp} \leftarrow (\mathcal{F}_e\backslash\{j\}) \cup \{i\}$; // $\mathcal{O}(1)$
7   *temp_tcost* $\leftarrow$ *TotalCost*$(C, \mathcal{F}_{temp})$; // $\mathcal{O}(N|\mathcal{F}_{temp}|)$
8   **if** *temp_tcost* $<$ *old_tcost* **then**
9   *min_cost_f* $\leftarrow$ i; // $\mathcal{O}(1)$
10   *old_tcost* $\leftarrow$ *temp_tcost*; // $\mathcal{O}(1)$
11   *flag* $\leftarrow$ 1; // $\mathcal{O}(1)$
12   **end**
13   **end**
14   **if** *flag* $= 1$ **then**
15      $\mathcal{F}_e \leftarrow (\mathcal{F}_e\backslash\{j\}) \cup \{min\_cost\_f\}$; // $\mathcal{O}(1)$
16   **end**
17 **end**
18 $\mathcal{F}_n \leftarrow \mathcal{F}_e$; // $\mathcal{O}(|\mathcal{F}_e|)$
19 **return** $\mathcal{F}_n$

3.4 Deterministic BFR

The deterministic BFR (i.e., Algorithm 4) takes the cost matrix C as its input and gives the set of opened facilities $\mathcal{F}$ and the corresponding total cost as its outputs. Algorithm 4 starts with opening all the facilities as shown in line 2. Then, the following steps are repeated to reduce the total cost. In line 6, the backward phase is used to close some opened facilities (if possible) and this is followed by the replacement phase in line 7 that may replace some opened facilities. Then, the forward phase is used in line 8 to open new facilities (if possible) which is again followed by the replacement phase in line 9. The steps at lines 6 to 9 are repeated as long as the total cost improves.

Algorithm 4: Deterministic BFR

Input: Cost matrix C
Output: Set of opened facilities $\mathcal{F}$, total cost $\mathcal{T}$
1 $n \leftarrow$ no. of potential facilities;
2 $\mathcal{F} \leftarrow \{1, 2, \ldots, n\}$;
3 $old\_tcost = TotalCost(C, \mathcal{F})$;
4 $improve \leftarrow True$;
5 **while** $improve = True$ **do**
6     $\mathcal{F} = BackwardPhase(C, \mathcal{F})$;
7     $\mathcal{F} = ReplacementPhase(C, \mathcal{F})$;
8     $\mathcal{F} = ForwardPhase(C, \mathcal{F})$;
9     $\mathcal{F} = ReplacementPhase(C, \mathcal{F})$;
10     $new\_tcost = TotalCost(C, \mathcal{F})$;
11   **if** $new\_tcost < old\_tcost$ **then**
12        $old\_tcost \leftarrow new\_tcost$;
13   **else**
14        $improve \leftarrow false$;
15   **end**
16 **end**
17 $\mathcal{T} \leftarrow TotalCost(C, \mathcal{F})$;
18 **return** $\mathcal{F}$, $\mathcal{T}$

3.5 Randomized BFR

It is likely that the deterministic BFR may stuck into a local optima. The randomized BFR tries to overcome this problem by the help of randomness. The randomized BFR given in Algorithm 5 also takes the cost matrix C as its input and gives the set $\mathcal{F}$ and the corresponding total cost as its outputs. The deterministic BFR is called in line 2 to produce a solution F_1. In line 3, the solutions F_2, F_3, and F_4 are generated randomly from the solution F_1 by arbitrarily changing the elements in F_1 by the elements in $\overline{F}_1$ keeping the cardinality same as F_1. We now modify each of the solutions F_i, $1 \leq i \leq 4$, in the following way. Each opened facility in F_i is swapped with a randomly chosen non-opened facility in $\overline{F}_i$ with probability 0.5. If any swapping occurs at all then both the sets F_i and $\overline{F}_i$ are modified. At the end of the *for* loop in line 13, the total cost for each F_i is computed in line 14 and the solution with the maximum total cost is replaced by the solution with the minimum total cost of the previous iteration. Here, the variables *iterate* and *count* are used to terminate Algorithm 5. The value of *count* is incremented by one if the minimum total cost for two consecutive iterations remains same, otherwise it is set to zero. Algorithm 5 terminates if the value of either *iterate* or *count* exceeds *max_iterate* or *max_count* respectively.

Algorithm 5: Randomized BFR

Input: Cost matrix C
Output: Set of opened facilities F, total cost T
1 $n \leftarrow$ no. of potential facilities;
2 Create a set of opened facilities F_1 using the deterministic BFR (Algorithm 4);
3 Create other three solutions F_2, F_3, F_4 randomly using F_1;
4 *iterate* $\leftarrow 1$, *count* $\leftarrow 1$;
5 **while** *iterate* $\leq$ *max_iterate* and *count* $\leq$ *max_count* **do**
6 $\quad$ Find the total cost for each of the solutions $F_1, \dots, F_4$;
7 $\quad$ Let F_{best} be the solution with the minimum total cost;
8 $\quad$ **for** $i \leftarrow 1$ *to* 4 **do**
9 $\quad\quad$ $r \leftarrow$ a random number in $[0, 1]$;
10 $\quad\quad$ **if** $r \geq 0.5$ **then**
11 $\quad\quad\quad$ Swap each of the opened facilities in F_i with a randomly chosen non-opened facility in $\overline{F}_i$;
12 $\quad\quad$ **end**
13 $\quad$ **end**
14 $\quad$ Find the total cost for each of the solutions $F_1 \dots F_4$;
15 $\quad$ Replace the solution having the maximum total cost value with F_{best};
16 $\quad$ **if** *iterate* $\neq 1$ **then**
17 $\quad\quad$ **if** *the minimum total cost for two successive iterations are same* **then**
18 $\quad\quad\quad$ *count* $\leftarrow$ *count* + 1;
19 $\quad\quad$ **else**
20 $\quad\quad\quad$ *count* $\leftarrow 0$;
21 $\quad\quad$ **end**
22 $\quad$ **end**
23 $\quad$ *iterate* $\leftarrow$ *iterate* + 1;
24 **end**
25 $F \leftarrow$ the solution from the set $F_1, \dots, F_4$ with minimum total cost;
26 $T = TotalCost(C, F)$;
27 **return** F, T

Table 1 Results obtained for OR-library benchmark data (uncapacitated)

Data file [21]	Size of data file	Optimal value [21]	Deterministic BFR			Randomized BFR		
			Total cost	Gap%	Time (s)	Total cost	Gap%	Time (s)
Cap71	16 × 50	932615.75	932615.75	0.0	0.087	–	–	–
Cap72	16 × 50	977799.4	977799.4	0.0	0.099	–	–	–
Cap73	16 × 50	1010641.45	1010641.45	0.0	0.118	–	–	–
Cap74	16 × 50	1034976.975	1034976.975	0.0	0.120	–	–	–
Cap101	25 × 50	796648.437	796648.4375	0.0	0.206	–	–	–
Cap102	25 × 50	854704.2	854704.2	0.0	0.242	–	–	–
Cap103	25 × 50	893782.112	893782.112	0.0	0.257	–	–	–
Cap104	25 × 50	928941.75	928941.75	0.0	0.276	–	–	–
Cap131	50 × 50	793439.562	793439.562	0.0	0.931	–	–	–
Cap132	50 × 50	851495.325	851495.325	0.0	1.031	–	–	–
Cap133	50 × 50	893076.712	893782.112	0.079	1.033	893076.712	0.0	2.524
Cap134	50 × 50	928941.75	928941.75	0.0	1.107	–	–	–
Capa	100 × 1000	17156454.47830	17156454.47830	0.0	12.827	–	–	–
Capb	100 × 1000	12979071.58143	12979071.58143	0.0	10.259	–	–	–
Capc	100 × 1000	11505594.32878	11535265.915	0.258	13.118	1505594.32878	0.0	35.572

4   Experimental Results and Discussion

The efficiency of the proposed two algorithms is tested on 15 benchmark instances of
Beaslay's OR-Library [20]. Here, all the benchmark instances and the corresponding optimal costs are taken from UflLib [21]. The proposed algorithms are coded
with MATLAB R2013a and all the computations are performed in a machine with
Intel Core i3 2.30 GHz processor having Ubuntu 14.40 LTS with 4 GB of RAM. To
evaluate the performance of the proposed algorithms, we define two performance
metrics *Gap%* and *Quality%* which are defined as follows:

$$Gap\% = \left(\frac{\text{Total Cost} - \text{Optimal Value}}{\text{Optimal Value}} \right) \times 100,$$

$$Quality\% = 100 - Gap\%.$$

At first, we run the deterministic BFR on the instances of UFLP. If optimal results
given in UflLib [21] are obtained for these instances then we do not run the randomized BFR. The randomized BFR is applied on UFLP instances only when the
deterministic BFR fails to give optimal results. In our experiments, the values of
max_iterate and *max_count* are set to 10 and 4 respectively. The results for the
benchmark instances are given in Table 1. Out of the total 15 instances, the deterministic algorithm achieves optimal results for 13 instances and for the remaining
two instances, the randomized algorithm achieves the optimal results. In Table 2, the
results of large size OR-Library instances [20] are compared with the Lagrangian-
type relaxation algorithm proposed by Monabbati [18]. It is observed from Table 2
that the proposed algorithms perform better.

Table 2   Comparison for OR-library benchmark

Data file [21]	HDA [18]		CPLEX time (s) [18]		Deterministic BFR		Randomized BFR	
	Time (s)	*Quality%*	With HDA	Without HDA	Time (s)	*Quality%*	Time (s)	*Quality%*
Capa	4.391	96.45	65.88	28.02	12.827	**100**	–	–
Capb	4.625	99.29	74.58	30.84	10.259	**100**	–	–
Capc	3.704	98.14	106.5	100.8	13.118	**99.742**	35.572	**100**

5 Conclusion

In this paper, two heuristic algorithms are proposed to solve the Uncapacitated Facility Location Problem (UFLP). For most of the instances, the deterministic BFR gives optimal or near-optimal results. The randomized BFR has been found to provide optimal results for all the instances where the deterministic BFR fails to give optimal results. It is to be noted that the result found by the randomized BFR is always better or at least same as the result obtained by the deterministic BFR. For future work, the effects of the three phases used in the proposed algorithms on the final result can be analyzed and more experiments on other UFLP instances available in the literature can be performed.

References

1. Krarup, J., Pruzan, P.M.: The simple plant location problem: survey and synthesis. Eur. J. Oper. Res. **12**(1), 36–81 (1983)
2. Balinski, M.: On finding integer solutions to linear programs. Technical Report, DTIC Document (1964)
3. Erlenkotter, D.: A dual-based procedure for uncapacitated facility location. Oper. Res. **26**(6), 992–1009 (1978)
4. Al-Sultan, K., Al-Fawzan, M.: A tabu search approach to the uncapacitated facility location problem. Ann. Oper. Res. **86**, 91–103 (1999)
5. Sun, M.: Solving the uncapacitated facility location problem using tabu search. Comput. Oper. Res. **33**(9), 2563–2589 (2006)
6. Kratica, J., Tošic, D., Filipović, V., Ljubić, I.: Solving the simple plant location problem by genetic algorithm. RAIRO Oper. Res. **35**(01), 127–142 (2001)
7. Khumawala, B.M.: An efficient branch and bound algorithm for the warehouse location problem. Manag. Sci. **18**(12), B–718 (1972)
8. Garey, M.R., Johnson, D.S.: Computers and intractability: a guide to NP-completeness (1979)
9. Lenstra, J., Kan, A.R.: Complexity of Packing, Covering and Partitioning Problems. Econometric Institute (1979)
10. Akinc, U., Khumawala, B.M.: An efficient branch and bound algorithm for the capacitated warehouse location problem. Manag. Sci. **23**(6), 585–594 (1977)
11. Bilde, O., Krarup, J.: Sharp lower bounds and efficient algorithms for the simple plant location problem. Ann. Discrete Math. **1**, 79–97 (1977)
12. Shmoys, D.B., Tardos, É., Aardal, K.: Approximation algorithms for facility location problems. In: Proceedings of the Twenty-Ninth Annual ACM Symposium on Theory of Computing, pp. 265–274. ACM (1997)
13. Guha, S., Khuller, S.: Greedy strikes back: improved facility location algorithms. J. Algorithms **31**(1), 228–248 (1999)
14. Ghosh, D.: Neighborhood search heuristics for the uncapacitated facility location problem. Eur. J. Oper. Res. **150**(1), 150–162 (2003)
15. Resende, M.G., Werneck, R.F.: A hybrid multistart heuristic for the uncapacitated facility location problem. Eur. J. Oper. Res. **174**(1), 54–68 (2006)
16. Beltran-Royo, C., Vial, J.P., Alonso-Ayuso, A.: Solving the uncapacitated facility location problem with semi-Lagrangian relaxation. Stat. Oper. Res., Rey Juan Carlos University, Mostoles, Madrid, España (2007)

17. Lazic, N., Frey, B.J., Aarabi, P.: Solving the uncapacitated facility location problem using message passing algorithms. In: International Conference on Artificial Intelligence and Statistics, pp. 429–436 (2010)
18. Monabbati, E.: An application of a Lagrangian-type relaxation for the uncapacitated facility location problem. Jpn. J. Ind. Appl. Math. **31**(3), 483–499 (2014)
19. Ardjmand, E., Amin-Naseri, M.R.: Unconscious search-a new structured search algorithm for solving continuous engineering optimization problems based on the theory of psychoanalysis. In: Proceedings of Advances in Swarm Intelligence, pp. 233–242. Springer (2012)
20. Beasley, J.E.: OR-library: distributing test problems by electronic mail. J. Oper. Res. Soc. 1069–1072 (1990)
21. Hoefer, M.: UflLib, benchmark instances for the uncapacitated facility location problem (2003). http://www.mpi-inf.mpg.de/departments/d1/projects/benchmarks/UflLib

Group Search Optimization Technique for Multi-area Economic Dispatch

Chitralekha Jena, Swati Smaranika Mishra and Bhagabat Panda

Abstract Group search optimization to solve multi-area economic dispatch (MAED) problem is presented in this paper with transmission losses, constraints in the tie-line with different fuels, the valve point loading effect, and the prohibited operating zones. The method proposed has been examined on two different test systems, large and small, considering a changing degree of complexity. Then, the comparison has been made with evolutionary programming, differential evolution, and real-coded genetic algorithm where the solution quality is considered. The method which is proposed here gives an alternative approach which is very promising solution for solving one of the power system problems like MAED.

Keywords Group search optimization · Tie-line constraints · Prohibited operating zone · Multi-area economic dispatch · Valve point loading

1 Introduction

The problem of economic dispatch (ED) [1] is a very vital in the power system. The extension of economic dispatch problem is multi-area economic dispatch. It finds the level of generation and changes the power between the areas so that the gross cost of fuel in the areas is minimized, while fulfilling the generating limits of power, tie-line capacity, and the constraints in power balance.

Due to the high efficiency of GSO, it is being applied in many fields.

Here, GSO is applied for solving MAED problem [2]. Here, three different types of MAED problem are taken for consideration. (1) MAED considering valve point

C. Jena (✉) · S. S. Mishra · B. Panda
School of Electrical Engineering, KIIT University, Bhubaneswar, India
e-mail: chitralekha.jenafel@kiit.ac.in

S. S. Mishra
e-mail: swatismaranika@gmail.com

B. Panda
e-mail: bpanda.fel@kiit.ac.in

© Springer Nature Singapore Pte Ltd. 2018   217
S. C. Satapathy et al. (eds.), *Information and Decision Sciences*,
Advances in Intelligent Systems and Computing 701,
https://doi.org/10.1007/978-981-10-7563-6_23

loading (2) MAED with prohibited operating zones transmission losses and valve point loading (3) MAED considering valve point loading effect, multiple sources of fuel, and losses in transmission. The GSO approach has been applied to two different test systems and the validation of the results has been done.

2 Problem Formulation

The objective of Multi-area ED [3] is to get minimum cost of production of the loads supplying to all the areas keeping in consideration the different constraints like power balance, constraints of the limits in generation and the constraints in the tie-line capacity.

Here, two types of multi-area economic dispatch problems have been taken into consideration.

2.1 MAED with Prohibited Operating Zones, the Losses in Transmission and Loading Due to Valve Point

F_t, which the objective function, is the gross cost of all the generators which are committed in all areas. Thus, the problem of MAED is given as

$$F_t = \sum_{i=1}^{N} \sum_{j=1}^{M_i} F_{ij}(P_{ij}) = \sum_{i=1}^{N} \sum_{j=1}^{M_i} \left(a_{ij} + b_{ij}P_{ij} + c_{ij}P_{ij}^2 \right) \qquad (1)$$

Real power balance constraint

$$\sum_{j=1}^{M_i} P_{ij} = P_{Di} + P_{Li} + \sum_{k, k \neq i} T_{ik} \quad i \in N \qquad (2)$$

The loss in transmission P_{Li} of the area i can be given using the B-coefficients,

$$P_{Li} = \sum_{l=1}^{M_i} \sum_{j=1}^{M_i} P_{ij}B_{ilj}P_{il} + \sum_{j=1}^{M_i} B_{0ij}P_{ij} + B_{00i} \qquad (3)$$

Capacity constraints of the tie-line

T_{ik} is the tie-line real power transfer from the area i to k, and does not exceed the transfer capacity of the line for considering the security.

$$-T_{ik}^{max} \leq T_{ik} \leq T_{ik}^{max} \qquad (4)$$

Generation of real power capacity constraints [4]

Real power of the generator should not exceed P_{ij}^{min} and P_{ij}^{max}, which are the lower and upper limit, respectively, hence

$$P_{ij}^{min} \leq P_{ij} \leq P_{ij}^{max} \quad i \in N \quad \text{and} \quad j \in M_i \tag{5}$$

The limits of the prohibited operating zone [4]

The limits of the prohibited operating zones feasible for the system are expressed as:

$$\begin{aligned} &P_{ij}^{min} \leq P_{ij} \leq P_{ij,1}^{l} \\ &P_{ij,m-1}^{u} \leq P_{ij} \leq P_{ij,m}^{l}, \quad m = 2, 3, \ldots, n_{ij} \\ &P_{ij,n_{ij}}^{u} \leq P_{ij} \leq P_{ij}^{max} \end{aligned} \tag{6}$$

2.2 MAED Considering the Valve Point Loading Effect [5]

The cost function of the fuel after taking into consideration the valve point loading effect is given as

$$F_t = \sum_{i=1}^{N} \sum_{j=1}^{M_i} F_{ij}(P_{ij}) = \sum_{i=1}^{N} \sum_{j=1}^{M_i} \left[a_{ij} + b_{ij}P_{ij} + c_{ij}P_{ij}^2 + \left| d_{ij} \times \sin\left\{ e_{ij} \times \left(P_{ij}^{min} - P_{ij} \right) \right\} \right| \right] \tag{7}$$

2.3 MAED with Valve Point Loading Effect [6], Multiple-Fuel Sources, and Transmission Losses

The cost function of the fuel of the ith generator with N_F different fuel types after consideration of the valve point effect is given as:

$$F_{ij}(P_{ij}) = a_{ijm} + b_{ijm}P_{ij} + c_{ijm}P_{ij}^2 + \left| d_{ijm} \times \sin\left\{ e_{ijm} \times \left(P_{ijm}^{min} - P_{ij} \right) \right\} \right| \tag{8}$$

if $P_{ijm}^{min} \leq P_{ij} \leq P_{ijm}^{max}$ $m = 1, 2, \ldots, N_F$ and is the types of fuel

F_t is given by,

$$F_t = \sum_{i=1}^{N} \sum_{j=1}^{M_i} F_{ij}(P_{ij}) \tag{9}$$

F_t has to get minimized by considering the above constraints which are given in Eqs. (2), (4), and (5).

3 Determining the Level of Generation of the Slack Generator

Taking assumption that the loading of power of the $(M_i - 1)$ generators is known, the power level of the slack generator (M_ith) is given by

$$P_{iM_i} - P_{Dl} + P_{Ll} + \sum_{k, k \neq i} T_{ik} \quad \sum_{j=1}^{M_i - 1} P_{ij} \qquad (10)$$

After expansion and again rearrangement of Eq. (10), it becomes

$$B_{iM_iM_i} P_{iM_i}^2 + \left(2 \sum_{j=1}^{M_i - 1} B_{iM_ij} P_{ij} + B_{0iM_i} - 1 \right) P_{iM_i} + \left(P_{Di} + \sum_{k, k \neq i} T_{ik} + \sum_{j=1}^{M_i - 1} \sum_{l=1}^{M_i - 1} P_{ij} B_{ilj} P_{il} + \sum_{j=1}^{M_i - 1} B_{0ij} P_{ij} \right.$$
$$\left. - \sum_{j=1}^{M_i - 1} P_{ij} + B_{00i} \right) = 0$$

$$(11)$$

By solving the Eq. (11), we can find the value of M_ith, loading of the slack generator.

4 Group Search Optimization

GSO, which is a stochastic algorithm and is population-based, has come from the idea of living theory in group and animal searching behavior [7, 8].

5 Simulation Results

Here, the validation of the GSO method which is proposed is examined by using two variety test systems. To verify the effectiveness of the method, the different optimization techniques like evolutionary programming (EP) [9], differential evolution (DE), and RCGA have been applied to the two test systems. The technique which is used in this paper to solve MAED problem has been applied using MATLAB 7.0 on a PC (Pentium-IV, 80 GB, 3.0 GHz).

Test System 1[10]: Here, the system consists of two areas. All the areas have three generators. Loss in transmission and prohibited operating zones is also considered. The data of the generator and the loss coefficients are shown in Appendix-1. Area 1 is 60% and in Area 2, it is 40% of the total load demand. 1263 MW is the total load demand and 100 MW is the limit of the power flow of the system.

Table 1 Results of the test system 1

	GSO	DE	EP	RCGA
$P_{1,1}$ (MW)	500.0000	500.0000	500.0000	500.0000
$P_{1,2}$ (MW)	200.0000	200.0000	200.0000	200.0000
$P_{1,3}$ (MW)	150.0000	150.0000	149.9919	149.6328
$P_{2,1}$ (MW)	204.3345	204.3341	206.4493	205.9398
$P_{2,2}$ (MW)	154.7030	154.7048	154.8892	155.8322
$P_{2,3}$ (MW)	67.5784	67.5770	65.2717	65.2209
T_{12} (MW)	82.7731	82.7731	82.7652	82.4135
P_{L1} (MW)	9.4269	9.4269	9.4267	9.4193
P_{L2} (MW)	4.1890	4.1890	4.1754	4.2064
Cost ($/h)	12255.38	12255.42	12255.43	12256.23
CPU time (s)	5.0324	5.9219	8.8906	9.6094

GSO is used for solving the problem. For this test system, the value of N_P and the number of iterations are taken as 100 and 50, respectively [11].

For validation of the GSO which is proposed, the test system has been solved by differential evolution (DE), real-coded genetic algorithm (RCGA), and evolutionary programming (EP). The scaling factor, population size, and crossover constant are taken as 1.0, 100, and 1.0, respectively, in DE method. The scaling factor and population size are taken as 0.1 and 100, respectively. In RCGA, the crossover, probability of mutation, and population size are taken as 0.9, 0.2, and 100, respectively. **N** number of iterations has been taken 50 for all the above algorithms.

The results from the proposed methods are summarized in Table 1 and the convergence characteristic of cost is given in Fig. 1.

Fig. 1 Characteristics of cost convergence for test system 1

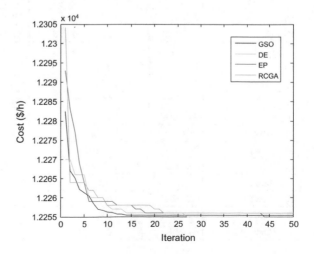

Test System 2[10]: Here, the system consists of 10 generators having valve point loading and multiple-fuel sources having three kinds of fuel options. Loss in transmission is also considered. Here, we have a entire load demand of 2700 MW which has 10 generators and those are splitted into three parts/areas. In the first area, there are four units then the second area has adjacent three units and third area consists of the rest three units. We have taken 50% of total load demand in the first area. Whereas in second area, the load demand is considered as 25% and in third area it is considered 25%. The limit of power flow from first area 1 to area 2 or from area 2 to area 1 is 100 MW. The limit of power flow from area 1 to 3 or from to third or third area to first is the same, i.e., 100 MW. Similarly, the power flow limit from second area to third area or from third area to second is the same, i.e., 100 MW. B-coefficients are shown in the Appendix 2.

GSO method is applied to the problem. The value of N_P and the iteration number are taken as 100 and 100, respectively, for this system.

For validation of GSO which is proposed, the test system 2 is solved by using different optimization techniques. The number of iterations is taken as 100 for all the optimization techniques used. The results found from the different techniques are presented in Table 2. The convergence of the cost characteristic of the system is given in Fig. 2.

Table 2 Results of the test system 2

	GSO		DE		EP		RCGA	
		Fuel		Fuel		Fuel		Fuel
$P_{1,1}$ (MW)	225.7002		220.2200	2	223.8491	2	239.0958	2
$P_{1,2}$ (MW)	212.1994		212.1510	1	209.5759	1	216.1166	1
$P_{1,3}$ (MW)	487.3917		493.2287	2	496.0680	2	484.1506	2
$P_{1,4}$ (MW)	242.1126		242.3742	3	237.9954	3	240.6228	3
$P_{2,1}$ (MW)	250.8376		251.5901	1	259.4299	1	259.6639	1
$P_{2,2}$ (MW)	234.9589		234.5500	3	228.9422	3	219.9107	3
$P_{2,3}$ (MW)	264.0710		268.1343	1	264.1133	1	254.5140	1
$P_{3,1}$ (MW)	236.7312		234.9833	3	238.2280	3	231.3565	3
$P_{3,2}$ (MW)	332.0932		328.5371	1	331.2982	1	341.9624	1
$P_{3,3}$ (MW)	249.4448		250.1525	1	246.6025	1	248.2782	1
T_{21} (MW)	99.7121		99.4945		100		93.1700	
T_{31} (MW)	100		99.9849		100		93.8739	
T_{32} (MW)	34.5573		30.0535		32.5231		43.7824	
P_{L1} (MW)	17.1160		17.5000		17.4884		17.0297	
P_{L2} (MW)	9.7127		9.8334		10.0085		9.7010	
P_{L3} (MW)	8.7118		8.6345		8.6056		8.9408	
Cost ($/h)	654.0572		654.0811		655.1716		657.3325	
CPU time (second)	64.0387		65.0351		78.0625		83.8438	

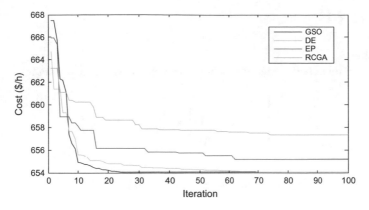

Fig. 2 Convergence characteristic cost of test system 2

Table 3 Data for two-area system

Generator ij	a_{ij} ($/ h)	b_{ij} ($/ MWh)	c_{ij} ($/ (MW)^{2}h)	P_{ij}^{min} (MW)	P_{ij}^{max} (MW)	Prohibited zones (MW)
$G_{1,1}$	550	8.10	0.00028	100	500	[210 240] [350 380]
$G_{1,2}$	350	7.50	0.00056	50	200	[90 110] [140 160]
$G_{1,3}$	310	8.10	0.00056	50	150	[80 90] [110 120]
$G_{2,1}$	240	7.74	0.00324	80	300	[150 170] [210 240]
$G_{2,2}$	200	8.00	0.00254	50	200	[90 110] [140 150]
$G_{2,3}$	126	8.60	0.00284	50	120	[75 85] [100 105]

6 Conclusion

Here, the group search optimization technique has been applied to resolve MAED problems. The validation of the method which is proposed here is shown by taking two test systems and the test results have been compared with the results obtained from other optimization technique. It was observed after comparison that the GSO converges to a much better result than DE, EP and RCGA. Due to these characteristics, in future, the GSO method may be used for solving variety power system problems in future.

Appendix 1

See Table 3.

The formula coefficients of transmission loss of two-area system are

$$B_1 = \begin{bmatrix} 17 & 12 & 7 \\ 12 & 14 & 9 \\ 7 & 9 & 31 \end{bmatrix} \times 10^{-6}$$

$$B_{01} = [\, -0.3908 \quad -0.1297 \quad 0.7047\,] \times 10^{-3}$$
$$B_{001} = 0.045$$

$$B_2 = \begin{bmatrix} 24 & -6 & -8 \\ -5 & 129 & -2 \\ -8 & -2 & 150 \end{bmatrix} \times 10^{-6}$$

$$B_{02} = [\,0.0591 \quad 0.2161 \quad -0.6635\,] \times 10^{-3}$$
$$B_{002} = 0.056$$

Appendix 2

Formula coefficients of transmission loss of three-area system:

$$B_1 = \begin{bmatrix} 8.70 & 0.43 & -4.61 & 0.36 \\ 0.43 & 8.30 & -0.97 & 0.22 \\ -4.61 & -0.97 & 9.00 & -2.00 \\ 0.36 & 0.22 & -2.00 & 5.30 \end{bmatrix} \times 10^{-5}$$

$$B_{01} = [\, -0.3908 \quad -0.1297 \quad 0.7047 \quad 0.0591\,] \times 10^{-3}$$
$$B_{001} = 0.056$$

$$B_2 = \begin{bmatrix} 8.60 & -0.80 & 0.37 \\ -0.80 & 9.08 & -4.90 \\ 0.37 & -4.90 & 8.24 \end{bmatrix} \times 10^{-5}$$

$$B_{02} = [\,0.2161 \quad -0.6635 \quad 0.5034\,] \times 10^{-3}$$
$$B_{002} = 0.045$$

$$
B_3 = \begin{bmatrix} 1.20 & -0.96 & 0.56 \\ -0.96 & 4.93 & -0.30 \\ 0.56 & -0.30 & 5.99 \end{bmatrix} \times 10^{-5}
$$

$$
B_{03} = \begin{bmatrix} -0.3216 & 0.4635 & 0.3503 \end{bmatrix} \times 10^{-3}
$$
$$
B_{003} = 0.055
$$

References

1. Manoharan, P.S., Kannan, P.S., Baskar, S., Iruthayarajan, M.Willjuice: Evolutionary algorithm solution and ICICT based optimality verification to multi area economic dispatch Int. J. Electr. Power Energy Syst. **31**(7–8), 365–373 (2009)
2. Wang, L., Singh, C.: Reserve-constrained multiarea environmental/economic dispatch based on particle swarm optimization with local search. Eng. Appl. Artif. Intell. **22**(2), 298–307 (2009)
3. Sharma, M., Pandit, M., Srivastava, L.: Reserve constrained multi-area economic dispatch employing differential evolution with time-varying mutation. Int. J. Electr. Power Energy Syst. **33**(3), 753–766 (2011)
4. Gaing, Z.-L.: Particle swarm optimization to solving the economic dispatch considering the generator constraints. IEEE Trans. Power Syst. **18**(3), 1187–1195 (2003)
5. Walter, D.C., Sheble, G.B.: Genetic algorithm solution of economic dispatch with valve point loading. IEEE Trans. Power Syst. **8**, 1325–1332 (1993)
6. Chiang, C.-L.: Improved genetic algorithm for power economic dispatch of units with valve-point effects and multiple fuels. IEEE Trans. Power Syst. **20**(4), 1690–1699 (2005)
7. He, S., Wu, Q.H., Saunders, J.R.: A novel group search optimizer inspired by animal behavioral ecology. In: IEEE Congress on Evolutionary Computation, CEC (2006)
8. He, S., Wu, Q.H., Saunders, J.R.: Group search optimizer: an optimization algorithm inspired by animal searching behavior. IEEE Trans. Evol. Comput. **13**(5), 973–990 (2009)
9. Sinha, N., Chakrabarti, R., Chattopadhyay, P.K.: Evolutionary programming techniques for economic load dispatch. IEEE Trans. Evol. Comput. **7**(1), 83–94 (2003)
10. Jain, K., Pandit, M.: Discussion of reserve constrained multi-area economic dispatch employing differential evolution with time-varying mutation by Manisha Sharma et al. international journal of electrical power and energy systems, 33 March (2011) 753–766. Int. J. Electr. Power Energy Syst. **39**(1), 68–69 (2012)
11. Shen, H., Zhu, Y., Zou, W., Zhu, Z.: Group search optimizer algorithm for constrained optimization. Comput. Sci. Environ. Eng. Ecoinform. (2011)

Adaptive Control of Aircraft Wing Oscillations with Stiffness and Damping Nonlinearities in Pitching Mode

L. Prabhu and J. Srinivas

Abstract This paper presents an adaptive control strategy for aircraft wing structure based on a nonlinear aeroelastic model with plunge and pitch degrees of freedom. System nonlinearities in terms of pitching degree of freedom are accounted in stiffness and damping terms of the model. The closed-loop response of the model is studied under two cases: (i) polynomial form of nonlinearities and (ii) combined free play and polynomial form of nonlinearities. The adaptive control strategy with wing flap based on partial feedback linearization is designed to suppress the instabilities occurring at certain freestream velocities. Objective of controller is to stabilize the system within the flutter boundary. A neural network based observer is used to estimate the uncertain parameters in control law. The designed control system with neural network estimator is effective in suppressing the limit cycle oscillations considerably.

1 Introduction

In aeroelastic studies, the interactions of various forces such as aerodynamics, elastic, and inertia are considered using simple mathematical models. A combination of these forces leads to an aircraft instability resulting in a direct consequence of an oscillatory instability known as a flutter which eventually leads to catastrophic failure due to the loss of a system damping. Aeroelastic instability region is identified by assuming system as linear one [1], but in real practice, the nonlinearities are inevitable. Woolston [2] accounted different types of structural nonlinearities and studied the influence of the initial conditions. The LCO of an aeroelastic model with hysteresis nonlinearity was controlled by sliding mode controller and effects of time delay were studied by Xu et al. [3].

L. Prabhu · J. Srinivas (✉)
Department of Mechanical Engineering, National Institute of Technology,
Rourkela 769008, Odisha, India
e-mail: srin07@yahoo.co.in

© Springer Nature Singapore Pte Ltd. 2018
S. C. Satapathy et al. (eds.), *Information and Decision Sciences*,
Advances in Intelligent Systems and Computing 701,
https://doi.org/10.1007/978-981-10-7563-6_24

To extend the flutter region, active flutter suppression is carried out using various control techniques. Efficiency of linear controller in a nonlinear system comes down when the system nonlinearity effects are aggressive. The fully linearized control system was designed to make the system globally stable with two flaps [4]. In several other works [5–7], adaptive controllers were employed to overcome the dynamic instabilities. The controller was designed to guarantee the stability of structurally nonlinear system with a single flap using explicit parameterization of structural nonlinearity [8, 9]. To improvise the controllability of a nonlinear system, control surfaces at trailing edge and leading edge were used [10]. Block and Strganac [11] used unsteady formulation with optimal controller and Kalman filter as an observer to enhance the flutter boundary. Effectiveness of various types of controllers such as artificial intelligence, robust and adaptive on flutter suppression was discussed in Refs. [12–14]. In earlier work, authors [15] employed a linear quadratic regulator with neural networks estimator to control the instabilities occurring in aeroelastic system.

In all the above works, parametric uncertainty is considered in pitch stiffness only. However, in few works [16], an aeroelastic system was studied with parametric uncertainties, similar kind of nonlinearities in stiffness and damping terms were used. In the present work, an adaptive feedback linearization controller is designed to control the aeroelastic model subjected to different structural nonlinearities in both stiffness and damping in pitch degree of freedom. First, the model is analyzed with polynomial nonlinearity in both stiffness and damping terms and then with free play and cubic nonlinearity in stiffness and damping terms, respectively. Finally, the neural network observer is employed as an estimator for the controller to predict the estimated uncertain parameters in the control law for further suppressing the nonlinearity effect. Additionally, the influence of initial conditions on stable region with polynomial nonlinearities and influence of free play region on system stability is presented.

2 Mathematics Modeling of Nonlinear Aeroelastic System

A two-dimensional aeroelastic system is illustrated by a lumped parameter model shown in Fig. 1, where the system has two degrees of freedom namely plunge translation h and the pitching rotation α with trailing edge surface angle β.

By defining elastic axis at E, and b as semi-chord, ba, bx_α as distance from airfoil mid-chord to elastic axis and distance from airfoil elastic axis to center of mass, respectively, the dynamic equation of aeroelastic system in its standard form is given by [16]:

$$\begin{bmatrix} m_T & m_w x_\alpha b \\ m_w x_\alpha b & I_\alpha \end{bmatrix} \begin{bmatrix} \ddot{h} \\ \ddot{\alpha} \end{bmatrix} + \begin{bmatrix} c_h & 0 \\ 0 & c_\alpha(\dot{\alpha}) \end{bmatrix} \begin{bmatrix} \dot{h} \\ \dot{\alpha} \end{bmatrix} + \begin{bmatrix} k_h & 0 \\ 0 & k_\alpha(\alpha) \end{bmatrix} \begin{bmatrix} h \\ \alpha \end{bmatrix} = \begin{bmatrix} -L \\ M \end{bmatrix}. \quad (1)$$

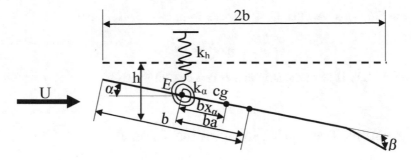

Fig. 1 Aeroelastic model with trailing edge

Here, m_T, m_w, c_h, k_h, L, and M are respectively indicating: total mass of the system, mass of wing section, plunge damping, plunge stiffness, aerodynamic lift and moment. The L and M have nonlinear terms in stiffness and damping in pitch degree of freedom. In real practice, different forms of nonlinearities occur in combination with other. In this work, two cases are considered.

Case 1: Both nonlinear terms are approximated in polynomial form as:

$$c_\alpha(\dot{\alpha}) = \sum_{i=1}^{m} c_i \dot{\alpha}^{i-1}.$$

$$k_\alpha(\alpha) = \sum_{i=1}^{n} k_i \alpha^{i-1}.$$
(2)

Case 2: Free play nonlinearity is considered in stiffness and polynomial form in damping. That is

$$k_\alpha(\alpha) = k_\alpha F(\alpha)/\alpha.$$
(3)

Here, $F(\alpha)$ is a function assigned to represent the free play nonlinearity given by:

$$F(\alpha) = \begin{cases} \alpha + \delta, & \text{if } \alpha < -\delta \\ 0, & \text{if } |\alpha| \leq \delta \\ \alpha - \delta, & \text{if } \alpha > \delta \end{cases}.$$
(4)

For quasi-steady aerodynamics, lift L and moment M are given by [16]:

$$L = \rho U^2 bsc_{l\alpha}\left(\alpha + \frac{\dot{h}}{U} + \left(\frac{1}{2} - a\right)b\frac{\dot{\alpha}}{U}\right) + \rho U^2 bsc_{l\beta}\beta$$

$$M = \rho U^2 b^2 s\bar{c}_{m\alpha}\left(\alpha + \frac{\dot{h}}{U} + \left(\frac{1}{2} - a\right)b\frac{\dot{\alpha}}{U}\right) + \rho U^2 b^2 s\bar{c}_{m\beta}\beta.$$
(5)

where $\bar{c}_{m\alpha}$ and $\bar{c}_{m\beta}$ represent the moment derivative coefficients per unit angle of attack and trailing edge angle, respectively.

The Eq. (1) is rewritten in state-space form as

$$\dot{X} = A(X)X + B\beta. \tag{6}$$

where $A(X)$ is system matrix and B is a control matrix.

3   Adaptive Feedback Linearization Control

The adaptive control scheme reforms the controller online depending on the system performance and changes its dynamics accordingly. Two important steps of adaptive controller are

1. Online parameter estimation.
2. Control law-redesign based on Step 1.

The feedback linearization is a method to transform the nonlinear equations of motion to an equivalent linear system by deriving a suitable control law to cancel the nonlinear terms. The output function is defined as $y = \alpha = x_2$. Before designing the controller, transformation of the equations of motion is carried out.

When parameter estimations $\{ \hat{c}_i \quad \hat{k}_i \}$ are unknown, then the control law of the control surface is given by [16]

$$\beta = \frac{-1}{B_4} \left(G(\theta) + \sum_{i=1}^{m} \hat{c}_i M_1(\dot{\alpha}^i) + \sum_{i=1}^{n} \hat{k}_i M_2(\alpha^i) - v \right). \tag{7}$$

where $\theta = [\theta_1, \theta_2]$ is a state vector, while M_1 and M_2 are nonlinear damping and stiffness terms and v is design input. Substituting the control law in θ, and simplifying, it is rewritten as

$$\dot{\theta}_1 = \theta_2, \dot{\theta}_2 = \sum_{i=1}^{m} (c_i - \hat{c}_i) M_1(\dot{\alpha}^i) + \sum_{i=1}^{n} (k_i - \hat{k}_i) M_2(\alpha^i) + v. \tag{8}$$

The control input $v = -a_1\theta_1 - a_2\theta_2$ must be selected such that the resulting linear subsystem is stable when the nonlinearities are eliminated via partial feedback linearization. The update law for parameter estimation is defined as [16]

$$\left[\dot{\hat{C}} \quad \dot{\hat{K}} \right]^T = \left[\dot{\hat{C}} \quad \dot{\hat{K}} \right]^T = \theta_2 M. \tag{9}$$

3.1 Parameter Estimation by Neural Network

The artificial neural networks are designed based on the human neuron system and successfully used in many engineering fields. The feedforward, backpropagation (BP) network [17] is a well-known model and the inputs are passed forward from the input to output layer via one/several hidden layers. The calculated error between actual and target values are propagated back in order to update the weights. The network is constructed based on the following cost function minimization:

$$Error = \frac{1}{2} \sum_p (y_p^d - y_p)^2,$$
(10)

where y_p and y_p^d are the pth neural network output and desired (target) values. The backpropagation algorithm minimizes the above cost function with the following output weights update law:

$$w_{new} = w_{old} - \eta \frac{\partial Error}{\partial w},$$
(11)

where $\eta \in (0, 1)$ is the learning rate. Likewise, hidden layer weights are also updated from the error in that cycle. After training the model, neural network can be utilized to predict the parameters. Now Eq. (8) is rewritten as

$$\theta_2 = \sum_{i=1}^{m} (c_i - \hat{y}_{ci}) M_1 (\dot{\alpha}^i) + \sum_{i=1}^{n} (k_i - \hat{y}_{ki}) M_2 (\alpha^i) + v.$$
(12)

where $\hat{y}_{ci}$ and $\hat{y}_{ki}$ are the parameters estimated by the neural network.

4 Results and Discussion

The analysis and control modules are implemented with MATLAB program. Numerical experiments are carried out to verify the performance of the controller discussed in this paper. Parameters employed in this work are taken from earlier work [16]. The flutter velocity of the aeroelastic model without considering nonlinearities is 11.57 m/s.

Polynomial nonlinearities in stiffness and damping terms are considered from [16]. The flutter boundary of the model with the polynomial nonlinearity is identified with various initial conditions and shown in Fig. 2.

The plunge "h(0)" is varied from 0 to 0.05 m, pitch "$\alpha(0)$" = −0.2 to 0.2 rad, $\dot{h}(0) = 0$ and $\dot{\alpha}(0) = 0$. The flutter boundary shrinks indirectly proportional to the plunge initial condition. For further analysis, the initial conditions considered are h(0) = 0.01, $\alpha(0)$ = 0.1, $\dot{h}(0) = 0$ and $\dot{\alpha}(0) = 0$. As an indication of instability, flutter velocity is first obtained from nonlinear responses. The flutter velocity of the

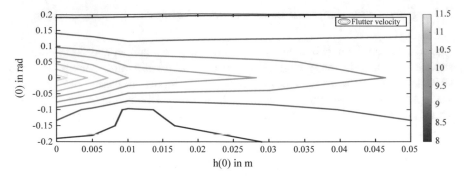

Fig. 2 Influence of initial conditions on flutter boundary

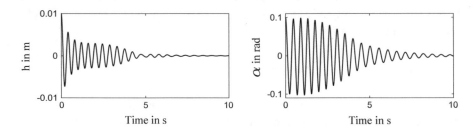

Fig. 3 Time response plot at freestream velocity 7 m/s

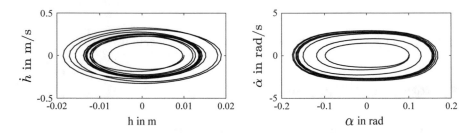

Fig. 4 Phase plot at freestream velocity 16 m/s

system for above initial conditions is found to be 7.93 m/s and the time responses are shown in Fig. 3 for the velocity 7 m/s and clearly shows the system as stable. Further the model is simulated at a velocity 16 m/s and is shown in Fig. 4, wherein an LCO is observed due to the presence of the polynomial nonlinearity making the system to oscillate periodically instead of becoming unstable.

To study the effect of the adaptive partial feedback linearization controller, the simulations are carried out at velocity 16 m/s with nonlinearity and Fig. 5 shows the time response of plunge and pitch degree of freedom and flap deflection. It is observed that the system is stable and it shows the effectiveness of the controller in suppressing the LCO. The pitching response settles at 1.5 s comparatively quicker

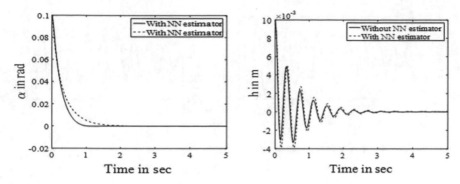

Fig. 5 Closed-loop response at freestream velocity 16 m/s

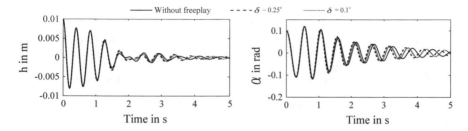

Fig. 6 Time histories at freestream velocity 8 m/s

to plunge response which takes 3.5 s to settle. This is because of the control law which is framed based on the pitch angle as output function and with damping nonlinearity. Now, the uncertain parameters in adaptive control law are identified by the three-layer feedforward, backpropagation neural network which acts as the estimator with learning rate $\eta = 0.7$ and α and $\dot{\alpha}$ as inputs. The system responses with such neural network estimator are also shown in Fig. 5. However, there is no marked variation observed by incorporating neural network based estimator.

In the next study, the free play nonlinearity in stiffness and polynomial structural nonlinearity in damping in pitching degree of freedom is added to a system to study their effect on the flutter boundary. The system is simulated with the same initial conditions and free play nonlinearity is not affecting the flutter boundary but the amplitudes of the pitch and plunge responses are higher as δ increases as seen in Fig. 6. The response at post flutter velocities is divergent in this case and the time history at the freestream velocity of 12.5 m/s is shown in Fig. 7.

Figure 8 shows the system as stable at freestream velocity 16 m/s with the controller active in the system. The time taken to converge is high in pith response compared to the previous case, where both stiffness and damping nonlinearities are of polynomial type. When the neural network estimator is added, the system becomes stable.

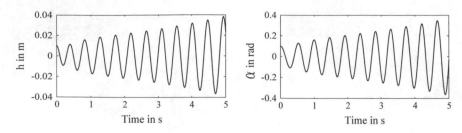

Fig. 7 Time histories at freestream velocity 12.5 m/s

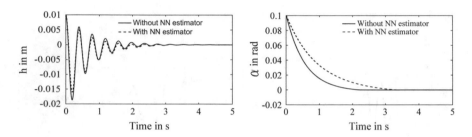

Fig. 8 Closed-loop response at freestream velocity 16 m/s

5 Conclusions

In this paper, the effect of a neural network based estimator with adaptive controller
was studied using two different cases on aeroelastic model with nonlinearities in
pitch direction. The adaptive control law was designed based on partial feedback
linearization and three-layer feedforward neural network is used as an estimator to
identify the estimated uncertainty parameters in control law. The effect of initial
conditions on flutter boundary with polynomial nonlinearities was studied. As the
initial conditions increase, the stable region got minimized. The free play nonlinearity effect on flutter boundary is small and the response is divergent in post flutter
operation. The effect of controller on unstable region was studied with and without
neural network estimator in both the cases. It practicality extends the flutter
boundary. For more detailed analysis, the control surface dynamics may be
included in the system of equations.

References

1. Fung, Y.C.: An Introduction to the Theory of Aeroelasticity. Dover Publications, Inc (2008)
2. Woolston, D.S.: An investigation of effects of certain types of structural nonlinearities on
   wing and control surface flutter. J. Aeronaut. Sci. **24**, 57–63 (1957)

3. Xu, X., Gao, Y., Zhang, W.: Aeroelastic dynamic response and control of an aeroelastic system with hysteresis nonlinearities. J. Control Sci. Eng. (2015)
4. Ko, J., Kurdila, A., Strganac, T., Ko, J., Kurdila, A., Strganac, T.: Nonlinear control theory for a class of structural nonlinearities in a prototypical wing section. In: 35th Aerospace Sciences Meeting and Exhibit. AIAA Paper 97–0580. American Institute of Aeronautics and Astronautics, Reno, Nevada (1997)
5. Ko, J., Kurdila, A., Stragnac, T., Ko, J., Kurdila, A., Stragnac, T.: Nonlinear dynamics and control for a structurally nonlinear aeroelastic system. In: 38th Structures, Structural Dynamics, and Materials Conference. AIAA Paper 97–1024. American Institute of Aeronautics and Astronautics, Kissimmee, FL, USA (1997)
6. Monahemi, M.M., Krstic, M.: Control of wing rock motion using adaptive feedback linearization. J. Guid. Control Dyn. 19, 905–912 (1996)
7. Li, N., Balas, M.J.: Aeroelastic vibration suppression of a rotating wind turbine blade using adaptive control. In: 32nd ASME Wind Energy Symposium. American Institute of Aeronautics and Astronautics (2014)
8. Ko, J., Strganac, T.W., Kurdila, A.J.: Adaptive feedback linearization for the control of a typical wing section with structural nonlinearity. Nonlinear Dyn. 18, 289–301 (1999)
9. Strganac, T.W., Ko, J., Thompson, D.E.: Identification and control of limit cycle oscillations in aeroelastic systems. J. Guid. Control Dyn. 23, 1127–1133 (2000)
10. Platanitis, G., Strganac, T.W.: Control of a nonlinear wing section using leading- and trailing-edge surfaces. J. Guid. Control Dyn. 27, 52–58 (2004)
11. Block, J.J., Strganac, T.W.: Applied active control for a nonlinear aeroelastic structure. J. Guid. Control Dyn. 21, 838–845 (1998)
12. Wang, Z., Behal, A., Marzocca, P.: Adaptive and robust aeroelastic control of nonlinear lifting surfaces. Int. J. Aeronaut. Space Sci. 11, 285–302 (2010)
13. Chen, C.-L., Chang, C.-W., Yau, H.-T.: Terminal sliding mode control for aeroelastic systems. Nonlinear Dyn. 70, 2015–2026 (2012)
14. Gujjula, S., Singh, S.N., Yim, W.: Adaptive and neural control of a wing section using leading- and trailing-edge surfaces. Aerosp. Sci. Technol. 9, 161–171 (2005)
15. Prabhu, L., Srinivas, J.: Robust control of a three degrees of freedom aeroelastic model using an intelligent observer. In: 2015 International Conference on Robotics, Automation, Control and Embedded Systems (RACE), pp. 1–5, Chennai, India (2015)
16. Li, D., Xiang, J., Guo, S.: Adaptive control of a nonlinear aeroelastic system. Aerosp. Sci. Technol. 15, 343–352 (2011)
17. Schalkoff, R.J.: Artificial Neural Networks. Tata McGraw-Hill Education, New Delhi (2011)

Application of Total Least Squares Version of ESPRIT Algorithm for Seismic Signal Processing

G. Pradeep Kamal and S. Koteswara Rao

Abstract Estimation of frequency with high resolution is a crucial task in signal processing. Raw seismic signals consist of huge noise which can be removed only by using some signal processing methods. In this paper, the ESPRIT algorithm is implemented in order to process the signal. A time series data is taken and the frequency is estimated by total least squares version of ESPRIT algorithm. ESPRIT employs a basic rotational invariance in the subspaces of the signal. The detailed implementation of the algorithm is greatly presented in the following sections.

Keywords Seismology · Power spectral density · Least square estimation
Digital signal processing · Frequency estimation

1 Introduction

Seismology is the study of earthquakes, according to some researches in olden times, the earthquakes are caused by the volcanic explosions that take place under the earth's crust and the waves travel to the earth's surface causing the tremors which cause lots of destruction to the mankind and according to some researches, the earthquakes are caused [1] by the drifting of continents which causes the landmass to move and create mass earthquakes. These waves are of two types, one is the transverse waves and the other is the longitudinal waves, the transverse waves travel parallel to the epicenter of the earthquake while the longitudinal waves travel perpendicular to the epicenter of the earthquake. Generally, these earthquakes are detected by a device known as seismograph, it simply records the data of the earthquake like duration, magnitude, etc. [2], this data is further converted,

G. Pradeep Kamal (✉) · S. Koteswara Rao
Department of Electronics and Communication Engineering,
K L University, Vaddeswaram, Guntur, Andhra Pradesh, India
e-mail: gollamudipradeep@gmail.com

S. Koteswara Rao
e-mail: rao.sk9@gmail.com

© Springer Nature Singapore Pte Ltd. 2018
S. C. Satapathy et al. (eds.), *Information and Decision Sciences*,
Advances in Intelligent Systems and Computing 701,
https://doi.org/10.1007/978-981-10-7563-6_25

simplified, etc., or simply it is called as processing of seismic signal which is briefly explained below.

1.1 ESPRIT Algorithm

ESPRIT stands for estimation of signal parameters via rotational invariance techniques which is developed on the similar values just like the other subspace procedures but additionally exploits a deterministic connection among subspaces [3]. It is a frequency estimation technique. This method differs from the other subspace methods in that the signal subspace is estimated from the data matrix **A** rather than the estimated correlation matrix. The essence of ESPRIT lies in the rotational property between staggered subspaces that is invoked to produce the frequency estimates. In the case of a discrete-time signal or time series, this property relies on observations of the signal over two identical intervals staggered in time [4]. This condition arises naturally for discrete-time signals, provided that the sampling is performed uniformly in time. Extensions of the ESPRIT method to a spatial array of sensors, the application for which it was originally proposed, the original, least squares version of the algorithm are described in first place and then the derivation to total least squares ESPRIT was extended [5], which is the preferred method for use, as the derivation of the algorithm requires an extensive amount of formulation and matrix manipulations.

2 Mathematical Modeling

Let us take a complex exponential $S_0 = e^{j2\pi fn}$ which has a complex amplitude α and a frequency f. The property of the signal which we have taken is shown below [6].

$$S_0(n+1) = \alpha\, e^{j2\pi f(n+1)} = S_0(n)e^{j2\pi f} \tag{1}$$

Hence, the phase-shifted version of the present value is the succeeding sample value. The rotation on the unit circle $e^{j2\pi f}$ is a representation of this phase shift.

$$x(n) = \sum_{p=1}^{P} \alpha_p V(f_p) e^{j2\pi n f_p} + W(n) = V\varnothing^n \alpha + W(n) = S(n) + W(n), \tag{2}$$

where the P columns of matrix U are length N interval frequency vectors of the complex exponentials.

$$U = [U(f_1) \, U(f_2) \ldots U(f_p)].$$ (3)

The complex exponentials α_p amplitudes are present in the vector α. The diagonal matrix of phase shifts among the adjacent time samples of the individual is the matrix $\emptyset$ complex exponential elements of $S(n)$ [7].

$$\emptyset = \text{diag}\{\emptyset_1, \ldots, \emptyset_p\} = \begin{bmatrix} e^{j2\pi n f_1} & 0 & \cdots & 0 \\ 0 & e^{j2\pi n f_2} & \cdots & 0 \\ \vdots & \vdots & \ddots & \vdots \\ 0 & \cdots & 0 & e^{j2\pi n f_p} \end{bmatrix}$$ (4)

For $p = 1, 2, \ldots, P$, $\emptyset_p = e^{j2\pi n f_p}$

This rotation matrix is entirely expressed by the complex exponential frequencies f_p. If $\emptyset$ can be acquired, then frequency estimates can be acquired. Take two overlaying sub-windows of length $N - 1$ with the length N time window vector and signal which has the sum of complex exponentials.

$$S(n) = \begin{bmatrix} S_{N-1}(n) \\ S(n+N-1) \end{bmatrix} = \begin{bmatrix} S(n) \\ S_{N-1}(n+1) \end{bmatrix},$$ (5)

where $S_{N-1}(n)$ is the length $(N-1)$ sub-window of $S(n)$, hence $S_{N-1}(n) = U_{N-1}\emptyset^n\alpha$.

Matrix U_{N-1} is built in the equivalent way as U other than its time window frequency vectors are of length $N - 1$, represented as $U_{N-1}(f)$ [8].

$$U_{N-1} = [U_{N-1}(f_1) \, U_{N-1}(f_2) \ldots U_{N-1}(f_p)]$$ (6)

Remember that $S(n)$ is a scalar signal which is shaped up of the sum of complex exponentials at time n.

$$U_1 = U_{N-1}\emptyset^n \text{ and } U_2 = U_{N-1}\emptyset^{n+1}$$ (7)

U_1 and U_2 relate to the unstaggered and staggered windows, which is

$$U\emptyset = \begin{bmatrix} U_1 \\ * * \ldots * \end{bmatrix} = \begin{bmatrix} * * \ldots * \\ U_2 \end{bmatrix}$$ (8)

The two matrices with vectors having intervals are expressed as

$$U_2 = U_1\emptyset$$ (9)

Observe that both matrices spaces a distinct, however related, $(N - 1)$ dimensional subspace. Assume that we possess a data matrix A with M data records of the length N interval vector signal x(n). By singular value decomposition (SVD), data matrix is

$$A = L\sum V^H \tag{10}$$

L is a M × M matrix of left singular vectors and v is a N × N matrix of right singular vectors. Each of these matrices is a unit matrix, hence $L^H L = I$ and $V^H V = I$. Dimensions of the matrix $\sum$ are M × N which contains singular quantities on the main diagonal which is ordered in a magnitude of decreasing value [9]. The singular valued magnitudes are squared and are equivalent to the eigenvalues of $\hat{R}$ scaled with a M factor and the V's columns are their related eigenvectors. Hence, v shapes an orthogonal and normalized foundation for the underlying N-dimensional vector space. The signal and noise subspaces are formed by dividing this subspace as $V = [V_s/V_n]$.

Relating to the p largest magnitudes of the singular values, V_s is a matrix of right-hand singular vectors [10]. All of these frequency vectors for $f = f_1, f_2, \ldots, f_p$ should lie in the signal subspace since the sum of complex exponentials formed as time window frequency vectors U(f) is contained in the signal portion. Hence, U and V_s matrices occupy the identical subspace [11]. Hence, there lies an invertible transformation T that draws V_s into $U = V_s T$.

In this derivation, T transformation is never elucidated, in the other way, it is only constructed just like a mapping inside the subspace of the signal among these two matrices. Divide the subspace of the signal into two tiny subspaces of dimensions (N − 1).

$$V_s = \begin{bmatrix} V_1 \\ * * \ldots * \end{bmatrix} = \begin{bmatrix} * * \ldots * \\ V_2 \end{bmatrix}, \tag{11}$$

where V_1 and V_2 relate to the staggered and unstaggered subspaces because U_1 and U_2 related to the equivalent subspaces.

$$U_1 = V_1 T \quad U_2 = V_2 T \tag{12}$$

The rotation Ø subspaces are being corresponded by the matrix U's staggered and unstaggered elements. A same, though unlike, rotation should be present that associates V_1 to V_2 because the matrices V_1 and V_2 also spaces these subspaces as $V_2 = V_1 \Psi$ [10]. Where Ψ is the matrix of rotation. Remember that the estimation of frequency arrives below for summarizing the rotation matrix Ø subspace. Rotations among the subspaces of staggered signal and the relations altogether combined can be made use of the estimation of Ø [12]. From the data matrix A's SVD, the matrices V_1 and V_2 are known from the procedure. Primarily solve Ψ by utilizing the technique of least squares.

$$\Psi = \left(V_1^H V_1 \right)^{-1} V_1^H V_2 \tag{13}$$

Substituting $V_2 = V_1 \Psi$, $U_2 = V_2 T = V_1 \Psi T$ is acquired. In the same way, solve U_2, utilizing the relation $U_2 = U_1 \emptyset$ and substituting $U_1 = V_1 T$ and $U_2 = V_2 T$ for U_1, $U_2 = U_1 \emptyset = V_1 T \emptyset$. Hence, by equating both the right-hand sides of $U_2 = V_2 T = V_1 \Psi T$ and $U_2 = U_1 \emptyset = V_1 T \emptyset$. The relation among the two subspaces rotations is

$$\Psi T = T \emptyset \text{ or}$$

Similarly,

$$\Psi = T \emptyset T^{-1} \tag{14}$$

Equations $\Psi T = T \emptyset$ and $\Psi = T \emptyset T^{-1}$ [13] must be realized as the association among the matrix Ψ's eigenvectors and eigenvalues. Hence, elements of the diagonal of $\emptyset$, $\emptyset_p$ for p = 1, 2, 3, ..., P are commonly the Ψ's eigenvalues. Finally, the frequency estimates are

$$\widehat{f_p} = \frac{\angle \emptyset_p}{2\pi}, \tag{15}$$

where the phase of $\emptyset_p$ is $\angle \emptyset_p$. Even though the utilization of rotational subspaces is the property of the ESPRIT algorithm is very easy. Pay heed that only matrix simple relationships are utilized by us. Primarily, provide an algorithm which has a version of total least squares, a best technique to utilize [10]. Pay heed that V_1 and V_2 subspaces are the original subspace's only estimates that relate to U_1 and U_2, naturally acquired through the data matrix A. The subspace rotation's estimate was acquired by solving $V_2 = V_1 \Psi$ utilizing the least square criterion.

$$\Psi_{ls} = \left(V_1^H V_1 \right)^{-1} V_1^H V_2 \tag{16}$$

This least square result is acquired by reducing the errors in least square perception from the formulation as given below [10].

$$V_2 + E_2 = V_1 \Psi \tag{17}$$

Since E_2 is a matrix which has errors among V_2 and the original subspace relating to U_2. The least square formulation presumes that errors exist especially on V_2 estimation and on the other side, it presumes that there exist no errors among V_1 and the original subspace that it is trying to estimate relating to U_1 [14]. Hence, V_1 is an estimated subspace too, an extremely accurate formulation is

$$V_2 + E_2 = (V_1 + E_1)\Psi \tag{18}$$

Errors among V_1 and the original subspace relating to U_1 are expressed by the matrix E_1. Minimizing the Frobenius norm of the two error matrices can acquire the result to this problem, which is known as total least squares (TLS).

$$\| E_1 \quad E_2 \|_F \tag{19}$$

As the properties of TLS are far away from the expectation, normally lend the process to acquire the TLS solution of Ψ. Primarily, prepare a matrix constructed by the staggered signal subspace matrices V_1 and V_2 located adjacent to each other and execute an SVD.

$$[V_1 \quad V_2] = \tilde{L}\tilde{\sum}\tilde{V}^H \tag{20}$$

Later we work on 2P × 2P matrix $\tilde{V}$ of right singular vectors which are divided as P × P quadrants.

$$\tilde{V} = \begin{bmatrix} \tilde{V}_{11} & \tilde{V}_{12} \\ \tilde{V}_{21} & \tilde{V}_{22} \end{bmatrix} \tag{21}$$

The subspace rotation matrix Ψ Total least square solution is $\Psi_{tls} = -\tilde{V}_{12}\tilde{V}_{22}^{-1}$. The estimation of frequencies is then acquired by $\Psi = T\emptyset T^{-1}$ and $\tilde{f}_p = \frac{\angle \emptyset_p}{2\pi}$ by utilizing Ψ_{tls} from $\Psi_{tls} = -\tilde{V}_{12}\tilde{V}_{22}^{-1}$ [10].

3   Simulation and Results

Step 1: The data utilized for the observation is acquired from Book_Seismic_Data.mat of East Texas [3] landmine is the file name. We have taken the source as a dynamite blast which took place at a depth of around 100 ft, one trace has 1501 samples of 0.002 s sampling interval.

Step 2: The algorithm's functioning is assessed with known synthetic signal and then ESPRIT algorithm is applied to calculate the seismic signal's tonal. Synthetic signal's frequencies are taken as 0.2π, 0.3π, 0.8π, and 1.2π and are shown as complex exponentials.

Step 3: The normalized frequencies are 0.2π and 0.7π. The signal generated is shown in Fig. 1.

Step 4: In Fig. 2, power spectral density of the synthetic signal is shown. The figure shows peaks are at 0.2 and 0.7 normalized frequencies. That means ESPRIT algorithm is working fine.

Step 5: The raw seismic signal is shown in Fig. 3, which is a single shot taken from [3].

Fig. 1 Synthetic signal with
and without noise

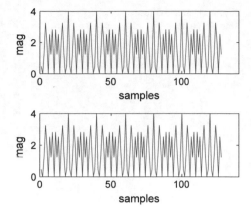

Fig. 2 PSD using
ESPIRIT TLS Method for
synthetic signal

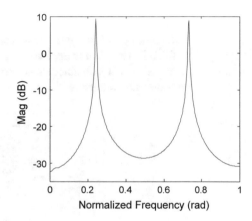

Fig. 3 Raw seismic signal

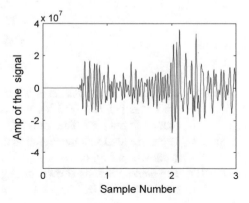

Fig. 4 Detrended raw
seismic signal

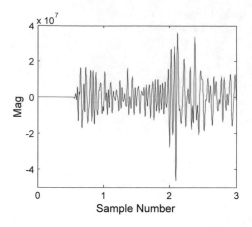

Step 6: The raw seismic signal is detrended which is shown in Fig. 4. Detrending is done in order to remove bias and baseline drift.

Step 7: ESPRIT algorithm is applied on detrended seismic signal and the power spectral density obtained is shown in Fig. 5. The max peak is at 0.958 normalized frequency.

$$w = \frac{2\pi f}{fs} = 0.0958\pi$$

$$= \frac{2\pi f}{500}$$

$$= \frac{2\pi}{fs}f = 0.0958\pi$$

$$f = \frac{500}{2} * 0.0958$$

$$= 250 * 0.0958$$

$$= 25 * 0.0958$$

$$= 23.950\,\text{Hz}$$

Step 8: In the reference book, it is written that the data is band-pass filtered in the range [15 Hz, 60 Hz]. For ensuring purpose, a BP filter with FIR order 8 is realized. The transfer function of the same is shown in Fig. 6.

Step 9: The detrended seismic signal is convolved with FIR BPF and the output is shown in Fig. 6.

Step 10: The same PSD, as shown in Fig. 7 is obtained. So, the seismic signal tonal is 23.95 Hz.

Fig. 5 FIR band-pass filtered
frequency spectrum

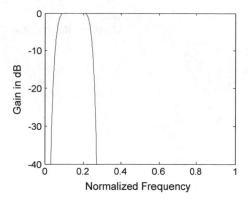

Fig. 6 FFT of band-pass
filtered seismic signal

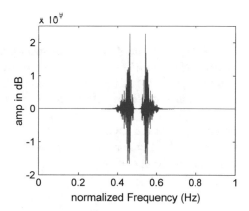

Fig. 7 ESPRIT spectrum of
BPF detrended seismic signal

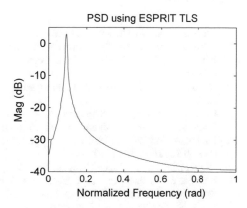

$$\text{Another insignificant tonal is } w_1 = \frac{2\Pi f}{f_s} = 0.119\pi$$

$$\frac{2\Pi f}{500} = 0.119$$

$$f = \frac{0.119\pi * 500}{2\pi}$$

$$\therefore f = 250 * 0.119$$

$$= 29.75\,\text{Hz}$$

4 Conclusion

In this paper, signal parameter estimation with high resolution is obtained using ESPRIT algorithm. The step-by-step process of the seismic signal analysis has been perfectly presented in the results section, the actual signal strengths in the seismic signal from time series data taken are clearly estimated with peaks of high resolution, and power spectral density for both synthetic and raw seismic signal is obtained. From the results obtained, it is concluded that the ESPRIT algorithm is the best technique for frequency estimation in seismic signal processing which requires less computation.

References

1. Mousa, W.A., Al-Shuhail, A.A.: Processing of seismic reflection data using Matlab. In: Synthesis Lectures on Signal Processing. Morgan & Claypool Publishers
2. Kirlin, R.L., Done, W.J.: Covariance Analysis of Seismic Signal Processing. Seismic Library (1999)
3. Mousa, W.A., Al-Shuhail, A.A.: Processing of Seismic Reflection Data Using MATLAB (2011)
4. Roy, R., Ottersten, B., Swindlehurst, L., Kailath, T.: Multiple invariance ESPRIT. IEEE Trans. Acoust. Speech Signal Process. (in preparation)
5. Paulraj, A., Roy, R., Kailath, T.: A subspace rotation approach to signal parameter estimation. Proc. IEEE 1044–1045 (1986)
6. Hayes, M.H.: Statical Digital Signal Processing and Modeling. Wiley (1996)
7. Lemma, A.N., van der Veen, A.-J., Deprettere, E.F.: Analysis of joint angle frequency estimation using ESPRIT. IEEE Trans. Signal Process. 55(5), 1264–1283 (2003)
8. Yuen, N., Friedlander, B.: Performance analysis of higher order ESPRIT for localization of near field sources. IEEE Trans. Signal Process. 46(3) (1998)
9. Vaseghi, S.V.: Advanced Signal Processing and Noise Reduction. Wiley (2008)
10. Manolakis, G., Ingle, V.K.: Statistical and Adaptive Signal Processing. McGraw-Hill (2000)
11. Stoica, P., Moses, R.: Spectral Analysis of Signals. Prentice Hall Inc. (2005)
12. Rouqutte, S., Najim, M.: Estimation of frequency and damping factors by two dimensional ESPRIT type methods. IEEE Trans. Signal Process. 49(1) (2002)

13. Roy, R., Paulraj, A., Kailath, T.: ESPRIT-a subspace rotation approach to estimation of parameters of Cisoids in noise. IEEE Trans. Acoust. Speech Signal Process. **ASSP-34**, 1340–1342 (1986)
14. Roy, R., Kailath, T.: ESPRIT-estimation of signal parameters via rotational invariance techniques. IEEE Trans. Acoust. Speech Signal Process. **37**(7), 984–995 (1989)

Metadata-Based Semantic Query in Multilingual Databases

Ch. V. S. Satyamurty, J. V. R. Murthy and M. Raghava

Abstract The retrieval of semantically equivalent information from various databases having internationalization feature is a major challenge for the data scientists. The criticality of this problem revolves around translation of the query in the native language into some target language despite the structural differences of the schema. The metadata generated and maintained by any relational database technology offers a scope for realizing such a seamless mapping with the help of industry offered translators. This paper proposes a model with such a capability and demonstrates the equivalence of forward and backward translations across various languages using WordNet, Microsoft Translator tools while presenting the results.

Keywords Semantic query · WordNet · RDBMS · Metadata

1 Introduction

The information systems are playing a major role to automate many tasks of the enterprise. Especially, the repetitive activities can be implemented through well-designed software tools and technologies leading to a burst in the amount of data. The generated data will be stored in database systems. Last 30 years witnessed the evolution in relational databases that are helpful in storing and retrieving data effectively and efficiently in various languages. This capability is possible because of the standards specified by Unicode Consortium to encode the characters. Eventually, nowadays, software gains the internationalization and localization

Ch. V. S. Satyamurty (✉) · M. Raghava
CVR College of Engineering, Hyderabad, India
e-mail: Satyamurtycvs@yahoo.co.in

M. Raghava
e-mail: raghava.m@cvr.ac.in

J. V. R. Murthy
JNTU Kakinada, Kakinada, India
e-mail: mjonnalagedda@gamil.com

© Springer Nature Singapore Pte Ltd. 2018
S. C. Satapathy et al. (eds.), *Information and Decision Sciences*,
Advances in Intelligent Systems and Computing 701,
https://doi.org/10.1007/978-981-10-7563-6_26

capabilities and can cater to users with plurality of languages. This has led to various departments of large organizations to implement local databases with language as a choice. The schemas of the databases created by different departments might differ in structure but can contain semantically equivalent data. Retrieval of the sensible information from these databases and presenting a consolidated and comprehensive view are not a straightaway query generation step [1–3]. The metadata available in the repositories of the database software is vital for retrieving the information efficiently. The graphs are helpful to map these tables onto proper indexing structures wherever design document is available. To simplify the task of information retrieval from different language databases with preserved semantics is a topic of interest to researchers. This paper proposes a framework which accepts a keyword from the user and translates it into the target language using the industry standard tools. Then, the proposed framework scan the metadata files of the target database in search a table with the translated word and tries to interpret the structure developed. Once a suitable table is selected then it automatically generates a query and executes the same to fetch the information of interest. The results of the automatically constructed query by the centralized model prove to compensate the naïve knowledge of the user on the language and data model.

2 Related Work

The authors in [4] presented a model on extraction of information from semantically equivalent databases of English language and presented the results at the centralized location. In [5], the authors considered a single query that can be abstracted in many ways by extracting the metadata of query logs and expanded the semantics of the metadata using WordNet [6]. The authors in [7] developed a framework which creates a sports ontology using the protégé tool in the first step. Next, it prompts the user to enter the query and parses it into a tree using a Stanford parser. Then, synonyms for the keywords of the parse tree are generated using the WordNet that overall helps to expand the query semantically and extract the information from the sports domain.

The authors in [8] exploited the keyword-based query applied on relational databases by employing the techniques of natural languages processing using the key-mantic. Key-mantic explores the semantically equivalent words using the intrinsic weight matrix and proximity of the words using the aggregate function to rebuild the query. In [9], the authors extensively reviewed the various query optimization techniques and how they are applicable for centralized databases and distributed databases. The optimization specifically discusses selection and sequence of operations performed to efficient retrieval of information from the database.

3 Proposed Model

The transaction databases are developed based on relational model which are free from many operational inconsistencies across multiple manipulations. While designing any database, the emphasis is on structures and based the structures it is expected to derive some inference. Hence, the language of schema specification does not have any bearing in information extraction as long as the schema is the same. This notion led the database designers to incorporate the internationalization and localization issues in the databases. However, the end user can enjoy this multi-language feature only if a third-party agent software is configured. In the proposed model, an end user works only in his native language without any concern to the language of database implementation and the corresponding query language. Once the language of user is mapped onto the language of databases, the query building is a straightforward step which involves replacing each keyword of source query with a semantically equivalent word in the target language. In this regard, the development in the allied areas in natural language processing helps the application developers to enable end users to use their native language while working on databases. WordNet is one such software that throws semantically equivalent words for the given input word and caters to multiple languages. Functionally, WordNet is a thesaurus for the language chosen. It is based on psycholinguistic studies. The WordNet is a lexical database of the language. Nouns, verbs, adverbs, and adjectives are grouped into cognitive word called as synsets. The synsets are helpful to obtain similar words having same meaning. It also consists of homonymy, hypernym, holonomy, etc.

3.1 Framework

Figure 1 presents the model view and the detailed description of the model is given in this subsection. The user interacts with the graphical user interface and enters his query in the form of keywords, which is the first layer of the framework. The entered keywords are translated into different target languages using the Microsoft Translator in the second layer of the model. There is a chance that the translated words may not map onto any table designed in the target database. In such cases, the framework queries the WordNet to workout with synonyms of the keyword. Once all keywords are resolved in the target languages, then SQL statements are built. This aspect contributes to the framework as the third step. These SQL statements, in turn, access the various data models of target languages and the retrieved information is presented to the end user.

a. Graphical User Interface

This interface prompts the user to enter the keywords for referring to the information to be extracted and the results of the query execution are displayed.

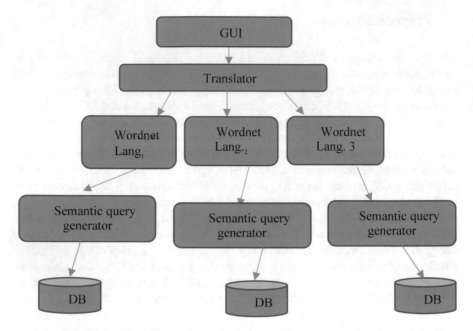

Fig. 1 The complete framework

b. Translator

This layer will translate the input keywords into multilingual words with the help of Microsoft Translator. It finds the equivalent word in target languages and maintains all of them in a linear data structure. Then, the framework checks the metadata maintained by the databases for the availability of a table with the translated keyword corresponding to the name of the table. It is very likely that the database schema represents the same information with a semantically equivalent word. Then the next step is involved.

c. WordNet

This step identifies the lexical data relationships of languages chosen and extracts the synonyms of the given keyword. The WordNet acts as mediator to convert the raw tokens of SQL query into a semantically equivalent multiple synonyms.

d. Query Generator

Then the framework checks if the recreated query can be executed in the target language by checking the database metadata for the availability of a tables with the synonyms. If the search against all keywords is a success, then the framework generates the query corresponding to the word.

e. **Database(s)**

The databases of different institutions contain semantic data of the keyword in different formats.

4 Experimental Results

We are interested in educational domain, so we considered educational domain here owned by different colleges across the globe. In a simple instance, if the user is interested in student information, created and maintained in local languages by e different colleges. The translator tool maps the keyword student in the local language into two target languages Spanish and Japanese. In Fig. 2, the input string is "student", the synonyms for the student keyword in English are "scholarly person", "scholar", "pupil", "bookman", and "educatee". The keyword student translated into Spanish and Japanese languages and extracted synonyms for the keyword student in Spanish are "alumna" "estudinate" and similarly in Japanese language. The synonyms extracted from the WordNet are helpful for finding the table existence in the database of any college using the metadata available in the database. On successful search of tables, the framework generated the query and extracted the information related to students and the results are presented to the end user as shown in latter part of Fig. 2.

Here, in Fig. 3, the word "alumno" from the Spanish language is semantically equivalent to "student" in English language. The output extracted from these tables is presented.

Figure 4 represents the output extracted from the "student" table and "scholar" table in English language and extracted information from the Japanese tables equivalent to "student" keyword.

```
Enter the string to be searched
student
en
estudiante student 学生
spanish
{'alumna', 'estudiante', 'alumno'}
english
{'scholarly_person', 'scholar', 'pupil', 'bookman', 'educatee', 'student'}
japan
{'学生', '生徒', '角帽', '大学生', '教え子', '学徒', '学習者'}
```

Fig. 2 Input string and equivalent synonyms in Spanish and Japanese languages

```
['alumno', 'scholar', 'pupil', 'student', '学生', '大学生']
alumno
select * from alumno
['NOMBRE', 'RAMA', 'CURSO']
('ashwini', 'ordenadores', 'Ingenieria')
('madhumitha', 'electrónica', 'microelectrónica')
('kisan', 'mecánico', 'Ingenieria')
scholar
select * from scholar
```

Fig. 3 Output from Spanish table "alumno"

```
select * from student
['NAME', 'ROLLNO', 'GRADE']
('a', 1, 'a')
('b', 2, 'b')
('as', 3, 'c')
('ashwini', 4, 'd')
学生
select * from 学生
['名', '数']
('アシュウィニ', '零')
('サイ', '四')
('マッドハブ', '六')
大学生
select * from 大学生
['名称', '数', '学科']
('アシュビニ', '零', '英语')
('塞', '四', '数学')
('马达夫', '六', '科学')
['alumno', 'scholar', 'pupil', 'student', '学生', '大学生']
finished printing
```

Fig. 4 Output extracted from English and Japanese

5 Conclusion

This paper presented an architectural framework for extraction of reliable information from multi-language databases. The structural stability of the SQL is exploited by this framework to translate the native language query into the target language SQL query through an exhaustive search mechanism. This paper considered English as the native language and Spanish and Japanese as target languages while translating one SQL statement into another. The results of the generated query are presented at a centralized location.

References

1. Saini, M., Sharma D., Gupta, P.K.: Enhancing information retrieval efficiency using semantic based-combined-similarity-measure. In: International Conference on Image Information Processing (ICIIP 2011). IEEE Computer Society (2011)
2. Mohan Kumar, P., Vaideeswaran, J.: Semantic based efficient cache mechanism for database query optimization. Int. J. Comput. Appl. **43**(23), 14–18 (2012)
3. Hsu, C., Knoblock, A.: Semantic query optimization for query plans of heterogeneous multidatabase systems. Knowl. Data Eng. **12**(6), 959–978 (2000)
4. Satyamurty, Ch.V.S., Murthy, J.V.R., Raghava, M.: Metadata based semantic query in relational databases. India-2017 Danang (2017)
5. Christodoulakis, C., Kandogan, E., Terrizzano, I.G.: VIQS-visual interactive exploration of query semantics. In: ESIDA. ACM, pp. 4503–4903, March 2017
6. Miller, G.: Nouns in WordNet: a lexical inheritance system. Int. J. Lexicogr. **3**(4) (1990)
7. Umadevi, M., Gandhi, M.: WordNet and ontology based query expansion for semantic information retrieval in Sports domain. J. Comput. Sci. **11**(2), 361–371 (2015). https://doi.org/10.3844/jcssp.2015.361.371
8. Ramada, M.S., da Silva, J.C., de Sa Leitão-Junior, P.: Data extraction from structured databases using keyword-based queries. In: SBBD Proceedings, Oct 2014. ISSN 2316-5170
9. Khan, M., Khan, M.N.A.: Exploring query optimization techniques in relational databases. Int. J. Database Theory Appl. **6**(3), 11–20 (2013)

A Step Towards Internet Anonymity Minimization: Cybercrime Investigation Process Perspective

Shweta Sankhwar, Dhirendra Pandey and R. A. Khan

Abstract Nowadays, people are heading towards an era where the use of personal devices such as Personal Digital Assistant, laptops and wireless networks is increasing. Users operate their personal devices to gain benefit from resources and services offered by the Internet. Sometimes, these Internet activities are susceptible to cybercrimes and their consequences can be as harmful as common physical crime. Cyber-criminals use fake geographical locations to commit frauds and easily get away without eroding their anonymity. The geographical location information should be mandatory to gain access control. Authentication of user's geographical location (geolocation) can be helpful in enhancing network security and control access to resources. In this paper, IP address is used to authenticate user's graphical location and some of its extended properties which can be used as a weapon to avoid users from entering fake geographical locations while using Internet services, so as to improve safety and decreasing cybercrimes. The proposed model takes proactive investigations to uncover cybercrimes and cyber-criminals who are actively engaged in cybercrime.

Keywords Cybercrimes · Internet Anonymity · Cyber investigation
Geolocation · IP address

S. Sankhwar (✉) · D. Pandey · R. A. Khan
Department of Information Technology, Babasaheb Bhimrao Ambedkar University,
Lucknow, Uttar Pradesh, India
e-mail: shweta.sank@gmail.com

D. Pandey
e-mail: prof.dhiren@gmail.com

R. A. Khan
e-mail: khanraees@yahoo.com

© Springer Nature Singapore Pte Ltd. 2018
S. C. Satapathy et al. (eds.), *Information and Decision Sciences*,
Advances in Intelligent Systems and Computing 701,
https://doi.org/10.1007/978-981-10-7563-6_27

257

1   Introduction

Cyber frauds are increasing day by day. Users often enter a fake geolocation and attempt to perform malicious activities and lure the victims which leads to hacking of bank or website accounts, fraudulent transactions or scams. IP address is the first asset which is used in a cyber investigation to solve a cybercrime. It provides some specific details which help in tracing the geolocation of the user. However, in most of the cases, it becomes difficult to retrieve the original location of criminal. This has created a threat to network security. That is why we need to come up with a better way to stop cybercrime. The proposed model will illustrate how IP address can be used in a way so as to put a halt on cybercrimes in such a way that a criminal can be stopped even before attempting a cybercrime.

Use of smartphones, Personal Digital Assistant (PDAs) and wireless networks is increasing and has become a part of our daily lives. They are extremely useful in reviewing documents, managing contact information, responding via electronic mail and instant messaging, delivering presentations, accessing as well as handling data. As the person continues to use these devices, certain sensitive information gets acquired on them. This information can be accessed by organizational resources through wireless or wired communication. Such conditions can create a security threat for a computing device which may include unauthorized access. Thus, we need to have a mechanism which could authenticate whether the geolocation given by user is valid or not.

Usually, all the smartphones have Global Positioning System (GPS) on it and it is most commonly used for location-based services. GPS tracking could be used to track mobile phones, equipments, vehicles, animal, people. GPS offers more security, however, it uses satellite navigation system to collect geolocation information due to which sometimes it fails in locating position and may not be able to provide accurate directions due to some obstacles to the signals. So, we need to promote security over Internet and authenticate location of user without relying on GPS [1]. For fulfilling this purpose, we propose a model that will be suitable for location authentication. Here, in this research paper, it illustrates how IP address can be used to determine geolocation information and how it would be used in lessening cybercrimes. The remainder of the paper is organized as follows: Sect. 2 provides some information about cybercrime and case study, Sect. 3 explains the concept of IP Address, Sect. 4 identifies the problem statement, Sect. 5 proposes the solution of cybercrime investigation, Sect. 6 compares the proposed models with other existing selected models. Last section concludes the paper with future scope.

2   IP Address and Geolocation

An IP address assigns a numerical label to a device (e.g. computer, mobile) which participates in computer network and uses Internet protocol for communication. IP is short for Internet protocol which refers to a set of rules that govern the activities over the Internet and facilitate in completing a wide range of actions on WWW (World Wide Web). Therefore, Internet Protocol Address (IP address) is an inter-connected grid which is systematically laid out and regulates online communication by identifying initiating devices and varied Internet destinations, thus attaining a two-way communication. It has two principle functions: Host or network interface identification and location addressing. IP address are written and displayed in such notations which could be read by humans. It looks like this: 78.135.0.245 [2].

Internet Assigned Numbers Authority (IANA) oversees the allocation of global IP address. IANA is responsible for allocating IP address blocks to Regional Internet Registries (RIR). There are five RIRs around the world, i.e. African Network Information Center (AFRINIC) for Africa, American Registry for Internet Numbers (ARIN) for the United States, Canada, several parts of the Caribbean region, and Antarctica, Asia-Pacific Network Information Centre (APNIC) for Asia, Australia, New Zealand, and neighbouring countries, Latin America and Caribbean Network Information Centre (LACNIC) for Latin America and parts of the Caribbean region, Réseaux IP Européens Network Coordination Centre (RIPE NCC) for Europe, Russia, the Middle East, and Central Asia. Each RIR is allocated with addresses of particular areas around the world. These RIR manages IP addresses within particular regions of world. RIR then allocates block of IP address to Local Internet Registry (LIR) which assigns IP address to the customers within a particular locality. Most LIR include Internet Service Provider (ISP), academic institutions or enterprises [2].

In the context of India, IANA delegates allocations of IP address blocks to APNIC (RIR) which then divides the allocated address pools into smaller blocks and delegates them to ISP or any other organization in their operating regions as shown in Fig. 1.

Fig. 1 IP address allocation process

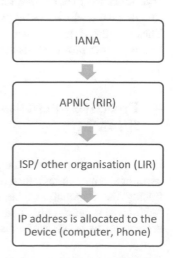

Public and Private IP Address

IP address can be public or private. A public IP address can be accessed on the Internet. It is a unique IP address that is assigned to a computing device. We can find our public IP address at www.whatismyipaddress.com, www.whatismyip.com, www.whatsmyip.org, etc. On the other hand, private IP address is assigned to the computers within a private space so as to avoid the devices from getting directly exposed to Internet [2]. For example, if a person has multiple computers at his company, then he might want to use private IP address for each computer within his company. In this case, his router will get a public IP address and each of the smartphones, computers and tablets connected to the router will get a private IP address from router via DHCP (Dynamic host configuration protocol) protocol. Private IP address is mostly used within internal network or LAN like company network or home whereas public IP address is used to communicate over the Internet [3].

3 Problem Identified: Internet Anonymity

IP address plays an important role in tracking locations. When a message is sent over the Internet, it breaks into small pieces, known as packets. Each packet contains the IP address of source computer and the receiving computer (destination). This IP address can be read by the router [3]. The message sent gets to the destination computer via many routers. So, when a person requests access to a Internet service his geographic location and personal information could be identified via IP address. However, a website cannot get all the information of a person solely from his IP address and a person can remain anonymous unless he/she registers to the website and provides his/her personal information. However, if the perpetrator is smart enough then an investigator could not rely on IP address alone and tracing the criminal can become difficult. Some cyber-criminals use advanced methods of cryptic actions over the Internet which includes hiding the IP address and pretending to be someone else by using proxy servers. They send traffic via another IP address. These methods are known as spoofing, masking and redirecting. Many times fraudsters use proxy IP address which could mislead cyber investigation [4].

4 Proactive Internet Anonymity Minimization Model (PIAMM)

IP address could provide an expeditious exposure of identity. However, if the perpetrator is smart enough then an investigator could not rely on IP address alone and tracing the criminal can become difficult. Some cyber-criminals use advanced methods of cryptic actions over the Internet which includes hiding the IP address and pretending to be someone else. They send traffic via another IP address. These methods are known as spoofing, masking and redirecting. This often misleads the investigation and prevents identification of user. In our proposed idea, the IP address is assigned to a device. As soon as the user connects to the Internet, ISP/service provider assigns an IP address. Everytime the user tries to access any service from the system, this IP address will be used to authenticate user's location [5].

Whenever the user tries to access any protected service from the system (e.g. login into his/her account), along with their username and password, the user will have to provide his/her geolocation to prove the authenticity. In this method, location of the user will be determined by matching his/her entered location with the location obtained by the IP address of their mobile. IP address is matched through reverse lookup and IP Address pooling. If the location does not match with the location obtained by IP address of the device then the user's request will get a warning notification as shown in Fig. 2. This process will be undertaken everytime the user tries to access any service for example creating a new account on social websites, making online transactions, registering on dating websites, generating one-time password (OTP) or issuing e-certificate. When the user connects to the Internet, his/her public IP address will be recorded by the server and his/her geolocation will be estimated based on their IP address information.

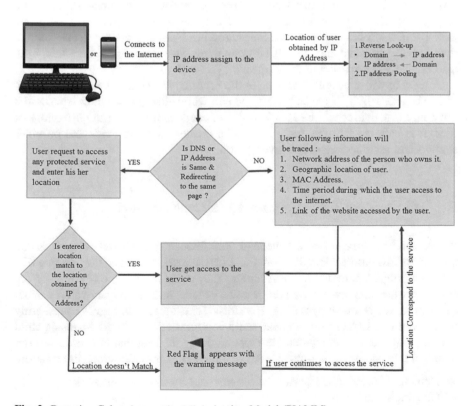

Fig. 2 Proactive Cyber Anonymity Minimization Model (PIAMM)

Fig. 3 Location verification
in PIAMM

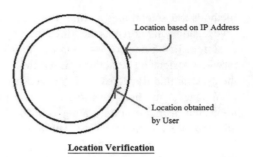

Location Verification

While user requests to access the service, he/she will be asked to enter location. The geolocation described by the user should be contained within the location area calculated by IP address [6]. If both locations coordinate as described in Fig. 3, then the user will be allowed to get further access to the service however, if both the geolocation does not coordinate then a notification will appear as 'the geolocation entered is incorrect as shown in Fig. 3. Do you still want to continue?', if the user still continues then the details of the user (IP address, MAC Address, Geographic Location, Time and website Link) will be recorded. In computer networking, a physical address is the hardware address embedded into network adapters. A physical address is not always a MAC address. A MAC address is only one example of a physical address. The timestamp is a basic component to measure the duration of events and the intervals between them. Successful logon timestamps are also critical when reviewing the suspect's system [3].

A red flag will appear as a warning to the user that someone is keeping an eye on him. The user will be listed as harmful person who is likely to commit crime and a track on his activities will be kept in order to check and monitor the behaviour of user. It is believed that this proposed model creates inclusive guide and provides support and assistance to crime investigators.

5 Comparative Analysis with Live Case Scenario

IP address has come up as a beneficial and frequently deployed weapon in the armoury of law enforcement division. For a case based on Internet, the first step of a cyber investigator is to find the IP Address. They could retrieve the IP address from ISP only when they have a warrant or court order. There have been various cyber cases which are resolved by tracing IP address, for example, 'In May 2010, an army officer who was stationed in Mumbai, India was arrested as he was dispensing child pornography from his computer. He was caught by the Mumbai Police, India after tracing the IP address he was using'. An IP address can provide certain information:

- Network address of the person who owns and operates the network
- Geographic location

- Associated computer name/domain name
- Location service provider identifier

Fraudsters of cyberspace have reared its ugly head, the first of its kind this year, by launching a phishing attack. The alleged mastermind registered himself in cyberspace with his wrong geolocation and fake details. He is using proxy IP address to access the target and steal the sensitive information of customers/clients from a commercial website (username, password, transaction password, debit card number and PIN, mother's maiden name, etc.) and misused it. Here, The investigation carried out with help of Generic Process Model for Network Forensics—2010 [7], A Digital Forensic Investigation Model for Online Social Networking—2010 [8], Cyber Crime Investigation Model—2014 [9] and Domain Specific Cyber Forensics Investigation Process Model—2015 [10] individually. The effectiveness of all aforesaid model is calculated on four parameters, as listed in Table 1. These four parameters are defined as follows:

1. Detectability: The ability to detect the geographical location of the suspected host or cyber-criminal.
2. Time consumption: The total time in the process of detecting the suspected host or cyber-criminal.
3. Accuracy: Detection of exact geographical location of suspected host.
4. Effort: The rate of effort by law enforcement agent or in cybercrime investigation process.

The proposed model will uncover the cybercrimes and cyber-criminals. It will store the user IP address, geolocation, MAC address, etc. If fraudsters commit cybercrime then investigators first investigate the database of the suspected user (shown in Red Flag).

Here, a comparative analysis is done on Proactive Cyber Anonymity Minimization Model (PIAMM) with existing cyber investigation models individually on the basis aforesaid of four parameters, i.e. high detectability, less time consumption,

Table 1 Effectiveness of existing cybercrime investigation models

S. No.	Model	High detectability	Less time consumption	High accuracy	Less effort
1	Generic Process Model for Network Forensics—2010 [7]	✓	✗	✗	✗
2	A Digital Forensic Investigation Model for Online Social Networking—2010 [8]	✓	✗	✗	✗
3	Cyber Crime Investigation Model—2014 [9]	✓	✗	✗	✗
4	Domain specific cyber forensics investigation process model—2015 [10]	✓	✗	✗	✗

Table 2 Comparison of Proactive Cyber Anonymity Minimization Model (PIAMM) with selected cyber investigation models

S. No.	Model	High detectability	Less time consumption	High accuracy	Less effort
1	PIAMM + Generic Process Model for Network Forensics—2010	✓	✓	✓	✓
2	PIAMM + A Digital Forensic Investigation Model for Online Social Networking—2010	✓	✓	✓	✓
3	PIAMM + Cyber Crime Investigation Model—2014	✓	✓	✓	✓
4	PIAMM + Domain specific cyber forensics investigation process model—2015	✓	✓	✓	✓

high accuracy, less effort. An environment is developed to run PIAMM. As the PIAMM is maintaining the suspected data on the basis of geolocation verification, it becomes easier to do cyber investigation with existing cyber investigation models as listed in Table 2.

It has been observed from above Table 2 that the PIAMM is capable of increasing the accuracy and detectability with less time consuming as compared to all other existing cyber investigation models.

6   Future Scope

The proposed model is proactive in nature. Researchers can do the implementation of proposed model is with experimental results. This would increase the effectiveness of cyber investigation models and make the task easier and comfortable for cyber investigator.

7   Conclusion

An approach is illustrated to the cybercrime investigation based on geographical location identification. In this research paper, we have proposed an idea to improve geolocation-based authentication through IP address. This will provide more secured access to Internet services and accurate geolocation could be tracked to improve safety. The cyber-criminals will get aware that their original geolocation could disclosed, which would erode their anonymity and chances of getting caught will increase. This proposed model could act as a deterrent, resulting in drastic decrease in cybercrime rate, thereby improving security and cyber investigation process. It is a time saving, accurate process with less effort as well as it will help in monitoring the behaviour of user and could avoid cyber frauds at a larger scale. It is believed that the research can efficiently assist law enforcement officials. This proposed model creates inclusive guide and provides support and assistance to crime investigators.

References

1. Sitaraman, A., Dos Santos, M.A., Lou, S., Zhang, S., Sthothra Bhasham, S.K.: Integrated IP address management. U.S. Patent 6,427,170, issued 30 July 2002
2. Schmidt, G.T.: INS/GPS technology trends. Massachusetts Institute of Technology, Lexington, MA (2010)
3. Kao, D.-Y., Wang, S.-J.: The IP address and time in cyber-crime investigation. Polic.: Int. J. Police Strateg. Manag. **32**(2), 194–208 (2009)
4. Denning, D.E., MacDoran, P.F.: Location-based authentication: grounding cyberspace for better security. Comput. Fraud Secur. **1996**(2), 12–16 (1996)
5. Padmanabhan, V.N., Subramanian, L.: Determining the geographic location of Internet hosts. In: SIGMETRICS/Performance, pp. 324–325 (2001)
6. Thorvaldsen, Ø.E.: Geographical location of internet hosts using a multi-agent system. Master's thesis, Institutt for telematikk (2006)
7. Pilli, E.S., Joshi, R.C., Niyogi, R.: Network forensic frameworks: survey and research challenges. Digit. Investig. **7**(1), 14–27 (2010)
8. Zainudin, N.M., Merabti, M., Llewellyn-Jones, D.: A digital forensic investigation model for online social networking. In: Proceedings of the 11th Annual Conference on the Convergence of Telecommunications, Networking and Broadcasting, Liverpool, pp. 21–22 (2010)
9. Poonia, A.S.: Analysis of existing models and proposed cyber crime investigation model. Analysis **10**(11), 77–81 (2014)
10. Satti, R.S., Jafari, F.: Reviewing existing forensic models to propose a cyber forensic investigation process model for higher educational institutes. Int. J. Comput. Netw. Inf. Secur. **7**(5), 16 (2015)

Comparison of Different Fuzzy Clustering Algorithms: A Replicated Case Study

Tusharika Singh and Anjana Gosain

Abstract Fuzzy clustering partitions data points of a dataset into clusters in which one data point can belong to more than one cluster. In the literature, a number of fuzzy clustering algorithms have been proposed. This paper reviews various fuzzy clustering algorithms such as Fuzzy C-Means (FCM), Possibilistic C-Means (PCM), Possibilistic Fuzzy C-Means (PFCM), Intuitionistic Fuzzy C-Means (IFCM), Kernel Fuzzy C-Means (KFCM), and Density-Oriented Fuzzy C-Means (DOFCM). We have demonstrated the experimental performance of these algorithms on some standard and synthetic datasets which include—Bensaid, Square (DUNN), D15, and D45 dataset. Then, the results are analyzed and compared to see the effectiveness of these algorithms in presence of noise and outliers.

Keywords Fuzzy clustering · FCM · PCM · PFCM · IFCM KFCM · DOFCM · Outliers

1 Introduction

The concept of separating data elements into groups or clusters in a way that elements in the same cluster are homogeneous and elements held by different clusters are disparate is called as clustering [1]. Clustering can be either hard clustering or fuzzy clustering (soft clustering). When data points of a dataset are divided into different clusters and each data point can belong to only one cluster, then this type of clustering is called hard clustering, whereas fuzzy clustering allows each data point to belong to multiple clusters [2]. Fuzzy clustering uses the concept

T. Singh (✉) · A. Gosain
University School of Information and Communication Technology,
Guru Gobind Singh Indraprastha University, Dwarka 110078, Delhi, India
e-mail: tusharikasingh170@gmail.com

A. Gosain
e-mail: anjana_gosain@hotmail.com

© Springer Nature Singapore Pte Ltd. 2018
S. C. Satapathy et al. (eds.), *Information and Decision Sciences*,
Advances in Intelligent Systems and Computing 701,
https://doi.org/10.1007/978-981-10-7563-6_28

of fuzzy logic where a membership grade is associated with each data point, between a range of 0 and 1 [3].

Mostly used algorithm for fuzzy clustering was proposed by Bezdek [4], called Fuzzy C-Means (FCM) [5]. FCM algorithm works well on most noise-free data but fails to detect data and segment images corrupted by noise and outliers [5]. Algorithms like Possibilistic C-Means (PCM) and Possibilistic Fuzzy C-Means (PFCM) perform better in presence of noise as compared to FCM but PCM fails to find global optimal cluster and leads to the generation of coincident clusters. PFCM lacks to give accurate results for datasets consisting of two clusters which are highly unequal in size with outliers given in it [6].

FCM, PCM, and PFCM are not effective in clustering nonspherical clusters [7]. Hence, a new algorithm, KFCM (Kernel Fuzzy C-Means) was proposed to deal with these types of nonspherical data. When the data sets contain one or more very large outliers, KFCM is more desirable than FCM and PCM [7].

In order to improve accuracy and effectiveness of clustering algorithms, a new algorithm called Intuitionistic Fuzzy C-Means (IFCM) was proposed by Chaira [8]. IFCM uses intuitionistic fuzzy set theory and converges to a more desirable location as compared to the cluster centers obtained using FCM, PCM, PFCM, and KFCM. IFCM works well for hyper-spherical data but is not suitable to cluster nonlinearly separable data. Another algorithm DOFCM [9] was proposed. Density-Oriented Fuzzy C-Means (DOFCM) is a robust technique which uses density of dataset to identify outliers before applying clustering.

In this paper, we have analyzed and compared the performance of these algorithms to see their effectiveness in presence of noise and outliers.

Paper is organized as follows: Sect. 2 briefly discusses various algorithms used in our work. Section 3 presents experimental result of datasets with respect to algorithms used in the form of figures and tables. Section 4 concludes this paper with a short summary.

2 Literature Review

2.1 Fuzzy C-Means (FCM)

FCM fragments a set of data points into a number of clusters "c", which are assumed to be known for a dataset and minimizes the objective function as expressed in Eq. (1) [5]:

$$J_{FCM} = \sum_{k=1}^{c} \sum_{i=1}^{n} u_{ik}^{m} d_{ik}^{2} \qquad (1)$$

With respect to membership function u_{ik} of a data point x_i in cluster k. d_{ik} is the Euclidean distance between data point, x_i and cluster center, v_k. "m" is fuzzifier.

2.2 Possibilistic C-Means (PCM)

PCM proposed by Krishnapuram and Keller [10] overcomes the FCM's problem of noise points which are equidistant from two clusters. Equation (2) shows objective function:

$$J_{PCM} = \sum_{k=1}^{c} \sum_{i=1}^{n} u_{ik}^{m} d_{ik}^{2} + \sum_{k=1}^{c} \sigma_{k} \sum_{i=1}^{n} (1 - u_{ik})^{m} \qquad (2)$$

σ_{k} are suitable positive numbers.

2.3 Possibilistic Fuzzy C-Means (PFCM)

The fuzzy approach of FCM and possibilistic approach of PCM was combined by Pal et al. [11]. Hence, it has two types of memberships, i.e., a fuzzy membership (u_{ik}) and possibilistic membership (t_{ki}). Objective function is given in Eq. (3):

$$J_{PFCM} = \sum_{k=1}^{c} \sum_{i=1}^{n} \left(a u_{ik}^{m} + b t_{ki}^{n} \right) d_{ik}^{2} + \sum_{k=1}^{c} \Upsilon_{k} \sum_{i=1}^{n} (1 - t_{ki})^{n} \qquad (3)$$

2.4 Intuitionistic Fuzzy C-Means (IFCM)

IFCM uses Intuitionistic Fuzzy Set (IFS) theory. IFS theory considers both membership and nonmembership functions [12]. This algorithm merged the hesitation degree (an uncertainty factor) with membership degree as shown in Eq. (4):

$$J_{IFCM} = \sum_{k=1}^{c} \sum_{i=1}^{n} u_{ik}^{*m} d_{ik}^{2} + \sum_{k=1}^{c} \eta_{k}^{*} e^{1 - \eta_{k}^{*}}, \qquad (4)$$

where u_{ik}^{*} denotes the intuitionistic fuzzy membership of the ith data in kth cluster.

2.5 Kernel Fuzzy C-Means (KFCM)

KFCM uses a new kernel-induced metric in the data space, instead of conventional Euclidean norm metric in FCM, in order to deal with high-dimensional data set. By swapping the conventional distance measure with a suitable "kernel" function, without increasing the number of parameters, a nonlinear mapping can be performed to a high-dimensional feature space [7]. Objective function is determined by Eq. (5):

$$J = \sum_{k=1}^{c} \sum_{i=1}^{n} u_{ik}^{m} \|\Phi(x_i) - \Phi(v_k)\|^2, \tag{5}$$

where $\|\Phi(x_i) - \Phi(v_k)\|^2 = K(x_i, x_i) + K(v_k, v_k) - 2K(x_i, v_k)$.

2.6 Density-Oriented Fuzzy C-Means (DOFCM)

DOFCM separates noise into different clusters, i.e., it finds "n" noiseless clusters and one cluster which is not valid consisting of all the outliers present in a dataset, resulting in total of "$n + 1$" clusters [13]. Neighborhood membership is defined as in Eq. (6):

$$M_{neighborhood}^{i}(X) = \frac{\eta_{neighborhood}^{i}}{\eta_{max}} \tag{6}$$

3 Result and Simulation

We have implemented FCM, PCM, PFCM, IFCM, KFCM, and DOFCM in MATLAB Version 7.10 on core i3 processor, 1.70 GHz with 4 GB RAM. For all datasets, we have considered m = 2, $\varepsilon = 0.0001$ and maximum number of iterations as 100.

3.1 Bensaid Dataset

Dataset: Bensaid [14]
Number of clusters: 3
Number of data points in respective clusters: 6, 25, 16
Number of outliers: 2

Figure 1 depicts the clustering result of discussed algorithms. Symbols "x", "o" and ">" represent the three clusters. Centroids of three clusters are plotted by "*" and outliers are plotted using symbol "o" in blue. We examined that FCM is slightly affected with the presence of outliers. PCM fails to give appropriate result due to unequal sized clusters in the dataset and hence only two clusters are obtained. PFCM results into two overlapping clusters with three centroids.

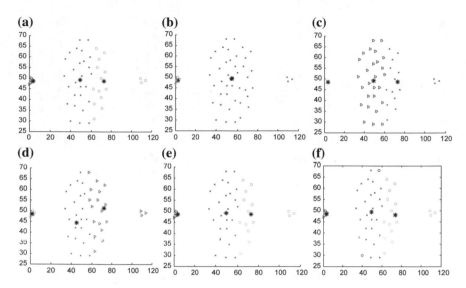

Fig. 1 Clustering result of **a** FCM, **b** PCM, **c** PFCM, **d** IFCM, **e** KFCM, **f** DOFCM on Bensaid dataset

Similar to FCM, presence of noise and outliers affects the performance of IFCM and KFCM. IFCM could not detect original clusters whereas KFCM detects clusters but results in faulty centroid locations. DOFCM detects outliers, revealing original clusters.

3.2 Square Dataset

Dataset: Square [15]
Number of clusters: 2
Number of data points in respective clusters: 53, 81
Number of outliers: 21

Figure 2 depicts the clustering result of discussed algorithms. Symbols "x" and "o" represent the two clusters. Centroids of two clusters are plotted by "*" and outliers are plotted using symbol "o" in blue. We examined that outliers affect the performance of FCM, PCM, PFCM, and IFCM. However, unlike Bensaid dataset, PCM and PFCM are able to detect actual number of clusters. KFCM obtains centroid of the clusters but fails to find outliers. DOFCM results in original clusters as it distinguishes outliers from the cluster's data points.

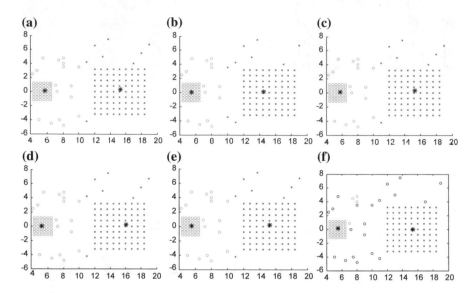

Fig. 2 Clustering result of **a** FCM, **b** PCM, **c** PFCM, **d** IFCM, **e** KFCM, **f** DOFCM on Square dataset

3.3 D15 Dataset

Dataset: D15 [15]
Number of clusters: 2
Number of data points in respective clusters: 6, 5
Number of outliers: 4

Figure 3 depicts the clustering result of discussed algorithms. We examined and observed that PCM cannot find appropriate number of clusters due to its unequal size (11 and 4 data points in two clusters) and hence provide only one cluster. FCM and PFCM produce centroids which are more attracted toward the outliers. IFCM could not detect original clusters and its performance is badly affected by noise. KFCM and DOFCM give centroid locations which preclude attraction of centroids toward outliers but compared to KFCM, DOFCM gives more accurate result and detects outliers, which KFCM lacks.

3.4 D45 Dataset

Dataset: D45 [16]
Number of clusters: 2

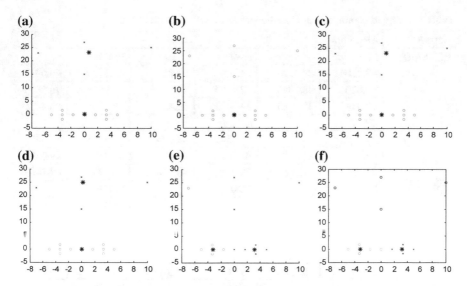

Fig. 3 Clustering result of **a** FCM, **b** PCM, **c** PFCM, **d** IFCM, **e** KFCM, **f** DOFCM on D15 dataset

Number of data points in respective clusters: 18, 18
Number of outliers: 9

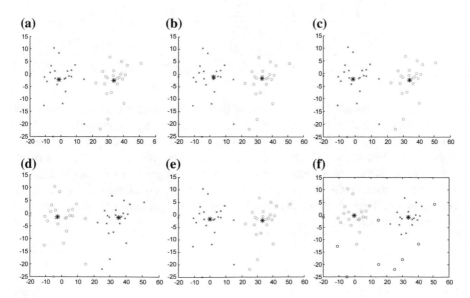

Fig. 4 Clustering result of **a** FCM, **b** PCM, **c** PFCM, **d** IFCM, **e** KFCM, **f** DOFCM on D45 dataset

Table 1 Centroid coordinates produced on standard and synthetic datasets

Dataset	Algorithms					
BENSAID	*FCM*		*PCM*		*PFCM*	
	49.3104	49.0970	54.5093	49.2560	49.3100	49.0910
	3.6201	48.4843	2.4704	48.5296	72.9868	48.5630
	72.9866	48.5565	54.3423	49.4217	3.6202	48.4851
	IFCM		*KFCM*		*DOFCM*	
	1.8103	48.5128	4.3509	48.4742	73.7985	48.0879
	45.2492	44.4176	49.7620	49.1125	49.5533	49.4391
	72.1888	51.2069	73.2312	48.5238	3.3955	48.5460
SQUARE	*FCM*		*PCM*		*PFCM*	
	15.3149	0.3322	14.5058	0.1195	15.3093	0.3286
	5.7652	0.1165	5.4570	0.0456	5.7667	0.1186
	IFCM		*KFCM*		*DOFCM*	
	5.2717	0.0418	15.2015	0.1724	5.4870	0.1719
	16.0559	0.2383	5.4841	0.0784	15.3848	0.0086
D15	*FCM*		*PCM*		*PFCM*	
	0.6757	23.1738	0.0033	0.0222	0.0040	0.1050
	0.0047	0.1227	−0.0016	0.1454	0.6756	23.1720
	IFCM		*KFCM*		*DOFCM*	
	0.0123	−0.0141	−3.2205	0.0033	−3.1672	0.0000
	0.1796	25.0757	3.1256	0.0026	3.1675	0.0000
D45	*FCM*		*PCM*		*PFCM*	
	−1.2111	−2.2085	2.3730	−1.1031	−1.1982	−2.2039
	33.8600	−2.6449	33.4284	−1.6086	33.8562	−2.6481
	IFCM		*KFCM*		*DOFCM*	
	−2.5559	−1.4287	33.3498	−2.2699	−0.4359	−0.3126
	35.3518	−1.6149	0.0353	−1.7038	33.8011	−0.9544

Figure 4 depicts the clustering result of discussed algorithms. D45 is a synthetic dataset consisting of two clusters which are represented by symbols "x" and "o". Centroids are plotted by "*" and outliers are plotted using symbol "o" in blue. FCM, PCM, PFCM, and IFCM performance is highly affected by outliers. KFCM detects centroids which are more attracted toward data points of the actual clusters instead of outliers. DOFCM provides original clusters, excluding the outliers from consideration. Table 1 shows the centroid coordinates produced by each algorithm on various datasets.

4 Conclusion

In this paper, we have analyzed some of the fuzzy clustering algorithms on standard and synthetic datasets considering noise and outliers in the datasets. We observed that FCM does not perform well in presence of noise and outliers whereas performance of PCM and PFCM improves over FCM but not significantly. IFCM could not detect original clusters and KFCM exhibits attraction of centroid toward outlier so the clustering accuracy is affected. DOFCM, compared to other algorithms, gives foremost cluster centroid location as it first detects outliers and then applies clustering technique on outlier-free clusters. In future, we will try to come up with an algorithm which optimizes existing clustering results.

References

1. Hung, C.C., Kulkarni, S., Kuo, B.: A new weighted fuzzy c-means clustering algorithm for remotely sensed image classification. IEEE J. Sel. Top. Signal Process. **5**(3), 543–553 (2011)
2. Grover, N.: A study of various fuzzy clustering algorithms. Int. J. Eng. Res. (IJER) **3**(3), 177–181 (2014)
3. Gosain, A., Dahiya, S.: Performance analysis of various fuzzy clustering algorithms: a review. Proc. Comput. Sci. 100–111 (2016)
4. Bezdek, J.: Pattern Recognition with Fuzzy Objective Function Algorithms. Plenum (1981)
5. Gong, M., et al.: Fuzzy c-means clustering with local information and kernel metric for image segmentation. IEEE Trans. Image Process. **22**(2), 573–584 (2013)
6. Sharma, S., Goel, M., Kaur, P.: Performance comparison of various robust data clustering algorithms. Int. J. Intell. Syst. Appl. **5**(7), 63–71 (2013)
7. Zhang, D., Chen, S.C.: Kernel-based fuzzy and possibilistic c-means clustering. In: Proceedings of the International Conference Artificial Neural Network (2003)
8. Chaira, T.: A novel intuitionistic fuzzy c means clustering algorithm and its application to medical images. Appl. Soft Comput. **11**, 1711–1717 (2011)
9. Kaur, P., Lamba, I.M.S., Gosain, A.: DOFCM: a robust clustering technique based upon density. IACSIT Int. J. Eng. Technol. **3**(3), 297–303 (2011)
10. Krishnapuram, R., Keller, J.M.: The possibilistic c-means algorithm: insights and recommendations. IEEE Trans. Fuzzy Syst. **4**(3), 385–393 (1996)
11. Pal, N.R., et al.: A possibilistic fuzzy c-means clustering algorithm. IEEE Trans. Fuzzy Syst. **13**(4), 517–530 (2005)
12. Kaur, P., et al.: Novel intuitionistic fuzzy C-means clustering for linearly and nonlinearly separable data. WSEAS Trans. Comput. **11**(3), 65–76 (2012)
13. Gosain, A., Singh, T.: DKFCM: kernelized approach to density-oriented clustering. In: Accepted in 4th International Conference on Computational Intelligence in Data Mining (ICCIDM-2017), Odisha, India, Nov 2017
14. Bensaid, A.M., et al.: Validity-guided (re) clustering with applications to image segmentation. IEEE Trans. Fuzzy Syst. **4**(2), 112–123 (1996)
15. Kaur, P., Soni, A.K., Gosain, A.: Robust kernelized approach to clustering by incorporating new distance measure. Eng. Appl. Artif. Intell. **26**(2), 833–847 (2013)
16. Rehm, F., Klawonn, F., Kruse, R.: A novel approach to noise clustering for outlier detection. Soft. Comput. **11**(5), 489–494 (2007)

Passive Object Tracking Using MGEKF Algorithm

M. Kavitha Lakshmi, S. Koteswara Rao, K. Subrahmanyam
and V. Gopi Tilak

Abstract This paper is mainly about the underwater object (Submarine) tracking as it plays a crucial role in maritime environment. Earlier many methods have been developed by using only bearing measurement which requires great computation time. The proposed method in the paper Modified Gain Extended Kalman Filter (MGEKF) focuses on the use of elevation measurement also in addition to bearing for tracking. This reduces the complexity in the detection of the object which is presented in the simulated results.

1 Introduction

In underwater, passive object tracking is generally followed to track an object [1]. The observer object is assumed to be standstill to reduce self-noise for tracking of the object. The proposed system considers both the bearings and elevation measurements to reduce the noise in estimates rather than considering the bearings measurements only used by other Kalman filtering algorithms earlier. Also MGEKF Algorithm makes the system linear. These days, object with Sonars is coming up having the facility to get object elevation measurements. In this paper, research is

M. Kavitha Lakshmi (✉) · S. Koteswara Rao · K. Subrahmanyam · V. Gopi Tilak
Koneru Lakshmaiah Education Foundation, Vaddeswaram, Guntur 522502,
Andhra Pradesh, India
e-mail: kavithalkshm67@gmail.com

S. Koteswara Rao
e-mail: rao.sk9@gmail.com

K. Subrahmanyam
e-mail: smkodukula@kluniversity.in

V. Gopi Tilak
e-mail: gopitilak7@gmail.com

© Springer Nature Singapore Pte Ltd. 2018   277
S. C. Satapathy et al. (eds.), *Information and Decision Sciences*,
Advances in Intelligent Systems and Computing 701,
https://doi.org/10.1007/978-981-10-7563-6_29

278 M. Kavitha Lakshmi et al.

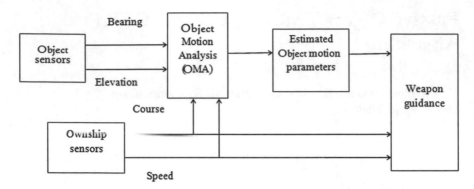

Fig. 1 Block diagram of passive object tracking using bearing and elevation measurements

toward the observer to tracking another object using elevation and bearing measurements. As angle measurements are only available, the process is highly nonlinear and hence modified gain extended Kalman filter (MGEKF), a nonlinear filter is explored for this application, as shown in Fig. 1 [1–4]. The contribution of this paper is the utilization of elevation measurements in passive object tracking and hence observer need not maneuver always. The estimated object range, course bearing, and speed (RCBS) are utilized in weapon guidance algorithm [5] (which is not discussed here).

2 Mathematical Modeling [3, 4]

2.1 Measurements, State Vector and Measurement Equation

State vector $X_s(k)$ is defined as

$$X_S(k) = \left[\dot{x}(k)\dot{y}(k)\dot{z}(k)R_x(k)R_y(k)R_z(k)\right]^T, \tag{1}$$

where, $\dot{x}(n), \dot{y}(n), \dot{z}(n)x(n), y(n), \dot{z}(n), R_x(n), R_y(n)$ and $R_z(n)$ are velocity and range components in x, y, and z. The state equation is given by

$$X_s(k+1) = \emptyset(k+1/k)X_s(k) + b(k+1) + \omega(k), \tag{2}$$

where $\omega(n)$ having zero mean white Gaussian power spectral density is noise. Transient matrix is $\emptyset(k+1/k)$

$$\emptyset(k+1/k) = \begin{bmatrix} 1 & 0 & 0 & 0 & 0 & 0 \\ 0 & 1 & 0 & 0 & 0 & 0 \\ 0 & 0 & 1 & 0 & 0 & 0 \\ t & 0 & 0 & 1 & 0 & 0 \\ 0 & t & 0 & 0 & 1 & 0 \\ 0 & 0 & t & 0 & 0 & 1 \end{bmatrix}, \tag{3}$$

where t is measurement interval.

$b(k + 1)$, deterministic vector, and it is

$$b(k+1) = [0 \quad 0 \quad 0 - [x_0(k+1) + x_0(k)] - [y_0(k+1) + y_0(k)] \\ - [z_0(k+1) + y_0(k)]]^T, \tag{4}$$

where $x_0(k)$, $y_0(k)$, $z_0(k)$ position components of ownship position.

For clarity of concepts, the observer and object encounter in horizontal plane as shown in Fig. 2. The observer is initially at the (0, 0, 0) and standstill. The object is moving at a consistent speed. To reduce the mathematical complexity, all angles are measured w.r.t to reference (True North).

$Z(k)$ is measurement vector and it is given by

$$Z(k) = \begin{bmatrix} B_m(k) \\ \theta_m(k) \end{bmatrix} \tag{5}$$

where $B_m(k)$ and $\theta_m(k)$ are bearing and elevation measurements respectively and are given by

Fig. 2 Object and observer encounter

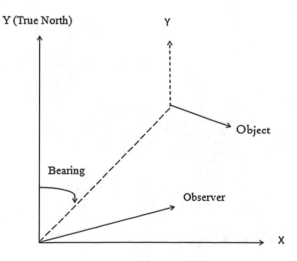

$$B_m(k) = B(k) + \gamma(k) \tag{6}$$

$$\theta_m(k) = \theta(k) + \eta(k) \tag{7}$$

Bearing and elevation angles are *B(k)* and *θ(k)*, respectively, and these are

$$B(k) = tan^{-1}\left(\frac{R_x(k)}{R_y(k)}\right) \tag{8}$$

$$\theta(k) = tan^{-1}\left(\frac{R_{xy}(k)}{R_z(k)}\right) \tag{9}$$

The noises γ(k) and *η(k)* are uncorrelated Gaussian noises. The equation of measurement is given as

$$Z(k) = H(k)X_s(k) + \xi(k), \tag{10}$$

where,

$$H(k) = \begin{bmatrix} 0 & 0 & 0 & \frac{\cos(\hat{B}(k))}{\hat{R}_{xy}} & \frac{-\sin(\hat{B}(k))}{\hat{R}_{xy}} & 0 \\ 0 & 0 & 0 & \frac{\cos(\theta(k))*\sin(\hat{B}(k))}{\hat{R}} & \frac{\cos(\theta(n))*\cos(\hat{B}(k))}{\hat{R}} & \frac{-\sin(\theta(k))}{\hat{R}} \end{bmatrix}, \tag{11}$$

where $\hat{B}(k), \hat{R}, (k), \theta(k)$ are estimated bearing, range and elevation respectively.
Algorithm flow is shown in Fig. 3 [4, 6–12].

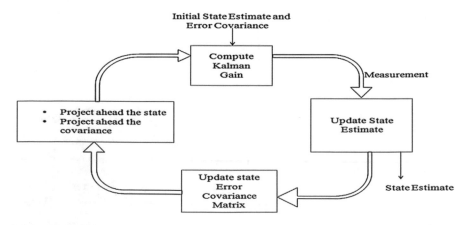

Fig. 3 MGEKF loop

2.2 MGEKF Algorithm: [1–4, 6] MGEKF Algorithm is Given in Table 1

Table 1 MGEKF algorithm

1. To start with estimation, *P(0/0)* and *X(0/0)* which are covariance matrix and its initial state vector respectively are chosen.
2.Kalman gain is given as
$$K((k+1)/k) = P((h((k+1)/k)(k+1)/k) P((k+1/k)h^T ((k+1)/k) + r(k))^{-1} \qquad (12)$$
here, *r(k)* is covariance matrix measurement
3.Updated state vector is
$$X(k+1/k+1) = K(k+1)(z(k+1) - h(k+1)X\left(k + \frac{1}{k}\right) + X(k+1/k)) \qquad (13)$$
4.Updated state covariance matrix is
$$P\left(k + \frac{1}{k} + 1\right) = K(k+1)r(k)G(k+1)^T + (I - K(k+1)g(z(k+1),X(k+1/k)).P(k+1/k)(I - (k+1)g(z(k+1),X(k+1/k))^T \qquad (14)$$
$$G(k) = \begin{bmatrix} 0 & 0 & 0 & \frac{\cos(\hat{B}(k))}{R} & \frac{-\sin(\hat{B}(k))}{R} & 0 \\ 0 & 0 & 0 & \cos(\phi(n))*\sin\left(\dfrac{\hat{B}(k)+B^{xy}(k)}{2}\right)_m & \cos\left(\dfrac{\hat{B}(n)+B^{xy}(k)}{2}\right)_m *\cos(\phi(k))_m & -\sin(\phi(k))_m \\ & & & \hat{R}*\cos\left(\dfrac{B(k)-\hat{B}(k)}{m}\right) & \hat{R}*\cos\left(\dfrac{B(k)-\hat{B}(k)}{2}\right) & \frac{m}{\hat{R}} \end{bmatrix} \qquad (15)$$
5.For next iteration,
$$X(k/k) = X(k+1/k+1)) \qquad (16)$$ $$P(k/k) = P(k+1/k+1)) \qquad (17)$$
6.Predicted state vector is
$$X(k+1/k) = X(k/k)\phi(k+1/k) \qquad (18)$$
7.Predicted state covariance matrix is
$$P(k+1/k) = \phi(k+1/k)\phi^T(k+1/k) + Q(k+1/k) \qquad (19)$$
Here *Q(k)* is plant noise covariance matrix

3 Generalized Simulator

Let initial position of the object be (x_t, y_t, z_t) and the object moves with velocity v_t. After time t seconds, observer position changes and the change in the observer position is given by

$$dx_0 = v_0 * \sin(ocr) * \sin(oph) * t \tag{20}$$

$$dy_0 = v_0 * \cos(ocr) * \sin(oph) * t \tag{21}$$

$$dz_0 = v_0 * \cos(oph) * t \tag{22}$$

where ocr and oph are observer course and pitch, respectively.
   Now, the new observer position becomes

$$x_0 = x_0 + dx_0 \tag{23}$$

$$y_0 = y_0 + dy_0 \tag{24}$$

$$z_0 = z_0 + dz_0 \tag{25}$$

From Fig. 4

$$x_t = R_{xy} * \sin(B) \tag{26}$$

$$y_t = R_{xy} * \cos(B) \tag{27}$$

$$\sin(\theta) = R_{xy}/R \tag{28}$$

Substituting Eq. (28) in (26) and (27)

$$x_t = R * \sin(\theta) * \sin(B) \tag{29}$$

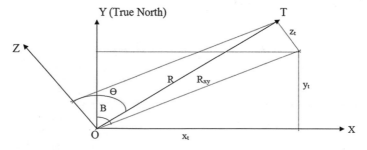

Fig. 4 Object and observer positions

$$y_t = R * \sin(\theta) * \cos(B) \tag{30}$$

$$z_t = R * \cos(\theta) \tag{31}$$

Change in object position after t seconds due to its speed v_t is shown in Fig. 5.

$$dx_t = v_t * \sin(tcr) * \sin(tph) * t \tag{32}$$

$$dy_t = v_t * \cos(tcr) * \sin(tph) * t \tag{33}$$

$$dz_t = v_t * \cos(tph) * t, \tag{34}$$

where tcr and tph are object course and pitch, respectively.
   Now the new object position is (Fig. 6)

$$x_t = x_t + dx_t \tag{35}$$

$$y_t = y_t + dy_t \tag{36}$$

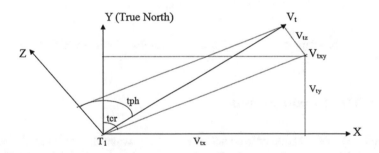

Fig. 5 Object and observer velocities

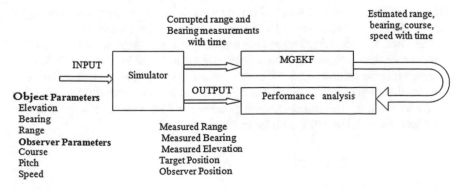

Fig. 6 Block diagram of OMA in simulation mode

$$z_t = z_t + dz_t \tag{37}$$

Target true bearing, range and elevation are

$$true\ bearing = tan^{-1}\left(\frac{x_t - x_0}{y_t - y_0}\right) \tag{38}$$

$$true\ range = \sqrt{(x_t - x_0)^2 + (y_t - y_0)^2 + (z_t - z_0)^2} \tag{39}$$

$$true\ elevation = tan^{-1}\left(\frac{R_{xy}}{z_t - z_0}\right) \tag{40}$$

Since the measurements are affected by noise in real situations, noise is added to these measurements.

Measured bearing = true bearing + sigma b

Measured range = true range + sigma r

Measured elevation = true elevation + sigma e

where sigma b, sigma r, and sigma e are 1σ values of white Gaussian process.

4   Simulation and Results

In simulation mode, estimated and actual values are available and hence the validity of the solution based on certain acceptance criterion is possible. The following acceptance criterion is chosen. The solution is converged when course in error estimate $<=3^0$ and speed in error estimate $<=5$ m/s and range estimate $<=8\%$. The errors in estimated range, speed, and course for scenario 1 are represented in Fig. 7, Fig. 8, and Fig. 9, respectively, for clarity of the concepts (Table 2).

The solution is converged when the course, speed, and range are converged. The convergence time (seconds) for the scenarios is given in Table 3. In simulation, it is observed that the estimated course of the object, speed, and range of the object are converged at 66th sample, 3rd sample and 97th sample, respectively, for scenario1. So, for scenario 1, the total solution is obtained at 97th sample. Similarly for the other scenario, the convergence time is shown in Table 3.

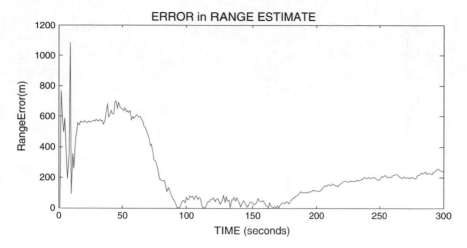

Fig. 7 Error in range estimate

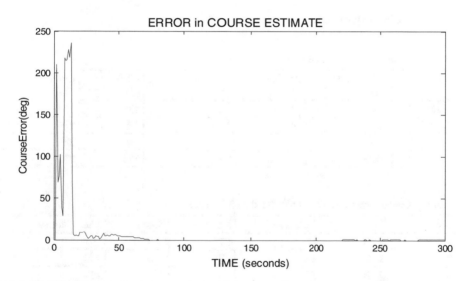

Fig. 8 Error in speed estimate

5 Conclusion

From the results, it is clear that Modified Gain Extended Kalman Filter (MGEKF) reduces the noise in measurements and hence the convergence time is extremely low. So, this method is highly recommended for the tracking of object easily in underwater environment.

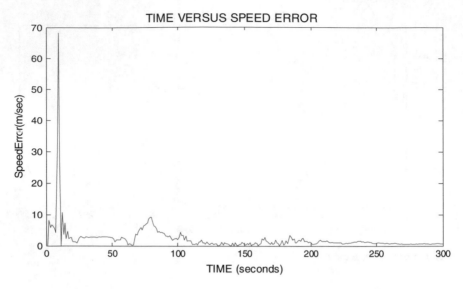

Fig. 9 Error in course estimate

Table 2 Input parameters chosen for the UKF algorithms

Scenario	Initial range (m)	Bearing (deg)	Elevation (deg)	Pitch (deg)	Course (deg)
Scenario1	3000	45	135	135	255
Scenario2	6000	45	135	135	255
Scenario3	19000	350	45	45	320
Scenario4	6000	45	135	135	275

Table 3 Convergence time in samples for the chosen scenarios

Scenario	Course	Elevation	Range	Speed	Total solution
1	66	3	97	3	97
2	90	3	77	290	290
3	250	3	3	169	250
4	65	3	33	209	209

References

1. Song, T.L., Speyer, J.L.: A stochastic analysis of a modified gain extended kalman filter with application to estimation with bearings only measurements. IEEE Trans. Autom. Control **30** (10), 940–949 (1985)
2. Galkowski, P.J., Islam, M.A.: An alternative derivation of the modified gain function of Song and Speyer. IEEE Trans. Autom. Control **36**(11), 1323–1326 Nov' 1991

3. Koteswara Rao, S.: Modified gain extended Kalman filter with application to angles only underwater passive target tracking. In: Proceedings of ICSP, pp. 1439–1442, Mar' 1998
4. Wan, E.A., Van Der Merwe, R.: The unscented Kalman filter for non-linear estimation. In: Proceedings of Symposiums 2000 on Adaptive System for Signal Processing, Communication and Control. Canada Oct 2000
5. Omkar Lakshmi Jagan, B., Koteswara Rao, S., Jawahar, A., Karishma, S.K.B.: Application of Bar-Shalom and Fortmann's input estimation for underwater target tracking. Indian J. Sci. Technol. (2016)
6. Ni, J., Wang, C., Fan, X., Yang, S.X.: Abioinspired neural model based extended Kalman filter for robot slam. (Research Article) (Report). Mathematical Problems in Engineering, Annual 2014 Issue
7. Koteswara Rao, S., RajaRajeswari, K., Lingamurthy, K.S.: Unscented Kalman filter with application to bearings-only target tracking. IETE J. Res. 55(2), 63–67 (2009)
8. Koteswara Rao, S., Sunanda Babu, V.: Unscented Kalman filter with application to bearings-only passive maneuvering target tracking. In: IEEE International Conference on Signal processing, Communications and Networking, pp. 219–224 Jan' 2008
9. Lalchbadriasl, kutluyildoganicay. Three-dimensional target motion analysis using azimuth/Elevation Angles. IEEE Trans. Aerosp. Electr. Syst. 50(4), 3178–3194 Oct' 2014
10. Omkar Lakshmi Jagan, B., Koteswara Rao, S., Lakshmi Prasanna, K., Jawahar, A., Karishma, S.K.B.: Novel estimation algorithm for bearings-only target tracking. Int. J. Eng. Technol. 8 (1), 238–246 Mar' 2016
11. Omkar Lakshmi Jagan, B., Koteswara Rao, S., Jawahar, A., Karishma, S.K.B.: Unscented Kalman filter with application to bearing-only passive target tracking. Indian J. Sci. Technol. 9(19), 1–10 (2016)
12. Ravi Kumar, D.V., Koteswara Rao, S.: Underwater bearings-only passive target tracking using estimate fusion technique. Adv. Mil. Technol. 10(2), 31–44, Dec' 2015

Instantaneous Time Smoothing in GPS Receivers Using Kalman Filter

R. Revathi, K. S. Ramesh, S. Koteswara Rao and K. Uday Kiran

Abstract Positioning precision aids in calibration of baseline estimation using Global Positioning Systems (GPS) receiver. Analysis of instantaneous epoch measurements improves coordinate positioning than estimation of position using 24-h batch processing of GPS data. In the present work, smoothing of the epoch measurements is done by using Kalman filter to achieve accurate positioning of GPS receivers. Using the above algorithm, the variations in the epoch time are minimized. These variations are caused by receiver loss of lock and cycle slips. These studies will lead to applications in other research areas such as earthquake geodesy and GPS seismology.

Keywords Precise positioning · Epoch measurements · Kalman filter

1 Introduction

For the past few decades, GPS has become a very important tool for geodetic research. Baseline estimation of GPS receivers is useful to infer global plate motions, boundary deformation, and permanent deformation caused before and after the occurrence of earthquakes. Precise positioning estimation by multi-epoch measurements and batch processing methods results in positioning errors. The errors can be reduced by processing epoch-by-epoch for position estimation.

R. Revathi (✉) · K. S. Ramesh · S. Koteswara Rao · K. Uday Kiran
Department of ECE, KL University, Vaddeswaram, Guntur (DT) 522502
Andhra Pradesh, India
e-mail: revathimouni@gmail.com

K. S. Ramesh
e-mail: dr.ramesh@klunivesity.in

S. Koteswara Rao
e-mail: skrao@gmail.com

K. Uday Kiran
e-mail: meemithrudu@gmail.com

© Springer Nature Singapore Pte Ltd. 2018
S. C. Satapathy et al. (eds.), *Information and Decision Sciences*,
Advances in Intelligent Systems and Computing 701,
https://doi.org/10.1007/978-981-10-7563-6_30

289

Generally, in 24-h data used for precise positioning, the parameters are estimated for every 30 s or by least squares or similar multi-epoch measurements [1].

GPS satellites transmit navigation and observation file in Receiver Independent Exchange Format (RINEX). The navigation file consists of different parameters such as time of epoch (t_{oe}), mean anomaly, corrections for the elliptical satellite orbits, etc. The observation file consists of pseudoranges, carrier phases, and Doppler data of L_1 and L_2, respectively. In the observation file, the information is recorded per second. This time information is called as t_{oee}. Instantaneous time is measured as

$$t_k = t_{sec} - t_{oe} \tag{1}$$

This instantaneous time is calculated for the GPS receivers on the baseline and also for the receiver position estimation of the same satellites. If the value of t_k is less than 302400 s of GPS week, the value of t_k is calculated as

$$t_k = t_k - 604800 \tag{2}$$

and if it is greater than 302400 s of GPS week, the value of t_k is calculated as

$$t_k = t_k + 302400 \tag{3}$$

This t_k is considered to calculate the position information of the receiver [2]. The variations in "t_k" cause errors in calculating the receiver position on the baseline. This instantaneous time is to be smoothed before the calculation of receiver position. In this work, an attempt is made to correct epoch time using Kalman filter.

Kinematic and rapid static processing of GPS data needs data for several minutes for precise site positioning. These were used to calibrate GPS arrays installed over the fault lines [3, 4]. The method of instantaneous approach reduces the necessity of GPS data rearrangement since receiver lock and cycle clips are irrelevant at single epoch level [5]. This method also reduces the complexity of integer cycle phase ambiguities, basic requirement for precise positioning services [6, 7]. It is easier to recognize erroneous data than to correct it. Elimination of such data leads to more robust results that are attainable in batch or filter-oriented processing techniques.

Smoothing of epoch time data recorded from GPS receivers over a baseline gives a good estimation of the precise positioning for geodetic applications. Kalman filter is used to smooth the raw epoch data obtained from GPS receivers. It uses Bayesian inference and estimates the joint probability distribution of variables for every epoch time. It is a real-time recursive algorithm using only present input measurements, previous state, and its uncertainty matrices. The smoothing of epoch data using Kalman filter has provided better results for precise position estimation.

2 Methodology

Kalman filter, a recursive filter gives a best estimate of the present state by considering the previous estimate of the variable. It thus helps in reduction of storage of data. This filter can be applied to all stationary and nonstationary problems. This filter has a set of mathematical equations which optimize the state variable with minimum error covariance. This method uses the observed variables (t_{k1}, t_{k2}, t_{k3},..., t_{kn}) to estimate the state (x_i) with a minimum value. Kalman filter has two steps.

1. Prediction
2. The correction

The procedure of Kalman filter is given in Fig. 1.

2.1 Nomenclature

$\hat{x}(k\|k)$	State vector at estimate for time t(k), including a measurement at time t(k)
$\hat{x}(k+1\|k)$	State vector at estimate for time t(k + 1) based upon the state vector estimate at time t(k)
$\hat{x}(k+1\|k+1)$	Corrected state vector estimate for time t(k + 1), including a measurement at time t(k + 1)
ts	Time sample, t(k + 1) − t(k)
F(k + 1|k)	Transition matrix
p(k|k)	Covariance matrix of $\hat{x}(k\|k)$
p(k + 1|k)	Covariance matrix of $\hat{x}(k+1\|k)$
p(k + 1|k + 1)	Covariance matrix of $\hat{x}(k+1\|k+1)$
H(k + 1)	Measurement matrix at time t(t + 1)
w(k + 1)	Input covariance matrix at time t(k + 1)

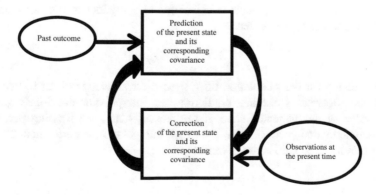

Fig. 1 Procedure for Kalman filter

K$(k+1)$ Kalman gain at time t$(k + 1)$
$b_m(k+1)$ Epoch measurement at time t$(k + 1)$
$b(k+1)$ True epoch value at time t$(k + 1)$

The algorithm is implemented in MATLAB as follows:
It is implemented in five steps.

Step 1: Calculate the pre-estimation of state.

First, we calculate the post-estimation of state $\hat{x}_{t-1}^{\,|}$. Now, pre-estimation state is calculated by multiplying the post-estimation state with the transition matrix. The equation to find pre-estimation state is given in Eq. 12.

Step 2: Calculate the pre-error covariance of the epoch time.

Here, the probability of the epoch time given at the instant $k - 1$ is multiplied with the transposed transition matrix and the standard deviation of the probability distribution of pre-estimation error is added to it. The equation to find pre-error covariance is given in Eq. 13.

Step 3: Calculation of Kalman gain.

The pre-estimation error is multiplied with the transposed measurement matrix. The product is multiplied with measurement matrix and the standard deviation of the normal distribution function. The equation to find the Kalman gain is given in Eq. 14.

Step 4: Calculation of post-estimated state.

First, we calculate the product of pre-error estimation state with the measured value. It is then subtracted from the present epoch value t_k. This is multiplied with the Kalman gain and is added to pre-error estimation state to find out post-estimation state of the epoch time. The equation to find out post-error estimation is given in Eq. 15.

Step 5: Calculation of the post-error covariance.

Finally, the post-error covariance is calculated as a product of the pre-estimation covariance and the S. S is given as

$$S = (1 - k_t H)$$

The equation for the calculation of post-error covariance is given in Eq. 15.
The three operations filtering, prediction, and interpolation are done depending on the value of the estimated state. If the present state $i = t$ filtering performed, $i > t$ predicted value is considered, $i < t$ then interpolation is performed. The state and measurement equations are given as

$$x_{t+1} = Fx_t + w_t \tag{4}$$

$$t_k = Hx_t + v_t, \tag{5}$$

where w_t and v_t are white noise vectors (P(w): N(0,Q) and P(v): N(0,R)), "P" is the probability distribution function and N is the normal distribution function with standard deviation parameters Q and R, respectively. The transition matrix F takes values from time t to t + 1 and H is called the measurement matrix. Let us consider "x_i" the real state at time t to t + 1. So, the pre-estimation error and pre-error covariance are given as

$$e_t^- = x_t - \hat{x}_t^- \tag{6}$$

$$p_t^- = E(e_t^- e_t'^-) \tag{7}$$

The post-estimation error and post-error covariance are given as

$$e_t^+ = x_t - \hat{x}_t^+ \tag{8}$$

$$p_t^+ = E(e_t^+ e_t'^+) \tag{9}$$

The Kalman filter mainly estimates the post-error state '$\hat{x}_t^+$ (post-estimation of state)' using linear integration of $\hat{x}_t^-$ (pre-estimation of state). It calculates the measured error $(t_k - H\hat{x}_t^-)$ as

$$\hat{x}_t^+ = \hat{x}_t^- + k_t(t_k - H\hat{x}_t^-), \tag{10}$$

where k_t is the Kalman coefficient. It is defined based on the minimum of post-error covariance given as

$$k_t = p_k^- H^T (Hp_k^- H^T + R)^{-1} \tag{11}$$

These filter measurements are good when the covariance of error measurements is close to zero. These filter equations are catalogued into time update and measurement update equations. The time update equations calibrate the state and covariance matrices depends on pre measurements. The measurement matrix equations are used for feedback purpose and the time update is used to study system effects in the leading toward optimum state measurements. These are given as

$$\hat{x}_t^- = F\hat{x}_{t-1}^+ \tag{12}$$

$$p_t^- = F_{p_{t-1}} F^T + Q \tag{13}$$

$$k_t = p_t^- H^T (H p_t^- H^T + R)^{-1} \tag{14}$$

$$\hat{x}_t^+ = \hat{x}_t^- + k_t (t_k - H\hat{x}_t^-) \tag{15}$$

$$p_t^+ = (1 - k_t H) p_t^- \tag{16}$$

Hence, the process of prediction is done initially. The predicted value is corrected depending on the observations and thus the prediction process replicated [8].

The available measured epoch data contains noise. The state vector consists of only one component, i.e., smoothed epoch data. State vector and its covariance vector are initiated with average epoch value of 25, and its variance as 0.001. After the mean is available, Kalman gain is calculated. The state and its covariance vectors are updated. The updated state and its variance indicate the standard deviation of the error in estimated or smoothed epoch. The state and its covariance vector are extrapolated for the next epoch time. Again, once the measurement is available, Kalman gain is calculated. The process is repeated for every epoch arrival.

3   Results and Discussion

In the present work, the epoch time data of a GPS receiver over a baseline is considered for precise positioning of receiver on the baseline. The epoch data for 125 data points is represented in Fig. 2.

The epoch data is corrected using Kalman filter. The smoothed epoch data given in Figs. 3 and 4 represents the Kalman gain and the smoothing of the epoch data using Kalman filter. From Fig. 2, it is clearly understood that the Kalman filter has smoothed the variations in the epoch measurements. It has given a good estimate of the input data. Thus, the analysis of position estimation using the smoothed data

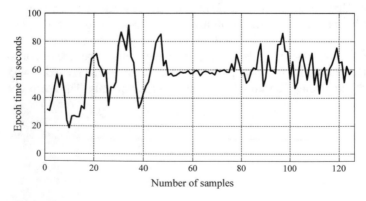

Fig. 2   Epoch data for 125 data points

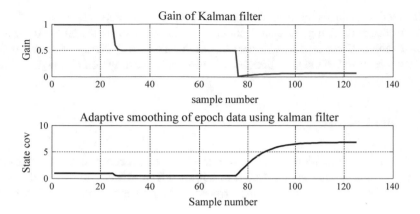

Fig. 3 Gain of Kalman filter and probability distribution function of state covariance vector

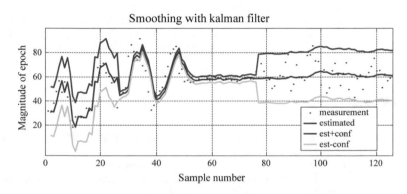

Fig. 4 Epoch smoothing using Kalman filter

will result in precise GPS receiver positions on the baseline. If the positions of the GPS receivers are calculated similarly by calibrating the epoch measurements over the baseline, then it is possible to study the baseline deformations caused by earthquakes.

The variations observed in the initial part of the data for the 50 epochs are caused due to the removal of erroneous data recorded remain unchanged and the rest of the data have been smoothened. This is mainly because of cycle slips or loss of lock of the receiver. But the variation seen from 50 to 125 epoch measurements has been smoothed to give a good estimate of the epoch data thus reducing the variations in the calibration of the position estimation of the GPS receiver.

The analysis clearly shows that the epoch data has been smoothed. Thus, it has reduced the errors in the calibration of the GPS receiver position on the baseline. The above analysis used helps in studying the plate tectonic motions, crustal deformations caused by coseismic, preseismic motions, and earthquakes.

This method reduces the risk of cycle clips, lock of lock, and also the correction of the erroneous data. It gives better estimation of the position of the GPS receivers on the baseline. These studies will lead for the better understanding of crustal deformations occurred in the fault line and will also help in estimation of new fault lines.

4 Conclusions

Implementation of Kalman filter on the epoch measurements has reduced the error caused by inaccurate time measurement in the GPS receivers. This also reduces the stringent problem of repairing the erroneous data which is not possible in the traditional processing techniques. The processed method reduces the inaccuracies in the baseline estimation of the GPS receivers. This new technique provides better opportunities in the research areas of geodesy, GPS seismology, etc.

Acknowledgements This present research is supported by the Department of Science and Technology (DST), Government of India via sponsored research projects SR/AS-04/WOS-A/2011 and SR/S4/AS-9/2012. The authors would like to thank the Atmospheric Sciences panel of DST and President, Koneru Lakshmaih University for their continuous support and assistance.

References

1. Bock, Y., Williams, S.: Integrated satellite interferometry in southern California. Eos Trans. A GU **78**(293), 2s–a00 (1997)
2. Mishra, P.: Global Positioning Systems, 2nd en. Ganga-Jamuna Press
3. Genrich, J., Bock, Y.: Rapid resolution of crustal motion at short ranges with the global positioning system. J. Geophys. Res. **97**, 3261–3269 (1992)
4. Genrich, J.F., Bock, Y., Mason, R.: Crustal deformation across the imperial fault: results from kinematic GPS surveys and trilateration of a densely-spaced, small aperture network. J. Geophys. Res. **102**, 4985–5004 (1997)
5. Blewitt, G.: An automatic editing algorithm for GPS data. Geophgs. Res. Left. **17**, 199–202 (1990)
6. Blewitt, G.: Carrier phase ambiguity resolution for the global positioning system applied to geodetic baselines up to 2000 km. J. Geophys. Res. **9**, 10,187–10,203 (1989)
7. Dong, D., Bock, Y.: Global positioning system network analysis with phase ambiguity resolution applied to crustal deformation studies in California. J. Geophys. Res. **9J**, 3949–3966 (1989)
8. Akhoondzadeh, M.: Anomalous TEC variations associated with the powerful Tohoku earthquake of 11 March 2011. Nat. Hazards Earth Syst. Sci. **12**, 1453–1462 (2012), © Author(s) 2012. CC Attribution 3.0 License

Application of Least Squares Algorithm for Precise GPS Receiver Positioning

R. Revathi, K. S. Ramesh, S. Koteswara Rao and K. Uday Kiran

Abstract Precise positioning measurements of Global Positioning Systems (GPS) have revolutionized scientific research in the area of geodynamics. Nowadays, it aids in the study of the crustal deformations caused by earthquakes, the unavoidable natural disasters. Calibration of the GPS data over fault lines, epoch-by-epoch results in precise positioning of the receiver. Least squares method is implemented in the calculation of the receiver coordinates obtained from the pseudoranges for every epoch. Application of this method has given a good approximation of the receiver coordinates at a given location.

Keywords Signal processing algorithms · GPS receiver coordinates Pseudoranges

1 Introduction

Inaccurate GPS receiver position coordinates are caused by errors in satellite ephemerides, ionospheric modeling, calibration of local equipment, etc. These errors are minimized by calibration of the GPS receiver epoch-by-epoch [1]. Baseline estimation using GPS receivers is useful for the analysis of the crustal deformation caused by earthquakes. Instantaneous processing of GPS data epoch-by-epoch reduces the error observed in multi-epoch processing.

R. Revathi (✉) · K. S. Ramesh · S. Koteswara Rao · K. U. Kiran
Department of ECE, KL University, Vaddeswaram Guntur (DT) 522502,
Andhra Pradesh, India
e-mail: revathimouni@gmail.com

K. S. Ramesh
e-mail: dr.ramesh@kluniversity.in

S. Koteswara Rao
e-mail: rao.sk9@gmail.com

K. U. Kiran
e-mail: meemithrudu@gmail.com

© Springer Nature Singapore Pte Ltd. 2018
S. C. Satapathy et al. (eds.), *Information and Decision Sciences*,
Advances in Intelligent Systems and Computing 701,
https://doi.org/10.1007/978-981-10-7563-6_31

In the conventional processing techniques (batch processing and 24 h data), the receiver has to re-initialize carrier phase ambiguities occurring due to loss of lock. The methods provide inaccurate position coordinates over the baselines [2–4]. Epoch-by-epoch processing resolves the complexity arising due to loss of lock, initialization of the carrier phase ambiguities as it considers the first available time information [5–7].

GPS receivers record the data at regular intervals of time, for example, every 30 s. The receiver clock time at which the data is recorded is the measurement time given by T, and also, the observation file is recorded with the receiver measurement time. Thus, the actual observation time of the satellite is

$$\rho^s = (T_1 - T_1^s)c \tag{1}$$

where "T_1^S" is the satellite clock time when the signal is transmitted, and "c" is the speed of light in vacuum. The receiver and the satellite clocks have bias. These clock biases have to be added with the measured and satellite times. The corrected measured and satellite times are given as

$$T = t + \tau \tag{2}$$

$$T_1^s = t^s + \tau^s \tag{3}$$

where "τ" and "τ^s" are the receiver and satellite clock biases, respectively. Thus, pseudorange "$\rho^s(t)$" is given as

$$\begin{aligned} \rho_1^s(t) &= ((t + \tau) - (t^s + \tau^s))c \\ \rho_1^s(t) &= (t - t^s)c + c\tau - c\tau^s \\ \rho_1^s(t) &= \rho^s(t, t^s) + c\tau - c\tau^s \end{aligned} \tag{4}$$

where $\rho^s(t, t^s)$ is the range from the receiver to the satellite. In the above equation, we assume that the speed of light in the atmosphere is "c" ignoring the theory of relativity. If we know the satellite position in the orbit (x^s, y^s, z^s) and the satellite clock bias τ^s, then the pseudorange is given as

$$\rho^s(t, t^s) = \sqrt{V_1 + V_2 + V_3} \tag{5}$$

where $V_1 = (x^s(t^s) - x(t))^2$, $V_2 = (y^s(t^s) - y(t))^2$

$V_3 = (z^s(t^s) - z(t))^2$

The satellite coordinates can be calculated from the navigation message information transmitted from the GPS satellite. Thus, we have four unknown values, i.e., the receiver coordinates and the receiver clock bias. The maximum satellite range from the time the signal has been transmitted is 60 m, i.e., 0.07 s lagging when it reaches the receiver. If we know the receiver clock bias, we can compute the transmit time.

The receiver clock bias is usually few milliseconds, i.e., relatively 50 m when S/A is switched on. Thus, it can be cautiously ignored in error computation of pseudoranges of different satellites. This error can be corrected by using more precise carrier phase observables [8]. Let us assume four different satellites with pseudoranges $\rho_1, \rho_2, \rho_3, \rho_4$ given as

$$\rho_1 = ((x_{s_1} - x_r)^2 + (y_{s_1} - y_r)^2 + (z_{s_1} - z_r)^{1/2} + c\tau - c\tau_{s_1} \tag{6}$$

$$\rho_2 = ((x_{s_2} - x_r)^2 + (y_{s_2} - y_r)^2 + (z_{s_2} - z_r)^{1/2} + c\tau - c\tau_{s_2} \tag{7}$$

$$\rho_3 = ((x_{s_3} - x_r)^2 + (y_{s_3} - y_r)^2 + (z_{s_3} - z_r)^{1/2} + c\tau - c\tau_{s_3} \tag{8}$$

$$\rho_4 = ((x_{s_4} - x_r)^2 + (y_{s_4} - y_r)^2 + (z_{s_4} - z_r)^{1/2} + c\tau - c\tau_{s_4} \tag{9}$$

In the present work, least squares method is implemented to find out the precise receiver coordinates epoch-by-epoch. This method is easier to implement on a large amount of data and get a good prediction of the input variables. Least square method is simpler to analyze mathematically, and its solutions can easily be interpreted. It works in a limited number of points and reduces the processing time, prediction time, and computer memory. The implementation of least squares method at every epoch results in the precise estimation of the receiver coordinates over fault lines aiding in the study of geodynamics caused by earthquakes.

2 Methodology

The accurate estimation of the receiver coordinates is solved by linearizing the pseudoranges. Then, the method of least squares analysis is implemented. Let us assume the physical observation ρ_o is the sum of modeled observations ρ_m, plus an error term given as

$$\rho_o = \rho_m + \text{noise} \tag{10}$$

$$\rho_o = \rho(x_s, y_s, z_s, \tau_s) + v \tag{11}$$

Now, by considering the initial values $\left(x_{s_o}, y_{s_o}, z_{s_o}, \tau_{s_o}\right)$, we expand the above equation using Taylor's theorem and ignore the second and higher order terms given as

$$\rho(x, y, z, \tau) \cong \rho(x_{s_o}, y_{s_o}, z_{s_o}, \tau_{s_o}) + (x - x_{s_o})\frac{\partial\rho}{\partial x}$$

$$+ (y - y_{s_o})\frac{\partial\rho}{\partial y} + (z - z_{s_o})\frac{\partial\rho}{\partial z} + (\tau - \tau_{s_o})\frac{\partial\rho}{\partial\tau} \qquad (12)$$

$$= \rho_{computed} + \frac{\partial\rho}{\partial x}\Delta x + \frac{\partial\rho}{\partial y}\Delta y + \frac{\partial\rho}{\partial z}\Delta z + \frac{\partial\rho}{\partial\tau}\Delta\tau$$

The partial derivatives in the above equation are computed using the initial values. The residual pseudorange is given as the difference between the observed and provisional parameter values as

$$\Delta\rho = \rho_o - \rho_{computed} \qquad (13)$$

$$\Delta\rho = \frac{\partial\rho_c}{\partial x}\Delta x + \frac{\partial\rho_c}{\partial y}\Delta y + \frac{\partial\rho_c}{\partial z}\Delta z + \frac{\partial\rho_c}{\partial\tau}\Delta\tau + v \qquad (14)$$

This can be written in the form

$$\Delta\rho = \begin{pmatrix} \frac{\partial\rho_c}{\partial x} & \frac{\partial\rho_c}{\partial y} & \frac{\partial\rho_c}{\partial z} & \frac{\partial\rho_c}{\partial\tau} \end{pmatrix} \begin{pmatrix} \Delta x \\ \Delta y \\ \Delta z \\ \Delta\tau \end{pmatrix} + v \qquad (15)$$

Such approximation is drawn for the satellites in view. The above equation can also be written as

$$b_1 = A_1 x + v_1 \qquad (16)$$

which shows a linear relation between the residual observations and the unknown correction to the parameters x. The design matrix in the above equation for four satellites is given as

$$A = \begin{pmatrix} \frac{x_0 - x_{s_1}}{\rho} & \frac{y_0 - y_{s_1}}{\rho} & \frac{z_0 - z_{s_1}}{\rho} & c \\ \frac{x_0 - x_{s_2}}{\rho} & \frac{y_0 - y_{s_2}}{\rho} & \frac{z_0 - z_{s_2}}{\rho} & c \\ \frac{x_0 - x_{s_3}}{\rho} & \frac{y_0 - y_{s_3}}{\rho} & \frac{z_0 - z_{s_3}}{\rho} & c \\ \frac{x_0 - x_{s_4}}{\rho} & \frac{y_0 - y_{s_4}}{\rho} & \frac{z_0 - z_{s_4}^4}{\rho} & c \end{pmatrix}$$

The least square solution is given below.

The solution for the linearised observation is considered as $\hat{x}$. Thus by using the linearised equation given above, we can write

$$\hat{v} = b_1 - A\hat{x} \tag{17}$$

where the estimated residual is given as difference of the original observations and the estimated model observations. Thus, the solution of the least squares can be written as

$$g(x) \equiv \sum_{i=1}^{4} v_i^2 = v^T v = (b_1 - Ax)^T (b_1 - Ax) \tag{18}$$

From the above equation, we are minimizing the estimated residuals. The following equations describe the application of the above method:

$$\delta g(\hat{x}) = 0 \tag{19}$$

$$\delta \left\{ (b_1 - A_1\hat{x})^T (b_1 - A_1\hat{x}) \right\} = 0 \tag{20}$$

$$\delta(b_1 - A_1\hat{x})^T (b_1 - A_1\hat{x}) + (b_1 - A_1\hat{x})^T \delta(b_1 - A_1\hat{x}) = 0 \tag{21}$$

$$(-A_1(\delta x)^T (b_1 - A_1\hat{x}) + (b_1 - A_1\hat{x})^T (-A_1\delta\hat{x}) = 0 \tag{22}$$

$$(-2A_1\delta x)^T (b_1 - A_1\hat{x}) = 0 \tag{23}$$

$$(\delta x^T A_1^T)(b_1 - A_1\hat{x}) = 0 \tag{24}$$

$$\delta x^T (A_1^T b_1 - A_1^T A_1\hat{x}) = 0 \tag{25}$$

$$A_1^T A_1\hat{x} = A_1^T b_1 \tag{26}$$

Equation (26) represents normal equations. The solution is given by

$$\hat{x} = \text{inverse}((A_1^T A_1)) A_1^T b \tag{27}$$

In the present case, data from the ohiosat is considered for analysis. For a single epoch, the algorithm is implemented consisting of five satellites. The pseudoranges are taken from the RINEX observation file collected on March 30th,1996.

3 Results and Discussion

In the present work, we have implemented the least square method for precise positioning of the GPS receiver epoch-by-epoch. Calibration of receiver position epoch-by-epoch will lead to a better estimation of the receiver coordinates. The receiver position for a particular epoch is represented in Fig. 1.

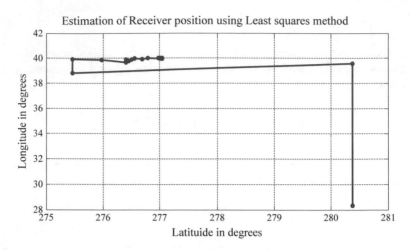

Fig. 1 Estimation of receiver position using least squares method

It is shown that the instantaneous positioning provides good results than that of batch processed data. The studies carried out over the baselines situated in the earthquake-prone area will aid in the analysis of coseismic, preseismic, and crustal motions in that region. The single-epoch analysis has provided a trade-off between the computation times, precision when compared with the conventional techniques like static and kinematic processes.

Epoch time smoothing has to be done before the calculation of the accurate receiver coordinates as this analysis overcomes the problems associated with carrier phase ambiguities.

Calibration of GPS receivers by the implementation of conventional techniques initializes the integer-cycle phase ambiguities at the first instant of its occurrence. Once it is fixed, it will not be changed further even though there are other incidents of their occurrence. In addition to the provision of providing accurate positioning, the method of instantaneous processing of GPS data also resolves integer-cycle phase ambiguities before proceeding further. Thus, in this method, the receiver is reinitialized for every occurrence of loss of lock. This re-initialization introduces a delay of 30–40 s. In this process, once the receiver comes out of this problem and the hardware reacquires the GPS satellites, the software will fully reclaim on the first available epoch.

In the present work least squares method is applied on the all the pseudoranges obtained for the number of satellite seen at every epoch. This approach gives good approximation of the receiver position using simple mathematical equations.

4 Conclusions

The least square method presented for the calibration of instantaneous receiver coordinates has given more accurate results compared to that of the batch processing methods. This method avoids the risk of cleaning the data and re-initialization of carrier phase ambiguities at time of loss of lock. It also provides better parameter estimates at high frequencies and reduces the effect of flicker noise seen in the coordinate estimation caused due to atmospheric delays for longer baselines.

Acknowledgements This investigation is financially funded by DST, Government of India through sponsored projects SR/AS-04/WOS-A/2011 and SR/S4/AS-9/2012. The authors would also thank President, K L University for their assistance.

References

1. Genrich, J., Bock, Y.: Rapid resolution of crustal motion at short ranges with the global positioning system. J. Geophys. Res. **97**, 3261–3269 (1992)
2. Genrich, J.F., Bock, Y., Mason, R.: Crustal deformation across the imperial fault: results from kinematic GPS surveys and trilateration of a densely-spaced, small aperture network. J. Geophys. Res. **102**, 4985–5004 (1997)
3. Blewitt, G.: An automatic editing algorithm for GPS data. Geophys. Res. Lett. **17**, 199–202 (1990)
4. Blewitt, G.: Carrier phase ambiguity resolution for the global positioning system applied to geodetic baselines up to 2000 km. J. Geophys. Res. **94**, 10187–10203 (1989)
5. Dong, D., Bock, Y.: Global positioning system network analysis with phase ambiguity resolution applied to crustal deformation studies in California. J. Geophys. Res. **9J**, 3949–3966 (1989)
6. Bock, Y., Williams, S.: Integrated satellite interferometry in southern California. Eos Trans. AGU **78**(293), 2s–a00 (1997)
7. Akhoondzadeh, M.: Anomalous TEC variations associated with the powerful Tohoku earthquake of 11 March 2011. Nat. Hazards Earth Syst. Sci. **12**, 1453–1462 (2012) © Author(s) 2012. CC Attribution 3.0 License
8. Basics of the GPS Technique: Observation Equations, "Geoffrey Blewitt", Department of Geomatics, University of Newcastle, Newcastle upon Tyne, NE1 7RU, United Kingdom

Application of Parametric Methods for Earthquake Precursors Using GPS TEC

R. Revathi, K. S. Ramesh, S. Koteswara Rao and K. Uday Kiran

Abstract Total Electron Content (TEC) perturbations observed for event happened on 15th January 2014 in Indonesia with the magnitude of M ∼ 4.5 on Richter scale are investigated. The earthquake occurred at 9:26 h Greenwich Mean Time (GMT). Yule–Walker, Covariance, Modified Covariance and Burg parametric methods are used to find out the change in the spectrum of signals with and without disturbances in the signal. From the analysis, it is observed that ionosphere was disturbed thrice due to the impending earthquake resulting in upraise of energy in the ionosphere at 10:13 h LTC, 10:40 h LTC and finally at 10:52 h LTC, which is coinciding with that of the earthquake occurrence. These studies in future may help in developing early-warning systems for earthquakes.

Keywords Total electron content · Parametric methods · Earthquake

1 Introduction

The analysis of electromagnetic perturbations, prior to the occurrence of the earthquake, has often been considered to be a promising tool for their short-term prediction [1]. The seismogenic electromagnetic signals cover a wide range of frequencies from DC to VHF [4]. Even though the cause for the seismo-ionospheric perturbations is not yet clearly known, these disturbances have been established by

R. Revathi (✉) · K. S. Ramesh · S. Koteswara Rao · K. Uday Kiran
Department of ECE, KL University, Vaddeswaram, Guntur (DT)
522502, Andhra Pradesh, India
e-mail: revathimouni@gmail.com

K. S. Ramesh
e-mail: dr.ramesh@kluniversity.in

S. Koteswara Rao
e-mail: rao.sk9@gmail.com

K. Uday Kiran
e-mail: meemithrudu@gmail.com

© Springer Nature Singapore Pte Ltd. 2018   305
S. C. Satapathy et al. (eds.), *Information and Decision Sciences*,
Advances in Intelligent Systems and Computing 701,
https://doi.org/10.1007/978-981-10-7563-6_32

many scientists statistically by removing the possible sources like solar flares, geomagnetic storm activity using running median values in Total Electron Content (TEC) [6].

Mechanical transformations with active geochemical processes are the most prevalent phenomena in the area of earthquake occurrence leading to the emanations of radon, noble gases. The radioactive emissions expend their energy in ionization and excitation of the elementary particles in the atmosphere [8]. The near ground plasma collides with the atmospheric constituents such as CO_2, SO_x, NO_x, etc. leading to the generation of ions and free electrons, which quickly attach with the oxygen in a three-body reaction [2].

The free electrons attach to the metal atoms generating negative ions. The activated ions participate in ion-molecular reactions with the H_2O, which is present in large amount in troposphere. Thus, H_2O in the troposphere plays a major role in the formation of long living ions [3]. Hydrated ions including radon become centres of water condensations. During this process, latent heat is released and the atmosphere gets heated up resulting in convective activity of the ions. The substantial dipole movement of the water molecule $p = 1.87D$ prevents these molecules from recombination. Coulomb force of attraction results in the formation of positive, negative and quasi-neutral clusters.

Intense movements in air damage the neutral clusters due to weak Coulomb force of attraction. Thus, the atmosphere near ground becomes rich in long-living ion clusters. The different mobility of the ion clusters results in charge separation [7]. The increase in ion hydration reduces the columnar conductivity of the air column. The above-discussed charge clusters are assembled from about 4–5 m above the ground. These assembled clusters develop an electric field in surface near earth vertically [4, 5]. Thus, the anomalous vertical electric field acts as an electrical source, connecting the near earth surface and the upper atmosphere, leading to change in the upper atmosphere dynamo currents resulting in seismo-ionospheric perturbations.

Pulinets and Boyarchuk developed a simulation model for penetration of the anomalous vertical electric field. According to it, the value of the electric field changes with height, and it has a rapid initial increase at minimum and slow decay after reaching the maximum value. The penetration of the electrical field is larger at night-time than in day-time [6]. The existing lithosphere–atmosphere–ionosphere coupling mechanism is persistent with these changes [7].

Advancement in the technology of the Global Positioning Systems (GPS) paved a way to the study of seismo-ionosphere perturbations. The ionosphere being a dispersive medium causes a delay in electromagnetic signals passing through ionosphere. The time delay is calculated by using principal of 'time of arrival'.

The ground-based GPS receiver calculates the Slant Total Electron Content (STEC) which is defined as the integral no. of electrons present in line of sight path from satellite to receiver. The change in speed of the GPS signal travelling through ionosphere leads to the change in the signal transit time due to the changes in refractive index of medium. The pseudorange in metres for GPS receiver of single frequency 'ρ_{L_1}' is given by

$$\rho_{L_1} = (40.3 * STEC)/f_{L_1}^2 \tag{1}$$

For GPS receivers with two frequencies

$$\rho_{L_1} - \rho_{L_2} = (40.3 * STEC) * [(1/f_1^2) - (1/f_2^2)] \tag{2}$$

where f_1 and f_2 correspond to frequencies of GPS L_1 and L_2 signals. STEC for GPS receiver with two frequencies is

$$STEC = (1/40.3) * [(f_{L_1}^2 * f_{L_2}^2)/(f_{L_1}^2 - f_{L_2}^2)] * (\rho_{L_1} - \rho_{L_2}) \tag{3}$$

The STEC values change with the movement of the satellite. At a particular time, many satellites are seen by the ground-based receiver. Vertical Total Electron Content (VTEC(V)) is calculated by

$$V = STEC * \cos(\xi) \tag{4}$$

where $\xi' = 90°$—satellite zenith angle at ionospheric pierce point. Generally, it is considered to be a height of 350 km from the surface of the earth [8]. The anomaly detection of seismo-ionospheric perturbations is carried on GPS TEC data on global scale (GIM data, GDGPS and NASA's data) [9–11]. Global scale TEC values are taken from a group of GPS receivers.

The demerits in analysing seismo-ionospheric perturbations using a network of ground-based GPS receivers (global TEC data) are as follows. 1. The global TEC data includes modelling errors, and it may not be possible to study the specific phenomena leading to the occurrence of earthquake. 2. A network of GPS stations, where all receivers are not situated in a particular geometry, estimation of TEC values for points nearer and farther distances leads to validation of the data at a specific point in the upper atmosphere. 3. The global TEC data used in the analysis has different sampling times. For example, the sampling period for EURF network situated in Italy is 2.5 min, and Global Ionospheric Maps (GIM) data has a sampling period of 2 h, etc.

Implementation of signal processing algorithms results in accurate signal estimation. Parametric methods are helpful in identification of underlying stationary signal structure, described by using a limited number of parameters. These methods are useful in detecting and tracking changes in the data set. In this paper, Autoregressive AR(4) methods are used for anomaly detection regarding earthquakes. Yule–Walker, Burg, Covariance and Modified Covariance are used for the analysis. The analysis is carried out for the catastrophic event occurred on 15th January 2014 in Indonesia (6.33° S, 106.896° E) at 9:56 h, Greenwich Mean Time with a depth of 125 km.

Fig. 1 Location map of the earthquake

Fig. 2 VTEC plot of satellite PRN 18

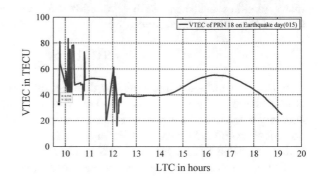

2 Data

GPS VTEC is staked by the International GNSS Service (IGS). It is noticed that the VTEC of satellite 18 is perturbed on the earthquake day. The location map of the earthquake is shown in Fig. 1, taken from http://earthquaketrack.com/quakes/2014-01-15-09-26-11-utc-4-5-125, and the VTEC plot of satellite 18 is shown in Fig. 2.

3 Methodology

Parametric methods use the knowledge of data generation in the estimation procedure. The investigation was carried out using fourth-order Autoregressive (AR (4)) models [12, 13].

Yule–Walker, Covariance, Modified Covariance and Burg algorithms are synthesized, and their respective normalized frequencies are 0.2002 and 0.298. Power Spectral Density (PSD) of Yule–Walker is different from the remaining PSDs. It does not effectively represent the frequencies present in the data. PSDs of Covariance and Modified Covariance are scaled so as to clearly represent them in

Fig. 3 PSD's of Yule–
Walker, Covariance, Modified
Covariance and Burg methods
for the synthetic signal

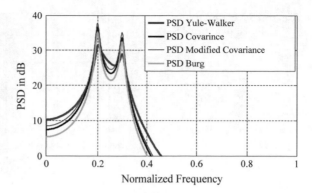

Fig. 4 Detrended R_1 data

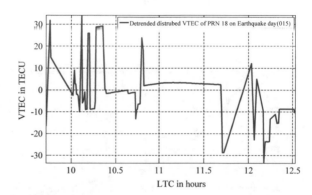

Fig. 3. Burg PSD has predominantly represented the normalized frequencies in the data. After the implementation of the algorithms on the synthetic signal these are applied on the disturbed and undisturbed VTEC for identificaiton of earthquake precursors.

The TEC perturbation in satellite number 18 is not available before occurrence time of the disturbance, and it is divided into disturbed and undisturbed parts on the earthquake day. The diurnal and seasonal variations in VTEC are removed by detrending the data. The detrended disturbed (R_1) and undisturbed (R_2) data sets are

Fig. 5 Detrended R_2

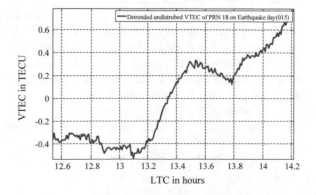

Fig. 6 PSD of disturbed
VTEC for all four methods

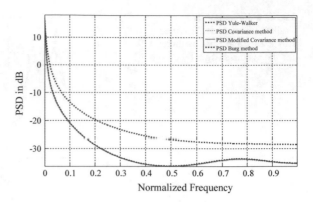

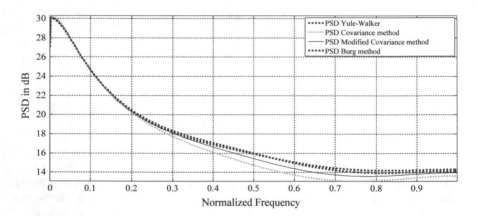

Fig. 7 PSD of undisturbed VTEC for all four methods

represented in Figs. 4 and 5. The TEC data of PRN 18 consists of 1042 points. R_1
and R_2 segments of data consist of 240 and 202 data points, respectively. The
methods are applied on both R_1 and R_2 sets to identify the underlying salient
features of non-stationary TEC data.

It is observed that the PSD of R_1 data has significant positive value (14 dB)
when compared to that of the undisturbed data. PSD plots characterize the energy in
the ionosphere due to impending earthquake. PSDs of above methods for both R_1
and R_2 sets are shown in Figs. 6 and 7. Among the four spectrum estimation
methods applied on GPS VTEC, Burg method has given better results. The PSDs of
R_1 and R_2 sets are given in Table 1 (Fig. 8).

The disturbed VTEC data segment is subjected to multi-resolution analysis to
identify earthquake precursors. Burg algorithm is implemented on those segments
of data to understand the conjuncture of earthquake on the ionosphere. The first
bisection of disturbed data segment resulted in upraise of energy at normalized

Table 1 PSD of the disturbed and undisturbed data using Yule–Walker, Covariance, Modified Covariance and Burg methods

S. No	Method	Disturbed approx. range (dB)	Undisturbed approx. range (dB)
1	Yule–Walker	14.19	−28.22
2	Covariance	13.19	−28.22
3	Modified Covariance	13.64	−33.64
4	Burg	13.95	−33.64

Fig. 8 PSD of R_1 for 120 points each

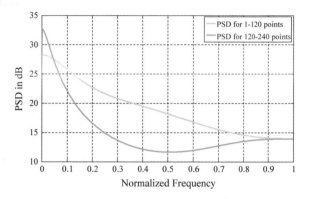

Fig. 9 PSD of R_1 for 60 points each

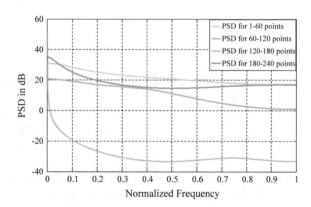

frequencies 0.9247 with PSD 13.98 dB in the second data segment. The PSD plot of two segments of R_1 of 120 points is shown in Fig. 9.

As spectral resolution increases, certainty in time of occurrence of perturbations in VTEC also increases. So R_1 is bisected into segments of 60, 30 and 15 points each. Application of Burg algorithm on second bisection of the data resulted in a small peak in the third segment of the data. Figure 10 represents the PSD for data segments of 60 data points each. Further, R_1 is sampled into sets of 30 and 15

Fig. 10 PSDs of R_1 for first
four sets with 30 points

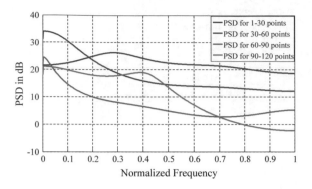

Fig. 11 PSDs for second
four sets of R_1 with 30 points

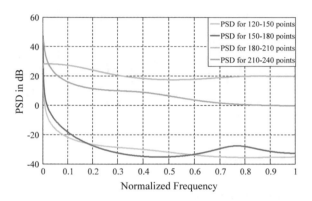

points to notice the clear upraise in the disturbance. In first and fourth sets of 30 points, peaks are identified at normalized frequencies of 0.2815 and 0.4066 with PSDs of 26.24 dB and 18.83 dB, respectively. In the third and the second sets, the energy is abating.

A peak is again seen in the sixth set with a normalized frequency of 0.7742 with a PSD of −27.62 dB. The PSDs for disturbed data sets with 30 points each are given in Fig. 11. The PSDs were also drawn for data frames of 15 points each. The PSDs of data with 15 points each are shown in Figs. 12 and 13. It is clearly observed from Fig. 12 that first, third and fourth data sets have peaks at a normalized frequency of 0.6432, 0.2893 and 0.6197 with a PSDs of 27.03 dB, 29.01 dB and −28 dB.

In the fifth to eight data sets containing 15 points, it is observed that the energy has abated in the seventh set. In the eight set, a peak is observed at normalized frequency of 0.6315 with a PSD of −28.83 dB. In the remaining sets of data with 15 points, each no significant change had been noticed. From the analysis, it is clearly identified that the ionosphere is disturbed thrice at 10:13 h LTC, 10:40 h LTC and finally at 10:52 h LTC. As the GPS receiver is located near to the epicentre (58 km), these perturbations may be considered to represent the impending earthquake.

Fig. 12 PSDs for first four
sets of R_1 with 15 points

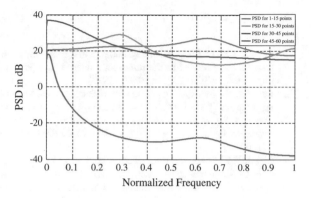

Fig. 13 PSDs for second
four sets of R_1 with 15 points

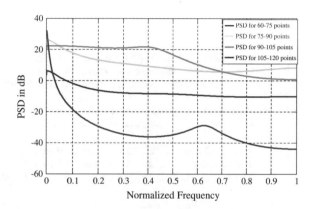

4 Conclusion

Anomalies in GPS VTEC on the earthquake day clearly show that use of Burg
algorithm could recognize co-seismic disturbances in ionosphere. From the analysis, the ionosphere is disturbed thrice at 10:13 h LTC, 10:40 h LTC and finally at
10:52 h LTC. According to the available literature, the ionosphere is disturbed few
days before the earthquake. In this analysis, we can also identify the latitude and
longitude of these perturbations. A thorough analysis of GPS VTEC for
seismo-ionospheric perturbations using spectrum estimation methods, ahead of the
occurrence, may lead to identification of possible precursors of earthquakes in
ionosphere. In future, these studies may be used for the development of
early-warning systems for earthquakes.

Acknowledgements Author's sincere acknowledgments are given to DST, Government of India
for finically funding this work under the projects SR/AS-04/WOS-A/2011 and SR/S4/AS-91/2012.

References

1. Su, Y.C., et al.: Temporal and spatial precursors in ionospheric total electron content of the 16 October 1999 Mw7.1 hector mine earthquake. J. Geophys. Res. Space Phys. **118**, 6511–6517. https://doi.org/10.1002/jgra.50586 (2013)
2. Pulinets, S.: Ionospheric precursors of earthquakes; recent advances in theory and practical applications. TAO **15**(3), 413–435 (2004)
3. Singh, A.K., et al.: Electrodynamical coupling of earth's atmosphere and ionosphere: an overview. Int. J. Geophys. **2011**, Article ID 971302, 13 pages https://doi.org/10.1155/2011/971302
4. Lin, J.W.: Ionospheric total electron content (TEC) anomalies associated with earth quakes through Karhunen-Loéve trans form (KLT). Terr. Atmos. Ocean. Sci. **21**(2), 253–265 (2010)
5. Gosh, D., Midy, S.K.: Associating an ionospheric parameter with major earthquake occurence throughout the world. J. Earth Syst. Sci. **123**(1), 63–71
6. Boyarchuk, K.A.: Estimation of the concentration of complex negative ions resulting from radioactive contamination of the troposphere. Tech. Phys. **44**(3)
7. Pulinets, S.A., Boyarchuk, K.A.: Ionospheric Precursors of Earthquakes. Springer Publ (2004)
8. Mishra, P.: Global Positioning Systems, 2nd edn. Ganga-Jamuna Press
9. Molchanov, O., et al.: lithosphere-atmosphere-ionosphere coupling as governing mechanism for pre-seismic short term events in atmosphere and ionosphere. Natural Hazards Earth Syst. Sci. **4**, 757–767 (2004)
10. Lin, J.W.: Ionospheric precursor for the 20 April, 2013, Mw = 6.6 China' Lushan earthquake: two-dimensional principal component analysis (2DPCA). German J. Earth Sci. Res. (GJESR), **1**(1), 1–12 (2013)
11. Contadakis, M.E., et al.: TEC variations over the mediterranean during the seismic activity period of the last quarter of 2005 in the area of Greece. Nat. Hazards Earth Sys. **8**, 1267–1276
12. Hayes M.H.: Statistical digital processing and modeling. Georgia Institute of Technology, Wiley, Inc
13. Revathi, R., Lakshminarayana, S., Koteswara Rao, S., Ramesh, K.S., Uday Kiran, K.: Observation of ionospheric disturbances for earthquakes (M > 4) occurred during june 2013 to july 2014 in Indonesia using wavelets. Proc. SPIE **9876**, 98763E. © 2016 SPIE · CCC code: 0277-786X/16/$18 · https://doi.org/10.1117/12.2227301

A Fuzzy-Based Modified Gain Adaptive Scheme for Model Reference Adaptive Control

A. K. Pal, Indrajit Naskar and Sampa Paul

Abstract In this paper, a fuzzy-based adaptive scheme for model reference adaptive control (MRAC) is proposed. In MRAC, the choice of proper adaptive gain (γ) is a cumbersome job, and it is usually done by trial and error method. To eliminate this shortcoming, here fuzzy logic is incorporated in the control loop to tune the adaptive gain (γ). In design of model reference adaptive control, MIT rule is followed, where a cost function is defined as a function of error between the outputs of the plant and the reference model, and the controller parameters are adjusted in such a way so that this cost function is minimized. The experiments on the different second-order linear/nonlinear systems are illustrated to show the merits of the proposed fuzzy-based model reference adaptive control (FMRAC) scheme over the MRAC. The performances of the proposed control algorithms are evaluated and shown by means of simulation on MATLAB and Simulink.

Keywords Adaptive control · Fuzzy control · MIT rule · Fuzzy-based MRAC
Adaptation gain

1 Introduction

Conventional controllers are the most used controllers in industry to control the dynamics of any process. Different control strategies are tried to control the process, and among them, adaptive control is one of the widely used control methods for the betterment of system performance and accuracy [1]. PID fails to deliver good results in the system where certain system parameters are not known or change over

A. K. Pal (✉) · I. Naskar · S. Paul
Department of AEIE, Heritage Institute of Technology, Kolkata, India
e-mail: arabindakumarpal@gmail.com

I. Naskar
e-mail: indrajit.naskar@heritaheit.edu

S. Paul
e-mail: sampa.paul@heritageit.edu

© Springer Nature Singapore Pte Ltd. 2018　　　　　　　　　　　　　315
S. C. Satapathy et al. (eds.), *Information and Decision Sciences*,
Advances in Intelligent Systems and Computing 701,
https://doi.org/10.1007/978-981-10-7563-6_33

time; in such cases, adaptive control techniques may be used as an alternative [2]. One particular way of handling this problem is the technique of "model reference adaptive control" (MRAC). In MRAC, a reference process model whose dynamics in response to a reference input should be followed by the actual process [3]. Model reference adaptive control (MRAC) is a direct adaptive strategy with adjusting mechanism that enables it to achieve the desired result even in system parameter variations [4, 5]. An adaptive controller consists of two loops, an outer loop or normal feedback loop and an inner loop or parameter adjustment loop [6, 7].

In MRAC, the process output response is forced to track the response of a reference model irrespective of plant parameter variations. The controller parameters are adjusted to give a desired closed-loop performance. It is observed that often MRAC is unable to control the higher order, complex, and nonlinear system [8]. However, fuzzy logic is knowledge-based which can track the system parameter variation easily and even able to deal with problem of uneven load, inertia and set point variation [9–11]. In this paper, a modified version of MRAC is proposed using fuzzy logic which is capable to adjust the system parameter variations by changing the adaptive gain (γ) as required.

2 Adaptive Schemes

The adaptive control process is one that continuously and automatically measures the dynamic behavior of a plant, compares it with the desired output, and uses the difference to vary adjustable system parameters or to generate an actuating signal in such a way so that optimal performance can be maintained regardless of system changes. The nature of the adaptation mechanism for controlling the system performance is greatly affected by the value of adaptation gain. The control law usually features time-variant controller parameter (θ), which reflects the algorithm's adaptation to the given system.

2.1 MRAC

In MRAC, a reference model is used to compare the system's performance. Here, the process output (y) is compared with the reference output (y_m), and their difference (e) is used to adjust the controller parameters as shown in Fig. 1. The inner and outer loops of the MRAC adjust the parameters of the regulator in such a way that it drives the error between the process output and reference output to zero.

MRAC is developed based on the principle of MIT rule. In this rule, the cost function or loss function $F(\theta)$ is defined as

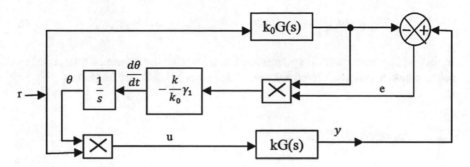

Fig. 1 MRAC scheme

$$F(\theta) = e^2/2 \tag{1}$$

where e is the difference of the output of the actual process and reference model, while θ is the adjustable parameter known as the control parameter.

In this rule, the parameter θ is adjusted in such a way that the loss function is minimized. Therefore, it is reasonable to change the parameter in the direction of the negative gradient of F and adaptation gain (γ_1), i.e.,

$$d\theta/dt = -\gamma_1 \partial F/\partial \theta = -\gamma_1 e \, \partial e/\partial \theta \tag{2}$$

Let the process transfer function is $kG(s)$ and $k_0G(s)$ is the reference transfer function.

From Fig. 1, it is observed that the controller output, $u = r\theta$, where r is the reference or command signal.

The process output y in then denoted by

$$y = u \cdot KG(s) = r \cdot \theta \cdot KG(s) \tag{3}$$

The MIT rule is used to obtain a way for adjusting the parameter θ when k is unknown. From Fig. 1, it is observed that error (e) is obtained from Eq. (4):

$$e = y - y_m = r\theta KG(s) - rK_0G(s) \tag{4}$$

The sensitivity derivative is calculated by

$$\partial e/\partial \theta = rKG(s) = K \cdot y_m/K_0 = (K/K_0)y_m \tag{5}$$

The adaptation law is derived using Eqs. (2) and (5),

$$d\theta/dt = -\gamma_1 e\, \partial e/\partial\theta = -(\gamma_1 K/K_0)y_m e = -\gamma y_m e \tag{6}$$

In this paper, under damped systems are taken both for the reference model and the process model and considered $k/k_0 = -1$ for both MRAC and FMRAC.

2.2 Proposed FMRAC

In MRAC, the control parameter θ is obtained by integrating $d\theta/dt$, which is proportional to $-\gamma y_m e$. In "$-\gamma y_m e$", γ ($\gamma = \gamma_1 \cdot k/k_0$) is adjusted by trial and error method to obtain the optimal performance. To overcome the difficulty, here, in this paper a fuzzy-based auto-tuning scheme for adaptive gain is proposed. In Fig. 2, a fuzzy-based tuning scheme for adaptive gain (γ) is proposed. Same like MRAC, here error variable (e) is obtained from Eq. (4),

The output of the fuzzy controller is generated from the two input variables error (e) and change of error (Δe) as shown in Fig. 2. The adaptation gain (γ') is a function of (e) and (Δe), i.e., $\gamma' = f(e, \Delta e)$.

In fuzzy control, scaling factor (SF) is a very important parameter as it is directly related to the system stability and system performance. The SF "G_u" is applied at the output of the fuzzy controller to obtain modified adaptive gain (γ_F), $\gamma_F = Gu \cdot \gamma'$. In this paper, γ represents an adaptive gain of MRAC, and γ_F

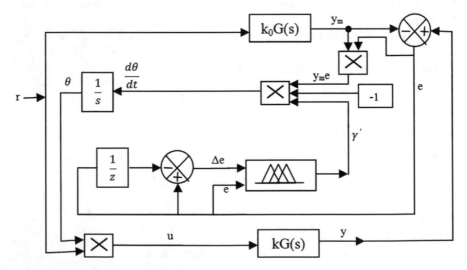

Fig. 2 Proposed FMRAC scheme

Fig. 3 MFs for e and Δe

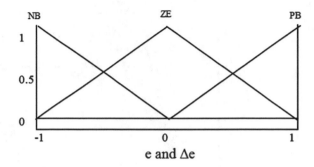

Fig. 4 MF for adaptation gain (γ')

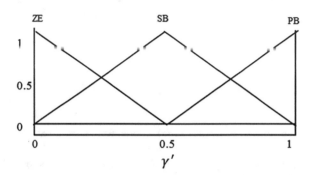

represents and adaptive gain of FMRAC. As shown in Fig. 2, the change of control parameter dθ/dt is obtained from Eq. (7), γ and γ_F

$$d\theta/dt = -\gamma_F y_m e \tag{7}$$

In this scheme, control output "θ", obtained by integrating $d\theta/dt$, is automatically adjusted accordingly to the change of process variable e and Δe, and no human intervention is required for this.

Figure 3 represents the triangular membership function used for e and Δe in the span of $[-1, 1]$, whereas a span of $[0, 1]$ is used for adaptation gain (γ') as shown in Fig. 4. The developed fuzzy-based model reference adaptive control (FMRAC) consists of nine fuzzy *if-then* rules as shown in Table 1. The linguistic membership functions are depicted here for input variables (e, Δe) are NB-negative big, ZE–zero and PB–positive big. Similarly, the variable used for gain adaptation output are ZE-zero, SB-small big and PB-positive big.

Table 1 Rules for computing fuzzy output (γ')

Δe/e	NB	ZE	PB
NB	PB	SB	ZE
ZE	SB	ZE	SB
PB	ZE	SB	PB

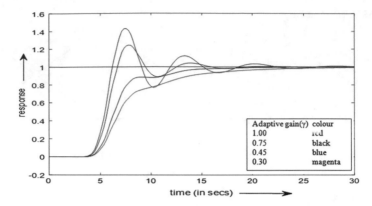

Fig. 5 Process response with different adaptive gains (γ) with MRAC for $1/s^2 + s + 1$ as reference and $0.5/s^2 + s + 1$ as process

3 Results

The proposed fuzzy-based model reference adaptive control (FMRAC) is investigated to control different types of second-order linear/nonlinear processes with step input. The output of the proposed controller is investigated under step input as shown in (Figs. 5, 6, 7, 8, 9 and 10). Simulation results on the second-order linear/nonlinear process are illustrated the merits of the proposed scheme over the others as shown in (Figs. 6, 8 and 10). The performance of MRAC with different adaptive gains (0.5–5) is investigated in linear and nonlinear processes. The effects of adaptive gain in MRAC are observed in Figs. 5, 7 and 9, respectively, and corresponding performance criteria are presented in Tables 2, 4, and 6. In case of MRAC, the choice of proper adaptive gain is a cumbersome job. Whereas the fuzzy-based modified version of MRAC (FMRAC) outperforms the MRAC as

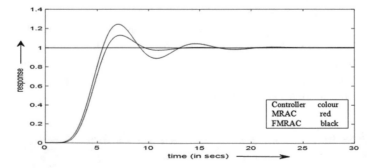

Fig. 6 Process response with MRAC and FMRAC for $1/s^2 + s + 1$ as reference and $0.5/s^2 + s + 1$ as process

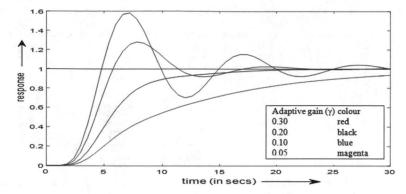

Fig. 7 Process response with different adaptive gains (γ) with MRAC for $1.6/s^2 + s + 1.6$ as reference and $1/s^2 + s + 0.6$ as process

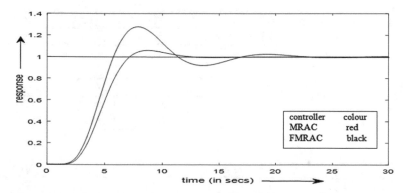

Fig. 8 Process response with MRAC and FMRAC for $1.6/s^2 + s + 1.6$ as reference and $1/s^2 + s + 0.6$ as process

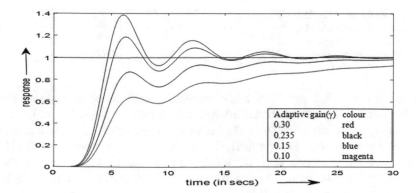

Fig. 9 Process response with different adaptive gains (γ) with MRAC for $\ddot{y} + 0.3y\,\dot{y} = u(t-1)$ as reference and $\ddot{y} + \dot{y} + 0.25y^2 = u(t-1)$ as process

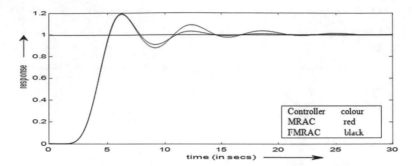

Fig. 10 Process response with MRAC and FMRAC for $\ddot{y} + 0.3y\,\dot{y} = u(t - 1)$ as reference and $\ddot{y} + \dot{y} + 0.25y^2 = u(t - 1)$ as process

Table 2 Performance indices with different adaptive gains (γ) with MRAC for $1/s^2 + s + 1$ as reference and $0.5/s^2 + s + 1$ as process

Adaptive gain (γ)	%OS	t_r (s)	t_s (s)
1.00	42.14	4.90	22.40
0.75	24.37	5.51	16.01
0.45 and 0.30	...	...	...

Table 3 Performance indices with MRAC and FMRAC for $1/s^2 + s + 1$ as reference and $0.5/s^2 + s + 1$ as process

Controller	%OS	t_r (s)	t_s (s)
FMRAC	12.40	5.80	11.90
MRAC	24.37	5.51	16.01

Table 4 Performance indices with different adaptive gains (γ) with MRAC for $1.6/s^2 + s + 1.6$ as reference and $1/s^2 + s + 0.6$ as process

Adaptive gain (γ)	%OS	t_r (s)	t_s (s)
0.3	58.10	4.78	29.90
0.2	27.56	5.77	20.16
0.10 and 0.05	...	...	...

Table 5 Performance indices with MRAC and FMRAC for $1.6/s^2 + s + 1.6$ as reference and $1/s^2 + s + 0.6$ as process

Controller	%OS	t_r (s)	t_s (s)
FMRAC	5.58	6.80	11.01
MRAC	27.56	5.77	20.16

referred in Tables 3, 5 and 7. A detail comparative analysis of the system performance in different processes under MRAC and FMRAC is evaluated in Table 8.

In any system design, the goals of minimizing the overshoot as well as rise time are very difficult, as they are conflicting in nature. However, the analysis of Table 8 shows a comparable rise time for MRAC and FMRAC, though overshoot is decreased in large proportion in case of FMRAC. In case of FMRAC, the system settles quickly compared to MRAC.

Table 6 Performance indices with different adaptive gains (γ) with MRAC for $\ddot{y} + 0.3y\ \dot{y} = u$ $(t - 1)$ as reference and $\ddot{y} + \dot{y} + 0.25y^2 = u(t - 1)$ as process

Adaptive gain (γ)	%OS	t_r (s)	t_s (s)
0.3	38.70	4.70	20.20
0.235	18.45	5.51	19.87
0.15 and 0.10	...	...	...

Table 7 Performance indices with MRAC and FMRAC for $\ddot{y} + 0.3y\ \dot{y} = u(t - 1)$ as reference and $\ddot{y} + \dot{y} + 0.25y^2 = u(t - 1)$ as process

Controller	%OS	t_r (s)	t_s (s)
FMRAC	19.88	5.06	13.27
MRAC	18.45	5.15	19.87

Table 8 Comparative study of all the processes with MRAC and FMRAC

Different types of systems	Controller	%OS	t_r (s)	t_s (s)
$1/s^2 + s + 1$ as reference and $0.5/s^2 + s + 1$ as process	FMRAC	12.40	5.80	11.90
	MRAC	24.37	5.51	16.015
$1.6/s^2 + s + 1.6$ as reference and $1/s^2 + s + 0.6$ as process	FMRAC	5.85	6.80	11.01
	MRAC	27.56	5.77	20.16
$\ddot{y} + 0.3y\ \dot{y} = u(t - 1)$ as reference and $\ddot{y} + \dot{y} + 0.25y^2 = u(t - 1)$ as process	FMRAC	19.88	5.06	13.27
	MRAC	18.44	5.15	19.87

4 Conclusion

The impact of variation of "γ" is studied in the last section, and it is observed that the slightest variation of "γ" effects the system performance of the investigated system. The proper selection of adaptation gain (γ) is very difficult, and there is no idea to find the suitable "γ" for optimal process performance. The development of fuzzy-based gain adaptive scheme for MRAC has removed the difficulty of choosing the proper adaptive gain. In FMRAC, the gain is automatically adapted/ varied for the variation of system parameter to achieve the optimal system performance. The proposed controller provides a useful alternative for controlling any such processes, which are difficult to control using conventional MRAC methods.

References

1. Benjelloun, K., Mechlih, H., Boukas, E.K.: A modified model reference adaptive control algorithm for DC servomotor. Second IEEE Conf. Control Appl. **2**, 941–946 (1993)
2. Tsai, P.-Y., Huang, H.C., Chen Y.-J., Hwang, R.-C.: The model reference control by auto tuning PID-like fuzzy controller. In: International Conference on Control Applications. Taipei, Taiwan, pp. 32–42 (2004)
3. Astrom, K.J., Bjorn, W.: Adaptive Control, 2nd edn, pp. 185–225. Pearson Education, Asia (2001)
4. Cirrincione, M., Pucci, M.: An MRAS-based sensorless high performance induction motor drive with a predictive adaptive model. IEEE Trans. Ind. Electr. **52**(2), 532–542 (2005)
5. Tayebi, A.: Model reference adaptive iterative learning control for linear systems. Int. J. Adapt. Control Signal Process. **20**, 475–489 (2006)
6. Ehsani, M.S.: Adaptive control of servo motor by MRAC method. In: IEEE International Conference on Vehicle, Power and Propulsion. Arlington, TX, pp. 78–83 (2007)
7. Kersting, S., Martin, B.: Direct and indirect model reference adaptive control for multivariable piecewise affine systems. IEEE Trans. Autom. Control 1–16 (2017)
8. Mushiri, T., Mahachil, A., Mbohwa, C.: A model reference adaptive control (MRAC) system for the pneumatic valve of the bottle washer in beverages using simulink. In: International Conference on Sustainable Materials Processing and Manufacturing, pp. 364–373 (2017)
9. Pal, A.K., Mudi Rajani K.: An adaptive fuzzy controller for overhead crane. In: IEEE International Conference on Advanced Communication Control and Computing Technologies (ICACCCT), pp. 328–332 (2012)
10. Koo, T.J.: Stable model reference adaptive fuzzy control of a class of nonlinear systems. IEEE Trans. Fuzzy Syst. **9**(4), 624–636 (2001)
11. Abid, H., Chtourou, M., Toumi, A.: An indirect model reference robust fuzzy adaptive control for a class of SISO nonlinear systems. Int. J. Control Autom. Syst. **7**, 982–991 (2009)

Early System Test Effort Estimation Automation for Object-Oriented Systems

Pulak Sahoo, J. R. Mohanty and Debabrata Sahoo

Abstract Quality and reliability are the two most important criterions used for judging a software product from customer's point of view. Software testing plays a critical role in delivering a high-quality software product. An early estimation of system test effort enables software organizations to plan and execute required test activities thoroughly. This results in the product meeting the required quality goals and improving customer acceptability. In this work, we propose a method for prediction of system test effort from Use Case models created in Requirement Analysis phase of software development. The estimation process includes automation of steps for extracting parameters from the system's Use Case models required for estimation of system test effort.

Keywords Use Case model · Use Case · Actor · Unified modeling language
Test effort estimator · CASE tool · UCPM

1 Introduction

Quality of a software product is a major concern for both customers and the organizations developing it. It is well known that software testing [1] is essential for producing quality software products. For conducting a thorough and effective testing, test planning is very much essential. Test planning based on early test effort estimations is an area of attention for software development organizations of late.

P. Sahoo (✉)
School of Computer Engineering, KIIT University, Bhubaneswar 751024, India
e-mail: sahoo_pulak@yahoo.com

J. R. Mohanty
School of Computer Applications, KIIT University, Bhubaneswar 751024, India
e-mail: jnyana1@gmail.com

D. Sahoo
Utkal University, Bhubaneswar 751004, India
e-mail: debabratasahoo@live.com

© Springer Nature Singapore Pte Ltd. 2018
S. C. Satapathy et al. (eds.), *Information and Decision Sciences*,
Advances in Intelligent Systems and Computing 701,
https://doi.org/10.1007/978-981-10-7563-6_34

If the effort required for testing a system is known early in Requirement Analysis phase, it will enable the project team to plan for the schedule and resources early. This will in turn help with effective conduction of test activities within required timeline.

Unified Modeling Language is currently the most popular standard for systems developed using object-oriented methodology [2]. Use Case models are created to capture the functional requirements, user interfaces, and scope of the system in the Requirement Analysis phase. Use Case models contain elements such as Use Cases, Actors, and their Associations. Since Use Case models are available early, these models can be used to provide inputs for an initial estimation of system test effort.

Our work proposes an estimation method for system test effort using Use Case models of the system. An automated parsing tool has been developed to extract essential information from Use Case models of the system. The Use Case models are created using a CASE tool called ArgoUML. This information is stored in a repository to be used for system test effort estimation.

ArgoUML is an open source Unified Modeling Language CASE tool written in Java. It supports exporting of Use Case models and essential parameters to XML metadata interchange (XMI) format, which then serves as inputs to the parsing tool.

The remainder of this paper is structured as follows. In Sect. 2 titled Related Work, we have described a number of relevant estimation approaches. Section 3, Proposed Method, contains the Use Case model-based test effort estimation method proposed by us. In Sect. 4, Implementation and Experimental Study, we describe the architecture, implementation steps, and experiments. Section 5, Conclusions and Future Work, provides the summary of our work and discusses the future scope.

2   Related Work

Over the years, experts have proposed a number of methods to estimate development and test effort for software products. In this section, we have discussed some relevant estimation methods which are widely used by software projects.

Function Point Analysis (FPA) method by Albrecht [3] is widely used to estimate the size of software products in terms of function points based on functionalities offered. In 1996, Caper Jones [4] proposed a method to calculate approximate number of acceptance test cases from function points using below formula.

$$\text{Number of Test Cases} = (\text{Function Points})^{1.2} \tag{1}$$

In 2000, Boehm proposed COCOMO [5] model, which estimates software development effort from the size of the system expressed in lines of code (LOC). Proposed in 2000, the Test Point Analysis (TPA) [6] method can estimate acceptance test effort of a system in test points based on its size in function.

Although FPA-based methods can provide early estimations, collecting the detailed inputs required for this process is time taking and can sometimes be costly.

The Use Case Points Methodology (UCPM) [7] was introduced to estimate system development effort from its Use Cases. The effort is expressed in Use Case Points calculated based on number of actors, number of transactions, and technical and environmental factors. In 2001, paper [8] proposed a refined UCPM-based method to estimate acceptance test effort. This method identified and classified Actors and Use Cases to assign points resulting in a total unadjusted Use Case Points for the system. Adjusted Use Case Points were obtained by taking into account nine technical and environmental factors. In paper [9], a more refined approach called N-Weighted method was proposed to estimate test activity effort based on systems Use Cases. This method separated Use Case scenarios into two types: normal and exceptional. Since a normal scenario has more steps than an exceptional scenario, the higher score was assigned to it.

In 2011, Paper [10] compared the accuracies of various UCPM-based estimation methods. It proposed a reduction in a number of environmental and technical factors from 21 to 6 by conducting factor analysis through experts. It also changed the Use Case Points calculation process by counting Use Case steps instead of transactions. But this method estimates development effort, not test effort. The estimation accuracy depends on expert's opinion and historical data of the organization.

In 2007, Paper [11] proposed an innovative approach to produce test cases from a system's Use Case and Sequence models. This method translated Use Case models to Use Case model Graph and Sequence models to Sequence model Graph. Both graphs were combined to create System Testing Graph which was traversed to produce test cases. Similarly, in 2010, Paper [12] proposed a method to produce test cases from State-Chart and Activity models. This method produced a combined state activity model and traversed all the basis paths to generate test cases. The abovementioned approaches can be expanded to produce test effort by classifying each test case by complexity and assigning scores. Unavailability of sequence, state-chart, and activity models early proves to be the hindrance for early estimation of test effort.

From above described methods, it is clear that there is a need for developing a simple and automated approach for estimating early test effort with reasonable accuracy. In this work, we propose an early system test effort estimation method using Use Case models and automation for extracting essential information from Use Case models required for the estimation process.

3 Proposed Method

For early system test effort estimation, we propose to take Use Case models of the system as inputs. Use Case models are the first UML model created in the requirement stage to capture functional requirements and user interfaces of

Fig. 1 Steps to determine
Use Case model weights

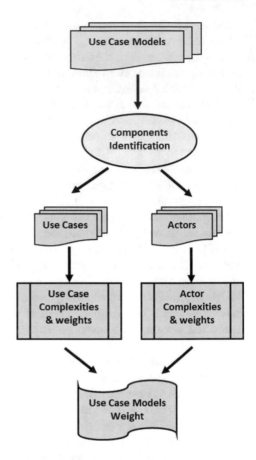

the system. Use Case models have components like Use Cases, Actors, and Associations between those [13]. Shown in Fig. 1 is the diagrammatic view of steps involved in determining Use Case model weights based on the complexity of the involved components.

3.1 Use Case and Actor Complexity

Use Cases present in Use Case models represent functional units of the system. A Use Case contains Use Case scenarios [14] of types main or exceptional. While the main scenario is made up of primary steps to achieve some functionality, the exceptional scenario contains steps for error handling. The scenarios include transactions. The complexity of a Use Case from testing perspective is calculated by counting the transactions in it. Weights assigned to normal scenario are higher than that assigned to exceptional scenarios due to the presence of greater number of checks. Apart from this, a Use Case's complexity depends on number of interacting Actors.

$$Usecase\_wt = F1(normal\_scenarios, excep\_scenarios, int\_actors) \qquad (2)$$

In Use Case models, Actors interacts with Use Cases to achieve some functionality. Actor may be an external user or an interfacing system. Actor's complexity depends on the mode of interaction with Use Cases. An Actor interacts with Use Cases via GUI, API, Protocol-Driven Interface, or Data Store. Apart from this, Actor complexity depends on number of interacting Use Cases.

$$Actor\_wt = F2(comm\_type, int\_usecases) \qquad (3)$$

3.2 Use Case Model Weight

To calculate Use Case model weight, we combine the Use Case and Actor weights obtained from Eqs. 2 and 3. A system may contain a number of Use Case models. Summing up weights for all Use Case models results in the total weight of the system.

$$Use\,Case\,Model\_wt = F1() + F2() \qquad (4)$$

After computing total weight of the system, adjustments will be applied by factoring in organization-specific technical and environmental factors relevant to testing of the software product. To this adjusted weight, organization-specific productivity factor will be applied to get system test effort.

4 Implementation and Experimental Study

The proposed architecture of UML test effort estimator containing steps of implementation is shown in Fig. 2. The major components are as follows: ArgoUML CASE tool, Use Case model Parser, Use Case model Classifier, Use Case model repository, test effort estimator, and classification setup interface. The estimator will produce Unadjusted Test effort, which will be adjusted by applying technical and environmental factors specific to the system and organization.

In the first step, Use Case models of the system are created with sufficient details by project team using ArgoUML CASE tool. The models are then exported to XMI format. In the second step, the Use Case model Parser extracts component information like Use Cases, Actors, and their associations from the XMI files and stores them in Use Case model repository. In the third step, the Use Case model component classifier categorizes the components based on complexity classification setup defined by subject matter experts. In step four, the test effort estimator computes the Unadjusted Test effort for the Use Case models. Later, relevant

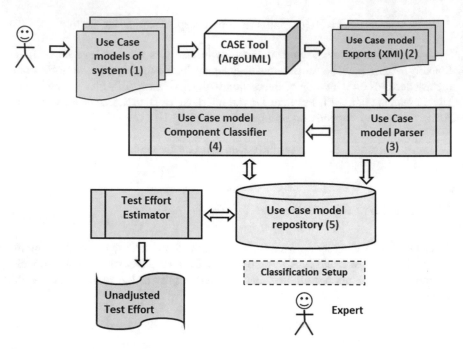

Fig. 2 Use Case model estimator architecture

technical and environmental factors can be applied to compute Adjusted Test efforts. In this work, we have implemented the first two steps on a real project. The project is titled "Cluster-based Agricultural Benefit Allocation (CABA)" which was executed by a reputed national IT organization. A brief description of CABA is given below.

For accelerated crop production, large clusters of agricultural lands are taken up for providing benefits (crop seeds) to farmers based on a cropping system. Using this system, the Admin allocates benefits to various districts according to crop production capacity. Then, the DDA (Deputy Director of Agriculture) distributes the allocated benefits among blocks and selects the crop varieties for production. Following this, the AAO (Assistant Agriculture Officer) selects the clusters and corresponding VAWs (Village Agriculture Worker) of that area to distribute the benefits. VAWs provide the seed requirements of the cluster to the registered seed agencies based on the production of crops. The seed agencies provide seeds to farmers according to the requirements and VAWs record the distribution of benefits among the clusters.

The Use Case models for CABA were created using ArgoUML and exported to XMI formats. Figure 2 shows the Use Case models for Crop Benefit Target Allocation (CBTA) and Crop Benefit Target Distribution (CBTD) modules. Figure 3 shows the Use Case model for Crop Benefit Cluster Distribution (CBCD) model and the XMI export file (Fig. 4).

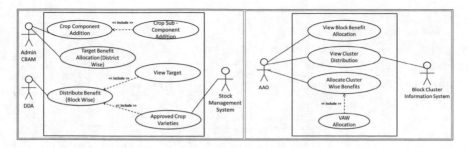

Fig. 3 Use Case models for CBTA and CBTD modules

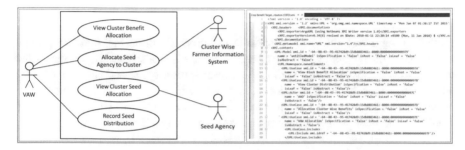

Fig. 4 Use Case model for CBCD module and XMI export

Module_ID	Use_Case_Model	Use_Case	No_of_Normal _Transactions	No_of_exception _Transactions	No_of_Interacting _Actors
CABA	CBTA	Crop_Component_addition	10	3	1
CABA	CBTA	Crop_Sub_Component_addition	8	2	0
CABA	CBTA	Target_Benefit_Allocation	14	3	1
CABA	CBTA	Distribute_Benefit	15	2	1
CABA	CBTA	View_Target	5	1	0
CABA	CBTA	Approved_Crop_Varteities	5	1	1
CABA	CBTD	View_block_Benefit_allocation	6	1	1
CABA	CBTD	View_Cluster_Distribution	8	2	2
CABA	CBTD	Allocate_Clusterwise_Benefits	15	3	1
CABA	CBTD	VAW_Allocation	7	1	0
CABA	CBCD	View_Cluster_Benefit_Allocation	6	1	1
CABA	CBCD	Allocate_Seed_Agency_to_Cluster	9	2	2
CABA	CBCD	View_Cluster_Seed_Allocation	7	1	1
CABA	CBCD	Record_Seed_Distribution	6	1	1

Fig. 5 Use Case model—Use Case data in repository

Module_ID	Use_Case_Model	Actor	Communication Type	No_of_Interacting _usecases
CABA	CBTA	Admin_CBAM	GUI	2
CABA	CBTA	DDA	GUI	1
CABA	CBTA	Stock_Management_System	Data Store	1
CABA	CBTD	AAO	GUI	3
CABA	CBTD	Block_Cluster_Information_system	Data Store	1
CABA	CBCD	VAW	GUI	3
CABA	CBCD	Clusterwise_farmer_Information_System	Data Store	2
CABA	CBCD	Seed_Agency	GUI	1

Fig. 6 Use Case model—actor data in repository

In the second step, the Use Case model Parser extracts relevant component information from Use Case models. It extracts all the Use Cases present in the models along with number of scenario transactions (both normal and exceptional) and number of interacting actors for each Use Case. It also extracts all the Actors present in the models along with the communication types and number of inter- acting Use Cases. This information is then stored in Use Case model repository to be used later for carrying out the remaining estimation steps. The two main data stores of Use Case model repository are shown in Figs. 5 and 6.

5   Conclusions and Future Work

In this work, we have proposed a simple method for system test effort estimation for object-oriented systems using Use Case models in early stage of software devel- opment. An early estimation of test effort will help project teams to plan ahead for the system testing phase, so that a quality product is delivered within the timeline. The estimation method and its steps are explained along with the estimator archi- tecture. We have implemented the steps to export Use Case models created using ArgoUML CASE tool into XMI format and extracting relevant component infor- mation from Use Case models using a Use Case model parser. The extracted details are stored in Use Case model repository and will be used to estimate system test effort required for the system. An experimental study was conducted on a real project named Cluster-based Agricultural Benefit Allocation (CABA) executed by a reputed IT organization.

References

1. Jorgensen, P.C.: Software testing: a craftsman's approach. CRC press (2016)
2. Binder, R.V.: Testing object-oriented systems: models, patterns, and tools. Addison-Wesley Professional (2000)
3. Albrecht, A.: Measuring application development productivity. Proc. Joint SHARE/GUIDE/IBM Appl. Develop. Symp. **10**, 83–92 (1979)
4. Capers, J.: Applied software measurement. McGraw-Hill (1996)
5. Boehm, B., Horowitz, E., Madachy, R., Reifer, D., Clark, B., Steece, B., Brown, W., Chulani, S., Abts, C.: Software Cost Estimation with COCOMO II. Prentice Hall (2000)
6. Van Veenendaal, E.P.W.M., Dekkers, T.: Test point analysis: a method for test estimation (1999)
7. Karner, G.: Metrics for objectory. Diploma Thesis. University of Linköping, Sweden (1993)
8. Nageswaran, S.: Test effort estimation using use case points, pp. 1–6. Quality Week (2001)
9. de Almeida, É.R.C., de Abreu, B.T., Moraes R.: An alternative approach to test effort estimation based on use cases. In: International Conference on Software Testing Verification and Validation, pp. 279–288. IEEE (2009)
10. Ochodek, M., Nawrocki, J., Kwarciak, K.: Simplifying effort estimation based on Use Case points. In: Information and Software Technology, vol. 53, pp. 200–213. Elsevier (2011)
11. Sarma, M., Mall, R.: Automatic test case generation from UML models. In: 10th International Conference on Information Technology, pp. 196–201. ICIT 2007, IEEE (2007)
12. Swain, S.K., Mohapatra, D.P., Mall, R.: Test case generation based on state and activity models. J. Obj. Technol. **9**, 1–27 (2010)
13. Sahoo, P., Mohanty, J.R.: Early test effort prediction using UML diagrams. Indones. J. Electr. Eng. Comput. Sci. **5**, 220–228 (2017)
14. Hussain, A., Nadeem, A., Ikram, M.T.: Review on formalizing use cases and scenarios: scenario based testing. In: International Conference on Emerging Technologies (ICET), pp. 1–6. IEEE (2015)

Identification of Coseismic Signatures by Comparing Welch and Burg Methods Using GPS TEC

K. Uday Kiran, S. Koteswara Rao, K. S. Ramesh and R. Revathi

Abstract Seismogenic ionospheric perturbations for a seismic hazard occurred on 23rd December 2013 in Indonesia are analyzed. The earthquake magnitude is 4.5 on Richter scale. The quake happened at 1:28 h local time coordinate. The total electron content data on occurrence day is collected from IGS System service station, BAKO Indonesia. In the present paper, Welch and Burg methods are implemented on earthquake day data. These two algorithms are implemented both on perturbed and unperturbed VTEC data. Enhancement in ionosphere is seen at the 13.84 h LTC in the ionosphere is identified by using these two methods.

Keywords Seismology · Stochastic signal processing · Ionospheric anomalies
Applied statistics

1 Introduction

Earthquakes are one of the vulnerable natural disasters occurred in nature. Abnormal changes in ionospheric electron density well before the occurrence have been reported for many large earthquakes around the world [1, 2]. Seismogenic perturbations in the ionosphere are simple and effective tool to understand the spectral characteristics of earthquakes.

K. Uday Kiran (✉) · S. Koteswara Rao · K. S. Ramesh · R. Revathi
Koneru Lakshmaiah Education Foundation, Vaddeswaram,
Guntur 522502, Andhra Pradesh, India
e-mail: meemithrudu@gmail.com

S. Koteswara Rao
e-mail: skrao@kluniversity.in

K. S. Ramesh
e-mail: dr.ramesh@klunivesity.in

R. Revathi
e-mail: revathimouni@gmail.com

© Springer Nature Singapore Pte Ltd. 2018
S. C. Satapathy et al. (eds.), *Information and Decision Sciences*,
Advances in Intelligent Systems and Computing 701,
https://doi.org/10.1007/978-981-10-7563-6_35

The physical explanation for the coupling of seismicity and ionosphere is explained by Pulinets [3]. There are various explanations like electromagnetic emissions stress changes in rocks which lead to piezoelectric effects. The investigation began from the Great Alaskan earthquake which occurred on 27th March 1964. The effect of induction is caused due to motion of electric charges in geomagnetic field, ionization of lower atmosphere due to the emission of radioactive gas and metal ions [5, 6]. Anecdotal studies show that ionospheric perturbations may be considered as short-term precursors for earthquakes.

The total number of electrons in line of sight of the satellite and GPS receiver is measured as time delay introduced in GPS receiver; it is represented as slant total electron content (STEC). STEC when expressed in meters is called as pseudorange. For single frequency GPS receiver, the pseudorange is "ρ_{L_1}" is given as

$$\rho_{L_1} = (40.3 * S)/f_{L_1}^2 \tag{1}$$

where "S" is the slant total electron content. For GPS receiver (dual frequency), it is given by

$$\rho_{L_1} - \rho_{L_2} = (40.3 * S) * [(1/f_{L_1}^2) - (1/f_{L_2}^2)] \tag{2}$$

where "f_{L_1}" and "f_{L_2}" frequencies correspond to GPS "L_1" and "L_2", signals. For a dual frequency GPS receiver, STEC is given by

$$S = \left[\frac{(\rho_{L_1} - \rho_{L_2})}{40.3} \right] * \left[\frac{f_{L_1}^2 * f_{L_2}^2}{f_{L_1}^2 - f_{L_2}^2} \right] \tag{3}$$

The VTEC values are calculated as

$$VTEC = S * \cos(\xi) \tag{4}$$

Here, ξ represents the difference value in between the 90° and zenith satellite angle. The analysis is carried out on the GPS TEC data. This global scale TEC values are obtained from the network of ground-based GPS receivers [7]. Statistical analysis is carried out for a large number of earthquakes [8–10]. All statistical parameters are used to analyze the seismo-ionospheric parameters. Wavelets, two-dimensional principal component analysis have been applied on global TEC values for the identification of the seismic anomalies in ionosphere [11, 12].

In the present work, single ground-based GPS receiver data. The data acquired is free from modeling errors with uniform sampling period. Spectral estimation techniques give better statistical and accurate estimate of the signal. Parametric methods suppose that the underlying stationary stochastic process has a certain structure which can be outlined with the help of minor amount of parameters. Nonparametric methods approximate the covariance without presuming that the process has a certain structure. In this paper, one of each of the above methods is used for the detection of the earthquake. Welch and Burg methods are applied for the analysis.

2 Details of Event Analyzed

The quake location map is in Fig. 1, taken from http://earthquaketrack.com/quakes/2013-12-23-07-58-10-utc-4-5-86 (Table 1).

Table 1 Details of the event analyzed

Date of the Earthquake	23rd December 2013
Time of occurrence	7.58 h universal time coordinate
Depth	86 km
Magnitude	4.5 on Richter scale

Fig. 1 Map of earthquake location

3 Data

International GNSS services (IGS) provides GPS VTEC. The VTEC values are collected from BAKO which is situated at Bakosurtanal, West Java. The VTEC data of the satellite PNR 18 is perturbed on earthquake day. VTEC plot of PNR number 18 is shown in Fig. 2. The satellite ray path is shown in Fig. 3.

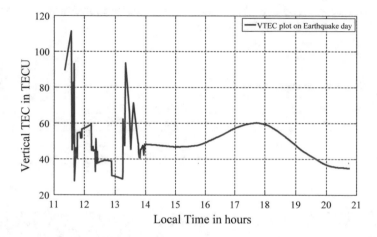

Fig. 2 VTEC plot of PRN 18 on earthquake day

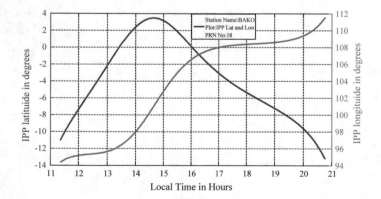

Fig. 3 Satellite ray path of PR

4 Methodology

To find out the perturbations in the signals, there are numerous statistical signal processing methods [13–16]. In this research work, Burg a parametric method and a nonparametric method Welch are applied on data of earthquake day. With the aid of these two methods, the power spectral density (PSD) of the undisturbed and disturbed VTEC data on the earthquake day is estimated. In Welch method, the signal is split into overlapping segments. The data is segmented into X number of data sets having length Y with Z points overlapped. In this method, if $Z = Y/2$ points, then the overlapping is said to be 50%. The overlapped segments are windowed to calculate the PSD.

Burg method uses autoregressive parameters by calculating forward and backward corrections. It gives stable results and also produces high-resolution PSD for short data records. The algorithms are simulated in Matlab for a synthetic signal. The PSD of the synthetic signal using both methods is represented in Fig. 4 and Fig. 5, respectively. The normalized frequencies in the synthetic signal are clearly seen in their PSD plots. Then, the algorithm has been applied to the VTEC data on the quake day [17].

The perturbed data had 1035 points (for satellite PRN number 18). The VTEC data is tailored into perturbed and unperturbed parts. These are represented Fig. 6 and Fig. 7, respectively. The data after 17:00 h LTC is not considered because it represents the post-sunset perturbations in that low latitude station. The disturbed data consists of 208 points, and the undisturbed data consists of 308 points, respectively.

The algorithms are implemented on the undisturbed and disturbed VTEC data. The PSDs of the disturbed and undisturbed data using Welch are represented in Fig. 8a, b. The PSDs of the perturbed and unperturbed data using Burg are represented in the Fig. 9a, b. The application of Welch method peak with a PSD of 14.14 dB at a frequency (normalized) of 0.6085 is observed. In the implementation

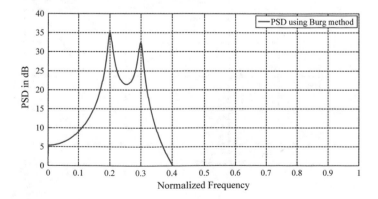

Fig. 4 PSD of the synthetic signal using Burg method

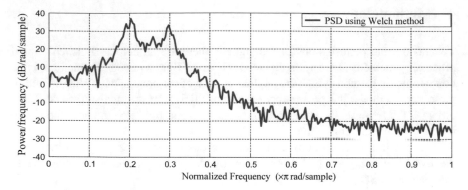

Fig. 5 PSD of the synthetic signal using Welch method

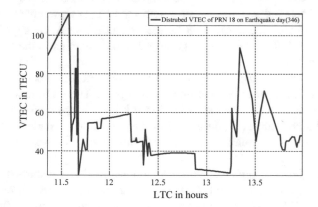

Fig. 6 Detrended disturbed VTEC of PRN 18

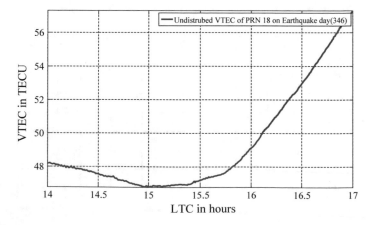

Fig. 7 Detrended undisturbed VTEC of PRN 18

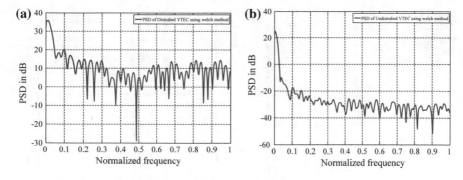

Fig. 8 PSD of the disturbed and undisturbed VTEC using Welch method

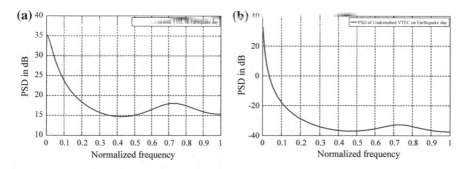

Fig. 9 PSD of the disturbed and undisturbed VTEC using Burg method

Table 2 PSD of disturbed and undisturbed VTEC using Welch and Burg methods

S. No	Method	Disturbed VTEC		Undisturbed VTEC
		Normalized frequency	PSD in dB	
1	Welch	0.6085	14.14	No peaks are observed
2	Burg	0.7351	18.03	No peaks are observed
		0.7312	−32.7	No peaks are observed

of Burg algorithm peak with a PSD of 18.03 dB at a frequency (normalized) of 0.7351 is seen. In the undisturbed parts, the PSDs had negative value. The PSDs of the two methods are tabulated in Table 2.

The application of the algorithms on the first bisection using Welch method resulted in peaks at normalized frequencies at 0.3851 and 0.4555 with PSDs of 4.51 dB and 9.15 dB in the first and second sets, respectively. In the application of Burg method on the first bisection, a peak is observed in the second set at normalized frequency of 0.6295 with PSD of 19.73 dB. The plot of their PSDs is given in Figs. 10 and 11.

Fig. 10 PSD of disturbed VTEC for 52 data using Welch method

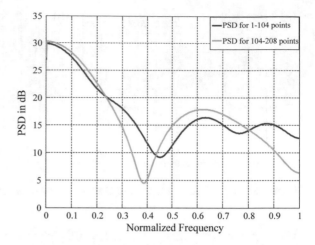

Fig. 11 PSD of disturbed VTEC for 52 data points using Burg method

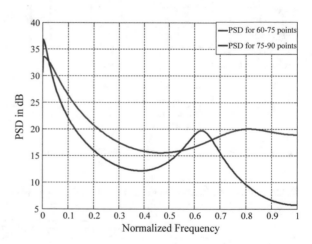

5 Discussion

The analysis of earthquake data using Welch and Burg methods has identified the seismic perturbations in the ionosphere during the earthquake. The above results are in accordance with coupling mechanisms of lower atmosphere and ionosphere. Enhancement in energy of ionosphere was observed during the occurrence of the quake. This enhancement is seen at 13.84 h LTC in the ionosphere in both the methods.

6   Conclusions

From the analysis, it may be concluded that the observed ionospheric perturbations represent the impending earthquake. The distance between earthquake epicenter and BAKO IGS station plays a prominent role in analysing the results obtained. Welch and Burg methods can represent the perturbations more precisely.

References

1. Akyol, A.A., Orhan, A., Feza, A., Necat Deviren, M.: Investigation on the reliability of EarthquakePrediction Based on Ionospheric Electron Content Variation. In: 16th IEEE International Conference on Information Fusion (FUSION), 2013, from 9–12 July 2013, pp. 1658–1663, ISBN; 978-605-86311-1-3
2. Komjathy, A.: Global Ionospheric Total Electron Content Mapping Using the Global Positioning System. Ph.D. dissertation, Department of Geodesy and Geomatics Engineering Technical Report No. 188, University of New Brunswick, Fredericton, New Brunswick, Canada, p. 248 (1997)
3. Pulinets, S.: Ionospheric precursors of earthquakes. Recent Advanc. Theor. Pract. Appl. TAO, 15(3), 413–435 (2004)
4. Thanassoulas, C., Tselentis, G.: Periodic variations in the Earth's electric field as earthquake precursors: results from recent experiments in Greece. Tectonophysics 224, 103–111, Elsevier Science Publishers B.V., Amsterdam (1993)
5. Freund, F.: Cracking the Code of Pre-earthquake signals. National Information Service for Earthquake Engineering, University of California, Berkeley (2005). http://solar-center. stanford.edu/SID/educators/earthquakes.html
6. Gosh, D., Midy, S.K.: Associating an ionospheric parameter with major earthquake occurrence throughout the world. J. Earth Syst. Sci. https://doi.org/10.1007/s12040-013-0372-1
7. Prathap, M.: Global Positioning Systems. 2nd ed., Ganga–Jamuna Press
8. Munawar Shaha, C., Shuanggen Jina, B.: Statistical characteristics of seismo-ionospheric GPS TEC disturbances prior to global Mw $\geq$ 5.0 earthquakes (1998–2014). J. Geodyn. 92, 42–49 (2015)
9. Liu, J.Y., Chen, Y.I., Pulinets, S.A., Tsai, Y.B., Chuo, Y.J.: Seismo-ionospheric signatures prior to M >= 6.0 Taiwan earthquakes. Geophys. Res. Lett. 27(19), 3113–3116 (2000)
10. Liu, J.Y., Chen, Y.I., Chuo, Y.J., Tsai, H.F.: Variations of ionospheric total electron content during the Chi-Chi earthquake. Geophys. Res. Lett. 28(7), 1383–1386 (2001)
11. Jyh-Woei, L.: Ionospheric precursor for the 20 April, 2013, M_W = 6.6 China' Lushan earthquake: two-dimensional principal component analysis (2DPCA). German J. Earth Sci. Res. (GJESR), 1(1), 1–12 (2013)
12. Song, Q., Ding, F., Yu, T., Wan, W., Ning, B., Liu, L., Zhao, B.: GPS detection of the co seismic ionospheric disturbances following the 12 May 2008 M7.9 Wenchuan earthquake in China. Sci. China Earth Sci. 58(1), 151–158 (2015). https://doi.org/10.1007/s11430-014-5000-7
13. Omkar Lakshmi, J., KoteswaraRao, S., Jawahar, A., Karishma, S.K.B.: Application of particle filter using bearing measurements. Indian J. Sci. Technol. 9(7) (2016). https://doi.org/10.17485/ijst/2016/v9i7/85557
14. Jawahar, A., KoteswaraRao, S.: Recursive multistage estimator for bearings only passive target tracking in ESM EW systems. Indian J. Sci. Technol. 8(26), 74932 (2015)

15. Annabattula, J., KoteswaraRao, S., SampathDakshina Murthy, A., Srikanth, K.S., RudraPrathap, D.: Multi-sensor submarine surveillance system using MGBEKF. Indian J. Sci. Technol. **8**(35) (2015). https://doi.org/10.17485/ijst/2015/v8i35/82088
16. Mylapilli, N.S., Koteswara R.S., RudraPrathatp, D., Lova, R.: Underwater target tracking using unscented Kalman filter. Indian J. Sci. Technol. **8**(31) (2015). https://doi.org/10.17485/ijst/2015/v8i31/77054
17. Hayes, M.H.: Statistical Digital Processing and Modeling, Georgia Institute of Technology, Wiley

Energy-Efficient GPS Usage in Location-Based Applications

Joy Dutta, Pradip Pramanick and Sarbani Roy

Abstract GPS is one of the most used services in any location-based app in our smartphone, and almost a quarter of all Android apps available in the Google Play store are using this GPS. There are many apps which require monitoring this for a continuous fashion because of the application's nature, and those kinds of apps consume the highest power from the smartphones. Because of the high-power draining nature of this GPS, we hesitate to take part in different crowd-sourced applications which are very much important for the smart city realization as maximum of these applications use GPS in real time or in a very frequent manner for the realization of participatory sensing in a smart city scenario. To resolve this, we have introduced an energy-efficient context-aware approach which utilizes user's mobility information from the user's context and as well smartphone's sensing values from the inbuilt accelerometer, magnetometer, and gyroscope of the smartphone to provide us a very close estimation of the present location of the user without using continuous GPS. It is an energy-efficient solution without sacrificing the accuracy compared to energy saving which will boost the crowd to take part in the smartphone-based crowd-sourced applications that depend on participatory sensing for the smart city environment.

Keywords Energy efficient · GPS · Location estimation · Sensor fusion Smart city

J. Dutta (✉) · P. Pramanick · S. Roy
Department of Computer Science and Engineering,
Jadavpur University, Kolkata 700032, India
e-mail: joy.dutta.in@ieee.org

P. Pramanick
e-mail: pradipcyb@gmail.com

S. Roy
e-mail: sarbani.roy@ieee.org

© Springer Nature Singapore Pte Ltd. 2018
S. C. Satapathy et al. (eds.), *Information and Decision Sciences*,
Advances in Intelligent Systems and Computing 701,
https://doi.org/10.1007/978-981-10-7563-6_36

345

1   Introduction

The fundamental thought behind location-based information systems is to connect information pieces to positions in outdoor and indoor space. To locate anything on earth, the importance of GPS is immense as of now there are not many alternatives available other than GPS in terms of location-based services. However, in some systems, GLONASS and DARPA are used along with GPS to precise the user location.

GPS has an assortment of uses ashore, adrift, and noticeable all around. Usually, GPS is usable wherever aside from where it is difficult to get the signal, for example, inside most structures, in caverns and other underground areas, and submerged. Its use is not bounded by only driving directions, which comes to our mind first, but there are plenty of other applications of GPS which we are using in our day-to-day life. Importance of GPS is endless, and in present days, maximum user-centric applications use the context of the user which also includes its location information to provide the user more optimal and valid feedback from the various smartphone-based applications. Some most recent applications in terms of smart city include various crowd-sourced and IoT-based applications where location-based participatory sensing is used. For example, GPS is used in various crowd-sourced environmental applications like air quality monitoring or even in opportunistic crowd sensing also for the realization of smart city [6, 7]. It is also used for various IoT-based smart city [8] related applications too. Practically, use of this GPS is very common and used widely across various types of applications.

However, GPS's battery depleting conduct is most recognizable amid the underlying procurement of the satellite's route message. During this, the phone is unable to enter a sleep state, which is sometimes essential to save energy. A-GPS (Assisted GPS) partially solves this, by sending the navigational message to the mobile device over a cellular data network or even Wi-Fi. As the bandwidth of either of these greatly reduces the 50 bps of the GPS satellites, the time spent powering the GPS antenna or avoiding sleep is greatly abridged. Also, the GPS signal is weaker than the other signals. It requires more amplification stage and other signal strength, increasing circuits than other sensors. Generally, mapping software needs more processing power, and it prevents the phone from going to sleep mode, thus increasing the energy consumption of the device.

As discussed in [4], the energy consumption of a GPS is the highest among all the sensors embedded in a smartphone. It is also a known fact that smartphones are battery-powered devices, and we cannot really afford the luxury to drain the battery. Also, it is now clear that it is a very important part, and we cannot skip using GPS. Also, we need to note that, if we use GPS continuously, the accuracy will be the highest but draining of the battery will be the highest as well, which is a serious issue. So, the solution is to use the sensor optimally without much compromise with the accuracy. We need to handle this one carefully such that we can do a balanced trade-off between the accuracy and the power consumption in location-based applications.

2 Related Work

There are works on using low power sensors such as accelerometer to minimize duty cycle time of GPS [1, 5]. In some sensing scenarios, the location information can be post-processed when the data are uploaded to a server; in such cases, CO-GPS [9] allows a sensing device to aggressively duty cycle its GPS receiver and log some raw GPS signal for post-processing. In [5], authors use the location-time history of the user to estimate the user velocity and adaptively turn on GPS only if the estimated uncertainty in position exceeds the accuracy threshold. Whereas in [10], methods to reduce the amount of GPS calculations and transmissions without sacrificing the tracking capabilities of the applications are based on a state machine within which modified Kalman filter is implemented to handle the presence of noise in GPS data.

From the above discussion, we found that there is a research challenge that how to optimize power consumption without sacrificing the accuracy much. Also, we found that inbuilt smartphone sensors are not utilized properly. Nowadays, even on low budget smartphones also we have all the sensors available. So here, we plan to utilize the same properly and for that, we are using the concept of sensor fusion [12] on the smartphone-based sensors to achieve the high level of accuracy. Here, we propose a solution to find a user's real-time location by merging sensors like GPS, accelerometer, gyroscope, and magnetometer data of the smartphone more accurately. Accelerometer requires much less power than GPS and using Kalman filter, the distance can be estimated by double integration of accelerometer values [2]. Using gyroscope and magnetometer orientation of the device can be found. Now, using the orientation and estimated distance by calculating the speed, next location can be predicted.

The main contribution of our work is twofold. First, we detect user's mobility using a low duty cycle accelerometer and turn on GPS only when user is having significant mobility and second, we periodically turn on GPS for trading off between accuracy and power efficiency. The technique of periodic GPS is described in [5] but the main difference in our work is that instead of only taking speed of the user as a period tuning parameter, we estimate the next location using accelerometer along with orientation sensors (gyroscope and magnetometer) and adaptively change the period taking estimation error into account. Our approach also improves the accuracy of the trajectory between two GPS points by location estimation and map-matching.

The rest of this paper is organized as follows: the proposed approach has been discussed next followed by evaluation section. This paper ends with conclusion.

3 Proposed Approach

In our proposed approach, we are detecting user's mobility to determine whether to keep GPS on and also instead of continuously using it to get location information, GPS is periodically turned on to re-synchronize estimated location to the actual location. The details of the proposed technique are described below.

Significant motion sensor is an ultralow power sensor; essentially a virtual sensor made from accelerometer values can detect a significant change in the user's activity, thereby indicating user's mobility [3]. This sensor can be effectively used to turn off or lower the sampling rate of GPS to save power when the user is not mobile. GPS can again be turned on or set to a higher sampling rate once this sensor triggers. We have used this one for our application development.

However, this sensor is not present in older devices. But user's mobility can again be detected by low power using accelerometer. Accelerometer reports acceleration values in three directions X, Y, and Z. Continuous positive acceleration values in X or Y axis (depending upon the orientation of the device) confidently indicate movement of the user using filters for noise removal [4].

If the position and horizontal orientation of the device are fixed, which is true for many application areas including vehicle-related ones, further optimization can be done to reduce GPS usage to a much greater extent using the inbuilt gyroscope sensor. We can measure the change in angle from the gyroscope directly. In our case, accelerometer along with gyroscope is playing the key role in our approach in predicting the user's path when the GPS is off.

To explain the scenario, let us consider that the user moves from point $P1(\varphi1, \lambda1)$ to point $P2(\varphi2, \lambda2)$ where φ is latitude and λ is longitude. Initially, GPS is turned on and used to get location samples with a high rate for a short amount of time. Then, distance between these two points given by the law of cosines which gives well-conditioned results down to distances as small as a few meters on the earth's surface as

$$D = cos^{-1}(sin\,\varphi1 \cdot sin\,\varphi2 + cos\,\varphi1 \cdot cos\,\varphi2 \cdot cos\,\Delta\lambda) \cdot R \qquad (1)$$

where R is the earth's radius (mean radius = 6,371 km).

Time interval T can be calculated from timestamp. Now average speed (S) during the synchronization interval from point P1 to P2 is S = D/T. Now the path between P1 and P2 is divided by small time interval t where t is the sampling period the application requires. Let d is the distance covered in time t. Again, intermediate location coordinates starting from point P1 over a smaller distance d can be estimated which essentially gives the GPS trajectory. The smaller distance interval, d can be found from accelerometer. Let a be the acceleration along the axis in which the user is moving, and v is the velocity assuming user goes straight in such a small interval. Then, v and d can be found by

$$v = \int a \, dt$$
$$d = \int v \, dt \tag{2}$$

By taking distance values from y axis of the accelerometer and orientation change in radians over small time interval t of the total synchronization interval T, and by summing up all these, we get the final estimated location $P2'(\varphi2', \lambda2')$ for a larger time interval T using Eq. (3) below:

$$\varphi2 = \{sin^{-1}(\sin(\varphi1) * \cos(d/R) + \cos(\varphi1) * \sin(d/R) * \cos(\theta)\}$$
$$\lambda2 = \lambda1 + tan^{-1}\{(sin(\theta) * sin(d/R) * cos(\varphi1)), (cos(d/R) - (sin(\varphi1) * sin(\varphi2)))\} \tag{3}$$

Initially, we are checking whether the user is mobile or not, based on the significant motion sensor or accelerometer, when the former is unavailable. When the user moves and the re-synchronization timeout occurs, then we are taking user's location based on GPS. If the user is not moving at that instant, then it waits for user's movement and then collects location information. And set the synchronization interval based on the context of the user. Between each synchronization, we take the accelerometer and gyroscope values and internally estimate distance in smaller time interval and velocity in that interval from Eq. (2). Using these calculated values, we are calculating next point from Eq. (3) until next re-synchronization interval occurs which can be further improved using opportunistic calibration method if you consider driving a car [11]. However, note that when the map-matching algorithm is applied along with our proposed approach for real-time location estimation, it is giving us significantly better result as shown in Fig. 1 (theoretically) which has been cross checked real time in the evaluation section. If the user is not within the synchronization interval, then we repeatedly check the status of the user and accordingly set the context for the user. After each re-synchronization interval, we are calculating actual distance traversed between two points using Eq. (1) which in turn helps the system to predict more accurately the next synchronization interval in the next turn.

If the user traverses in a straight line, then the difference between estimated locations P2' and actual location P2 is usually smaller than curved path. Here, the re-synchronization interval is very crucial. For a longer re-synchronization interval (T), the error is estimated with respect to the present location of the user tends to be high. So, choosing this interval is balancing the accuracy vs. battery consumption which requires attention. To make the system intelligent, we make this interval context-dependent, i.e., based on the user's speed and traveling mode, this interval changes internally in a dynamic way, which is again doing an excellent job in terms of accuracy. Here, in each synchronization phase, GPS is turned on to get actual

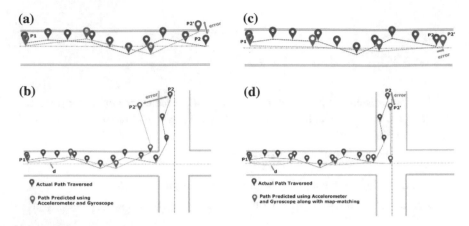

Fig. 1 Path estimation without map-matching in straight road segment (**a**), turning section of a road and with map-matching in (**b**), straight road segment (**c**), and turning section of a road (**d**)

location, and user's current location is set to this to correct the error. Also, new average speed is calculated after each T period.

If the path is not straight and a curved one, then the difference between P2 and P2' can be large due to the combined noise of accelerometer, gyroscope and magnetometer. The GPS trajectory estimated by Eq. (3) is further improved by interpolation.

4 Evaluation

We have tested the proposed approach in real-life traffic conditions in the Jadavpur University area and get the results as shown in Figs. 2, 3 and 4. Here, we have denoted real GPS traversal path to be in red color, and the path that we are calculating is set to be in blue color. Also note that P1, P2, P3, and P4 points shown in the map outputs are re-synchronization points, where estimated location synchronizes with actual GPS-based location.

Initially, we have taken straight road and tested how our proposed approach is performing in real life. We tested the synchronization interval to be 2 min, and we found the estimation is very close to the actual GPS whereas when we make the interval 4 min, the error in the estimation slightly increases but still very much accurate and valid for our everyday applications as shown in Fig. 2a, b. Next, when the proposed approach is merged with map-matching we get the result as shown in Fig. 2c, d which is actually improving system's accuracy.

Next, we have tested the efficiency by taking a left turn while walking. Here, also we have considered three different types of synchronization interval to validate our claim. It is clear from Fig. 3a, b that accuracy is high if the synchronization interval is small.

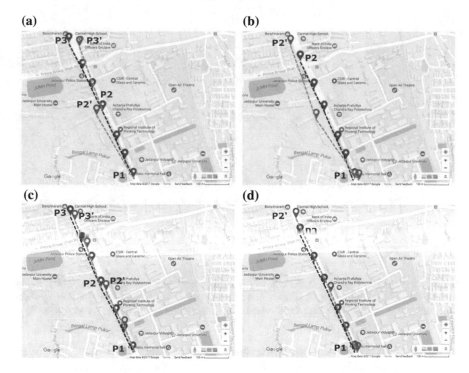

Fig. 2 Case studies: Path estimation without map-matching when user walking (**a**), biking in a straight line and path estimation with map-matching when user (**b**), walking (**c**), and biking in a straight line with 2 min synchronization interval (**d**)

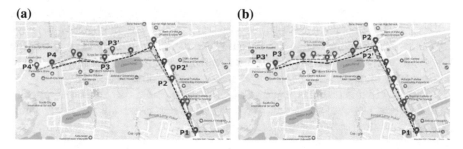

Fig. 3 Case studies: Path estimation without map-matching when user taking turn while walking with GPS synchronization interval 2 min (**a**), and 4 min (**b**)

Also note that in these types of cases, map-matching is not essential as the speed of the user is low. So, using only accelerometer and gyroscope, we can estimate the location of the user efficiently as shown above. As the walking speed of a human is not very fast, our estimated location is very much closer to the actual location point of any individual. Thus, our proposed scheme is very much valid from this

(a)   (b)

(c)   (d)

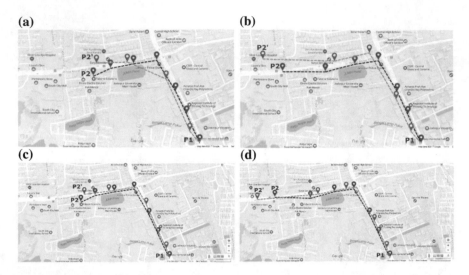

Fig. 4 Case studies: Path estimation without map-matching when user taking turn while biking with GPS synchronization interval 2 min (**a**), 4 min and path estimation with map-matching when user taking turn while biking with GPS synchronization interval (**b**), 2 min (**c**), and 4 min (**d**)

viewpoint also. However, even if we apply map-matching here, that will improve the accuracy which is of no harm. Practically, when the synchronization interval increases, we are becoming more dependent on the accelerometer and gyroscope values. However, it is clear from Figs. 2, 3 or from Fig. 4 also that without using map-matching also, the system is proving us very much descent result.

Now, the crucial part is predicting the location of a moving car and validating the same. This was the most challenging task. Here, due to the speed factor, the collected GPS points on the road are less as compared to walking because we have used the same sampling rate for all the cases. Now, we can see from Fig. 4 that the low synchronization interval gives better accuracy, as expected, when compared to all time GPS on a system which is acceptable in most of the applications.

According to this research result, turning the GPS periodically on/off along with intelligent sensing does not affect much the accuracy as we are gathering information from other energy-efficient sensors.

5 Conclusion

Here, we have used the GPS sensor optimally without affecting location accuracy of the user using the concept of sensor fusion. As GPS is a power-hungry sensor, our approach is to add intelligence in the system such that the system can read the user, understand the state, and work accordingly without affecting the performance. Here, we have introduced the context-aware sensing which, when assembles with accelerometer, magnetometer, and gyroscope, is providing us almost the same result as if you are using GPS continuously. Here, using different real-life test cases, we have validated the same by using the location estimation and map-matching for the trajectory between different GPS points.

Acknowledgements The research work of the first author is funded by "Visvesvaraya PhD Scheme, Ministry of Communications & IT, Government of India".

References

1. Abdesslem, F.B., Phillips, A., Henderson T.: Less is more: energy-efficient mobile sensing with senseless. In Proceedings of the 1st ACM workshop on Networking, Systems, and Applications for Mobile Handhelds (2009)
2. Singhal, T., Harit, A., Vishwakarma, D.N.: Kalman filter implementation on an accelerometer sensor data for three state estimation of a dynamic system. Int. J. Res. Eng. Technol. (2012)
3. Muthohar, M.F., Nugraha, I.G.D., Choi, D.: Exploring significant motion sensor for energy-efficient continuous motion and location sampling in mobile sensing application. Int. J. Technol. 38–49 (2016)
4. Kjærgaard, M.B., Langdal, J., Godsk, T., Toftkjær, T.: Entracked: energy-efficient robust position tracking for mobile devices. In: Proceedings of the 7th International Conference on Mobile Systems, Applications, and Services. ACM. New York. USA (2009) 221–234
5. Paek, J., Kim, J., Govindan, R.: Energy-efficient rate-adaptive GPS-based positioning for smartphones. In Proceedings of the 8th International Conference on Mobile Systems, Applications, and Services. ACM. New York. USA (2010) 299–314
6. Dutta, J., Gazi, F., Roy, S., Chowdhury, C.: AirSense: opportunistic crowd-sensing based air quality monitoring system for smart city. In: Proceedings of IEEE Sensors. Orlando, FL, USA (2016). https://doi.org/10.1109/icsens.2016.7808730
7. Dutta, J., Chowdhury, C., Roy, S., Middya, A.I., Gazi, F.: Towards smart city: sensing air quality in city based on opportunistic crowd-sensing. In: Proceedings of the 18th International Conference on Distributed Computing and Networking. Hyderabad, India. ACM. (2017). https://doi.org/10.1145/3007748.3018286
8. Dutta, J., Roy, S.: IoT-fog-cloud based architecture for smart city: prototype of a smart building. In: Proceedings of 7th International Conference on Cloud Computing, Data Science & Engineering. Noida, India, pp. 237–242 (2017). https://doi.org/10.1109/confluence.2017.7943156
9. Liu, J., et al.: CO-GPS: energy efficient GPS sensing with cloud offloading. IEEE Trans. Mob. Comput. **15**(6), 1348–1361 (2016). https://doi.org/10.1109/TMC.2015.2446461
10. Taylor, I. M., Labrador, M. A.: Improving the energy consumption in mobile phones by filtering noisy GPS fixes with modified Kalman filters. IEEE Wireless Communications and Networking Conference, Cancun, Quintana Roo, Mexico 2006–2011. (2011). https://doi.org/10.1109/wcnc.2011.5779437
11. Khaleghi, B., El-Ghazal, A., Hilal, A. R., Toonstra, J., Miners, W. B., Basir O. A.: Opportunistic calibration of smartphone orientation in a vehicle. In: IEEE 16th International Symposium on a World of Wireless, Mobile and Multimedia Networks. Boston, MA, USA (2015). https://doi.org/10.1109/wowmom.2015.7158210
12. Abyarjoo, F. et al.: Implementing a sensor fusion algorithm for 3D orientation detection with inertial/magnetic sensors. Innovations and Advances in Computing, Informatics, Systems Sciences, Networking and Engineering. Springer. Cham pp. 305–310 (2015)

Group Recommender Systems-Evolutionary Approach Based on Consensus with Ties

Ritu Meena

Abstract The issue regarding aggregation of multiple rankings into one consensus ranking is an interesting research subject in a ubiquitous scenario that includes a group of users. For minimizing the fitness value of Kendall tau distance (KtD), the well-known optimal aggregation method of Kemeny is used to generate an aggregated list from the input lists. A primary goal of our work is to recommend a list of items or permutation that can effectively handle the problem of full ranking with ties using consensus (FRWT-WC). Additionally, in real applications, most of the studies have focused on without ties. However, the rankings to be aggregated may not be permutations where elements have multiple choices ordered set, but they may have ties where some elements are placed at the same position. In this work, in order to handle problem of FRWT in GRS using consensus measure function, KtD are used as fitness function. Experimental result are presents that our proposed GRS based on Consensus for FRWT (GRS-FRWT-WC) outperforms well-knows baseline GRS techniques. In this work, we design and evolve an innovative method to solve the problem of ties in GRS based on consensus and results show that efficiency of group does not certainly reduce in which the group has similar-minded user.

Keywords Group recommender systems · Rank aggregation · Genetic algorithm · Kendall tau distance · Consensus

1 Introduction

Recommender systems (RSs) have evolved as a phenomenal mechanism which skillfully manages data excess issue which is generated by unmatched development of amenities accessible on the web. However, the most RSs [1] produce recommendations for single users, in many situations, the selected items (e.g., movies) are used by group

R. Meena (✉)
School of Computer & Systems Sciences, Jawaharlal Nehru University,
New Delhi, India
e-mail: meena.ritu@gmail.com

© Springer Nature Singapore Pte Ltd. 2018   355
S. C. Satapathy et al. (eds.), *Information and Decision Sciences*,
Advances in Intelligent Systems and Computing 701,
https://doi.org/10.1007/978-981-10-7563-6_37

of users. There has been much work done in group recommender systems (GRSs) with full ranking but full ranking with ties (list of items is ranked not clear) still remains a challenge [2, 3].

A recommendation system generates an item suggestion to a user focused on a study of interests. Such study is instructively developed from the single profile [4], built from the individual item evaluation made by this user, based on this interest profile [5]. For an example [6], situation related to recommendation for group's are-recommending repertories of songs for group of friends or online people, recommending a restaurant for group of people, a travel destination for family and movies for group of friends [7, 8].

Rank aggregation (RA) approach is being conveniently used in the domain of GRSs for aggregating group of users rankings [9]. Further, in actual approaches, the rankings which have to be aggregated are strictly ordered, but they do have ties explained in [10], where some elements are placed at the same position [1].

There is in GRS in general unique solution is not possible so optimization technique will be used [2]. And the genetic algorithm (GA) along with correlation is used to provide recommendations to the user. Optimization is the process to find that point or set of points in the search space and to making something better. The set of all possible solutions or values which the inputs can take make up the search space [11].

In specific circumstances where groups are formed randomly and thus the chances for heterogeneous random group results into consensus agreement failure.

Specifically, [12], this paper describes the notion of consensus measure which consists of two components, group ranking (Gr) and average pairwise dissimilarity (APD) between users for an item, and each of candidate item produce a single recommendation scores, higher the score means that items is for that particular group is highly consensus item and priority of that item should be first [13].

Section 2 reviews recommender systems for group ranking with ties and consensus strategy. In Sect. 3, we define formally the problem of recommendation for group ranking with ties and present an illustrative example of the functioning of the approach to a small number of users in a group; Sect. 4 discusses experimental results using consensus; and Sect. 5 finally shows some concluding states and future work.

2 Background and Related Work

Given a domain of choices (like books, movies, or CDs), user can express his preferences by ranking these choices, thus ranking serve as an approximate representation of users preferences, and the recommender system will match these rankings against rankings suggested by all other users.

There exist approaches [12, 13] which make use of consensus mechanism to reach a final item recommendation strategy accepted by the all users of group; recently, these approaches have also been called borderline and role-based strategy consensus used in Travel Decision form Collaborative Advisory Travel System [5].

A consensus ranking is not necessarily optimal solution of the problem; when a solution is optimal, it is explicitly signify as an optimal consensus. Different kinds of groups affect the way users evaluate the result of the adopted aggregation strategy.

2.1 Full Ranking with Ties (FRWT)

Full ranking means all users give their preferences for all the items in a group [10, 14]. It can be ties or without ties. A tie means a user give same preference of two items in a list so that items not clearly preferred to other. In this work, we expand a computationally efficient framework for ranking data which have same preference for more than one item. The framework starts by considering full ranking with Ties (FRWT) and for that we evaluate well-known notion of metrics, namely Kendall tau distance (KtD) [15].

Let σ, τ be two ranking with domain D and $G = \{\{i, j\} | i \neq j$ and i, j D$\}$ be the set of district pairs of discrete elements. The Kendall tau distance will be equal to the number of exchanges needed in a bubble sort to convert one list of items to the other [2, 13].

2.2 Genetic Algorithms (GAs)

The genetic algorithm (GA) is based on a set of feasible resolution of the optimization problem which is needed to be solved. The representation of the candidate solution plays a Euclid role as it determines which genetic operators are to be employed. That represent a solution of the optimization problem, which we want to solve [11, 16].

They can be represented by the sets of symbols or the list of values for the continuous values, they are called as vectors. In case of combinatorial problems, the solutions often consist of character that appears in a list [17, 18]. Following are the pseudocode for genetic algorithms [19]:

1. Initial population,
2. Crossover and mutation,
3. Fitness computation,
4. Go to step 2 **until** population complete,
5. Selection of parental population, and
6. **Go to step 2 until** termination condition.

2.3   Consensus Measure (CM)

The goal of consensus measure (CM) in GRS is to compute a group ranking
(Gr) for every item that reflects the interests and preferences of all members of the
group. A CM for every item needs to be carefully represented because, in general,
members of group may not have the same tastes [3, 4]. Intuitively, there are two
main aspects to the CM which are as follows:

Average Pairwise Dissimilarity (APD). The dissimilarity of a group of user U
over an item i, denote dis (U, i), indicate the score of consensus in the ranking score
for item i among group members. We consider the following dissimilarity consensus methods:

$$dis(U,i) = \frac{2}{|U|(|U|-1)} \sum (|r(u,i) - r(v,i)|) \tag{1}$$

where r is ranking of user u and v for item i, and u $\neq$ v where u, v $\in$ U.

Group Ranking (GR). The ranking of an item i to group of user U, denoted Gr.
There are several rank aggregation strategies used in group recommendation to
aggregate the group rating.

Average. In this aggregation strategy for item i, the group ranking (Gr) is
calculated as the average of the predicted ranking for the group of user U.

$$Gr(U,i) = \frac{1}{|U|} \sum (r(u,i)) \tag{2}$$

Least Misery. In this aggregation strategy for item i, the Gr is equal to the
smallest predicted ranking in the group for i.

$$Gr(U,i) = Min(r(u,i)) \tag{3}$$

Most Pleasure. In this strategy for item i the Gr is equal to the largest predicted
ranking for i in the group.

$$Gr(U,i) = Max(r(u,i)) \tag{4}$$

Borda_count. In this Strategy for each user u_j, the item with the highest satisfaction gets the rank 1, the next product gets rank 2, however, the satisfaction level
of the two products are equal, and the rank values are averaged and then assigned to
both the products.

$$Gr(U,i) = \sum_{j=1}^{m} (rank_{uj}^i) \tag{5}$$

The consensus measure function, symbolized CM (U, i), combines the group of users ranking Gr and the group dissimilarity of i for U into a single group recommendation score using the following fitness function for consensus measure (CM):

$$CM(U,i) = w_1 * Gr(U,i) + w_2 * (1 - dis(U,i)) \tag{6}$$

where w_1 and w_2 denote the relative importance of preferences and dissimilarity in the final decision, $w_1 + w_2 = 1$. Here, these two values for w_1 and w_2 (0.8 and 0.2, respectively) are chosen after observing the data that we have collected from our experiment [3].

So not only are the results stable across groups of different sizes for a single consensus list of recommended items but they also calculate the effectiveness of group of member using normalized discounted cumulated gain with different rank aggregation strategies [3].

3 Proposed Consensus-Based Recommendation

Consider a set G of all groups with at least two members that may be formed by group of users U. Consider, finally, $U \in G$ and |U| defined as the number m of group members U. If for instance, a group consists of user u_1, u_2, u_3, thus this can be expressed as $U = \{u_1, u_2, u_3\}$ and |U|.

Step 1. Group Generation.

First, we initiate synthetic groups of various set of different sizes [3]. We want to check the performance of proposed strategies change with varying group size. We randomly generated several groups and selected those set of groups which have ties with randomly generated different proportions. We have chosen a group with different data sets.

Let us consider a set of group of m users and n items.

$$U = \{u_1, u_2, \ldots \ldots u_m\}$$
$$I = \{i_1, i_2, \ldots \ldots \ldots, i_n\}$$

Matrix on full ranking with ties is defined as follows:

item =	i_1	i_2	i_3	i_4	i_5	i_6	i_7	i_8	i_9	i_{10}
u_i =	4	9	8	9	7	9	6	9	3	8
u_j =	10	5	2	7	9	7	5	6	7	6

Step 2. Fitness Function for GRS-FRWT (Sum-KtD).

The GRS problem is now to select group of n similar users and 15 items. We have to compare these matrix with full ranking with ties using fitness formula that is Kendall tau distance (KtD) that satisfies n users optimally.

Fitness Function. Let σ and τ both are full rankings with ties. Here, the fitness function is the *minimum sum of distance offer* which represents the sum of the distance for each individual in the group. We have to find the offer which is having minimum sum of the distance. In order to generate such an offer, sum of Kendall tau distance (Sum-KtD) formula is used. Similar to our definition we have to calculate KtD for every σ with $\tau_1, \tau_2, \ldots \ldots \tau_m$.

If $\sigma(i) \geq \sigma(j)$ and $\tau(i) \geq \tau(j)$, or $\sigma(i) \leq \sigma(j)$ and $\tau(i) \leq \tau(j)$, than $Ktd = 0$.
And If $\sigma(i) > \sigma(j)$ and $\tau(i) < \tau(j)$, or $\sigma(i) < \sigma(j)$ and $\tau(i) > \tau(j)$ than $Ktd = 1$.
And if $\sigma(i) \cong \sigma(j)$ and $\tau(i) = \tau(j)$, or $\sigma(i) = \sigma(j)$ and $\tau(i) \cong \tau(j)$ than Ktd $= 0.5$
Based on these cases, the *Kendall tau distance* is estimated as follows:

$$KtD(\sigma, \tau) = \sum_{\{i,j\} \in \mathcal{P}} \bar{k}_{i,j}^{(p)}(\sigma, \tau) \tag{7}$$
$$\sigma = 5 \quad 1 \quad 3 \quad 10 \quad 2 \quad 9 \quad 6 \quad 10 \quad 1 \quad 5$$

Finally, GRS-FRWT recommends the list of items (minimum Sum-KtD) that satisfied n group of users optimally (Table 1).

Genetic Algorithm (GA). Genetic operators create new solutions, combine them with existing solutions, and select between solutions in order to maintain diversity. Here, we have to apply Crossover and Mutation for GRS-FRWT to retain the best chromosome from generation to generation.

Crossover. There are many popular crossover techniques exist (e.g., single point, two point). In a single-point crossover, single point on both parents' a set of list is selected. All data beyond that point in either set of list is swapped between the two parents. The resulting are the offspring. In this paper, we are using two-point crossover where the suitable crossover point is randomly chosen from the two parents (Fig. 1).

Mutation. In our model, we first select two randomly generated numbers out of possibilities and replace it by a randomly generated number ranging from 1 to 10. For example (Fig. 2) of mutation, 3 and 10 genes are replaced by a randomly generated number 8 and 4.

Table 1 Fitness function for group with ties (Sum-KtD)

Ranking													KtD
τ_1	10	4	3	6	10	9	6	8	10	8			25
τ_2			4	9	8	9	7	9	6	9	3	8	14
τ_3	10	5	2	7	9	7	5	6	7	6			21
τ_4	7	6	9	8	6	6	10	9	9	9			19
	Total distance Sum-KtD												79

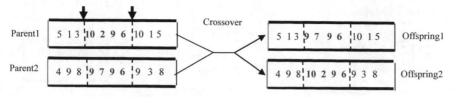

Fig. 1 A shadow of two-point crossover

Fig. 2 A shadow of two-point mutation

Stopping Standard. In order to best individual may retain from generation to generation, we are using elitist approach. When there is no improvement in the fitness value after 30 consecutive generations, the evolution process stops.

Step 3. Fitness Function for Consensus Measure (CM).

When the system generates the recommendations, we measure consensus score using formulas (1 to 7) which described in Sect. 2 to reach a consensus between items which have ties on the recommendations made for a group. Final recommendation for this matrix is as follows:

$$
\begin{array}{ccccccccccc}
I = & i_1 & i_2 & i_3 & i_4 & i_5 & i_6 & i_7 & i_8 & i_9 & i_{10} \\
\sigma = & 5 & 1 & 3 & 10 & 2 & 9 & 6 & 10 & 1 & 5
\end{array}
$$

Possible permutation for this matrix is as follows: [4–8], 6, 7, [1–10], 3, 5, [2–9].

Here, we can see that there are ties in recommendation in between items 4, 8, and 1, 10, and 2–9. Calculate consensus for this recommendation list for using CM fitness function is as follows:

$$
\begin{array}{ccccccccccc}
I = & i_1 & i_2 & i_3 & i_4 & i_5 & i_6 & i_7 & i_8 & i_9 & i_{10} \\
CM = & 3.6 & 0.6 & 1.8 & 7.9 & 1.4 & 7.2 & 4.6 & 7.9 & 0.5 & 4.0
\end{array}
$$

After consensus, permutation will be [4, 8, 6, 7, 10, 1, 3, 5, 2, 9].

Step 4. Effectiveness of a Recommend List of Ranking.

Using Consensus-based permutation can evaluate the effectiveness of recommendation of a ranked list and calculate the normalized discounted cumulative gain (nDCG) at rank k given below:

$$
DCG_k^u = r_{up1} + \sum_{i=2}^{k} \frac{r_{up_i}}{\log_2(i)} \tag{8}
$$

$$
nDCG_k^u = \frac{DCG_k^u}{IDCG_k^u} \tag{9}
$$

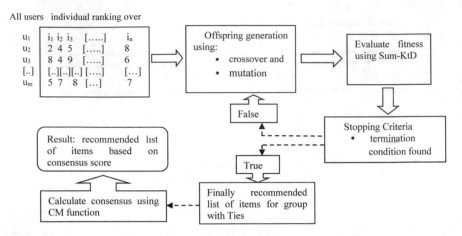

Fig. 3 A model recommendation aggregation scheme and consensus measure

where DCG is discounted cumulative gain, and IDCG is the maximum possible profit value for user u that is obtained from the optimal reorder of the k items in permutation p and n items I = 1, 2..... n. A model recommendation aggregation scheme and consensus measure is depicted in Fig. 3.

4   Experiments and Results

Data Set. The real data sets are not publicly available therefore we evaluate our proposed algorithms on a very large panel of carefully generated synthetic data set that has 15 items of different size of groups which have randomly generated ties. Experiments have been performed in order to compare the proposed approval with four states of art aggregation strategies.

Experiment 1. In this experiment, Sum-KtD is computed for different group sizes (G5, G10, G15, and G20). The results shown in Fig. 4 clearly indicates that for all group of different sizes GA meets near optimal solution after 200 generations.

Experiment 2. In this experiment, the proposed GRS-FRWT is compared with the different base line techniques. In order to compare performance of our proposed GRS-FRWT scheme with different baseline GRS techniques, we conducted experiments with groups of different sizes (G5, G10, G15, and G20). The results depicted in Fig. 5 clearly demonstrate that our scheme GRS-FRWT outperforms least misery, most pleasure, average, and Borda count.

Experiment 3. Here, we have compared the effectiveness of the group recommendation by our proposed scheme GRS-FRWT-WC with baseline techniques based on mean nDCG with varying group sizes. Results are shown in Fig. 6.

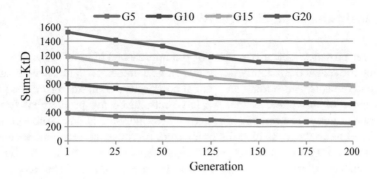

Fig. 4 The variation of Sum-KtD for group of different sizes

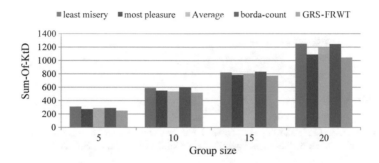

Fig. 5 The comparison of proposed GRS-FRWT to various baseline techniques for different group sizes

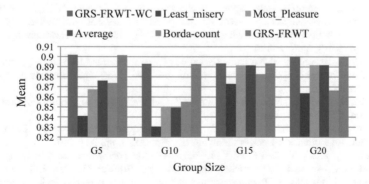

Fig. 6 The effectiveness of group recommendation with different rank aggregation techniques

5 Conclusions and Future Work

This paper provides a clear overview of the approached able to aggregate ranking with ties for selected randomly generated large number of data set with different size group of users, and finding an optimal consensus ranking in the context of ties. The purpose of this paper is that we have introduced the problem of ties using consensus in group recommender system where individuals have same preference for different items, how to solve the problem of ties in group recommender system. This system differs from normal personalized items to a group of users [18].

As a matter of feature work, we would like to experiment this strategy on real data (e.g., movie Lens, group Lens) set and to produce recommendations using trust-aware recommender systems and investigate incorporation of negotiation mechanism [20, 21].

Acknowledgements The work presented here has been supported partly by DST-PURSE and partly the RGNF-SRF for the scholar.

References

1. Adomavicius, G., Tuzhilin, A.: Toward the next generation of recommender systems: a survey of the state-of-the-art and possible extensions. IEEE Trans. Knowl. Data Eng. **17**, 734–749 (2005)
2. Meena, R., Bharadwaj, K.K.: Group recommender system based on rank aggregation–an evolutionary approach. In: Proceedings of the International Conference on Mining Intelligence and Knowledge Exploration (MIKE), LNCS 8284, Springer, pp. 663–676 (2013)
3. Salamo, M., McCarthy, K., Smyth, B.: Generating recommendations for consensus negotiation in group personalization services. Pers. Ubiquit. Comput. **16**(5), 597–610 (2012)
4. Baltrunas, L., Makcinskas, T., Ricci, F.: Group recommendations with rank aggregation and collaborative filtering. In: Proceedings of the 4th ACM Conference on Recommender Systems (RecSys 2010), pp. 119–126
5. Anand, D., Bharadwaj, K.K.: Utilizing various sparsity measures for enhancing accuracy of collaborative recommender systems based on local and global similarities. Exp. Syst. Appl. Elsevier (2010)
6. Sascha, H., Rösch, S., Beckmann, C., Gross, T.: Informing the design of group recommender systems. CHI Extend. Abst. (2012)
7. Baskin Jacob, P., Krishnamurthi, S.: Preference Aggregation in Group Recommender Systems for Committee Decision-Making. (RecSys 2009), pp. 337–340
8. Bharadwaj, K.K., Al-Shamri, M.Y.H.: Fuzzy-Genetic approach to recommender systems based on a novel hybrid user model. Exp. Syst. Appl. Elsevier **35**, 1386–1399 (2007)
9. Cantador, I., Castells, P.: Group recommender systems: new perspectives in the social web. In: J.J. Pazos Arias, A. Fernández Vilas, R.P. Díaz Redondo (Eds.): Recommender Systems for the Social Web. Springer, Intelligent Systems Reference Library, Vol. 32, ISBN: 978-3-642-25693-6 (2012)
10. Dwork, C., Kumar, R., Naor, M., Sivakumar, D.: Rank aggregation methods for the web. In: Proceedings of the Tenth International Conference on the World Wide Web, pp. 613–622, Hong Kong (2001)

11. Nguyen, H.D., Yoshihara, I., Yasunaga, M.: Modified edge recombination operators of genetic algorithm for the travelling salesman problem. In: Proceedings of the IEEE International Confeence on Industrial Electronics, Control, and Instrumentation (2000)
12. García, J.M., Tapia, Moral, M. J., del, Martínez, M.A., Herrera-Viedma, E.: A consensus model for group decision making problems with linguistic interval fuzzy preference relations. Expert Syst. Appl. **39**(11), 10022–10030 (2012)
13. Ioannidis, S., Muthukrishnan, S., Yan, J.: A Consensus-focused group recommender system. CoRR abs/1312.7076 (2013)
14. Brancotte, B., Yang, B., Blin, G., Boulakia, S.C., Denise, A., Hamel, S.: Rank aggregation with ties: experiments and analysis. PVLDB **8**(11), 1202–1213 (2015)
15. Fagin, R., Kumar, R., Mahdian, M., Sivakumar, D., Vee, E.: Comparing and Aggregating Rankings with Ties. Pods 47–58 (2004)
16. Lawrence, D.: Schedule optimization using genetic algorithms. In: Handbook of Genetic Algorithms, ed. Van Nostr, Reinhold, New York (1991)
17. Melanie, M.: An Introduction to Genetic Algorithms. MIT Press, ISBN 978–0-262-63185-3, pp. I–VIII, 1-208 (1998)
18. Salamo, M., McCarthy, K., Smyth, B.: Generating recommendations for consensus negotiation in group personalization services. Pers. Ubiquit. Comput. **16**(5), 597–610 (2012)
19. Chuan-Kang, T.: Improving edge recombination through alternate inheritance and greedy manner. Evo COP **2004**, 210–219 (2004)
20. Garcia, I., Pajares, S., Sebastia, L., Onaindia, E.: Preference elicitation techniques for group recommender systems. Informat. Sci. **189**, 155–175 (2012)
21. Onaindia, E., García, I., Sebastia, L.: A negotiation approach for group recommendation. In: Proceedings of the International Conference on Artificial Intelligence (ICAI-2009), CSREA Press, pp. 919–925

Cryptanalysis of Image Cryptosystem Using Synchronized 4D Lorenz Stenflo Hyperchaotic Systems

Musheer Ahmad, Aisha Aijaz, Subia Ansari,
Mohammad Moazzam Siddiqui and Sarfaraz Masood

Abstract Lately, a color image cryptosystem is suggested for secure wireless communication using 4D Lorenz Stenflo hyperchaotic systems. The proposition specified a nonlinear state feedback-based synchronization for master–slave Lorenz Stenflo chaotic systems. It presents seemingly successful application of synchronized chaotic systems for image encryption which is backed by simulations to assess the efficiency and stability of encryption. However, the image cryptosystem has the presence of certain loopholes. This paper aims to propose the cryptanalysis of this cryptosystem by exploiting existing vulnerabilities and loopholes. To prove that encryption algorithm is devoid of security, we mount the proposed attacks in the form of chosen-plaintext attack that recover the plaintext image from encrypted image without secret key. It is, therefore, shown through experimental simulations that the image cryptosystem is all insecure for use in practical applications of image-based secure wireless communication.

Keywords Lorenz Stenflo hyperchaotic system · Image cryptosystem
Synchronization · Cryptanalysis

1 Introduction

In the technologically advanced modern era, the communication sector has witnessed an evolution of transferred content, from signals to symbols, text, images, and moving frames. With the increasing complexity of content, the task of accomplishing a safe and secure transmission scheme has become no simpler [1]. Such is

M. Ahmad (✉) · A. Aijaz · S. Ansari · S. Masood
Department of Computer Engineering, Faculty of Engineering and Technology,
Jamia Millia Islamia, New Delhi 110025, India
e-mail: musheer.cse@gmail.com

M. M. Siddiqui
Subhash Institute of Software Technology, A.P.J. Abdul Kalam Technical University,
Lucknow 226031, Uttar Pradesh, India

© Springer Nature Singapore Pte Ltd. 2018   367
S. C. Satapathy et al. (eds.), *Information and Decision Sciences*,
Advances in Intelligent Systems and Computing 701,
https://doi.org/10.1007/978-981-10-7563-6_38

the case with image encryption systems. In the masterpiece work, Claude E. Shannon described the essential characteristics of a strong encryption method meant to secure plaintext data [2]. According to him, a security method should be able to possess a considerable amount of confusion which causes the encoded image to be highly uncorrelated to the plaintext image. Such factors may be achieved by exploring chaotic systems features [3]. In chaos-based encryption methods, they provide extreme sensitivity to initial conditions and control parameters, thus making them apt for constructing efficient and secure image encryption [4]. In past couple of decades, a number of studies have carried out by researcher worldwide for data like tests, image, audio, video, etc., encryption which utilized different types of chaotic dynamical systems to generate efficient and cryptographic random sequences of bits to be used as encryption keystreams in stream or block cryptosystems to resist any external threat from attackers [4, 5]. The field of cryptanalysis consists of mathematical tools and approaches which target toward security management, security systems, and their applications being a large fraction of it [3, 6].

In recent past, an image cryptosystem is suggested by Sonia Hammami in [7], which was based on hyperchaotic synchronization using state feedback control technique. In the work, two (master and slave) 4D Lorenz Stenflo systems were synchronized by application of aggregation techniques and arrow form matrix. The chaotic system has nonlinear dynamics which exhibits chaotic phenomenon. The cryptosystem comprises the usage of 4-D Lorenz Stenflo hyperchaotic system to change the pixel positions in highly random chaotic fashion to obtain encryption of plain-images. The parameters of master hyperchaotic system were chosen as transmitter system. The generation of chaotic behavior (which ensures encryption process) and the synchronization property (which serves during decryption) are maintained during the combination of states of chaotic system parameters with the image. Although the cryptosystem may be applied to basic or premature communication systems, it is not a feasible proposition for any sensitive image transmission. The cipher-image entirely depends on the keystream and makes the algorithm unsuitable for practical use. Also, the encryption process is not dependent on the pending plaintext image. In this paper, we focused on careful examination of security of Hammami's cryptosystem, aiming to find latent vulnerabilities and loopholes and to exploit them by applying proposed attack to regain plaintext image without incorporation of secret key. The cryptosystem is unqualified for secure image communication.

The outline of rest of this paper is as follows: In Sect. 2, a compendious review of Hammami image cryptosystem is provided. Section 3 describes the security analysis along with the entire break and its computer simulation. The work done in this paper is concluded in Sect. 4.

2 Hammami's Image Cryptosystem

Referring to [7], the Hammami's image cryptosystem is based on hyperchaotic system. The 4D Lorenz Stenflo hyperchaotic system employed in [7] is studied and analyzed in [8]. This hyperchaotic system is given in Eqs. (1) and (2), denoted by subscript m (for master), and by s for slave counterpart of same chaotic system. Where x_1, x_2, x_3, and x_4 denote the system state variables, and α, β, γ, and r are the different system parameters. The given system depicts chaotic phenomenon for $\alpha = 1$, $\beta = 0.7$, $\gamma = 1.5$, and $r = 26$.

The complete synchronization of two systems (1) and (2) is maintained using new nonlinear state feedback control law before any error-free encryption of images at the two ends during communication. The error dynamics of the coupled system, $e_{si}(t) = x_{si}(t) - x_{mi}(t)$ for $i = 1–4$, converges to zero after initial synchronization time of about 5 s. The interested readers are referred to Ref. [7] for more details.

$$
\begin{aligned}
\dot{x}_{m1}(t) &= \alpha(x_{m2}(t) - x_{m1}(t)) + \gamma x_{m4}(t) \\
\dot{x}_{m2}(t) &= x_{m1}(t)(r - x_{m3}(t)) - x_{m2}(t) \\
\dot{x}_{m3}(t) &= x_{m1}(t)x_{m2}(t) - \beta x_{m3}(t) \\
\dot{x}_{m4}(t) &= -x_{m1}(t) - \alpha x_{m4}(t)
\end{aligned}
\tag{1}
$$

The slave 4D hyperchaotic system is defined by

$$
\begin{aligned}
\dot{x}_{s1}(t) &= \alpha(x_{s2}(t) - x_{s1}(t)) + \gamma x_{s4}(t) \\
\dot{x}_{s2}(t) &= x_{s1}(t)(r - x_{s3}(t)) - x_{s2}(t) \\
\dot{x}_{s3}(t) &= x_{s1}(t)x_{s2}(t) - \beta x_{s3}(t) \\
\dot{x}_{s4}(t) &= -x_{s1}(t) - \alpha x_{s4}(t)
\end{aligned}
\tag{2}
$$

Let us take a color plain-image IMG with three dimensions (M × N × H), where M and N are the number of rows and columns, respectively, and H represents the color combination. Where H = 3 for RGB images, depicting the red, green, and blue components of image. Following are the various steps of image processing involved in the image cryptosystem:

Step 1. First, separate IMG matrix of the plain-image and convert it into three single array vectors (say R, G, and B) each of dimension (1 × MN).

Step 2. Now for encryption, generate elements from Lorenz Stenflo hyperchaos map equal to dimension of the matrix, i.e., (M × N × 3), say K, using the equations given as

$$X(1) = \alpha(y(i-1,2) - y(i-1,1)) + \gamma y(i-1,4)$$
$$X(2) = y(i-1,1)(r - y(i-1,3)) - y(i-1,2)$$
$$X(3) = y(i-1,1)y(i-1,2) - \beta y(i-1,3) \tag{3}$$
$$X(4) = -y(i-1,1) - \alpha y(i-1,4)$$
$$y(i,:) = y(i-1,:) + hX(:)$$

The initial values of different parameters are kept as it is in order to get a random sequence. The elements of K must be unique and distinct.

Step 3.   Divide elements obtained above into three blocks (say X, Y, and Z) each of size M × N and then sort the elements of each block (in ascending or descending order).

Step 4.   After sorting, we compare the disorder between original and sorted elements of each block and tabulate the index shuffle, say as T, as explained in Fig. 5 of Ref. [7]. This is repeated for all three color components.

Step 5.   According to the permuted indices vector T, the intensity positions of each plain-image components, i.e., R, G, and B are changed to get the encrypted image.

In order to decrypt, the encryption process is reversed. At the receiver side, the synchronized Lorenz Stenflo hyperchaotic map is used to generate a random sequence to obtain the sorted index elements and hence original color components, thus the plain-image IMG is regained.

3   Cryptanalysis and Its Simulation

The analysis of any cryptosystem is just as imperative as the process of building a security scheme. With minds working toward the essential protection of sensitive content, those who wish to conquer intellectual properties work just as hard to gain leverage over even the strongest encrypted systems. This is why experts opt to gauge and assess every new security primitive system before putting it out in market for practical usage [9–13]. The security of a cryptosystem depends on its secret key as it is the only hidden aspect of the system. It is assumed when scrutinizing any encryption–decryption system that the attacker knows everything about how the system works except the secret key, it is often termed as the Kerckhoffs's principle [14]. If the key itself is too fragile, the entire scheme can be broken with little effort. This is known as a "total break" and is an undesirable property for any cryptosystem. Under the said principle, the attacker may gain the access of cryptosystem as black box but without any knowledge of secret key used.

In Ref. [7], the image cryptosystem takes permutation sequences generated through initial values of Lorenz Stenflo hyperchaotic system and uses to encrypt color image. After security analysis, it is found that the entire security of system depends on merely the secret key, which if compromised may lead to complete unwanted regeneration of plain-images passing through the wireless channel. The loopholes inherent to cryptosystem are as follows: (1) the random permutation sequence obtained by the 4D Lorenz Stenflo system remains unchanged every time a plaintext image is encrypted, (2) the bit-sequence is completely independent of the information regarding the pending plaintext image, and (3) minute changes in the plaintext image result in almost no changes in the encoded image. In other words, there is actually no plaintext image sensitivity on any platform of the image cryptosystem which is of utmost significance. We explored these highlighted defects and cryptanalyze the cryptosystem in the subsequent paragraphs. Three components of color plain-image are treated and encrypted independently. The secret key used by authorized sender privately generates the permutation sequence T for each component; this is again the independent process. Thus, secret sequence of T corresponds to the secret key employed in the cryptosystem. Hence, recovering T to rearrange the encrypted image pixels for recovering plain-image pixels is as good as revealing the secret key. Therefore, in the proposed attack, we, instead of trying to find secret key, attempt to recover the permutation sequences T for each color component.

Let the attacker has access to Hammami's image cryptosystem and received an encrypted color image, say C, which is to be attacked to recover its corresponding color plain-image, say IMG. The attack proposed in this paper is chosen-plaintext attack and requires a special image Q whose first 255 pixels has gray value, in sequential order from 1 to 255, i.e., $Q(i) = i$ for $i = 1$–255 and rest of the pixels values are 0. The method $E = Hencryption(Q)$ implements the Hammami's image cryptosystem and performs encryption of input image Q and gives encrypted image E. The method $y = Find\text{-}position(D, j)$ returns the index of element j in array D, method $Q = Rotate\text{-}right(Q, 255)$ rotates the content of array Q in right direction circularly by 255 positions, and $max\_itr = ceil(M \times N/255)$. After getting all three recovered components PR, PG, PB corresponding to CR, CG, CB, we combine them to obtain the recovered color plain-image IMG $= (PR, PG, PB)$. The attack procedure for one red component is described.

Repeat the following procedure (given only for red component) for all three color components red (CR), green (CG), blue (CB) of received encrypted image C to recover the corresponding component of the plain-image.

In : chosen image Q and red component CR of received encrypted image C.
Out: recovered permutation indices $T = \{t_1, t_2, t_3, ..., t_{MN}\}$ and red component PR of IMG

```
begin
    for n = 1 to max_itr
            E = Hencryption(Q)
            D = Reshape(E, 1, MN)
            for j = 1 to 255
                    i = 255×(n – 1) + j
                    t_i = Find-position(D, j)
            end
            Q = Rotate-right(Q, 255)
    end
end

begin
    CR = Reshape(CR, 1, MN)
    for i = 1 to MN
        PR(i) = CR(t_i)
    end
    PR = Reshape(PR, M, N)
end
```

An example for the hypothetical color image of size $3 \times 3 \times 3$ $(M = N = 3)$ is provided in Table 1 to make understand the proposed attack procedure comprehensively. Whereas the computer simulation of the same attack is done for the standard color plain-image and shown in Fig. 1.

Furthermore, for any encryption scheme to be successful, a modification in one pixel of the plaintext image must capitulate at least a 50% change in the ciphertext image to oppose any statistical cryptanalysis [15–20]. In Hammami's image cryptosystem, however, only the positions of the pixels are changed in some random order to assess the plain-image sensitivity. For further understanding, consider two images with a disparity of no more than one pixel. Applying this cryptographic analysis to both these images results in two very similar ciphertext images (due to high dependency to initial parameters only and no dependency to change in plain-images). Thus, the difference obtained by applying XOR to both the encrypted image contents, with almost exactly the same datasets, gives us a black image. This is unlike the preferred case of distinct ciphertext generation which would have produced a much distorted image ideally for cryptographic sense. The difference image obtained by the Hammami's encryption approach has roughly all zeroes gray value pixels due to very little or no plaintext sensitivity. A single bit change which should have resulted in at least more than half the distortion of the

Table 1 An example description of proposed attack on a color image of size 3 × 3 × 3

$$IMG = \begin{bmatrix} 178,125,31 & 9,181,87 & 196,174,192 \\ 81,114,128 & 112,193,150 & 203,168,66 \\ 243,165,245 & 98,71,58 & 48,42,130 \end{bmatrix}$$

$$C = \begin{bmatrix} 203,71,87 & 243,181,130 & 9,42,66 \\ 98,165,31 & 112,114,245 & 178,193,192 \\ 196,125,58 & 81,174,150 & 48,168,128 \end{bmatrix}$$

$$Q = \begin{bmatrix} 1 & 2 & 3 \\ 4 & 5 & 6 \\ 7 & 8 & 9 \end{bmatrix}$$

recovered permutation sequences T

$$TR = [6 \ 3 \ 7 \ 8 \ 5 \ 1 \ 2 \ 4 \ 9]$$
$$TG = [7 \ 2 \ 8 \ 5 \ 6 \ 9 \ 4 \ 1 \ 3]$$
$$TB = [4 \ 1 \ 6 \ 9 \ 8 \ 3 \ 5 \ 7 \ 2]$$

$$PR = \begin{bmatrix} 178 & 9 & 196 \\ 81 & 112 & 203 \\ 243 & 98 & 48 \end{bmatrix}$$

$$PG = \begin{bmatrix} 125 & 181 & 174 \\ 114 & 193 & 168 \\ 165 & 71 & 42 \end{bmatrix}$$

$$PB = \begin{bmatrix} 31 & 87 & 192 \\ 128 & 150 & 66 \\ 245 & 58 & 130 \end{bmatrix}$$

$$recovered\_image = \begin{bmatrix} 178,125,31 & 9,181,87 & 196,174,192 \\ 81,114,128 & 112,193,150 & 203,168,66 \\ 243,165,245 & 98,71,58 & 48,42,130 \end{bmatrix}$$

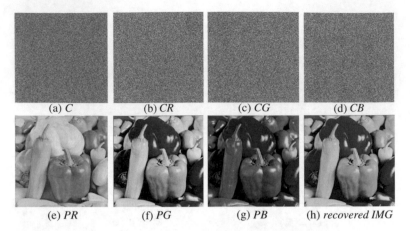

(a) C (b) CR (c) CG (d) CB

(e) PR (f) PG (g) PB (h) *recovered IMG*

Fig. 1 Simulation of recovery of color plain-image *IMG* from received encrypted image C under proposed attack

(a) IMG_1 (b) IMG_2 (c) J

Fig. 2 Simulation of no plaintext image sensitivity of cryptosystem under analysis

plaintext image by degree yields almost none. The simulation of this sensitivity test is shown in Fig. 2, where J is the difference image of C_1 and C_2, and $C_1 =$ *Hencryption*(IMG_1); $C_2 =$ *Hencryption*(IMG_2). It can be easily seen with naked eye that difference image J is all black except that middle changed pixel. Therefore, the Hammami's cryptosystem further fails in producing sufficient plaintext image sensitivity unlike a strong security system, this further evidence for its insecurity, inefficiency, and impracticability.

4 Conclusion

The security analysis of recent image cryptosystem suggested by Hammami is performed in this paper so as to bring into notice its inability to qualify as a practical encryption algorithm. The encryption procedure seems secure due to highly sensitive parameters of two 4D Lorenz Stenflo hyperchaotic systems. However, its sole dependency on the secret key brings flaws in the method which proves to be an easy advantage for a potential attacker to recover the plaintext content. These flaws and vulnerabilities have been studied, analyzed, and highlighted in this paper. The security analysis unveils and exploits basic vulnerabilities to attack the cryptosystem. Further, we have successfully broken the encryption scheme by proposed attack supported computer simulation outcomes. It is also found that encryption has poor sensitivity to plaintext images unlike key sensitivity. Thus, the cryptanalysis of the image cryptosystem under study is deemed not viable for use in real-life secure communication applications.

References

1. El-Samie, F.E.A., Ahmed, H.E.H., Elashry, I.F., Shahieen, M.H., Faragallah, O.S., El-Rabaie, E.S.M., Alshebeili, S.A.: Image Encryption: A Communication Perspective. CRC Press (2013)
2. Shannon, C.E.: Communication theory of secrecy systems. Bell Syst. Tech. J. **28**, 662 (1949)
3. Menezes, A.J., Van Oorschot, P.C., Vanstone, S.A.: Handbook of Applied Cryptography. CRC press (1996)
4. Kocarev, L., Lian, S. (eds.): Chaos-Based Cryptography: Theory, Algorithms and Applications, vol. 354. Springer (2011)
5. Kocarev, L., Galias, Z., Lian, S. (eds.): Intelligent Computing Based on Chaos, vol. 184. Springer (2009)
6. Bard, G.V.: Algebraic Cryptanalysis. Springer, Berlin (2009)
7. Hammami, S.: State feedback-based secure image cryptosystem using hyperchaotic synchronization. ISA Trans. **54**, 52–59 (2015)
8. Chen, Y., Wu, X., Gui, Z.: Global synchronization criteria for two Lorenz-Stenflo systems via single-variable substitution control. Nonlinear Dyn. **62**(1), 361–369 (2010)
9. Özkaynak, F., Özer, A.B., Yavuz, S.: Cryptanalysis of a novel image encryption scheme based on improved hyperchaotic sequences. Opt. Commun. **285**(2), 4946–4948 (2012)
10. Akhavan, A., Samsudin, A., Akhshani, A.: Cryptanalysis of an image encryption algorithm based on DNA encoding. Opt. Laser Technol. **95**, 94–99 (2017)
11. Norouzi, B., Mirzakuchaki, S., Norouzi, P.: Breaking an image encryption technique based on neural chaotic generator. Optik-Int. J. Light Electr. Opt. **140**, 946–952 (2017)
12. Singh, S., Ahmad, M., Malik, D.: Breaking an image encryption scheme based on chaotic synchronization phenomenon. In: 2016 Ninth International Conference on, Contemporary Computing (IC3), pp. 1–4. IEEE Aug 2016
13. Sharma, P.K., Kumar, A., Ahmad, M.: Cryptanalysis of image encryption algorithms based on pixels shuffling and bits shuffling. In: Proceedings of the International Congress on Information and Communication Technology, pp. 281–289. Springer, Singapore (2016)
14. Kerckhoffs's principle. http://crypto-it.net/eng/theory/kerckhoffs.html
15. Lambić, D.: Security analysis and improvement of a block cipher with dynamic S-boxes based on tent map. Nonlinear Dyn. **79**(4), 2531–2539 (2015)
16. Wu, J., Liao, X., Yang, B.: Cryptanalysis and Enhancements of image encryption based on three-dimensional bit matrix permutation. Sig. Process. **142**, 292–300 (2018). https://doi.org/10.1016/j.sigpro.2017.06.014
17. Ahmad, M., AlSharari, H.D.: On the security of chaos-based watermarking scheme for secure communication. In: Proceedings of the 5th International Conference on Frontiers in Intelligent Computing: Theory and Applications, pp. 313–321. Springer, Singapore (2017)

18. Ahmad, M., Shamsi, U., Khan, I.R.: An enhanced image encryption algorithm using fractional chaotic systems. Procedia Comput. Sci. **57**, 852–859 (2015)
19. Sharma, P.K., Ahmad, M., Khan, P.M.: Cryptanalysis of image encryption algorithm based on pixel shuffling and chaotic S-box transformation. In: International Symposium on Security in Computing and Communication, pp. 173–181. Springer, Berlin, Heidelberg, Sept 2014
20. Verma, O.P., Nizam, M., Ahmad, M.: Modified multi-chaotic systems that are based on pixel shuffle for image encryption. J. Inf. Process. Syst. **9**(2), 271–286 (2013)

Automatic Text-Line Level Handwritten *Indic* Script Recognition: A Two-Stage Framework

Pawan Kumar Singh, Anirban Mukhopadhyay, Ram Sarkar
and Mita Nasipuri

Abstract Script dependency of the Optical Character Recognition (OCR) systems is a huge obstacle for the digitalization of document images in a multi-script environment. Researchers around the world have developed various feature extraction and classification methodologies till date but mostly those are limited to bi-script and tri-script scenarios. The present work proposes an automatic two-stage framework for text-line based script recognition from the document images written in 12 *Indic* scripts. A misclassified text-line, at the first stage, is further examined by segmenting the same into its constituent words and the script recognition module is repeated on the obtained words. The pooled consequence of this two-stage framework helps to improve the overall accuracy of text-line level script classification.

Keywords Text-line level script identification · Handwritten documents
Two-stage framework · *Indic* scripts · Modified log-Gabor filter transform
Multi Layer Perceptron

P. K. Singh (✉) · A. Mukhopadhyay · R. Sarkar · M. Nasipuri
Department of Computer Science and Engineering, Jadavpur University,
188, Raja S.C. Mullick Road, Kolkata 700032, West Bengal, India
e-mail: pawansingh.ju@gmail.com

A. Mukhopadhyay
e-mail: anirbanmcse@gmail.com

R. Sarkar
e-mail: raamsarkar@gmail.com

M. Nasipuri
e-mail: mitanasipuri@gmail.com

© Springer Nature Singapore Pte Ltd. 2018   377
S. C. Satapathy et al. (eds.), *Information and Decision Sciences*,
Advances in Intelligent Systems and Computing 701,
https://doi.org/10.1007/978-981-10-7563-6_39

1 Introduction

The aim of research in document analysis for a multi-script environment is to conceive and establish a comprehensive model which can be capable of recognizing the actual script of a particular document image. Automatic script recognition from document images is a preprocessing step for applications such as sorting, searching, and indexing of multi-script data. There has been extensive research towards the digitization of the documents written in a specific script but for multi-script milieu, such OCR system would not be enough to solve the problem. Therefore, to make a successful multi-script document processing system, recognition of the scripts is very essential before running the OCR module [1]. Most of the published works on automatic script identification problem for *Indic* scripts deal with printed documents [2–7] whereas very few researchers have paid attention to its handwritten counterpart [8, 9]. Hangarge et al. [8] introduce Directional Discrete Cosine Transform (D-DCT) based word-level handwritten script identification. This is done in order to capture directional edge information, one by performing 1D-DCT along left and right diagonals of an image, and another by decomposing 2D-DCT coefficients in left and right diagonals. The classification is performed using Linear Discriminant Analysis (LDA) and k-Nearest Neighbor (k-NN) classifiers. Pardeshi et al. [9] present a word-level handwritten script identification technique for 11 different major Indian scripts (including *Roman*) in bi-script and tri-script scenarios. The features are extracted based on the combination of Radon transform, Discrete Wavelet transform, Statistical filters, and DCT which are tested using Support Vector Machine (SVM) and k-NN classifiers.

The problem of automatic script recognition has gained a boost since last decade. An exhaustive literature review on script recognition by Singh et al. [10] present the research work focused on the development of novel feature extraction approaches that work for some specific script classes [11]. Script dependent and independent feature sets have been evaluated at the page, block, text-line, and word-levels by Obaidullah et al. in [12] to determine the optimal level of use for each feature set. S. Chanda et al. have devised a two-stage approach for word-wise script recognition in a multi-script environment [13]. Ubul et al. in [14] provide a survey of the global and local features used in script identification in different multilingual environments.

But a generalized approach, which considers all handwritten *Indic* scripts texts under a common research platform, is still found to be missing. It is to be noted that script recognition can be performed at page/text-line/word/block-level. But none of the existing methods takes any feedback from other lower level (i.e., page to text-line or text-line to word) to alter the wrong decision made at higher level. This observation has motivated us towards developing a two-stage framework for handwritten script identification technique where any mistaken outcome obtained at text-line is further processed at word-level in order to improve the performance of the overall recognition system.

2   Proposed Work

The scripts, considered here, are as follows: *Tamil, Telugu, Kannada, Gurumukhi, Gujarati, Oriya, Devanagari, Bangla, Manipuri, Malayalam, Urdu,* and *Roman.* The proposed work follows a two-stage framework for handwritten *Indic* script recognition. In the first stage, Modified log-Gabor filter transform (described by Singh et al. [15]) is applied to extract features from text-line images scribed in 12 different scripts. The classification of the scripts is performed using Multi Layer Perceptron (MLP) classifier. Then, in the second stage, misclassified text-line images are forwarded to the next level where the words of the corresponding text-lines are extracted using a method described in [16]. After that, same feature extraction procedure is performed on the word images and the identification of the script in which the words belong is done using MLP classifier. If the classifier predicts a certain percentage of recognized words (say, α) belong to a particular script, then its parent text-line is assigned to that script. Otherwise, that text-line gets the label of the script class in which maximum number of words belong to the said text-line is voted. Block diagram of the proposed framework is shown in Fig. 1.

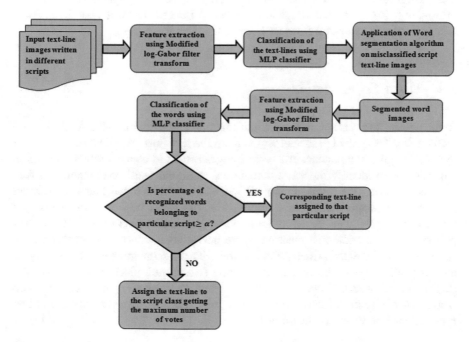

Fig. 1 Proposed two-stage framework for text-line-based script recognition

2.1 Modified Log-Gabor Filter Transform

In the recent past, Modified log-Gabor filter transform based features [15] were proven to be very effective in script classification task. Hence, in the present work, this is chosen as the feature descriptor in order to classify the text-lines and its constituent word images to its true script class. To preserve the spatial information, a Windowed Fourier Transform (WFT) is considered during the feature computation. WFT processes the image by multiplying it with the window function and the resultant output is followed by applying the Fourier transform. Since both spatial and frequency information are preferred while analyzing the texture, hence, this process tries to make a trade-off between these two. Basically, a Gaussian function is used here in the spatial as well as frequency domains [15]. Inverse Fourier transform is then applied to the obtained vector. Images, thus obtained, are provided as input to a function which computes the Gabor energy feature from them. Image representing magnitude and real part of the Gabor array pixels are retrieved from the function.

In the present work, both entropy and energy features based on Modified log-Gabor filter transform are extracted by taking 5 scales (n_s = 1, 2, 3, 4 and 5) and 12 different orientations (n_0 = 0°, 15°, 30°, ..., 165°) to capture discriminative information from different script text-line/word images. Here, a 120-dimensional feature vector is estimated from each input text-line/word image.

2.2 Word Segmentation

In the present work, a Spiral Run Length Smearing Algorithm (SRLSA) [16] based methodology is applied to extract words from the text-lines. Application of SRLSA over the disjoint components of a word images results in connectivity among them which helps to identify the word boundaries. A single word component can face two issues during the segmentation: (i) when it is broken down into two or more parts, an over-segmentation case occurs, and (ii) when two or more words get joined and formed a single word, an under-segmentation case occurs. But in this present work, no additional measures have been undertaken to correct these error cases because, here, the primary aim is to classify the segmented words (or may be part of a word) by which its parent text-line (which got misclassified in the first stage) can be classified correctly. Hence, exact segmentation of a text-line into words is not of prime concern here. Application of SRLSA on a sample handwritten *Bangla* text-line image is shown in Fig. 2.

Fig. 2 Output of application of SRLSA on a sample handwritten *Bangla* text-line image (components of each word are connected by smearing which are shown in yellow)

3 Experimental Results and Analysis

Due to the absence of publicly available database for handwritten script recognition, we have created an in-house database for conducting the current experiment. Document pages, written in 12 different scripts, are collected from different individuals belonging to varying sex, age and educational backgrounds. Scanning of the obtained pages is done at a resolution of 300 DPI which are then accumulated as 24-bit bitmap images. The noise present in these pages are removed using Gaussian filter [17]. A set of 2400 individual text-lines is prepared from these handwritten documents pages for training and testing the proposed framework by using neighborhood connected component analysis technique described in [18]. Figure 3 shows text-line sample images taken from our database. MLP classifier with Back Propagation learning algorithm has been used for the classification task. Values of its two important parameters, viz., Learning rate (η) and Momentum term (α) are experimentally set to 0.4 and 0.5 respectively. The configuration of MLP used here is 120-50-12, i.e., 120, 50 and 12 neurons are considered in its input, hidden, and output layers, respectively. The neurons are being trained for 100 epochs. Recognition accuracy of the handwritten script classification at text-line level is calculated as follows:

Fig. 3 Samples of handwritten text-line images taken from our database

382 P. K. Singh et al.

Table 1 Confusion matrix generated by MLP classifier after applying Modified log-Gabor filter transform features on handwritten text-line images

Class	A	B	C	D	E	F	G	H	I	J	K	L
A	**188**	3	0	1	0	1	0	2	2	3	0	0
B	2	**195**	0	0	0	0	0	0	2	0	0	1
C	6	0	**175**	2	0	0	0	7	3	0	2	5
D	0	4	3	**183**	0	5	2	0	1	0	1	1
E	1	0	0	0	**189**	4	0	1	1	2	2	0
F	3	0	2	3	2	**181**	0	3	4	0	2	0
G	0	2	1	3	0	1	**186**	0	0	0	7	0
H	3	0	2	0	2	1	0	**183**	5	1	1	2
I	6	0	0	0	1	0	0	0	**189**	1	3	0
J	4	0	2	2	4	2	0	0	0	**184**	0	2
K	0	0	0	0	0	4	1	1	1	2	**191**	0
L	2	2	3	0	0	0	1	0	0	1	0	**191**

$$Accuracy = \frac{\# text - lines\ correctly\ classified}{\# Total\ text - lines\ present} \times 100\% \qquad (1)$$

An average accuracy rate of 93.125% is realized through a 3-fold cross validation scheme. Table 1 shows the confusion matrix generated by MLP classifier. Scripts are numbered in the following order from *A* to *L* that denote *A*: Bangla, *B*: Devanagari, *C*: Gujarati, *D*: Gurumukhi, *E*: Kannada, *F*: Malayalam, *G*: Manipuri, *H*: Oriya, *I*: Tamil, *J*: Telugu, *K*: Urdu, and *L*: Roman scripts.

Table 1 shows that a total of 165 text-lines is misclassified at the first stage. Moreover, the highest classification (i.e., least confused script) is seen in case of *Devanagari* script and the lowest classification (i.e., highly confused script) is observed for *Malayalam* script. According to the proposed approach, in the second stage, the words from these misclassified text-lines are fed to the handwritten script recognition module to know the true script class of their parent text-lines. The outcome of the word segmentation procedure is recorded in Table 2.

The word images (obtained from the misclassified text-lines) undergo the same feature extraction procedure as described in Subsection 2.1. Classification of these segmented words is again done using MLP classifier whose configuration is taken as 120-50-12. The value of the threshold α is experimentally set to 0.7 where the best result has been achieved. After classification of the words, it is decided that if at least 70% of the words belonging to a particular text-line vote for a specific script, then the corresponding text-line is assigned to its voted script class. Otherwise, the text-line is assigned to the script class in which the maximum number words of that text-line are classified into, which occurs mostly for text-lines with less than five words. Proceeding in the following way, an average recognition accuracy of 97.75% is noted with 3-fold cross validation scheme. The final confusion matrix is shown in Table 3.

Table 2 Segmentation accuracy attained by the present word segmentation algorithm

# of actual text words (A)	# of text words found experimentally	Over-segmentation errors (O)	Under-segmentation errors (U)	$SR = \left[\dfrac{(A-(O+U))*100}{A}\right]$ $= 92.89\%$
929	863	27	39	

Table 3 Final confusion matrix generated after the application of two-stage framework

Class	A	B	C	D	E	F	G	H	I	J	K	L
A	193	3	0	1	0	0	0	2	0	1	0	0
B	0	199	0	1	0	0	0	0	0	0	0	0
C	4	0	192	0	0	0	0	3	1	0	0	2
D	0	3	1	194	0	2	0	0	0	0	0	0
E	0	0	0	0	199	0	0	0	0	1	0	0
F	1	0	1	0	2	192	0	2	2	0	0	0
G	0	1	1	2	0	1	193	0	0	0	2	0
H	2	0	1	0	1	0	0	194	2	0	0	0
I	0	0	0	0	1	0	0	0	198	1	0	0
J	1	0	1	1	2	1	0	0	0	194	0	0
K	0	0	0	0	0	0	0	0	0	0	200	0
L	1	0	1	0	0	0	0	0	0	0	0	198

It can be observed from the table that most of the confusing script text-lines (see Table 1) are now classified into their appropriate script classes. Here, the maximum confusing scripts are found as *Gujarati* and *Malayalam* whereas the least confusing script is *Urdu*. Performance enhancement of the overall system after applying the present two-stage framework is 4.62% which is quite impressive. That is, a total of 111 text-lines (previously misclassified) are correctly classified into their respective script classes. It is to be noted that in few cases, the proposed methodology also undergoes misclassification. Observing the misclassified images closely, it is found that where the number of words corresponding to a particular script text-line is less than three, the system has failed to classify the same. The reason behind this may be dearth of discriminate features to classify the words. Even, over-segmentation and under-segmentation of the words also sometimes cause the wrong estimation which we have ignored here.

4 Conclusion

The current methodology is devoted towards developing a two-stage framework for text-line based script recognition from the handwritten documents of *Indic* scripts. An encouraging outcome of this framework justifies the utility of considering the decision of lower level to rectify the erroneous result at the higher level. In future, the model would be applied on page-level where incorrect decision about the script class of a page can be altered by taking the feedback from both text-line and word-level. Another plan is to employ different features at different levels which may augment the recognition accuracy of the entire system. Size of the database will also be increased, in future, to prove the robustness of the proposed system in a diverse environment.

References

1. Singh, P.K., Sarkar, R., Das, N., Basu, S., Nasipuri, M.: Identification of Devnagari and Roman script from multiscript handwritten documents. In: Proceedings of 5th International Conference on PReMI, pp. 509–514. LNCS 8251 (2013)
2. Joshi, G.D., Garg, S., Sivaswamy, J.: Script identification from Indian documents. In: Lecture Notes in Computer Science: International Workshop Document Analysis Systems, pp. 255–267. Nelson, LNCS-3872, Feb 2006
3. Hiremath, P.S., Shivashankar, S.: Wavelet based co-occurrence histogram features for texture classification with an application to script identification in document image. Pattern Recogn. Lett. 29(9), 1182–1189 (2008)
4. Pati, P.B., Ramakrishnan, A.G.: Word level multi-script identification. Pattern Recogn. Lett. 29(9), 1218–1229 (2008)
5. Padma, M.C., Vijaya, P.A.: Wavelet packet based texture features for automatic script identification. Int. J. Image Process. 4(1), 53–65 (2010)
6. Obaidullah, S.M., Mondal, A., Roy, K.: Structural feature based approach for script identification from printed Indian document. In: Proceedings of IEEE Signal Processing and Integrated Networks (SPIN), pp. 120–124 (2014)
7. Obaidullah, S.M., Mondal, A., Das, N., Roy, K.: Script identification from printed Indian document images and performance evaluation using different classifiers. Appl. Comput. Intel. Soft Comput. Article ID: 896128, 1–12 (2014)
8. Hangarge, M., Santosh, K.C., Pardeshi, R.: Directional discrete cosine transform for handwritten script identification. In: Proceedings of 12th IEEE International Conference on Document Analysis and Recognition (ICDAR), pp. 344–348 (2013)
9. Pardeshi, R., Chaudhuri, B.B., Hangarge, M., Santosh, K.C.: Automatic handwritten Indian scripts identification. In: Proceedings of 14th IEEE International Conference on Frontiers in Handwriting Recognition (ICFHR), pp. 375–340 (2014)
10. Singh, P.K., Sarkar, R., Nasipuri, M.: Offline script identification from multilingual Indic-script documents: a state-of-the-art. Comput. Sci. Rev. (Elsevier) 15–16, 1–28 (2015)
11. Pal, U., Sinha, S., Chaudhuri, B.B.: Multi-script line identification from Indian documents. In: Proceedings of 7th International Conference on Document Analysis and Recognition (ICDAR), pp. 880–884, Aug 2003
12. Obaidullah, S.M., Santosh, K.C., Halder, C., Das, N., Roy, K.: Automatic Indic script identification from handwritten documents: page, block, line and word-level approach. Int. J. Mach. Learn. Cybern. 1–20 (2017)
13. Chanda, S., Pal, S., Franke, K., Pal, U.: Two-stage approach for word-wise script identification. In: Proceedings of 10th IEEE International Conference on Document Analysis and Recognition (ICDAR), pp. 926–930 (2009)
14. Ubul, K., Tursun, G., Aysa, A., Impedovo, D., Pirlo, G., Yibulayin, T.: Script identification of multi-script documents: a survey. IEEE Access 5, 6546–6559 (2017)

15. Singh, P.K., Chatterjee, I., Sarkar, R.: Page-level handwritten script identification using modified log-Gabor filter based features. In: Proceedings of 2nd IEEE International Conference on Recent Trends in Information Systems (ReTIS), pp. 225–230 (2015)
16. Sarkar, R., Malakar, S., Das, N., Basu, S., Kundu, M., Nasipuri, M.: Word extraction and character segmentation from text lines of unconstrained handwritten bangla document images. J. Intel. Syst. **20**(3), 227–260 (2011)
17. Gonzalez, R.C., Woods, R.E.: Digital Image Processing, vol. I. Prentice-Hall, India (1992)
18. Khandelwal, A., Choudhury, P., Sarkar, R., Basu, S., Nasipuri, M., Das, N.: Text line segmentation for unconstrained handwritten document images using neighborhood connected component analysis. In: Proceedings of 3rd International Conference on Pattern Recognition and Machine Intelligence (PReMI' 09). LNCS 5909, pp. 369–374 (2009)

Weight-Based Secure Approach for Identifying Selfishness Behavior of Node in MANET

Shoaib Khan, Ritu Prasad, Praneet Saurabh and Bhupendra Verma

Abstract Mobile adhoc networks (MANET) is an interesting concept, as all mobile nodes work like router. They can receive and transmit data packet any time in hostile environment even when the source and the destination mobile nodes are not directly connected to each other. In this circumstance, data packets are forwarded to the destination mobile node by relaying the transmission through intermediary mobile nodes. Due to selfish behavior, packets are intentionally dropped by intermediate nodes, also intermediate nodes continuously forward or broadcast control packets in network that consumes energy of node and lowers performance. This makes discovery of secure routing path and detection of selfish nodes and their activities as important challenges in MANET. This paper proposes a weight-based secure approach for identifying selfish behavior in MANET (WSISB). The proposed WSISB identifies the above-discussed activities and it will also discover a trusted path for secure data transmission. Experimental results show that WSISB performs much better than existing techniques and show significant improvement in terms of performance.

Keywords Selfish nodes · Trust · Packet drop · Secure · MANET

S. Khan (✉) · R. Prasad
Technocrats Institute of Technology (Excellence), Bhopal 462021,
Madhya Pradesh, India
e-mail: shoaib.khan005@gmail.com

R. Prasad
e-mail: rit7ndm@gmail.com

P. Saurabh · B. Verma
Technocrats Institute of Technology, Bhopal 462021, Madhya Pradesh, India
e-mail: praneetsaurabh@gmail.com

B. Verma
e-mail: bkverma3@gmail.com

© Springer Nature Singapore Pte Ltd. 2018
S. C. Satapathy et al. (eds.), *Information and Decision Sciences*,
Advances in Intelligent Systems and Computing 701,
https://doi.org/10.1007/978-981-10-7563-6_40

387

1 Introduction

Recent years have witnessed proliferation of different computing devices including mobile computing and various other communication devices imparting revolutionary change in information processing [1]. This pervasive communication is transforming the communication paradigm and bringing ease of life in all verticals. Consequently, the computing paradigm is shifting from the personal computing to ubiquitous computing where a user has the capability to access and utilize information through several electronic platforms whenever and wherever needed [2]. In this context, mobile ad hoc network (MANET) is a very fascinating concept due to its infrastructureless, IP-based network of mobile nodes architecture connected with radio. MANET also does not require a centralized administration and routers for execution and forwarding packets. Each node in MANET acts as a router to forward the packets either to the destination or to the next node in a specified path [3]. MANET is also blessed to have free or autonomous nodes which play a key and significant role in route discovery and then packet forwarding from source to destination. These nodes arrange themselves in new structures and operate without strict top-down network administration to facilitate packet delivery illustrated in Fig. 1 [4].

MANET witness more frequent changes in its structure, node placement, and communication links than fixed networks due to its dynamic structure. Environmental changes also affect the working of MANET in adverse manner [5]. One more key point is the assumption of trust between nodes in MANET. However, there exist chances that a node can demonstrate selfish behavior and might not act properly in either route creation or packet delivery. These selfish nodes utilize the

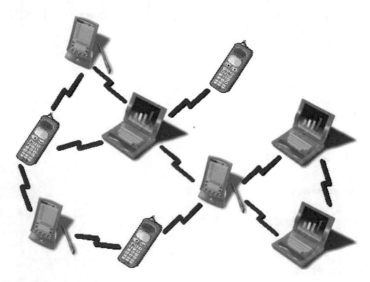

Fig. 1 Illustration of MANET

network and receive services from other nodes but they do not cooperate with other nodes of the network [6]. This behavior lowers the trust value of the network and lowers the performance. This paper proposes a Weight-based secure approach for identifying selfish behavior in MANET (WSISB) that will identify the selfish nodes and respective behaviors in order to discover a trusted path for secure data transmission. Rest of the paper is organized in the following manner; Sect. 2 presents the related work while Sect. 3 introduces the proposed work. Section 4 explains the results and analysis, while Sect. 5 concludes the paper.

2 Related Work

MANET is defined as a network which is composed of various autonomous nodes and demonstrates flexible and decentralized architecture [4]. These autonomous nodes act as a router to forward packets to the destination using multi-hop routing. Due to this fact, MANET does not have a centralized administrator and shows dynamic structure [6]. But, along with several benefits, MANET also demonstrates certain limitations like lack of centralized structure, limited bandwidth, limited power supply, routing overhead, and dynamic topology. MANET concept is based on trust that each node will cooperate in packet forwarding and delivery but in actuality some nodes act as selfish nodes [7]. This problem becomes more complex due to no centralized control and defined structure of the network [8]. In this scenario, detection of a selfish node becomes very difficult because there is no monitoring of the nodes. These selfish nodes remain in the network and utilize the services from different nodes; however, they do not collaborate with different nodes of the network in packet delivery [9]. Although, MANET uses multi-hop routing and routes for packet delivery but these selfish nodes in order to conserve energy do not demonstrate desired behaviors and eventually collapse the network.

Feng et al. [1] illustrated a game theory based cooperative incentive mechanism for detection of selfish nodes that included different utility functions based on three parameters—dependence, cooperation capability, and reputation. This cooperative incentive mechanism worked to maximize the utility function for selfish node detection. In an another work, Wu and Huey [2] created a MANET composed of "n" number of nodes linked through wireless links. This mechanism found the best route from source to the destination node. Thereafter, Wu and Yu [3] in their work defined AODV and suggested a mechanism to discover selfish nodes based on a threshold-based approach that achieved better detection rate. Buttyan and Hubaux [4] developed a system based on virtual currency, which focused on charge and reward. This system builds a trust-based hierarchy and offered incentives to the nodes that behaved in cooperative manner and nodes received reputation [6]. Around this time, Tootaghaj et al. [7] suggested a tit-for-tat mechanism for overcoming the limitations of presence of selfish nodes in MANET. Thereafter, Usha and Radha [8] introduced multi-hop acknowledgement scheme to detect the presence of misbehaving nodes in MANET. Agarwal and Motwani [10] and Bisen and

Sharma [11] represented other important contributions related to detection of malicious and selfish node in MANET. Recently, bio-inspired advances [12] also gained attention in realizing different goals on this domain [13–15].

3 Proposed Method

This section presents weight-based secure approach for identifying selfish behavior in MANET (WSISB) in AODV routing protocol in which weight-based node confidence factor is defined for identifying selfish behavior of node. The proposed algorithm make efforts to identify the above-discussed activities and it also tries to discover a trusted path for secure data transmission using three modules which include (i) monitoring module, (ii) identification module of selfish node, and (iii) elimination module of selfish node.

3.1 WSISB Working

This section describes the working of the proposed WSISB with algorithms employed at different modules. In the first step, **Monitoring Module** creates and monitors the mobile adhoc network. Detailed algorithm is given in Algorithm-1 given below:

Algorithm-1

Step 1: Create Mobile Ad hoc Network

Step 2: Initially, Mobile ad hoc network is configured with random assignment of mobility, mobile nodes, connections between source to destination and energy.

Step 3: Multiple sender (SN) and destination nodes (DN) define in network and AODV routing protocol configures to create route between them.

Step 4: In AODV routing, when sender node wants to communicate with the destination node then:

If destination node comes under the transmission range of source node then SN direct communicates with DN.
Otherwise source node uses intermediate (peer) nodes (IN) for packet forwarding to DN.

Step 5: To create effective routing for packet transmission, source node initially broadcasts Route request packet (RREQ) to its intermediate nodes.

    (a) If (Node == IN && Packet == RREQ)
         Send Acknowledgment to SN
    (b) If (IN == valid node)
         Use in routing process and SN can transmit data

    Otherwise
    IN may be selfish node and go to step 11 (Module II)

Step 6: Each intermediate node check following condition:

    If (Neighbor of IN == DN)

    IN won't transmits RREQ again

    Only data packet can be forwarded to DN and go to **Step 7**
    Otherwise

    IN again broadcasts the RREQ control packet to their neighbor & increase hop count and step 6 is repeated

Step 7: Finally destination node (DN) sends route reply packet (RREP) through IN to SN and connection is established.

    If (DN == True)
    Reply RREP (unicasting) to SN
    SN can send data packets
    Otherwise
    Repeat step 6

Step 8: Intermediate nodes are configured with random mobility therefore sometime communication link between source and destination can be changed.

    If (any IN moves from routing path or dead)
    Previous node broadcast Route Error packet to SN
    Otherwise Continue packet transmission (Repeat **Step 6**)

Step 9: Initially using previous steps simulation of MANET is done in monitoring module and analyzes performance of each node in the form of trace record.

Step 10: After analysis of trace record, performance of MANET is calculated in following terms:

(a) Packet delivery Ratio in variation of network size
(b) Throughput in variation of network size

3.2 Identification Module of Selfish Node

In this module, weight-based node confidence factor is defined for identifying selfish behavior of node. MANAT consists of different kinds of mobile nodes. Sometimes, the performance of network decreases in various conditions due to the presence of selfish nodes. Intermediate nodes can work like selfish node that performs the following activities:

- Due to selfish behavior, the packet is intentionally dropped by intermediate nodes.
- Intermediate nodes continuously forward or broadcast control packets in network that consumes more energy of node.
- Packets are dropped due to congestion and collision at the forwarding or intermediate node and destination nodes.

Therefore, this type of node reduces the performance of network. Finally, to identify selfish activities, weight-based node confidence factor ($WBCF_N$) has defined, which is calculated on the basis of node performance parameters like packet forwarding ratio of node (PFR_N), residual energy of node (RE_N), and packet dropped by node (PD_N).

(i) **Packet forwarding ratio of node (PFR_N)**: This parameter is the ratio of packets forwarded by a node to the packets received for forwarding (i.e., it is the intermediate node for that packet) by that node given in Eq. (1).

$$\text{Packet Forwarding Ratio of IN}(PFR_N) = \frac{\text{Packet Forwarding by node N}}{\text{Packet Received by node N}} \quad (1)$$

if node N forward all the packets it received for forwarding, then the value of the packet forwarding ratio will be unity; on the other hand, if it does not forward any packet or less than 50%, then the ratio will be zero.

(ii) **Residual energy of node (RE_N)**: Residual energy is the remaining energy of node at particular time of simulation which can be calculated by subtracting current energy at given time from initial energy of node stated in Eq. (2).

$$\text{Residual Energy of node}(\text{RE}_\text{N}) = \text{Initial Energy} - \text{Current Energy}(\text{in joule})$$
$$(2)$$

(iii) **Packet Dropped by Node (PD$_\text{N}$)**: Packet dropped is calculated by subtracting total packets received at the time of data communication by each node to total packet send by node. For any node N, if packet received equal to packet dropped, then that means trust value will be unity in Eq. (3).

$$\text{Packet Dropped by Node}(\text{PD}_\text{N}): \text{Packet Received} - \text{Packet Send} \quad (3)$$

Now, finally on the basis of these parameters, the confidence value of a node is defined which is estimated as follows in Eq. (4):

$$\text{WDCF}_\text{N} - \text{W1} \cdot (\text{PFR}_\text{N}) + \text{W2} * (\text{RE}_\text{N}) + \text{W3} * (\text{PD}_\text{N}) \quad (4)$$

The importance of a particular parameter is decided by assigning a unique weight to it. The weight assigned to a factor depends upon the system need. The flexibility in assigning weights to factors provides freedom to use this weight-based approach in different environments. The weights are chosen in such a way that the sum of three weights must be equal to unity given in Eq. (5).

$$\sum_{i=1}^{3} W_i = 1, \quad (5)$$

where PFR$_\text{N}$, RE$_\text{N}$, and PD$_\text{N}$ are the confidence factors related to data packets, respectively, and w1, w2, and w3 are weights or importance factors assigned to confidence parameters. In this calculation, values of PFR$_\text{N}$ and RE$_\text{N}$ should be maximum for best performance, while value of PD$_\text{N}$ should be minimum. At the end, overall estimated value of WBCF$_\text{N}$ should be maximum for best node and that node is participated in routing process after that repeat from step 5.

3.3 Elimination Module of Selfish Node

When the value of $WBCF_N$ is zero or under 25%, then that node may be declared as selfish node. Consider a case for best node, suppose trust value of PFR_N, RE_N, and PD_N are 1 and the values of weight (assign randomly W1, W2 and W3) are 0.5, 0.3, and 0.2 so that WBCF$_\text{N}$ will be 3 (three). In this scenario, calculated number is considered as a threshold value of $WBCF_N$. On the basis of that, value following confidence factors has been shown in Table 1.

In this condition, if (WBCF$_\text{N}$ == threshold_value ‖ WBCF$_\text{N}$ $\geq$ 25% of threshold), node will consider "Best Node" otherwise node will consider "Selfish node" and that will eliminate from the network. After elimination of selfish node,

Table 1 Confidence factor versus category of node

Confidence factor (WBCF$_N$)	Category of node
20	Best node
40	May be considered as intermediate node
80	Selfish node

again simulation is performed with same configurations and parameters, and result is compared with previous results and for this process repeat steps 9 and 10 of Algorithm 1.

4 Result Analysis

This section discusses the different experiments conducted on the developed WSISB under different scenarios to validate its efficiency. All the experiments are performed on NS-2.35 simulator on Ubuntu platform. The main objective of simulation is to ensure that proposed WSISB approach provides effective performance in the presence of selfish nodes. The experimental results and observations are also analyzed in this section. Performance results are compared with conventional AODV protocol because it is a highly used on-demand routing method for route establishment and maintenance. The performance has been analyzed in the following terms: (i) Packet delivery ratio in variation of network size and (ii) throughput in variation of network size.

These experiments measure the effect of network size on packet delivery ratio with the presence of 10% selfish node and compare the throughput in AODV and WSISB.

Table 2 and Fig. 2 show the effect of network size in packet delivery ratio. It is the ratio of total number of packets received by a node to the total number of packets sent from source. This experiment also states that as the number of nodes increases, the performance of conventional AODV degrades, which is due to that there is no mechanism in conventional protocol to detect selfish nodes so that each node like selfish node may participate in routing. On the other side, the proposed WSISB calculates the confidence factor of each node and on this basis on that factor only authorized nodes are used in routing process.

Table 2 Comparison of packet delivery (%)

Network size (nodes)	AODV (%)	WSISB (%)
20	61.35	95.35
40	63.45	96.48
60	64.99	97.65
80	72.89	98.37

Table 3 Comparison of throughput (kbps)

Network size (nodes)	AODV	WSISB
20	64	120
40	75	134
60	89	137
80	98	198

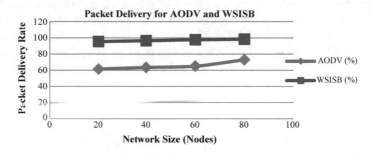

Fig. 2 Comparison of packet delivery ratio between AODV and WSISB

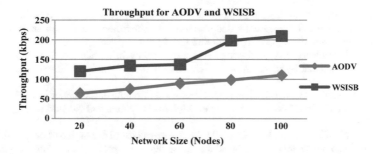

Fig. 3 Throughput comparison between AODV and WSISB

Table 3 and Fig. 3 represent the effect of number of nodes in throughput and compare the results of conventional AODV with proposed approach. Throughput defines total number of packets successfully delivered between source and destination in given simulation time period. In case of AODV, the performance is decreasing because of the presence of selfish nodes which drops the packets those are received by other nodes within simulation. But WSISB identifies these types of node using some confidence factor, and therefore nodes are successfully delivered packet from source to destination that will enhance the performance of network. Finally, improved results are observed in terms of packet delivery ratio and throughput as shown in Figs. 2 and 3.

5 Conclusion

In the area of MANET, finding secure link and detection of malicious activities are an important challenging issues. This paper mainly deals with the identification of selfish nodes. This paper has proposed a weight-based secure approach for identifying selfish behavior in MANET (WSISB) and it is also discovering a trusted path for secure data transmission. In the first phase of approach, monitoring algorithm has designed to trace the record of all nodes, and detection process of selfish nodes has defined on the basis of WBCF in the second module. Furthermore, simulation has performed in NS-2.35, and experimental results show that the proposed WSISB achieves much better results in terms of performance parameters such as packet delivery ratio and average throughput than existing techniques. In future point of view, more advanced security algorithm can be developed and analyzed some other attacks.

References

1. Feng, D., Zhu, Y., Luo, X.: Cooperative Incentive Mechanism based on Game Theory in MANET. In: ICNDS, pp. 201–204 (2009)
2. Wu, J.S., Huey, R.S.: A Routing Protocol Using Game Theory in Ad hoc Networks (2013)
3. Wu, L., Yu, R.: A Threshold-based method for selfish nodes detection in MANET. In: Computer Symposium (ICS), pp. 875–882 (2010)
4. Buttyn, L., Hubaux, J.P.: Enforcing service availability in mobile Ad-Hoc WANs. In: 1st ACM International Symposium on Mobile Ad hoc Networking and Computing, pp. 87–96 (2000)
5. Marti, S., Giuli, T.J., Lai, K., Baker, M.: Mitigating Routing Misbehavior in Mobile Ad hoc Networks 6th International Conference on Mobile Computing and Networking, pp. 255–265 (2000)
6. Xu, Z., Wu, G., Xia, Q., Ren, J.: GTFTTS: A generalized tit-for-tat based corporative game for temperature-aware task scheduling in multi-core systems. In: Third International Symposium on Parallel Architectures, Algorithms and Programming (PAAP), pp. 81–88 (2010)
7. Tootaghaj, D.Z., Farhat, F., Pakravan, M.R., Aref, M.: Game-theoretic approach to mitigate packet dropping in wireless Ad-hoc networks. In: Consumer Communications and Networking Conference (CCNC), pp. 163–165 (2011)
8. Usha, S., Radha, S.: Co-operative approach to detect misbehaving nodes in MANET using multi-hop acknowledgement scheme. In: Advances in Computing, Control, Telecommunication Technologies, pp. 576–578 (2009)
9. Komathy, K., Narayanasamy, P.: Study of co-operation among selfish neighbors in manet under evolutionary game theoretic model. In: Signal Processing, Communications and Networking, pp. 133–138 (2007)
10. Agarwal, R., Gupta, R., Motwani, M.: Performance optimisation through EPT-WBC in mobile Ad hoc networks. Int. J. Electron. 103(3), 355–371 (2015)
11. Bisen, D., Sharma, S.: An enhanced performance through agent-based secure approach for mobile Ad hoc networks. Int. J. Electron. (2017)
12. Saurabh, P., Verma, B.: An efficient proactive artificial immune system based anomaly detection and prevention system. Expert Syst. Appl. 60, 311–320 (2016)

13. Saurabh, P., Verma, B.: Cooperative negative selection algorithm. Int. J. Comput. Appl. (0975–8887) **95**(17), 27–32 (2014)
14. Saurabh, P., Verma, B, Sharma, S.: An immunity inspired anomaly detection system: a general framework. In: 7th BioInspired Computing: Theories & Applications, pp. 417–428. Springer (2012)
15. Saurabh, P., Verma, B., Sharma, S.: Biologically inspired computer security system: the way ahead. In: CICS, vol. 335, pp. 474–484. Springer (2011)

Assisting Vehicles Using Cyber-Physical Systems

Navid Anjum Munshi, Chandreyee Chowdhury and Sarmistha Neogy

Abstract Real-time vehicle control is one of the main applications of cyber-physical system. Modern smart vehicles contain different sensors and actuators and these execute physical processes, thereby making it a typical cyber-physical system. The exact operation of the vehicle depends on the correct interaction between its cyber and physical component. Any error or malfunctioning of any component in the vehicle may be fatal or cause property damages. Now-adays, accidents on roads are very frequent and in most of the cases, the main reason is over speeding. The main aim of this paper is to calculate average speed that uses real-time traffic information with historical traffic data. This may be used to develop traffic strategies that will improve and smoothen real-time traffic. Here, we propose a scheme to control vehicle speed using Cyber-Physical System (CPS) for driver assistance and traffic safety.

1 Introduction

Cyber-Physical Systems (CPS) have already taken a place in research in future systems. CPS has the promise of a near perfect technology as it has under itself frontiers of embedded systems, real-time systems, to name a few [1].

CPSs signify a class of next-generation engineered systems that integrate computing, physical, and communication components [1]. The integration of computations and communications as the cyber components with physical elements such as mechanical, electronic, and electromagnetic devices can provide better control of the physical system in addition to improved robustness, efficiency, and autonomy.

The use of CPS is evident in different areas like aeronautics, automotive, civil infrastructures, etc. and many more [2]. Figure 1 shows the entire system of CPS [3].

N. A. Munshi · C. Chowdhury · S. Neogy (✉)
Department of Computer Science and Engineering, Jadavpur University, Kolkata, India
e-mail: sarmisthaneogy@gmail.com

© Springer Nature Singapore Pte Ltd. 2018
S. C. Satapathy et al. (eds.), *Information and Decision Sciences*,
Advances in Intelligent Systems and Computing 701,
https://doi.org/10.1007/978-981-10-7563-6_41

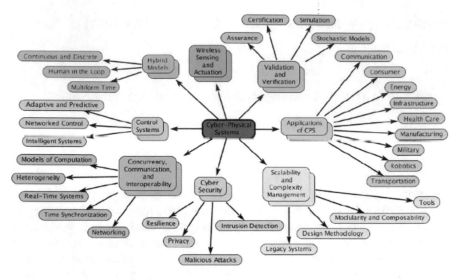

Fig. 1 Entire system of CPS [3]

With a view to enhance safety measures in smart transportation, accelerometers, GPS receivers, and other variety of sensor nodes can be used. This will also increase efficiency. For example, if bad road condition is detected by an accelerometer, the vehicle can use its GPS information and warn other vehicles of the same [4]. This will therefore increase efficiency as vehicles in that area will be prepared beforehand and act accordingly. Smart transportation is now becoming an essential part of daily human life.

The typical architecture of a CPS is shown in Fig. 2 that mainly consists of physical system, wireless router, and the cyber system.

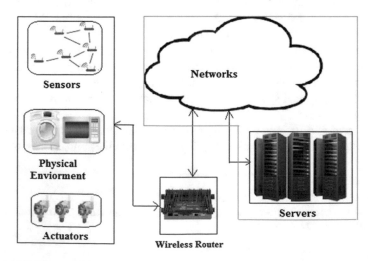

Fig. 2 Architecture of CPS

Physical system connects a number of different physical devices and components (such as sensors and actuators) by various communication protocols, such as Wi-Fi, Bluetooth, ZigBee, IEEE802.15.4, infrared, and so on. The networked embedded devices (physical system) work together toward sensing events and collecting data, transmitting it to the cyber system, thereby controlling the physical domain.

In cyber system, a decision-maker process will be responsible for receiving inputs from the physical system and analyzing the received data. Thereafter, it will activate the actuators in the physical system. This may be done with the help of a series of control processes. The entire process involves all layers of the network architecture.

The application layer abstracts the data received from the network layer and processes it according to some specified policies and rules. The control commands sent from this layer are directed to the physical components to be executed by the actuators.

Wireless router is a mediator between physical system and cyber system. One of the main objectives of wireless router is to switch the message structure as part of interfacing between physical and cyber world. It should provide seamless, high-quality service across different types of networks [5].

CPS can be designed in two ways depending on infrastructure. One is sever based; it is fitted for small architecture and requires individual maintenance. Another one is cloud-based; it is mainly effective for scalability, cost-effectiveness, and accessibility.

General CPS issues are as follows:

- Real-time operation: The system state can change rapidly [6].
- Distributed control: CPS is complex as it incorporates different systems, possibly each with own control mechanisms. So requirement is to have a decentralized control mechanism for overall functioning of the system [7].
- Security: Security is one of the main issues in CPS because physical system is accessed through the cyber system. For example, air traffic control mission-support systems of the U.S. Federal Aviation Administration is hacked several times in the recent past [8, 9].

There are additional issues like availability, reconfigurability, real-time operation (timeliness), fault-tolerance [10, 11], scalability, autonomy, reliability, heterogeneity, and geographic dispersion [12].

The rest of the paper runs as follows: Sect. 2 discusses state of the art. Present work is described in Sect. 3. Implementation of the scheme using R2016a is discussed along with the results in Sect. 4. Finally, Sect. 5 concludes the work.

2 Related Work

Over the last few years, a number of works on intelligent transport system using CPS, like car merging assistants [13], Cyber-physical traffic control systems [14], Smart traffic lights for traffic flow control [15], Traffic delay estimation [16], Traffic flow dynamics [17], Vehicle tracking systems [18], and many more, are mentionable.

In most cases, present-day smart road traffic habitually depends on the Google map. But Google map has some limitations:

- Users require all time internet connectivity in their smartphones or smart devices so that navigation is possible;
- Google map has limited offline support;
- Depending on the device, it might be harder to operate;
- Phone signal may be weak at certain locations; and
- Navigation can be interrupted when a call and other notification comes in.

Other problems like Global Positioning System (GPS) error may also creep in. The GPS can be responsible for one's location, altitude, and speed. It can be accurate as well. But error creeps in as a receiver has the signals from the orbiting satellites. So this has to be considered. In this paper, we introduce a new concept that does not use Google map and GPS; yet, our proposed technique provides an alternative smooth smart transportation system.

3 Our Work

Drivers care about their vehicles when they drive, so they want that the vehicle status is under their control and follow traffic rules (maintain the average speed) as well. Here, we want to control vehicle speed using CPS. The vehicle contains sensors, actuators, and controller. We want to maintain the vehicle speed, say, at 40 km/h or according to situation. In case of using Google map, Android phone users may turn on Google Maps app with GPS location enabled, in their phones. This allows location data to be updated continuously so that the authority knows the speed of their mobility. In fact, the apps read many data (including private) irrelevant to the current application. This raises privacy issues. Moreover, the driver him (her)self needs to maintain the speed limit recommended by Google Map.

In this work, we propose a novel scheme for vehicles in maintaining a fixed speed that is provided by an intelligent transport system. This system consists of wireless routers, computers, communication, and sensors. It is the application of sensing, analysis, control, and communications technologies to ground transportation (vehicles) in order to improve safety, mobility, and efficiency. The scheme is designed in such a way that even if a vehicle is driven at, say, 160 km/h and the intelligent transport system wants real-time traffic speed to be 60 km/h, then that will be maintained. So, even if the driver wants to accelerate, he will not be able to attain a speed greater than 60 km/h. Thus, the proposed system aims to provide safety and thereby avoid accidents considering real-time scenario.

Our proposed system architecture is shown in Fig. 3, described below. Here, we assume that multiple Wireless Routers (WRs) are kept at different locations, covering over 5,000 square feet. We consider that WR collects speed of a vehicle through a vehicle speed sensor. This speed sensor is attached with a special board like Arduino Uno [19] which is attached to the vehicle. WR is also responsible for giving the speed limit (average speed) to the sensor.

In this architecture, vehicle sensor is connected with WR, WR is connected with Local Server, and Local Server is connected with Central Server. The vehicle speed sensor is connected through wireless network, and the network is planned to meet the requirements of a particular application, such as smart transportation. WR acts as a gateway and the WR connects to a local server (LS) through the internet. If a vehicle is in the range of any WR, it takes average speed from the corresponding WR.

Here, we consider a four-lane road that is x km long. After each y km (calculated from the transmission range and terrain), there is a local server (LS). The main role

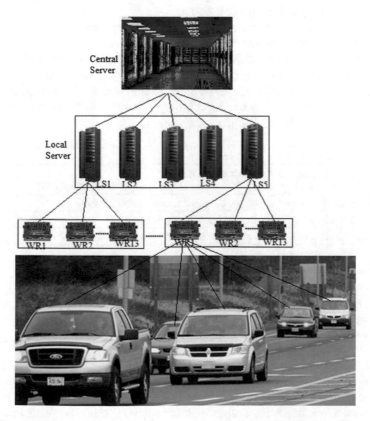

Fig. 3 Proposed system architecture

of the LS is to collect data from the vehicles through the WR and send it to the central server and vice versa (CS to WR). The LSs are connected to the CS over the Internet. When a vehicle is in the range of any WR, it takes necessary information from WR and takes appropriate action.

LS records contain two types of database, one is weekday's average speed and another is weekend average speed database with respect to time. After each 5 min interval, the LS updates information to the CS and deletes previous record. Then, LS creates new database from the fresh data (information) which is collected from the vehicle via WR. After collecting information from the LS (via WR), it is communicated to the central server to organize and analyze in order to take appropriate decisions (average speed). The CS updates the LS, if required. For example, if a procession is going on at a particular time that is known from the authorities, then the CS sends real-time delay information to all LS. The CS administrator is also linked with government traffic control authorities. The main role of the CS is to control the traffic. The vehicle speed to be maintained is then provided by the central server to the vehicle through the LS via WR.

In our scheme, if a vehicle is in range of any WR, the vehicle gets current average speed of that particular area spontaneously through the WR. So this scheme provides automatic vehicle driver assistance.

Delay in traffic is a primary cause of inefficiency, wasted fuel, and traveler frustration. Here, we introduce a technique so that each vehicle knows the current average speed and real-time delay when it is in range of any WR. In this scheme, each WR tracks the flow rate (traffic) according to time and stores it at the LS. The LS gives a threshold value (flow rate) to the WR. If the threshold value is exceeded, then LS instantly sends the value (real delay) to WR. The threshold value is calculated at the CS and CS gives the value to LS.

Average speed selection from WR algorithm:

1. Detect the WR region under which the vehicle is
2. If the vehicle is in range of any WR it gets information of that particular WR

---

3. If any update is required in the LS, CS takes instant action for those LS

---

4. a. If an LS fails at any time, the CS immediately backup

   b. If the CS crashes, then lost information would be collected from LSs

---

Real-time traffic delay calculation algorithm:

1. WR checks the vehicle flow rate at regular interval

---

2. If the flow rate is less than the predefined threshold (which is given by the LS)
   2.1 Then, LS calculates the delay and sends it to CS

---

3. CS broadcasts the information to all LS

Here, vehicle flow rate = the rate at which vehicles pass a fixed point (vehicles/h).

Traffic Delay = Current time to reach the destination—normal average time to reach the destination. Here, the threshold is defined according to average flow rate of a particular WR range. Basically, the traffic control system is dependent on travel behavior of vehicle. So the server must know the traffic (current speed) at a certain time and it gets information from the WR. The server also analyzes the collected data and takes some smart action. We also consider two types of databases: one for weekdays and another for weekend as we know that weekday traffic is typically much higher than weekend traffic. The database mainly consists of average speed according to time.

Here, we calculate average speed (S_{avg})

$$S_{avg} = \frac{L}{(T_f + C_j + D_i + D_r)PHF} \tag{1}$$

where L is the total distance, T_f is the time taken to cross the distance in freeway traffic, C_j is the average throughput of a queue, D_i is the specific delay (e.g., procession, accident, etc.), D_r is the real-time delay, and PHF is peak hour factor [20]. We have considered this PHF which denotes the relationship between the peak 15-min intervals of traffic compared to the total vehicle volume over the entire peak hour. It actually is a measure of finding vehicle volumes during the rush hours in a day.

The PHF represents the ratio of the traffic volume in an hour to the busiest 15-min interval in that hour. The peak 15-min interval is for the entire hour.

$$PHF = \frac{\text{total hourly volume}}{(\text{peak 15 minute volume within the hour} \times 4)} \qquad \text{from [20]}$$

$$= \frac{V}{(V_{peak} \times 4)} \text{ where: } 0 \langle PHF \rangle = 1$$

A PHF of 1 means that traffic volume in every 15-min interval is the same. This can be interpreted to be constant traffic flow in that hour. A decrease in PHF indicates that the traffic flow is variable, and the traffic volume has a spike during the peak 15-min interval.

$$C_j = \frac{No \ of \ vechiles \ in \ the \ junction}{Queue \ clerance \ rate(Q_c)},$$

where $Q_c = \frac{No \ of \ Vechiles}{\sum (No \ of \ lanes + Non \ of \ junction) Time}$ /hour

Here, number of lanes is four and total no of junctions is five. It is an infrastructure-based communication; so the communication delay is minimum.

4   Experimental Setup and Results

The traffic model is coded and simulated in Matlab (R2016a) m-file. Simulink is a block diagram environment for modeling and simulation of dynamic system. Sensors and actuators are directly connected to computing nodes. Here, these are

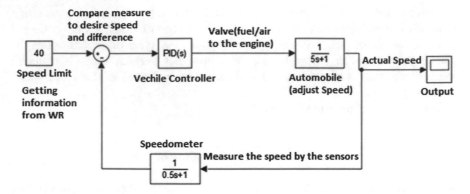

Fig. 4 Vehicles average speed control system

single-board computers like Arduino Uno board. It is an open-source computer hardware and software used for developing digital devices and interactive objects with activities of sensing and controlling objects in the physical world. The Arduino Uno catches the signal coming from WRs and sends the information to the vehicle control system. Here, the control system is PID controller. Here, we represent sensor as a speedometer, actuator as a fuel/air to the engine that is controlled by a valve, and controller as a computer in the vehicle which makes decision (maintain speed) that is shown in Fig. 4. Using R2016a, we create this model as is described below.

In Fig. 4, the constant will work as a set point (speed limit). The role of summation is to calculate the difference between desire and current speed. Automobile and speedometer both will work as a transfer function. The vehicle controller is to perform as a PID controller and finally scope will give us the output.

First, we set a speed limit. The summation block can add or subtract scalar, vector, or matrix inputs. It is also able to collapse the elements of a single input vector. The role of summation is to compare our set point and calculate the difference between measure and desired value. PID controller is also known as vehicle controller. Its main function is to take decision (average Speed). This block implements continuous- and discrete-time PID control algorithms. Advanced features of anti-windup, external reset, and signal tracking are included in this block. Main function of automobile (transfer function) is to adjust the vehicle speed. The numerator can be either a vector or any matrix expression. The denominator must be a vector. The output width denotes number of rows in the numerator. The coefficients are to be specified in descending order of powers of s. Here, numerator coefficient is 1 and denominator coefficient is (5, 1). Speedometer is also a transfer function. Here, numerator coefficient is 1 and denominator coefficient is (0.5, 1).

In Fig. 5, it is shown that after receiving information from WR the vehicle maintains maximum speed of 40 km/h. In Fig. 6, it is shown that after collecting information from WRs the CS analyzes the data and produces average speed for weekdays. Figure 7 shows that after collecting information from WRs the CS analyzes the data and produces average speed for weekend.

Fig. 5 Maintain speed 40 km/h

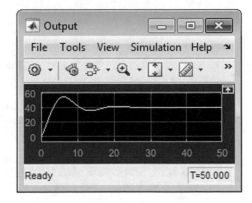

Fig. 6 CS weekday's average speed

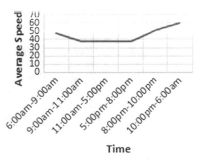

Fig. 7 CS weekend average speed

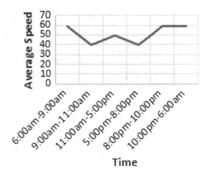

5 Conclusion

In this paper, an intelligent transport system providing automatic average speed to vehicles is proposed. This helps in reducing congestion or accident. We also create the average speed database from real-time scenario for certain time interval.

References

1. Kyoung, D.K., Kumar, P.R.: Cyber–physical systems: a perspective at the centennial. In: Proceedings of the IEEE 2012, pp. 1287–1308
2. http://www.ibm.com/smarterplanet
3. http://CyberPhysicalSystems.org
4. Lin, C.-Y., Zeadally, S., Chen, T.S., Chang, C.-Y.: Enabling cyber physical systems with wireless sensor networking technologies. Int. J. Distrib. Sens. Netw. Article ID 489794 (2012)
5. Lehpamer, H.: Transmission Systems Design Handbook for Wireless Networks. Artech House (2002)
6. Mitchell, R., Chen, I.-R.: A survey of intrusion detection techniques for cyber-physical systems. ACM (2014). ISSN: 0360-0300, EISSN: 1557-7341
7. Rajkumar, R.R., Lee, I., Sha, L., Stankovic, J.: Cyber-physical systems: the next computing revolution. In: Design Automation Conference 2010, Anaheim, California, USA, pp. 731–736. ACM
8. Mills, E.: Hackers broke into FAA air traffic control system. Wall Str. J. A6 (2009)
9. Biswas, S., Dey, P., Neogy, S.: Secure checkpointing-recovery using trusted nodes in MANET. In: IEEEXplore 2013 4th ICCCT, pp. 174–180 (2013)
10. Chowdhury, C., Neogy, S.: Checkpointing using mobile agents for mobile computing system. Int. J. Recent Trends Eng. 1(2), 26–29 (2009)
11. Chowdhury, C., Neogy, S.: Consistent checkpointing, recovery protocol for minimal number of nodes in mobile computing system. In: Lecture Notes in Computer Science, High Performance Computing—HiPC 2007, vol. 4873, pp. 599–611 (2007)
12. Mitchell, R., Chen, I.R.: A survey of intrusion detection techniques for cyber-physical systems. ACM Comput. Surv. 46(4) (2014)
13. Bell, M., Muirhead, L., Hu, F.: Cyber-physical system for transportation application. In: Hu, F. (ed.) Cyber-Physical Systems, pp. 253–265. CRC Press (2014)
14. Jianjun, S., Xu, W., Jizhen, G., Yanzhou, C.: The analysis of traffic control cyber-physical systems. In: Procedia Social and Behavioral Sciences, vol. 96, pp. 2487–2496 (2013)
15. Möller, D.P.F., Fidenco, A.X., Cota, E., Jehle, I.A., Vakilzadian, H.: Cyber-physical smart traffic light system. In: Proceedings of the IEEE EIT 2015 Conference
16. Astarita, V., Giofre, V.P., Guido, G., Festa, D.C.: Traffic delays estimation in two lane highway reconstruction. Proced. Comput. Sci. 32, 331–338 (2014)
17. Treiber, M., Kestling, A.: Traffic Flow Dynamics: Data, Models and Simulation. Springer Publishing (2014)
18. Möller, D.P.F., Deriyenko, T., Vakilzadian, H.: Cyber- physical vehicle tracking system-requirements for using radio frequency identification technique. In: Proceedings of the IEEE IET 2015 Conference
19. https://www.arduino.cc/
20. https://help.miovision.com/kb/peak-hour-factor-phf/

An Efficient Framework Based on Segmented Block Analysis for Human Activity Recognition

Vikas Tripathi, Durgaprasad Gangodkar, Monika Pandey
and Vishal Sanserwal

Abstract Video surveillance systems core component is human activity recognition. Analysis and identification of human activity emphasize on understanding human behavior in the video. Human activity recognition aims to automatically conjecture the activity being acted by a person. In this paper, we propose a novel feature description algorithm in which a segmented block of logarithm-based motion-generating frames is normalized for analysis of action being performed in the image sequences. The features obtained are classified using random forest classifier. We evaluated the framework on HMDB-51 and ATM datasets and achieved an average accuracy of 58.24 and 93.57%.

1 Introduction

Activity recognition is widely reviewed in the field of computer vision. Activity recognition is a systematic approach to examine the activities in a camera viewpoint. Human activity recognition (HAR) aims to automatically speculate the activity being acted by a person. For instance, recognition of actions like whether person is walking, running, and performing different activities. This usually emphasize the analysis and recognition of various motion patterns to outperform human activity description. Computer vision is significantly concerned with the analysis and understanding images in terms of features extant in the image frames. Computer vision also describes the transformation of visual images to the image data for vision perception [1]. The visual analysis-based human activity recognition

V. Tripathi (✉) · D. Gangodkar · M. Pandey · V. Sanserwal
Department of Computer Science and Engineering, Graphic Era University,
Dehradun, Uttarakhand, India
e-mail: vikastripathi.be@gmail.com

M. Pandey
e-mail: monikapandey234@gmail.com

V. Sanserwal
e-mail: vishuchaudhary28@gmail.com

© Springer Nature Singapore Pte Ltd. 2018
S. C. Satapathy et al. (eds.), *Information and Decision Sciences*,
Advances in Intelligent Systems and Computing 701,
https://doi.org/10.1007/978-981-10-7563-6_42

is the foremost application in computer vision. Identification of human activity recognition has conjured significance due to its prospective applications. A framework capable of identifying various activities of human has abundant significant applications like security system, healthcare systems, automated video surveillance system in public places [2], etc. to carry out activity analysis and data extraction from image sequences; series of frames are processed to depict the temporal motion information, which can be used as a feature to understand activity proficiently. In this paper, we show the effective approach for activity recognition using a feature extracting framework, which uses information from log-based motion-generating frames. To evaluate our framework, we use two HAR-based datasets, i.e., HMDB-51 [3] and ATM [4] datasets. The paper is organized in the following manner: Sect. 2 provides the previous work performed for activity recognition. In Sect. 3, methodology of the proposed framework has been explained. Section 4 contains the analysis of the result and Sect. 5 gives the conclusion of the paper.

2 Literature Survey

Lot of research has been accomplished in the field of computer vision, and numerous approaches have been projected to detect human actions. Computer vision enables to understand the behavior by analyzing and processing images. Human-action recognition is more complex in terms of slightly different actions [5, 7]. There are various approaches proposed by researchers for human activity analysis. Davis and Bobick [6] make use of different components of templates, i.e., Motion History Image and Motion Energy Image (MHI). MHI is a scalar-valued image in which intensity is a function showing recency of motion (Poppe [8]). Directional motion history image (DMHI) is an extension of MHI introduced by Ahad et al. [9, 10]. The temporal-based approaches utilize vector images where each vector point motion in the image [11, 12]. Descriptors like HOF and histograms of oriented gradients (HOG) give effective depiction and computation of actions. As per HOF descriptor, window is focused on the extracted features and after that sliced into a frequency histogram which is produced from each cell of the grid [n × n] is used by Hu et al. [13] to show the edge orientation in the cell. Laptev et al. [14] described the immobility in video for unlike actions. Representation based on histograms of oriented gradients is proficient as orientation information obtained is robust to changes in the camera view [15]. Motion using Optical flow is analyzed by the HOF, and it quantizes orientation of flow vectors [16]. HOG method is used to locate people in each frame of the surveillance video by A. Yussiff et al. [17]. A paper by Oreifej et al. [18] in which they used HOG to detect and analyze unstructured human activity in unstructured environment. Some approaches use combination of two or three different descriptors to enhance the accuracy. The fusion of the two descriptors, HOG with HOF, is used for the recognition of the objects from the frames and is used for the state-of-the-art performed, respectively [19, 20]. Optical flow model with new robust data obtained from HOG is proposed by Rashwan et al. [21].

In terms of optical flow method and frame sampling rate, the trade-off between accuracy and computational efficiency of descriptors such as HOF, HOG, and MBH is presented by Uijlings et al. [22]. A method called Hu moments is proposed by Hu which is invariant to scale, translation, and rotation. Tripathi et al. [23] used fusion of MHI with Hu moments to extract features from the video. Motion in the form of flow is shown by Mahbub et al. [24] using the approach of optical flow and give some effective results. An algorithm in which a fusion of HOG descriptor, Hu moments, and Zernike Moments is used to detect human activity from a single point of view is proposed by Sanserwal et al. [25]. This paper presents a novel feature descriptor-based algorithm using motion-generating frames to detect motion frames.

3 Methodology

In this method, frames are extracted from the video which is taken as an input. Subsequently, we add the three consecutive frames and perform logarithmic analysis on the results obtained. The resultant values are provided to our segmented block framework to extract features from the image sequences. Further, to classify the frames, random forest classifier is used. The framework is clearly illustrated in Fig. 1. Figure 2 represents the algorithmic representation of our framework.

In this proposed method, the video input is taken from two different datasets, i.e., HMDB and ATM datasets, Fig. 3 shows some classes of HMDB dataset, and Fig. 4 depicts the different classes of ATM dataset. For the generation of the motion image, we have used logarithmic analysis on the resultant sum of three consecutive frames. First, we calculate the summation of all the values of the three consecutive frames and after that we take the log of the summation. Further, a new feature extraction-based framework, i.e., segmented block descriptor is used to obtain motion features. The descriptor calculates x and y derivatives of image using convolution operation as shown in Eqs. 1 and 2.

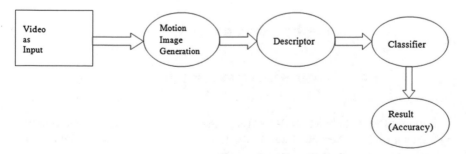

Fig. 1 Architecture of proposed framework

Input: Video	

1. Compute frames
2. Initialize x=0
3. Initialize n=3
4. While x < frames do
    a. Initialize y=0,sum=0
    b. While y < n
       i. sum=sum+y
      ii. y=y + 1
    c. Log(sum)
    d. x=x+1
5. Compute Descriptor
	1. Computing the frames of the video.
	2. Initialize x and sum by 0.
	3. Initialize n by 3.
	4. Frames = Total number of frames computed.
	    a. Initialize y and sum by 0.
	    b. Taking the sum of 3 frames only
	      i. Addition of frames.
	      ii. Increment y by 1.
	    c. Taking log of sum
	    d. Increment x by 1.
	5. Computing Descriptor.

Fig. 2 Algorithm for generation of descriptor

Fig. 3 Some classes of HMDB dataset

$$A_x = A * p_x \text{ where, } p_x = \begin{bmatrix} -1 & 0 & 1 \end{bmatrix} \tag{1}$$

$$A_y = A * p_y \text{ where, } p_y = \begin{bmatrix} 1 \\ 0 \\ -1 \end{bmatrix} \tag{2}$$

Random forest classifier used to classify the actions in the frames that functions by creating multiple decision trees during training. In our case, the model has been trained using random forest classifier which creates 100 trees.

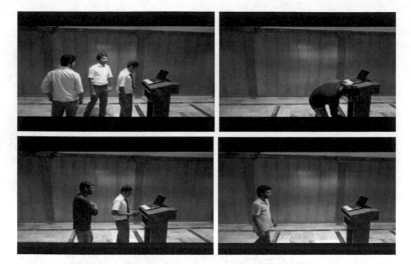

Fig. 4 Different classes of ATM dataset

4 Result and Analysis

The framework has been trained and tested using python and opencv at computer system having Intel i5, 2.4 GHz processor with 32 GB RAM on the videos for recognizing the human activities performed in the image sequences. To analyze the effectiveness of our method, we have performed analysis using two different datasets, i.e., HMDB dataset and ATM dataset. The HMDB dataset contains 51 classes and each class has 100 videos. Second dataset, i.e., ATM contains total 49 videos in all the four classes (10 single, 10 single abnormal, 20 multiple, and 9 multiple abnormal). Testing is done on different videos from the one used for the training purpose. For validating our result, we have tested our system with three different sets of videos in test and train in HMDB dataset. Thereafter, we consider that the accuracy of our proposed framework in HMDB dataset is 58 and 93% in ATM dataset, which can be analyzed by Table 1. Figure 5 shows a pictorial representation of the values of instances in test and train dataset of all the three train sets in terms of the frames. We have taken approximately 100,000 instances for the training purpose and approximately 40,000 for the testing purpose which shows the ratio of 60–40% as depicted in Fig. 5.

Table 1 Accuracy of descriptor on HMDB dataset

	Accuracy (%)
Train 1	56.8379
Train 2	57.7786
Train 3	60.1166
Average	58.2443

Fig. 5 Graph between
instances of test and train in
terms of frames in HMDB

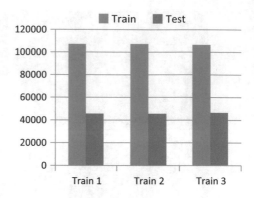

Table 2 Result analysis on HMDB dataset

	TP Rate	FP Rate	Precision	Recall	F-measure	ROC area
Train 1	0.568	0.012	0.593	0.568	0.562	0.908
Train 2	0.578	0.012	0.598	0.578	0.57	0.905
Train 3	0.601	0.011	0.617	0.601	0.594	0.918
Average	0.582	0.011	0.602	0.582	0.575	0.910
ATM average	0.936	0.022	0.942	0.936	0.933	0.998

Table 2 depicts the true-positive rate which shows the number of instances
predicted positive that are actually positive, false positive which is the number of
instances predicted positive that are actually negative, precision which is the
refinement in the calculations, recall means the fraction of the relevant instances
that are retrieved, f-measure takes the combination of precision and recall to provide
a single value of measurement, and ROC values of all 51 classes for three independent testing HMDB dataset and ATM dataset.

Figure 6 compares our framework with other HMDB-based methods and shows
that the improved dense trajectory method [24] gives highest accuracy rate,

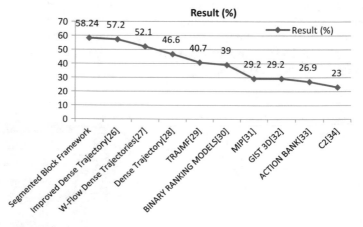

Fig. 6 Comparison with other algorithms (in percentage)

i.e., 57.2%, which is further improved by around 1% using our framework. The method represented outperforms proficiently from the other methods [26–34] and also use less computation for detecting the human activity from the videos. We also observed the performance of the framework on ATM dataset and analyzed that the ATM surveillance applications can be designed using the framework.

5 Conclusion

In this paper, we introduce an efficient logarithm-based framework to detect motion from video frames. The logarithm-based frames are used to capture the motion characteristics and these characteristics taken by segmented block descriptor to extract features. The features obtained are classified using random forest classifier. We validate our framework using two datasets HMDB and ATM dataset and achieved average accuracy of 58.24 and 93.57%, outperforming the existing HMDB-based algorithms. We analyzed that the proposed method performs well in controlled and uncontrolled environment and may be used as an application for ATM surveillance. Real-time analysis may be implemented to further improve the computation.

References

1. Pandey, M., Sanserwal, V., Tripathi, V.: Intelligent Vision Based Surveillance Framework for ATM Premises. Int. J. Control Theor. Appl. (2016)
2. Ryoo, M.S., Agarwal, J.K.: UT-Interaction dataset. In: IEEE International Conference on Pattern Recognition Workshops, vol. 2 (2010)
3. Kuehne, H., Jhuang, H., Stiefelhagen, R., Serre, T.: HMDB: a large video database for human motion recognition: In Proc. IEEE International Conference in Computer Vision (ICCV), Barcelona, Spain, pp. 2556–2563 (2011)
4. Tripathi, V., Mittal, A., Gangodkar, D., Kanth, V.: Real time security framework for detecting abnormal events at ATM installations. J. Real-Time Image Process. (2016)
5. Ramasso, E., Panagiotakis, C., Pellerin, D., Rombaut, M.: Human action recognition in videos based on the Transferable Belief Model. Pattern Anal. Appl. 11(1), 1–19 (2008)
6. Davis, J.W., Bobick, A.F.: The representation and recognition of human movement using temporal templates. In: Proceedings of the IEEE Computer Society Conference on Computer Vision and Pattern Recognition, June 1997, pp. 928–934
7. Marín-Jiménez, M., Pérez de la Blanca, N., Mendoza, M.: Human action recognition from simple feature pooling. Pattern Anal. Appl. 17(1), 17–36 (2014)
8. Poppe, R.: A survey on vision-based human action recognition. Image Vis. Comput. 28(6), 976–990 (2010)
9. Ahad, M.A.R., Ogata, T., Tan, J.K., Kim, H.S., Ishikawa, S.: Directional motion history templates for low resolution motion recognition. In: 34th Annual Conference on Industrial Electronics, pp. 1875–1880 (2008)

10. Ahad, M.A.R., Ogata, T., Tan, J.K., Kim, H.S., Ishikawa, S.: Template-based human motion recognition for complex activities. In: IEEE International Conference on Systems, Man and Cybernetics, pp. 673–678 (2008)
11. Garrido-Jurado, S., Munoz-Salinas, R., Madrid-Cuevas, F.J., Jimenez, M.J.: Automatic generation and detection of highly reliable fiducial markers under occlusion. Pattern Recogn. **47**(6), 2280–2292 (2016)
12. Holte, C.B., Moeslund, T.B., Gonzàlez, J.: Selective spatio-temporal interest points. Comput. Vis. Image Underst. **116**(3), 396–410 (2012)
13. Hu, R., Collomosse, J.: A performance evaluation of gradient field hog descriptor for sketch based image retrieval. Comput. Vis. Image Underst. **117**(7), 790–806 (2013)
14. Laptev, I., Lindeberg, T., Velocity adaptation of space-time interest points. In: Proceedings of the 17th International Conference on Pattern Recognition, Cambridge, vol. 1, pp. 52–56 (2004)
15. Freeman, W.T., Roth, M.: Orientation histograms for hand gesture recognition. In: International Workshop on Automatic Face and Gesture Recognition, vol. 12, pp. 296–301 (1995)
16. Wang, H., Schmid, C.: Action recognition with improved trajectories. In: Proceedings of the IEEE International Conference on Computer Vision, pp. 3551–3558 (2013)
17. Yussiff, A., Yong, S., Baharudin, B.B.: Detecting people using histogram of oriented gradients: a step towards abnormal human activity detection. In: Advances in Computer Science and its Applications, pp. 1145–1150. Springer, Berlin, Heidelberg (2014)
18. Oreifej, O., Liu, Z.: Hon4d: Histogram of oriented 4d normals for activity recognition from depth sequences. In: Proceedings of the IEEE Conference on Computer Vision and Pattern Recognition, pp. 716–723 (2013)
19. Laptev, I., Marszałek, M., Schmid, C., Rozenfeld, B.: Learning realistic human actions from movies. In: CVPR (2008)
20. Wang, H., Ullah, M., Klaser, A., Laptev, I., Schmid, C.: Evaluation of local spatio-temporal features for action recognition. In: BMVC (2009)
21. Rashwan, H.A., Mohamed, M.A., Angel Garca, M., Mertsching, B., Puig, D.: Illumination robust optical flow model based on histogram of oriented gradients. In: German Conference on Pattern Recognition, pp. 354–363. Springer, Berlin, Heidelberg (2013)
22. Uijlings, J., Duta, I.C., Sangineto, E., Sebe, N.: Video classification with densely extracted hog/hof/mbh features: an evaluation of the accuracy/computational efficiency trade-off. Int. J. Multimedia Inf. Retr. **4**(1), 33–44 (2015)
23. Tripathi, V., Gangodkar, D., Latta, V., Mittal, A.: Robust abnormal event recognition via motion and shape analysis at ATM installations. J. Elect. Comput. Eng. (2015)
24. Mahbub, U., Imtiaz, H., Ahad, M.A.R.: An optical flow based approach for action recognition: In: Computer and Information Technology (ICCIT), Dhaka, Bangladesh, pp. 646–651 (2011)
25. Sanserwal, V., Pandey, M., Tripathy, V., Chan, Z.: Comparative Analysis of Various Feature Descriptors for Efficient ATM Surveillance Framework (2016)
26. Wang, H., Schmid, C.: Action recognition with improved trajectories. In: ICCV (2013)
27. Jain, M., Jegou, H., Bouthemy, P.: Better exploiting motion for better action recognition. In: CVPR, (2013)
28. H. Wang, A. Klaeser, C. Schmid, and C-L Liu: Dense trajectories and motion boundary descriptors for action recognition. IJCV, (2013)
29. Jiang, Y., Dai, Q., Xue, X., Liu, W., Ngo, C.: Trajectory-based modeling of human actions with motion reference points. In: ECCV (2012)
30. Can, E.F., Manmatha, R.: Formulating action recognition as a ranking problem. In: International workshop on Action Similarity in Unconstrained Videos (2013)

31. Kliper-Gross, O., Gurovich, Y., Hassner, T., Wolf, L.: Motion Interchange Patterns for Action Recognition in Unconstrained Videos. In: ECCV (2012)
32. Solmaz, B., Assari, S.M., Shah, M.: Classifying web videos using a global video descriptor. Mach. Vis. Appl. (2012)
33. Sadanand, S., Corso, J.: Action Bank: a high-level representation of activity in video. In: CVPR (2012)
34. Kuehne, H., Jhuang, H., Garrote, E., Poggio, T., Serre, T.: HMDB: a large video database for human motion recognition. In: ICCV pp. 2556–2563 (2011)

Ranked Gene Ontology Based Protein Function Prediction by Analysis of Protein–Protein Interactions

**Kaustav Sengupta, Sovan Saha, Piyali Chatterjee,
Mahantapas Kundu, Mita Nasipuri and Subhadip Basu**

Abstract Computational function prediction of unknown protein is a challenging task in proteomics. As protein–protein interactions directly contribute to the protein function, recent efforts attempt to infer about proteins' functional group by studying their interactions. Recently, use of hierarchical relationship between functional groups characterized by Gene Ontology improves prediction ability compared to hierarchy unaware "flat" prediction methods. As a protein may have multiple functional groups with different degrees of evidences, function prediction is viewed as a complex multi-class classification problem. In this paper, we propose a method which assigns multiple Gene Ontology terms to unknown protein from its neighborhood topology using a ranking methodology showing different levels of association. This work achieves precision of 0.74, recall of 0.67, and F-score of 0.73,

---

K. Sengupta · M. Kundu · M. Nasipuri · S. Basu
Department of Computer Science and Engineering, Jadavpur University,
Kolkata 700032, India
e-mail: rish.kaustav@gmail.com

M. Kundu
e-mail: mahantapas@gmail.com

M. Nasipuri
e-mail: mitanasipuri@yahoo.com

S. Basu
e-mail: subhadip.basu@jadavpuruniversity.in

S. Saha (✉)
Department of Computer Science and Engineering, Dr. Sudhir Chandra
Sur Degree Engineering College, Dum Dum, Kolkata 700074, India
e-mail: sovansaha12@gmail.com

P. Chatterjee
Department of Computer Science and Engineering, Netaji Subhash
Engineering College, Garia, Kolkata 700152, India
e-mail: chatterjee_piyali@yahoo.com

© Springer Nature Singapore Pte Ltd. 2018
S. C. Satapathy et al. (eds.), *Information and Decision Sciences*,
Advances in Intelligent Systems and Computing 701,
https://doi.org/10.1007/978-981-10-7563-6_43

respectively, on 19,247 human proteins having 8,548,002 interactions in between themselves.

Keywords Gene ontology (GO) · Enrichment score · Edge weight
Shore protein · Bridge protein · Fjord protein · Gene ontology similarity
Protein–Protein interaction network (PPIN)

1 Introduction

The biological function of a protein is mainly governed by its three-dimensional structure which is encoded in its sequence. But finding a simple functional map from sequence to structure to its corresponding function is still a challenge. Moreover, protein does not perform its activity in isolation but in a group. Moreover, a given protein may have more than one function.

Protein function is a challenging task which can be characterized as follows: (1) Protein may have more than one functional groups with different reliabilities; (2) A large number of functional groups which are unbalanced; (3) They are not independent but exist in a hierarchy; and (4) Sometimes, true functional annotations are not available which can be denoted as "incomplete", "missing", or "unknown".

Inferring functions from protein–protein interactions are based on the fact that the unknown functions of a protein are related to its interacting partner. So, propagating annotation from "relevant" neighbors to unknown protein by analysis of its proximity and neighborhood properties is becoming a trend. But still this method faces challenge like protein interaction network may contain noisy or false data.

Protein interaction network-based approaches are classified into four groups: neighbor based, optimization based, Markov random field based, and clustering based [1]. Neighbor-based approach assumes that closer proteins in the network tend to have similar functions. Schwikowski et al. [2] and Hishigaki et al. [3] assigned function to target proteins primarily considering the probability of occurrence of functions among the neighboring proteins. One major drawback of this approach is that choice of the neighboring proteins is limited to only the level-1 neighbors of the target proteins. Chen et al. [4] have further extended the neighbor-based approach by labeling network motifs in protein interactomes for protein function prediction. Among optimization-based approaches, Vazquez et al. [5] introduced a global optimization principle by optimizing the constraints of making a functional assignment to an unclassified (target) protein consistent with the functional assignments to its unclassified (target) partners in the protein–protein interaction network (PPIN). Karaoz et al. [6] have mapped the functional linkage graph into a variant of a discrete-state Hopfield network to achieve a maximally consistent assignment by minimizing an "energy" function through a local search procedure guided by heuristics. In this work, edges in interaction networks are weighted using gene expression data. It considers more global properties of interaction maps but does not emphasize on local proximity of interacting proteins in the graph, as commented by Nabieva et al. [7]. To overcome this limitation and take care of distant effects of annotated proteins in the network, Nabieva et al. [7] have introduced an algorithm called functional flow using the idea of network flow and by treating each protein of known functional annotation as a "source" of "functional flow". Functional module detection [8] and graph clustering methods [9] are significant module-assisted approach. Along with interaction information, protein sequence, structure, and gene expression data can be combined to improve its accuracy as in the work of Zhao et al. [10] and Piovesan et al. [11].

From recent researches, it can be observed that protein interaction networks may contain false positive or negative data which are obstacles for obtaining higher recall and precision. Filtering protein interaction network by removal of these noisy data may be the key to improve the predictor's accuracy. This necessitates preprocessing of network by eliminating nonessential proteins or edges as found in the work of Peng et al. [12]. Another observation is that proteins have association with several functional groups of different degrees of reliability. Functional groups are not independent but exist in a hierarchy. Thus, taking into account the hierarchical relationship among labels organized by Gene Ontology (GO), protein function prediction can be extended to a multi-label classification problem. In the initial effort of another work [13] to predict protein's function, neighborhood score is computed for level-1 and level-2 neighbors of target proteins. Next, the score is empowered by incorporating functional similarity, edge-clustering coefficient, and path connectivity score [14, 15]. Finally, with neighborhood score thus empowered, this method identifies the most promising dense subnetwork where most interactions are expected to be found. Motivated by all these ideas, we have proposed a method here which creates a neighborhood graph of a target protein firstly, then it filters and prunes the neighborhood graph by removing nonessential proteins and insignificant interactions using graph-theoretic measures, and finally it propagates GO terms with a score or rank from neighbors to target proteins.

2 Materials and Methods

The dataset of protein interaction networks of Homo Sapiens is downloaded from STRING [16]. Corresponding GO terms are collected from UniProt [17] by mapping which consists of 156,710 proteins involving 46,410 number of protein interactions. Experimental studies show that significant inconsistencies are found if individual functional group is treated independently. Here, we use Gene Ontology (GO) [18] characterized by hierarchical relationship among functional groups which is better than "flat" prediction. So, in this method, functionally unannotated proteins and their interactions are filtered out from the dataset. The entire statistical

Table 1 The statistics of homo sapiens protein interaction network

Description	UniProt	STRING
#proteins	156,710	19,247
#Interactions	46,410	8,548,002
Total GO terms	16,915	–
Total unique GO terms/functions	12,366	7
Cellular component (CC)	1547	–
Molecular function (MF)	4105	–
Biological process (BP)	11,263	–
Proteins in top10 GO (Target proteins)	9141	6999
Randomly chosen targets (20%)	1828	1400
Proteins after filtering	1156	937

data of our input database in our proposed method is described in Table 1. The proposed method is outlined in three steps which are as follows:

2.1 Target Protein Selection

In this part, target proteins are selected whose functional annotation is to be predicted. The set of directly interacting pairs is collected from STRING database [19], and corresponding GO terms are retrieved from UniProt database [17]. If direct mapping from STRING ID to UniProt ID is not available, then the proposed method tries to find a similar entry with at least 90% sequence identity from UniProt [17]. Next frequency of GO terms is separately measured. According to frequency of GO terms, the top 10 frequently occurred GO terms are taken. The corresponding STRING IDs are obtained from these GO terms by reverse mapping. Randomly chosen 20% of these proteins are considered to be the target proteins (see Fig. 1).

2.2 Pruning and Filtering Neighborhood Graph

For each target protein, interaction information is retrieved from STRING database [16] and their neighborhood graph is formed. To remove nonessential protein, a test is performed for the neighbors of target protein whether it is of either Bridge or Fjord or Shore type [20]. If the neighbor is proven to be any one of this type, then it is not included. Thus, a pruned neighborhood graph is obtained. Next, to remove insignificant interactions, the edge weight [21] of every edge is computed, and if it does not exceed the threshold then that edge is not considered as valid interaction (see Fig. 2).

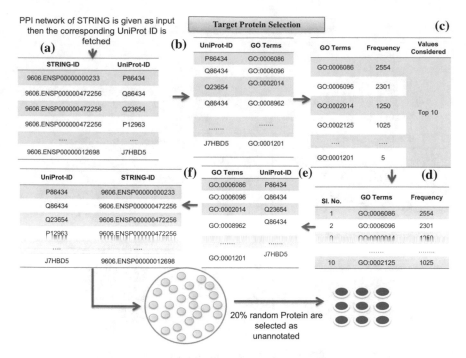

Fig. 1 Stepwise representation of target proteins selection (**a**: The first table lists the protein from STRING [16] and mapped to their UNIPROT ID, **b**: the second table lists their corresponding GO terms, **c**: the third one ranks GOs according to their frequency; Here, only 10 top-ranked GOs are considered; **d–f**: In the next three tables, UNIPROT Ids and STRING IDs are retrieved by reverse mapping for proteins corresponding to those top 10 GO and 20% random proteins are selected from them as target proteins)

2.3 GO Terms Propagation from Neighbors to Target Proteins Using Ranking Method

In this step, the corresponding UniProt ID is fetched for each level-1 protein of each target protein in the pruned and filtered neighborhood graph. Then, the GO [18] terms for each corresponding fetched UniProt ID are obtained from UniProt [17]. Then, proteins associated with these GO terms are considered to be its level-1 proteins. The enrichment of each GO term in the group of interacting proteins is then calculated by comparing it with the rest of annotated nodes in the entire STRING network of human. Enrichment is measured by calculating a P value with Fisher's exact test and used to rank GO terms. According to computed enrichment score [11], GO terms are ranked. Thus, GO terms with reliability measure are propagated from neighbors to target proteins. In this way, a target protein is annotated with multi-label functional groups along with their rank (see Fig. 3).

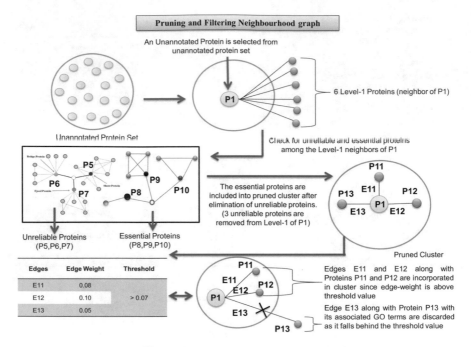

Fig. 2 Refinement of neighborhood graph of target protein (This figure shows removal of nonessential proteins and insignificant interactions from the level-2 and level-1 neighborhood graph of target proteins)

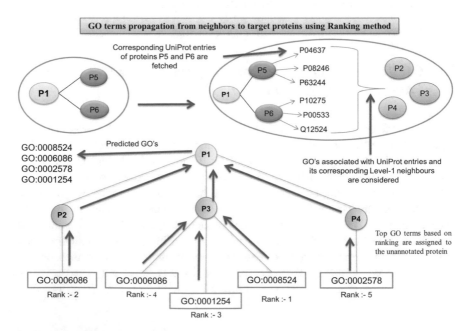

Fig. 3 GO terms propagation from neighbors to target proteins using ranking method (This figure shows the entire procedure of ranking each GO term and the ultimate propagation of top-ranking GO terms to the unannotated protein)

3 Result and Discussion

The performance evaluation of our algorithm has been estimated using precision (P), recall (R), and F-score (F) [14]. The method achieves an overall precision, recall, and F-score of 0.74, 0.67, and 0.73, respectively. Predicted GO terms are also validated with the online GO similarity tool, G-Sesame [22], using Lin's similarity. G-sesame usually reveals the similarity scoring matrix of the predicted GOs with respect to the original GOs which depicts the fact how close two GOs are. Few samples have been highlighted in Table 2. From Table 2, it can be said that both the proposed method and G-sesame yield more or less similar results except in the case of protein 9606.ENSP00000300935 where predicted result is satisfactory than others. The method has been compared with seven of the existing methods: neighborhood counting method [2], the chi-square method [3], a recent version of the neighbor relativity coefficient (NRC) [23], FS-weight based method [24], NAIVE method [11], BLAST method [11], and INGA [11].

From Table 3, it can be also concluded that our method outperforms the other methodologies in the same dataset of Homo Sapiens in the terms of precision–recall and F-score values due to several reasons which are as follows: (1) Highly connected proteins tend to be more important and are found to be associated with multiple functions. Target proteins are selected from proteins having top-tanked GO terms. (2) In order to remove false and noisy data, topologically nonessential nodes and their interactions are removed from neighborhood graph by identifying Bridge,

Table 2 Statistics of matched and unmatched GOs predicted by the proposed method by validating with G-sesame GO similarity tool

Sl. No.	Protein name (STRING)	Methods	Observed GO	Predicted GO	Matched GO	Unmatched GO
1	9606. ENSP00000300935	Proposed method	44	41	36	5
		G-Sesame [22]	44	41	34	7
2	9606. ENSP00000354876	Proposed method	17	13	10	3
		G-Sesame [22]	17	13	10	3
3	9606. ENSP00000289081	Proposed method	14	12	11	1
		G-Sesame [22]	14	12	13	−1
4	9606. ENSP00000364028	Proposed method	43	31	31	0
		G-Sesame [22]	43	31	31	0

Table 3 Comparison of performance evaluation metrics of the proposed method with other existing methods

Methods	Precision	Recall	F-score
The Proposed Method	0.74	0.67	0.73
INGA [11]	0.44	0.51	0.47
BLAST [11]	0.30	0.50	0.37
NAÏVE [11]	0.33	0.31	0.31
Chi-square #1&2 [3]	0.13	0.12	0.12
Chi-square #1 [3]	0.12	0.15	0.13
Neighborhood counting #1&2 [2]	0.21	0.25	0.18
Neighborhood counting #1 [2]	0.15	0.21	0.17
Fs-weight #1&2 [24]	0.24	0.22	0.22
Fs-weight #1 [24]	0.16	0.19	0.19
Nrc [23]	0.25	0.24	0.22

Shore, and Fjord proteins. (3) Every GO as well as non-GO terms are taken into account while prediction which provides a sort of balance in the methodology.

It can be concluded that these three factors boost up prediction accuracy of the proposed method. Incorporating protein domain or motif and sequence might help to enhance its accuracy. Moreover, the proposed method is not limited to PPI network of human but may also be applicable to PPI network of other organisms which may show a future direction in this regard.

References

1. Tiwari, A.K., Srivastava, R.: A survey of computational intelligence techniques in protein function prediction. Int. J. Proteomics. **2014**, 845479 (2014)
2. Schwikowski, B., Uetz, P., Fields, S.: A network of protein-protein interactions in yeast. Nat. Biotechnol. **18**, 1257–1261 (2000)
3. Hishigaki, H., Nakai, K., Ono, T., Tanigami, A., Takagi, T.: Assessment of prediction accuracy of protein function from protein–protein interaction data. Yeast **18**, 523–531 (2001)
4. Chen, J., Hsu, W., Lee, M.L., Ng, S.-K.: Labeling network motifs in protein interactomes for protein function prediction. In: 2007 IEEE 23rd International Conference on Data Engineering, pp. 546–555. IEEE (2007)
5. Vazquez, A., Flammini, A., Maritan, A., Vespignani, A.: Global protein function prediction from protein-protein interaction networks. Nat. Biotechnol. **21**, 697–700 (2003)
6. Karaoz, U., Murali, T.M., Letovsky, S., Zheng, Y., Ding, C., Cantor, C.R., Kasif, S.: Whole-genome annotation by using evidence integration in functional-linkage networks. Proc. Natl. Acad. Sci. USA **101**, 2888–2893 (2004)
7. Nabieva, E., Jim, K., Agarwal, A., Chazelle, B., Singh, M.: Whole-proteome prediction of protein function via graph-theoretic analysis of interaction maps. Bioinformatics **21**(Suppl 1), i302–i310 (2005)
8. Sharan, R., Ulitsky, I., Shamir, R.: Network-based prediction of protein function. Mol. Syst. Biol. **3** (2007)

9. King, A.D., Przulj, N., Jurisica, I.: Protein complex prediction via cost-based clustering. Bioinformatics **20**, 3013–3020 (2004)
10. Zhao, B., Wang, J., Li, M., Li, X., Li, Y., Wu, F.-X., Pan, Y.: A new method for predicting protein functions from dynamic weighted interactome networks. IEEE Trans. Nanobiosci. **15**, 131–139 (2016)
11. Piovesan, D., Giollo, M., Leonardi, E., Ferrari, C., Tosatto, S.C.E.: INGA: Protein function prediction combining interaction networks, domain assignments and sequence similarity. Nucleic Acids Res. **43**, W134–W140 (2015)
12. Peng, W., Wang, J., Wang, W., Liu, Q., Wu, F.-X., Pan, Y.: Iteration method for predicting essential proteins based on orthology and protein-protein interaction networks. BMC Syst. Biol. **6**, 87 (2012)
13. Saha, S., Chatterjee, P., Basu, S., Kundu, M., Nasipuri, M.: Improving prediction of protein function from protein interaction network using intelligent neighborhood approach. In: Proceedings of 2012 International Conference on Communications, Devices and Intelligent Systems, CODIS (2012)
14. Saha, S., Chatterjee, P., Basu, S., Kundu, M., Nasipuri, M.: FunPred-1: Protein function prediction from a protein interaction network using neighborhood analysis. Cell. Mol. Biol. Lett. **19** (2014)
15. Saha, S., Chatterjee, P., Basu, S., Nasipuri, M.: Gene Ontology Based Function Prediction of Human Protein Using Protein Sequence and Neighborhood Property of PPI Network. In: Proceedings of 5th International Conference on Frontiers in Intelligent Computing: Theory and Applications: FICTA 2016, vol. 2, pp. 109–118. Springer, Singapore (2017)
16. Szklarczyk, D., Franceschini, A., Wyder, S., Forslund, K., Heller, D., Huerta-Cepas, J., Simonovic, M., Roth, A., Santos, A., Tsafou, K.P., Kuhn, M., Bork, P., Jensen, L.J., von Mering, C.: STRING v10: protein-protein interaction networks, integrated over the tree of life. Nucleic Acids Res. **43**, D447–D452 (2015)
17. The UniProt Consortium: UniProt: a hub for protein information. Nucleic Acids Res. **43**, D204–D212 (2014)
18. Ashburner, M., Ball, C.A., Blake, J.A., Botstein, D., Butler, H., Cherry, J.M., Davis, A.P., Dolinski, K., Dwight, S.S., Eppig, J.T., Harris, M.A., Hill, D.P., Issel-Tarver, L., Kasarskis, A., Lewis, S., Matese, J.C., Richardson, J.E., Ringwald, M., Rubin, G.M., Sherlock, G.: Gene Ontology: tool for the unification of biology. Nat. Genet. **25**, 25–29 (2000)
19. Franceschini, A., Szklarczyk, D., Frankild, S., Kuhn, M., Simonovic, M., Roth, A., Lin, J., Minguez, P., Bork, P., Von Mering, C., Jensen, L.J.: STRING v9.1: Protein-protein interaction networks, with increased coverage and integration. Nucleic Acids Res. **41**, 808–815 (2013)
20. Hanna, E.M., Zaki, N.: Detecting protein complexes in protein interaction networks using a ranking algorithm with a refined merging procedure. **15**, 1–11 (2014)
21. Wang, S., Wu, F.: Detecting overlapping protein complexes in PPI networks based on robustness. Proteome Sci. **11**, S18 (2013)
22. Du, Z., Li, L., Chen, C.F., Yu, P.S., Wang, J.Z.: G-SESAME: Web tools for GO-term-based gene similarity analysis and knowledge discovery. Nucleic Acids Res. **37**, 345–349 (2009)
23. Moosavi, S., Rahgozar, M., Rahimi, A.: Protein function prediction using neighbor relativity in protein-protein interaction network. Comput. Biol. Chem. **43**, 11–16 (2013)
24. Chua, H.N., Sung, W.-K., Wong, L.: Exploiting indirect neighbours and topological weight to predict protein function from protein-protein interactions. Bioinformatics **22**, 1623–1630 (2006)

Investigation of Optimal Cyclic Prefix Length for 4G Fading Channel

Ch. Vijay, G. Sasibhushana Rao and Vinodh Kumar Minchula

Abstract The Orthogonal Frequency Division Multiplexing (OFDM) system plays a key role in 4G communication systems due to its ability to mitigate the frequency selective fading in a multipath channel. Channel fading causes Intersymbol Interference (ISI) and Intercarrier Interference (ICI) because of delay spread effect. These problems can be reduced by correlating the cyclic prefix with the OFDM symbol. This paper analyzes the OFDM performance with and without cyclic prefix of various lengths by considering the Bit Error Rate (BER) as performance metric. Results indicate that for an optimum cyclic prefix length (FFT length = 128 and cyclic prefix size = 16), the BPSK modulation technique is providing better BER performance when compared to other modulation techniques such as QPSK, 8PSK, 8QAM, 16QAM, 32QAM, and 256QAM.

1 Introduction

In modern wireless communication systems, Orthogonal Frequency Division Multiplexing (OFDM) is the most widely applicable in 4G technology. OFDM is a Frequency Division Multiplexing (FDM) contrivance implemented as a digital multicarrier modulation method [1]. It is also a technique that can be used to mitigate frequency selective channels. OFDM uses a large number of subcarriers to carry data on several parallel data streams [2]. OFDM is attracting considerable attention due to its excellent performance under intense channel condition.

Ch. Vijay (✉) · G. Sasibhushana Rao · V. K. Minchula
Department of Electronics and Communication Engineering, A.U.C.E(A),
Andhra University, Visakhapatnam 530003, Andhra Pradesh, India
e-mail: vchanamala@gmail.com

G. Sasibhushana Rao
e-mail: sasigps@gmail.com

V. K. Minchula
e-mail: vinodh.edu@gmail.com

© Springer Nature Singapore Pte Ltd. 2018   429
S. C. Satapathy et al. (eds.), *Information and Decision Sciences*,
Advances in Intelligent Systems and Computing 701,
https://doi.org/10.1007/978-981-10-7563-6_44

The advantages of OFDM are as follows:

- High spectral efficiency due to the absence of guard band.
- Simple and efficient hardware implementation using the FFT operation.
- Avoids intersymbol interference, thereby avoids the equalizers, and hence, the complexity of receivers is reduced.
- Each subcarrier can have different modulation or coding schemes leading to the design of highly robust adaptive transmission schemes.
- Enables frequency diversity by spreading the subcarriers across the usable spectrum.
- Provides good resistance against co-channel interference and impulsive noise.

Disadvantages:

- High sensitivity to Doppler shifts, requiring accurate frequency and time synchronization.
- Loss in spectral efficiency due to loss of guard interval.
- High PAPR due to overlap of a large number of modulated subcarrier signals which require the transmit power amplifier to be linear across the whole signal.

2 A Brief Study of Cyclic Prefix (CP)

The term CP refers to the prefixing of a symbol with a repetition of the end symbol. Different OFDM CP lengths are available in various systems, Normal and Extended [3]. Figure 1 shows the OFDM signal with cyclic prefix attached to the data payload.

The positive and negative effects of cyclic prefix within OFDM are as follows:
  Positive effects:

- The use of cyclic prefix gives robustness to OFDM signal. If necessary, the retransmitted data can be used again.
- ISI can be reduced due to the guard symbol initiated by the cyclic prefix.
- By using simple single tap equalizer, linear convolution is converted into cyclic convolution (CC) which mitigates the task of detection of the received signal.

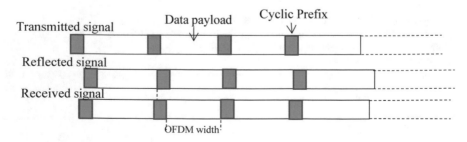

Fig. 1 OFDM signal with cyclic prefix

Negative effects:

• Reduction of the overall data rate as the retransmitted data uses the system capacity.

In wireless communication application, there is degradation in performance of OFDM due Channel Impulse Response (CIR) if the cyclic prefix length is too low. A synchronization mismatch or time of reference moves part of the impulse signal outside CP causing ISI and ICI, even for a CIR shorter than the CP length. In multipath environment for all the paths to arrive at the receiver, the addition of the CP allows time, before the demodulation of the symbol [4].

Delay dispersion occurs when the CIR h(t) is not equal to δ(t), which in turn causes ISI. If the orthogonality of the received signal is not achieved, it will cause the ICI. The ISI and ICI problems can be overcome by inserting CP to each symbol.

Generally, without CP the transmitted data is the K value as given in Fig. 2. $(K = K_{fft} = \text{length of FFT/IFFT})$ for a single OFDM symbol.

$$X = X_0, X_1, X_2, \ldots, X_{N-1} \tag{1}$$

Consider a cyclic prefix with K_{cp} length, which is formed by placing the last K_{cp} values from Eq. (1) and adding these K_{cp} values to the beginning of the X vector. Now the OFDM symbol constitutes the following values for the cyclic prefix length K_{cp} (where the length of $K_{cp} < K$).

If "S" is time span of an OFDM symbol in seconds, addition of cyclic prefix of length K_{cp} which causes the total time span as given by Eq. (2),

$$S_{OFDM} = S + S_{cp} \tag{2}$$

where S_{cp} indicates $K_{cp}*S/K$.

Therefore, the allocation of CP samples can be obtained from $K_{cp} = S_{cp}*K/S$, where K is the length of FFT/IFFT, S indicates time span of IFFT/FFT, and S_{cp} is the CP time span.

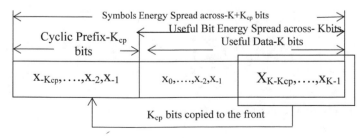

Fig. 2 The process of cyclic prefix

Assume "K_{cp}" is the length of CP which is affix to ODFM symbol, and channel (r) is the output given by CC of CIR (h) and CP (x).

$$r = h \circledast x \qquad (3)$$

Equation (3) can be translated to multiplication in frequency domain as follows:

$$R = HX \qquad (4)$$

"R" is frequency domain signal at the receiver which is used to estimate the transmitted signal (X) from the received signal R. From Eq. (4), the problem detected at the receiver end of the transmitting signal into a simplified equation is

$$\hat{X} = \frac{R}{H} \qquad (5)$$

After the performance of FFT at the receiver end, a single tap equalizer is utilized to evaluate the transmitted OFDM symbol, which also improves the performance of phase and amplitude.

From Fig. 3, time-domain symbol can be represented by small case letters whereas the frequency domain symbols with uppercase letters [5].

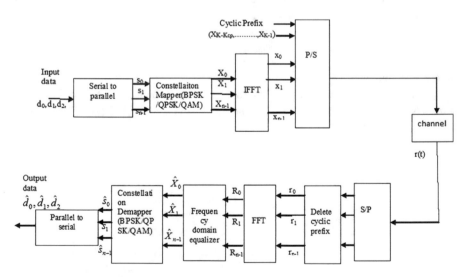

Fig. 3 An OFDM communication architecture with cyclic prefix

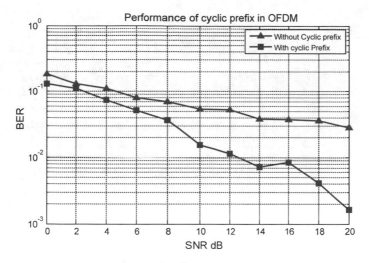

Fig. 4 Variation in BER and signal-to-noise ratio (SNR) in OFDM through the performance of with and without cyclic prefix

3 Results and Discussion

This section discusses the performance of BER and SNR in OFDM using cyclic prefix. All the simulations are done in MATLAB. Figure 4 shows the performance of BER using cyclic prefix and without using cyclic prefix in OFDM. And it is found that BER is better in cyclic prefix when compared to without cyclic prefix.

CP Length specifies the cyclic prefix mode for non-MBSFN (Multicast-Broadcast Single-Frequency Network) subframes and the non-MBSFN region of MBSFN subframes. The length of the cyclic prefix can be autodetected or can be specified as Normal in which CP Length = 7.03125% the length of the symbol or as Extended in which CP Length = 25% the length of the symbol.

Figure 5a shows the performance of cyclic prefix for the FFT of size 128 by applying various modulation techniques for long preamble of length 25% of the length of the symbol, and Fig. 5b shows the performance of cyclic prefix for the FFT of size 128 by applying various modulation techniques for short preamble of length 7.03% of the length of the symbol.

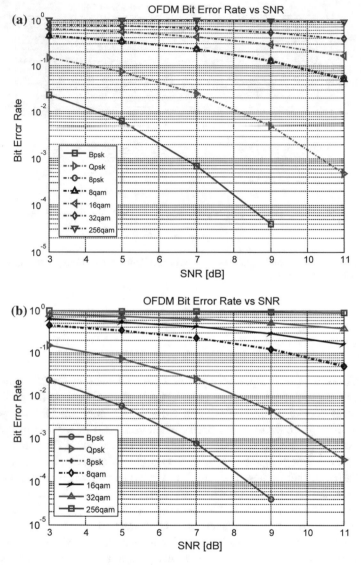

Fig. 5 **a** For FFT length = 128 and cyclic prefix size = 32. **b** For FFT length = 128 and cyclic prefix size = 16

4 Conclusion

OFDM system performance analysis is evaluated for determining the optimal CP length by considering the OFDM symbol with and without the inclusion of cyclic prefix of various lengths. Further, the optimum CP length performance is analyzed for different modulation techniques and in which the BPSK provides better performance when compared to other modulation techniques. It is found that for an optimum cyclic prefix length (FFT length = 128 and CP size = 16) the BPSK modulation technique is providing better BER performance (3.9×10^{-5} at 9 dB SNR) when compared to other modulation techniques (e.g., observed BER at 9 dB SNR are 0.9173 for 256QAM, 0.5157 for 32QAM, 0.1252 for 8QAM).

References

1. Sasibhushana Rao, G.: Mobile Cellular Communication. Pearson Education, New Delhi (2013)
2. Narasimhamurthy, A.B., Banavar, M.K., Tepedelenlioglu, C.: OFDM Systems for Wireless Communications. Morgan and claypool publishers (2010)
3. Batariere, M., Baum, K., Krauss, T.P.: Cyclic Prefix Length Analysis for 4G OFDM Systems. Motorola Labs Communication Systems Research Laboratory, 1301 E. Algonquin Road, Schaumburg, IL-60196USA
4. IEEE 802.11 specification: Orthogonal Frequency Division Multiplexing (OFDM) PHY specification for the 5 GHz band–chapter 17
5. Effect of Cyclic Prefix on OFDM over AWGN Channel, Wasiu lawal Adewuyi, s.o, Ogunti, e. o. Int. J. Innov. Res. Adv. Eng. (IJIRAE) (2014). ISSN: 2349-2163

High-Density Noise Removal Algorithm for Brain Image Analysis

Vimala Kumari G, Sasibhushana Rao G and Prabhakara Rao B

Abstract Noise is added to an image while acquiring or transmitting an image. One of the most commonly added noises is impulse noise. This work aims to remove this impulse noise from the high-density noise in medical images. The proposed algorithm is executed in two stages for removal of noise in an image. The first stage is for removing low-density noise which is cascaded to the second stage for removing high-density noise. First order neighborhood pixels are considered for identifying noisy pixels and cascaded filter is considered for replacement of the identified noisy pixel. This algorithm has given a good result when compared to other popular algorithms and has good noise removing capabilities. Different grayscale Magnetic Resonance Imaging (MRI) brain images are tested by using this algorithm and has given better results of the various performance metrics at different noise densities.

Keywords Impulse noise · Image de-noising · DBA · DMF
MDBUTMF · CDBNLF · DBTFOMF

Vimala Kumari G. (✉)
Department of Electronics and Communication Engineering,
M.V.G.R. College of Engineering, Vizianagaram 535005, India
e-mail: vimalakumari7@gmail.com

Sasibhushana Rao G.
Department of Electronics and Communication Engineering,
AU College of Engineering, Visakhapatnam 530003, India

Prabhakara Rao B.
Department of Electronics and Communication Engineering,
JNTUK, Kakinada 533003, Andhra Pradesh, India

© Springer Nature Singapore Pte Ltd. 2018
S. C. Satapathy et al. (eds.), *Information and Decision Sciences*,
Advances in Intelligent Systems and Computing 701,
https://doi.org/10.1007/978-981-10-7563-6_45

1 Introduction

Noise occurs generally in an image during transmission or reception. Noise is the irrelevant information that obscures the original information in an image. The most commonly occurring noise in the images is the Impulse Noise. Impulse noise maybe the noise of random values or noise of fixed values. The second type which is the fixed value impulse noise is generally referred as salt and pepper noise which replaces original with pixel values, either 0 or 255. 0 represents the pepper noise as it is a black dot and 255 represents salt noise as it is a white dot in image. Elimination of this noise is a vital processing step to achieve accurate and dependable information. For the elimination of impulse noise of fixed value from images, several nonlinear filters exist. In all of them, Standard Median Filter (SMF) is reliable to a great extent as this makes the median of the processing pixels in the window to replace current pixel, so that salt and pepper could be effectively removed. But, even if the density of noise in the image increases, SMF simply replaces the processing pixel with the median value irrespective of it being noisy or not, so there arises a chance that the non-noisy pixels might be replaced by the noisy pixels. So, this filter produces good results at low-density noise but as the noisy pixels in the image increases, this filter fails [1]. Adaptive Median Filter (AMF) [2] has been proposed to eliminate this problem, this filter increases the processing kernel size for eliminating the noisy pixels in the image, this filter also produces absolute results in the elimination of noise at lower densities but at higher densities, this algorithm produces the effect of blurring in images. For eliminating this problem, the Adaptive Decision-Based Median Filter (ADMF) [3] has also been introduced. To overcome the drawbacks and to remove the noise effectively at even higher densities Decision Based Algorithm (DBA) [4] has been brought up. DBA produces effective de-noising of images at low density of noise, this filter fails mainly whenever there is a higher noise density, and the neighborhood pixels replace the processing pixel and thereby producing the effect of Streaking. Hence, by using this algorithm the edges in the image cannot be reconstructed efficiently at higher densities of noise. To eliminate the problem of streaking, several filters have been introduced over the years like Decision Median Filter (DMF), Decision-Based Algorithm (DBA), Modified Decision-Based Unsymmetrical Trimmed Median Filter (MDBUTMF), and Cascaded Decision-Based Nonlinear Filter (CDBNLF) [5–8]. A Non-Local Means (NLM)-based filtering algorithm for de-noising Rician-distributed MRI and utilizes the concept of self-similarity for MRI restoration [9]. An approach involving minimum absolute difference criteria for identifying the noisy pixels and replacement by the mean value computed [10]. If the affecting noise is of low density the above-mentioned filters and other existing algorithms could be used, but if there is an effect of high-density noise on the images the above techniques fail. So, there is a requirement for efficient filters to be designed for de-noising. In this research work, a new algorithm is proposed to eliminate noise from images even at high-density level. The proposed algorithm employs two stages where the first stage removes low-density noise and the second stage removes the high-density noise. The proposed algorithm has been simulated for several Brain images. Simulation results clearly indicate that better results are produced at all noise density levels in terms of Mean Absolute Error (MAE), Mean Square Error (MSE) and Peak Signal-to-Noise Ratio (PSNR) when compared with several existing algorithms.

2 Proposed Algorithm

The algorithm is proposed for noise removal by processing first order neighborhood pixels by identifying the noisy pixel in the image. The main idea behind using this algorithm is to reduce the amount of noisy pixels in the image. A 3×3 window is considered. If the pixel is at the 2×2 position, it is assumed to be the processing pixel. This pixel is tested for noise. If the pixel is found to be a non-noisy pixel, it remains the same and if the pixel is found to be noisy then it is filtered. The algorithm can be described as

Stage 1: The first stage involves a Decision-Based Median Filter (DBMF). De-noising low-density salt and pepper noise is the primary focus of this stage. A 3×3 window is to be taken. The window is used to identify the noisy central pixel. If the pixel is a non-noisy pixel, it will not be filtered and if the pixel is a noisy one, it is filtered. The DBMF is executed as given below

Step 1: A 3×3 sized 2-D window is selected. The processing pixel is assumed to lie at the center of the window and is denoted as S_{ij} and it is surrounded eight of its neighboring pixels. If the processing pixel's gray level value lies in a range of $0 < S_{ij} < 255$, it is considered as non-noisy pixel and is not changed.

Step 2: If the intensity of the processing pixel is not in above range i.e., if $S_{ij} = 0$ or $S_{ij} = 255$: then it is declared as noisy and is filtered by the median.

Step 3: The 3×3 window is made to move to the next column along the row to process the next pixel and the process is repeated for processing the entire image. The result of this stage is passed on to the second stage to remove the unremoved noise.

Stage 2: Decision-Based Trimmed First-order Mean Filter (DBTFOMF) is used in this stage to de-noise medium and high-density noise. A 3×3 window is considered. The central pixel is considered as the processing pixel. If the gray scale value of processing pixel is 0 or 255 then it is noisy and is filtered, else the uncorrupted pixel is preserved. In this method, the pixels are processed as given below.

Step 1: The processing pixel which is in the 3×3 window considered as S_{ij} and is surrounded by eight of its neighbors. If the value of this pixel is in the range of $0 < S_{ij} < 255$, then it is an uncorrupted pixel and is not filtered.

Step 2:   If the grayscale value of the processing pixel does not lie in the above range, i.e., if $S_{ij} = 0$ or $S_{ij} = 255$, then this pixel is corrupted. Then two cases are possible: the central pixel is noisy and values of remaining pixels are 0 and 255 or central pixel is noisy and values of all other pixels may not be 0 and 255 but may have other values in between 0 and 255. The two cases are given below:

   Case i: If values of all the elements in the window are either 0 or 255, then replace S_{ij} with the mean value of elements of the window.
   Case ii: Even if at least one element in the considered window having pixel value in between 0 and 255, then all the other pixels having a value of either 0 or 255 are eliminated and mean of the remaining elements is calculated. Then the S_{ij} is replaced with this mean.

Step 3:   The 3×3 window is shifted to the next pixel for processing all the pixels in the entire image, repeat the above-mentioned steps till all pixels are processed.

3   Results and Discussion

Above proposed algorithms have been simulated on the MRI brain images. While simulating, salt and pepper noise will be added to the image to distort it. The density of noise applied to the image vary from 10 to 90% and the performance of above algorithm will be represented in case of PSNR, MAE, and MSE. MAE is given in Eq. 1, MSE is given in Eq. 2 and PSNR is given in Eq. 3.

$$MAE = \frac{\sum_{i,j} \left| Y(i,j) - \widehat{Y}(i,j) \right|}{M \times N} \tag{1}$$

$$MSE = \frac{\sum_{i,j} \left(Y(i,j) - \widehat{Y}(i,j) \right)^2}{M \times N} \tag{2}$$

$$PSNR = 10 \, \log_{10} \left(\frac{255^2}{MSE} \right) \tag{3}$$

Here, $M \times N$ is image dimensions, entire rows are represented with M and a total number of columns is represented with N, Y indicates the input image, η indicates noisy image and $\widehat{Y}$ indicates the reconstructed image. The obtained values of PSNR, MAE and MSE for a Huntington brain image for various noise densities from 10% to 90% in the steps of 10% have been tabulated and are shown in Tables 1, 2 and 3 respectively and it can be stated that the proposed algorithm

Table 1 PSNR values for a Huntington image at various densities of noise

PSNR in dB						
Noise density (in %)	SMF	DBA	DMF	MDBUT MF	CDBNLF	PA
10	36.67	37.52	44.68	44.74	44.81	45.28
20	35.37	36.07	41.28	41.36	41.87	41.93
30	35.02	35.02	39.24	39. 32	39.64	40.07
40	34.24	34.22	37.76	37.54	38.23	38.42
50	33.63	33.68	33.74	34.15	36.37	36.92
60	30.47	30.76	32.80	33.10	35.37	36.19
70	29.09	32.05	32.60	34.01	34.65	35.09
80	28.07	31.38	31.80	32.75	33.41	34.01
90	27.31	27.44	30.68	30.78	31.71	31.77

Table 2 MAE values for a Huntington image at various densities of noise

MAE values						
Noise density (in %)	SMF	DBA	DMF	MDBUT MF	CDBNLF	PA
10	1.93	1.47	0.28	0.28	0.24	0.26
20	2.33	2.07	0.62	0.62	0.52	0.50
30	2.82	2.78	1.02	0.96	0.88	0.78
40	4.04	3.50	1.48	1.52	1.25	1.15
50	8.74	7.81	5.39	4.19	2.17	1.71
60	23.9	21.95	8.36	5.23	3.08	2.02
70	34.3	11.97	6.38	4.88	3.36	2.7
80	43.6	16.41	8.14	7.79	4.92	3.67
90	49.9	49.84	21.59	12.8	12.5	8.89

Table 3 MSE values for a Huntington image at various densities of noise

MSE values						
Noise density (in %)	SMF	DBA	DMF	MDBUT MF	CDBNLF	PA
10	15.07	11.50	2.21	2.18	2.15	1.93
20	18.02	16.07	4.84	4.76	4.23	4.17
30	20.46	20.47	7.75	7.06	7.60	6.40
40	24.49	24.62	11.40	10.90	9.77	9.36
50	28.21	27.87	28.21	24.99	15.00	13.20
60	58.34	54.56	34.16	31.82	18.87	15.63
70	80.23	40.55	35.70	25.84	22.29	20.13
80	101.4	47.34	42.94	34.50	29.69	25.81
90	117.3	117.3	55.61	54.31	43.82	43.22

exhibits better PSNR, lesser MAE, and lower MSE than several existing algorithms namely SMF, DBA, DMF, MDBUTMF, and CDBNLF.

The comparison of PSNR, MAE, and MSE has been plotted in the Figs. (1, 2, and 3) respectively. From the figures, it can be stated that the proposed algorithm produces effective results than the other existing algorithms in all image quality factors considered.

Figures 4, 5 and 6 illustrate performance of various de-noising algorithms for brain images Huntington, Motor Neuron, and subacute stroke at noise density 40, 50, and 80 percentages of salt and pepper noise respectively. Figures 4, 5, and 6 (i) represent original brain image, Figs. 4, 5, and 6 (ii) represent the noisy image

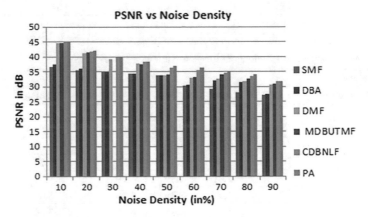

Fig. 1 Evaluation of the PSNR values for various algorithms applied on Huntington image for varying noise levels from 10 to 90% of noise

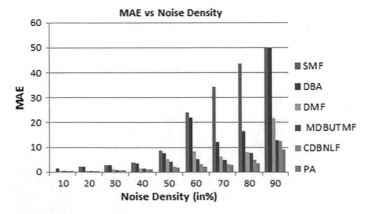

Fig. 2 Evaluation of the MAE values for various algorithms applied on Huntington image for varying noise levels from 10 to 90% of noise

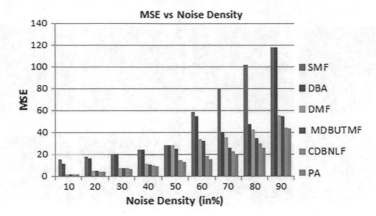

Fig. 3 Evaluation of the MSE values for various algorithms applied on Huntington image for varying noise levels from 10 to 90% of noise

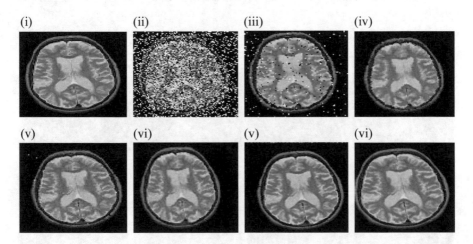

Fig. 4 Performance of various filters for Huntington image at a noise density of 40% salt and pepper noise **i** original image **ii** noisy image **iii** SMF **iv** DBA **v** DMF **vi** MDBUTMF **vii** CDBNLF **viii** PA

and Figs. 4, 5, and 6 (iii)–(viii) represent reconstructed images of various de-noising algorithms. From the Figs. 4, 5 and 6 (viii) it is observed that reformed images obtained by applying the proposed algorithm are superior to the results obtained by applying all previous algorithms.

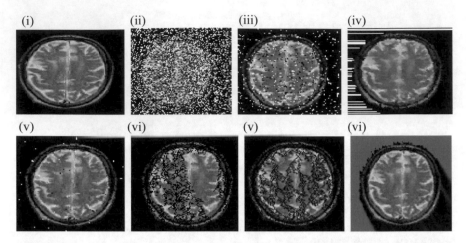

Fig. 5 Performance of various filters motor neuron image at a noise density of 50% salt and pepper noise **i** original image **ii** noisy image **iii** SMF **iv** DBA **v** DMF **vi** MDBUTMF **vii** CDBNLF **viii** PA

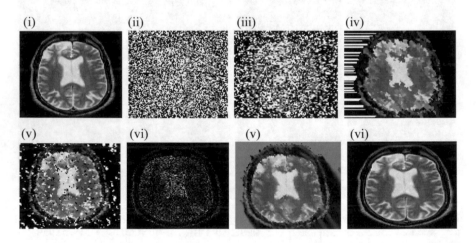

Fig. 6 Performance of various filters for subacute stroke image at a noise density of 80% salt and pepper noise **i** original image **ii** noisy image **iii** SMF **iv** DBA **v** DMF **vi** MDBUTMF **vii** CDBNLF **viii** PA

4 Conclusion

This proposed algorithm has been implemented for de-noising of fixed valued impulse noise from medical images. The algorithm has been tested on various MRI brain images and the algorithm has been found to produce better and accurate results. The performance of this algorithm has been computed in the case of MAE,

MSE, and PSNR and the results are compared with existing de-noising algorithms. From the output, it is evident that it has given excellent de-noising capabilities for high-density images also. This algorithm can also be used in tumor detection noise removal; this algorithm without disturbing the tumor can de-noise the noisy tumor images, so it would be very helpful in biomedical applications.

References

1. Balasubramanian, S., Kalishwaran, S., Muthuraj, R., Ebenezer, D., Jayaraj, V.: An efficient non-linear cascade filtering algorithm for removal of high density salt and pepper noise in images and video sequence. In: International Conference on Control, Automation, Communication and Energy Conservation, pp. 1–6 (2009)
2. Hwang, H., Hadded, R.A.: Adaptive median filter: new algorithms and results. IEEE Trans. Image Process. **4**, 499–502 (1995)
3. Suman, S.: Image denoising using new adaptive based median filter. Int. J. (SIPIJ) **5**(4), 1–13 (2014)
4. Srinivasan, K.S., Ebenezer, D.: A new fast and efficient decision based algorithm for removal of high density impulse noise. IEEE Signal Process Lett. **14**, 1506–1516 (2007)
5. Aiswarya, K., Jayaraj, V., Ebenezer, D.: A new and efficient algorithm for the removal of high density Salt and Pepper noise in images and videos. In Second International Conference on Computer Modeling and Simulation, pp. 409–413 (2010)
6. Esakkirajan, S., Veerakumar, T., Subramanyam, A.N., PremChand, C.H.: Removal of high density salt and pepper noise through modified decision based unsymmetric trimmed median filter. IEEE Signal process. Lett. **18**, 287–290 (2011)
7. Dash, A., Sathua, S.: High density noise removal by using cascading algorithms. In: International Conference on Advanced Computing and Communication Technologies, pp. 96–101 (2015)
8. Santhanam, T., Chithra, K.: A new decision based unsymmetric trimmed median filter using Eucledian distance for removal of high density salt and pepper noise from images, pp. 1–5. IEEE Conference Publications (2014)
9. Vikrant, B, Tiwari, H., Srivastava, A.: A non-local means filtering algorithm for restoration of Rician distributed MRI. In: Emerging ICT for Bridging the Future-Proceedings of the 49th Annual Convention of the Computer Society of India CSI, vol. 2. Springer, Cham (2015)
10. Awanish, K.S., Vikrant B, Verma, R.L., M. S. A.: An improved directional weighted median filter for restoration of images corrupted with high density impulse noise. In: 2014 International Conference on IEEE Optimization, Reliabilty, and Information Technology (ICROIT) (2014)

A Pragmatic Study and Analysis of Load Balancing Techniques in Parallel Computing

Varsha Thakur and Sanjay Kumar

Abstract Availability of economical systems and advancement in communication technicalities has galvanized appreciable attentiveness in a parallel architecture. Parallel architecture is materialized as a research area with the perspective of furnishing faster results. In a parallel architecture, balancing of load over engaged cores or nodes is a critical issue. Imbalance in cores or processors degrades the performance of systems. This paper presents a study and comparative anatomy of some load balancing approaches on parallel architecture with its effect on parallelism. The performances are analyzed on the basis of execution time, speed up, throughput, and efficiency. To implement load balancing approaches we have used OpenMP in Linux Environment.

Keywords Load balancing · Dynamic · Static · Parallel and distributed computing · OMP

1 Introduction

Load balancing is a terminology that emanates from applied science, it is a skill used by electric power stations to accumulate surplus electrical power during low need and rescue that power as requirements increase. Load balancing channelizes the total load to the individual processors for effective utilization of resources and improves the performance in terms of response time. Concurrently, eliminating a situation by which few processors or cores are heavily loaded while others may be lightly loaded [1]. The requirement of Load balancing occurs when load situation of processors has changed during execution or certain conditions which can alter the load of processors like the advent of any new task and assigning of that task to any

V. Thakur (✉) · S. Kumar
SoS CS & IT, PRSU, Raipur, Chhattisgarh, India
e-mail: varshathakur1308@gmail.com

S. Kumar
e-mail: sanraipur@rediffmail.com

© Springer Nature Singapore Pte Ltd. 2018
S. C. Satapathy et al. (eds.), *Information and Decision Sciences*,
Advances in Intelligent Systems and Computing 701,
https://doi.org/10.1007/978-981-10-7563-6_46

processor, finalization of any running process, the advent of new task and removal of any existing task. Load balancing optimizes the way the processing load is shared among the multiple processors. Load balancing migrates the work from the overloaded processor to underloaded processor. Migration of a process is a mechanism in which processes are transferred from one processor to another along with handling related messages during migration [2]. Various researchers admit that load balancing, with its aim of equalizing the load on all the processors, is an inappropriate goal. Due to the overhead included in collecting, the details of the state are usually more cumbersome, chiefly in the systems owning a huge number of processors. Furthermore, balancing of the load in the strictest way is not attainable since the number of tasks, in processors is perpetually changing that why materialistic imbalance exists among the processors at every instant. For genuine usage of the resources, it is not essential to balance the load on all the processors. Preferably, it is enough to avert the processors from being idle. In the heterogeneous environment term 'fairly' appears to be more justified than 'evenly'. A good load balancing algorithm is a difficult task due to the various issues involved in designing like estimating the load, transferring of the process, information of state, the location of the process, priority assignment and limitation of Migration [2].

2 Parallel Architecture

Parallel computing has turn out to be in demand as it provides a means to succeed in dealing with the limitations forced by the sequential systems. As a result, a different parallel system (multiprocessor systems) has been developed [3]. From, last few decades, the area of Parallel architectures has emerged as challenging task for discovering new architectures that can perform better. The basic objective of parallel architecture is improving the speed. Parallel computing employs a parallel system to lessen the time required to figure out a single computational problem. A Parallel computer means a system having multiple processors which can bear parallel programming. Parallel programming can be defined as a programming language that supports a different portion of the program to run concurrently. Parallel computers only became attractive to a wide range of customers with the emergence of very large-scale integration in the late 1970s. Supercomputers such as the Cray-1 were far too expensive for most organizations. Experimental parallel computers were less expensive than supercomputers, but they were still relatively costly. They were unreliable, to boot. VLSI technology allowed computer architects to reduce the chip count to the point where it became possible to construct affordable, reliable parallel systems. In roughly three decades between the early 1960s and the mid-1990s scientist and engineers explored a wide variety of parallel computer architectures. Development reaches a zenith in the1980s [4]. Parallel architecture is a group of processing elements that work together to compute larger problems quickly. Parallel architecture emphasizes on parallel processing between operations in some way. Parallel processing means dividing a problem into subproblem and executing these subproblems simultaneously. Parallel computing utilizes simultaneously running processes which are the parts of the larger computation, for this divide and conquer approach is usually preferred [5]. Conventionally, the software has been written for sequential systems in which only one instruction is executed at a time [6]. The basic reason for using parallel architecture is to save time and compute larger problem. The design of parallel architecture considered various factors like the number of processing elements, the arrangement of processing elements and memory, the capability of each processor, the way of addressing memory, mechanism of processor communication and synchronization, phenomena of connections of memory with the processing elements, kind of interconnections, and coinciding of programs. Parallel computers can be stratified into different types depending upon various factors such as classification based on architecture (Flynn's, Feng's, Handler, and Shores), classification based on granularity (fine, medium, and coarse), classification based on arrangement and communication among processing elements and memory (shared and distributed) based on interconnections and characteristic of processors. ArrayProcessor, Multicomputer, and Multiprocessors are the most popular parallel computer architectures [7].

3 Load Balancing

A load of the system can be defined as an assessment of computational work that a system performs [8]. Balancing deals with even distribution of load enabling something to remain upright [9]. For balancing of uneven workload, load distribution is required. Load distribution can be categorized in three ways first is task assignment, second load Sharing and last as load balancing. In the case of task assignment, each task is assigned to suitable processors for performance improvement. In the case of load sharing, no processor should be found idle, while in the case of load balancing it tries to equalize the load. Another method is there called Load leveling which comes between load balancing and load sharing [10]. Generally, when a large program is implemented on a system only one CPU is utilized and others are negligibly utilized which can be verified experimentally to avoid this type of situation load balancing is needed. In computer domain, Load balancing is an important term, one common feature is that number of processors are many in these types of computing. The basic reason for load balancing is that in parallel architecture the entrant of tasks and service time of processor for processing that tasks are random, so a situation may occur that at any instance some processors are more loaded, while the other may be idle. Instead of overloading a few processors with large numbers of tasks, it is better to allocate some of these tasks to the less loaded processor by which task can complete earliest. Load balancing is a challenging task in these types of computing. In recent years, load balancing and scheduling have received inclination from the researchers in parallel and distributed computing. In computer domain, Load balancing is an important term common features are that number of processors is many of these types of computing. In real life, whenever we have to join some queue, we always try to join the small queue or the queue which is served fast this is actually a load balancing. A parallel Architecture composed of groups of processing unit linked by some interconnections (mesh, hypercube). The redistribution of load enhances the performance by displacement of tasks from the extremely loaded processor, to the moderately loaded processor, where the tasks take the benefits of being processed [6]. Initially, load balancing can be broadly categorized as static and dynamic. Static will use information regarding the behaviors of the processors, paying no attention to the current state. Dynamic response to the current state and state that changes during runtime. Static load balancing algorithms are straightforward. The attraction of dynamic is the respondents to the change in system states. Due to this reason, dynamic load balancing achieves better performance. Although, complexity will be more [2]. Further, static can be classified as deterministic or probabilistic. The deterministic approach utilizes information regarding characteristics of the processes and the properties of processors. Probabilistic load balancing bears in mind, about the facts regarding static attributes of the system such as processing capability, topology, and number of processors. Dynamic load balancing is moreover stratified as centralized and distributed. In the case of centralized load balancing, the information related to nodes of the state is collected at a single processor while in distributed load balancing each processor involves in taking decisions regarding processes of their own. Distributed may be categorized as cooperative and non-cooperative. In cooperative load balancing processors cooperate with each other in making decisions while in the case of non-cooperative processor does not cooperate with each for making decisions. Load balancing on the basis of task transfer can be classified as Preemptive or non-preemptive Preemptive task transfers involve transferring of the task that has been executing while Non-preemptive transfer task that is in a queue means whose execution is not started. Depending on initiation of task transfer, it can be classified as sender, receiver or symmetrical. In sender initiated, the transfer of load is started by the sender, while in the case of receiver initiated the transfer of load is started by the receiver. Symmetric initiated incorporates the property of sender initiated and receiver initiated both. Major components of load distributing approaches are policy related to transferring of loads (from sender or receiver), policy related to the selection of process, policy related to finding the location of the destination processor and policy regarding information of processors. The index used to measure a load of system predicts the performance of a task if it is running on some processor. Load indexes that have been generally considered are the queue length of the processing elements, the average queue length of processing elements between some period of time, availability of memory, switching rate, service time, and processor utilization [11] (Fig. 1).

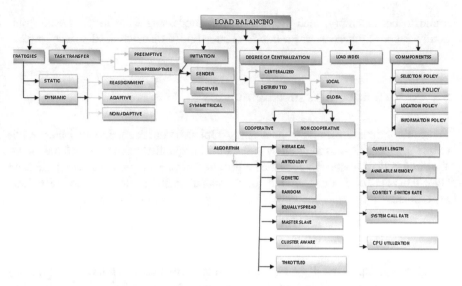

Fig. 1 Load balancing

4 Comparative Study

For comparative analysis, three load balancing algorithms in parallel architecture are considered namely random, cluster aware and hierarchical [12]. In random load balancing, the load is transferred randomly from overloaded node to any underloaded node. Hierarchical load balancing was proposed for divide and conquers approach. Here, systems are organized into a tree-like structure. Loads are transferred from one node to another. In the case of cluster aware, the load is transferred from the inside and outside cluster in parallel. At any time when processors become idle to take a load, it tries to take a load from a node in the remote cluster if not then in the local cluster.

Algorithm (Random)

1: Assume that α be the set of all processing elements that have work to transfer. 2: Randomly a victim (β) is selected from α. 3: Dispatch steal strive to β.

Algorithm (Hierarchical):—1: Assume that α be the set of all thief's child processing elements that have work to transfer. 2: If α is nonempty then victim (β) processing element are selected randomly. 3: If the thief has a parent then victim (β) processing element is thief's parents. 4: Otherwise value of β becomes NULL EndIF. 5: If value of β not equal to NULL then dispatch steal strive to β EndIF.

Algorithm (Cluster aware):—1: If outside Stealing then $\alpha 1$ be the set of all processing element outside of the thief's cluster that has work to transfer. 2: If $\alpha 1$ is nonempty, then the victim ($\beta 1$) processing element is selected randomly from $\alpha 1$ otherwise from outside of the thief's cluster. 3: EndIf dispatch steals strives to $\beta 1$. 4: Outside stealing = true Endif. 5: EndIF. 6: Assume that $\alpha 1$ be the set of all processing elements from the thief's the cluster that has work to transfer. 7: If $\alpha 2$ is found to be nonempty, then the victim ($\beta 2$) processing Element is selected randomly from $\alpha 2$ otherwise from thief's cluster EndIF dispatch steal strives to $\beta 2$.

5 Performance Measurements

These algorithms are implemented using OpenMP programming in Linux environment. OpenMP is used to analyze the effect of parallelization on load balancing algorithms. Performance metrics are helpful to describe the performance of different algorithms [13]. Some suitable performance metrics are selected for this experiment.

1.4.1 Parallel Run Time (T_n):—The time needed to execute the programs on parallel systems (having n number of processors). When the value of n is one it referred as sequential time.

1.4.2 Speed Up (S_n):—Speed up means time needed before improvement (serial) divided by the time needed after improvement (parallel).

$$\textbf{Speed up} = \textbf{t}(\textbf{1})/\textbf{t}(\textbf{n})$$

1.4.3 Efficiency (E_n):—While speed up measures how much faster a program runs on a parallel computer rather than on a parallel computer rather than on a single. Efficiency is a term which means speed up upon a number of processors that are used.

$$\textbf{Efficiency} = \textbf{Speedup}/\textbf{n}$$

1.4.4 Throughput:—Throughput is a measure of a number of processes that can be processed per time unit.

1.4.5 Cost of the parallel algorithm:—The cost of the parallel algorithm means the multiplication of the parallel runtime into the number of processors used.

$$\textbf{Cost} = \textbf{Running time X Number of processors}.$$

1.4.6 Schedule Length:—The time that all modules complete their execution and return result is called schedule length.

6 Observation

The performance of random, cluster aware and hierarchical Load balancing algorithms/approaches was observed in both sequential and parallel processor environment. This experiment carried out for n independent task and m

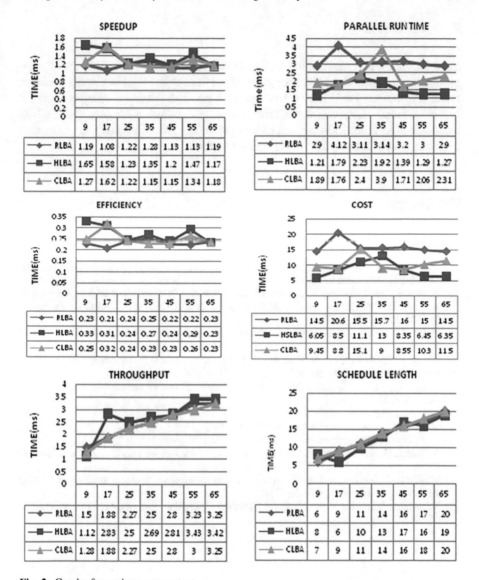

Fig. 2 Graphs for various parameters

heterogeneous processors (here we are considering m = 5). The 'n' independent task generated randomly and assigned to any available processor. In graphs, the horizontal axis represents a number of tasks and the vertical axis represents time. It has been analyzed through experiments that hierarchical load balancing performs better as compared to others (Fig. 2).

7 Conclusion

Load balancing is a challenging task in these (parallel architecture) types of computing. It was found through experiments that Hierarchical load balancing performs better than others. Performance wise (parallel runtime, speedup, efficiency, etc.) it has been found Hierarchical > Cluster aware > Random. Although various load balancing algorithms exist still an efficient technique for load balancing is needed by which much better performance can be achieved.

References

1. Thakur, V., Kumar, S.: Load balancing approaches, and recent computing trends. Int. J. Comput. Appl. **131**, 43–47 (2015)
2. Sinha, P.K.: Distributed Operating Systems concepts and design, pp. 356, 358, 367, 414. PHI Learning Private Limited (2007)
3. Kumar, S.: Mathematical modeling and simulation of a buffered fault tolerant double tree network. In: 15th International Conference on Advanced Computing and Communication, pp. 422–426. IEEE (2007)
4. Quinn, M.J.: Parallel Programming in C with MPI and OpenMP. McGraw Hill Inc., pp. 8–11 (2004)
5. Cheng, J., Grossman, M., McKercher, T.: Professional Cuda C Programming. Wiley (2014)
6. Parallel Computing: http://en.wikipedia.org/wiki/Parallel_computing
7. Thakur, V., Kumar, S.: Perspective study and analysis of parallel architecture. Int. J. Comput. Appl. **148**, 22–25 (2016)
8. https://en.wikipedia.org/wiki/Load_(computing)
9. htts://en.oxforddictionaries.com/definition/balance
10. http://shodhganga.inflibnet.ac.in/bitstream/10603/26264/5/chapter%203.pdf
11. Shivaratri, N.G., Krueger, P., Singhal, M.: Load distributing for locally distributed systems, 31–41
12. Vladimir, J.: Load balancing of irregular parallel applications on heterogeneous computing environments, thesis, pp. 93–100 (2012)
13. Rajaraman, V., Shiva Rama Murthy, C: Parallel Computing, pp. 441–422. PHI Publication (2009)

A Novel Segmentation Algorithm for Feature Extraction of Brain MRI Tumor

Ch. Rajasekhara Rao, M. N. V. S. S. Kumar
and G. Sasi Bhushana Rao

Abstract A new algorithm is projected in this paper for the identification and classification of tumors. For this, a set of MRI slices is considered from the database. As the images from electronic equipment contain noise, first the denoising of images is done using wavelets. Now, the identification of tumor is done by segmentation. Initially, the existing methods like expectation–maximization, histogram, and object-based thresholding are analyzed and implemented. But some of the features are missing in all these methods. So a new algorithm is proposed in which all the features from above methods are fused. The total analysis is done for 2D images, and the results obtained are in 2D. The performance analysis of the existing and proposed algorithms is compared in terms of size of the resultant tumor.

1 Introduction

The death rate due to tumor has been increasing enormously over the past three decades. Brain tumor is a pathology appearing in the intracranial anatomy due to abnormal and unstructured augmentation of cells. It is a very aggressive and life-threatening condition, which must be promptly diagnosed and cured to prevent mortality [1].

Brain tumors may be of different sizes and locations and may be even overlapped with regular tissues. The uneven growth of tissue may be benign or malignant and may arise in several elements of the brain which might not be prime tumors. So it is very essential to identify tumors before reaching uncontrollable stage. The mechanism used to identify the tumors is MRI.

Ch. Rajasekhara Rao · M. N. V. S. S. Kumar (✉)
Departemnt of ECE, AITAM, Tekkali 532201, India
e-mail: muvvala_sai@yahoo.co.in

G. S. B. Rao
Departemnt of ECE, Andhra University, Visakhapatnam 530003, India

© Springer Nature Singapore Pte Ltd. 2018
S. C. Satapathy et al. (eds.), *Information and Decision Sciences*,
Advances in Intelligent Systems and Computing 701,
https://doi.org/10.1007/978-981-10-7563-6_47

Fig. 1 Slice of brain

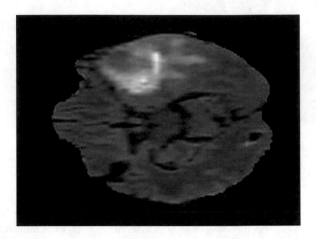

MRI is a sophisticated imaging method providing data relating to the human soft tissue structure which is largely engaged in radiology to observe the structure and behavior of the physical body. This will provide the intricate pictures of the body in all directions. MRI will be useful in medical imaging as it offers a large amount of dissimilarity among the different soft tissues of the body [2]. This MRI image contains lots of information along with tumor.

In this paper, an attempt is made to identify the tumor by developing a new hybrid algorithm based on expectation–maximization, histogram, and object-based thresholding methods. The result obtained from this algorithm is in 2D format. For higher understanding of the tumor stage, these 2D images are combined to create a 3D view of the tumor. The MRI slices considered for processing are shown in Fig. 1 [3].

2   Proposed Methodology

To reduce the complication and improve the accuracy of tumor 3D reconstruction, the below steps are implemented in the projected algorithm, which is presented in this section and is shown in Fig. 2. In this, first, the MRI images are denoised using wavelets as it contains noise. The segmentation for identification of tumor is done using EM segmentation, histogram thresholding, and object-based segmentation. To get missing edges in these three methods, all these methods are fused to segment the image as shown in Fig. 2.

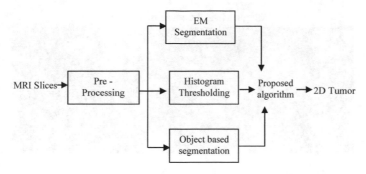

Fig. 2 Flowchart of proposed algorithm

3 Preprocessing

Before the presentation of the brain tumor segmentation methods, the MRI pre-processing operations are introduced because it is related to the traits of the segmentation results. The raw MRI images need to be preprocessed to realize the segmentation purposes. These preprocessing operations include denoising, skull-stripping, intensity normalization, etc., which will improve the results of brain tumor segmentation.

For clinical diagnosing, the visual quality of magnetic resonance images plays a crucial role. The MRI images are corrupted by noise while acquiring or transmitting them. Noise in MRI poses a plenty of problem to medical personnel by interfering with interpretation of MRI for diagnosing and treatment of human [4]. Noise in MRI image makes it complex to accurately outline areas of significance amid brain tumor and regular brain tissues. So, it is required to denoise the MRI images to remove noise and to improve distinction between regions [5].

Image denoising is a standard preprocessing task for MRI. Denoising is nothing but the removing of noise from image by retaining the initial quality of the image [1]. Since the brain images are more sensitive than other medical images, MRI images are typically degraded by noises like Gaussian and Poisson. Generally, in almost all denoising algorithms, the noise may be considered as additive white Gaussian noise [6].

By using wavelet transforms, the noise can be reduced to a large extent. The denoising of an image degraded by Gaussian noise is a regular problem image processing. To solve this problem wavelet transform is used in practice because of its energy compaction property. It decomposes the signal into wavelets and the coefficients are picked using thresholds and the signal is synthesized [7].

The discrete wavelet transform transforms the image information into an estimated sub-band and a group of detail sub-bands at different orientations and resolution scales. Typically, the bandpass content at every scale is split into three orientation sub-bands characterized by horizontal, vertical, and diagonal directions. The approximation sub-band consists of the so-called scaling coefficients, and the

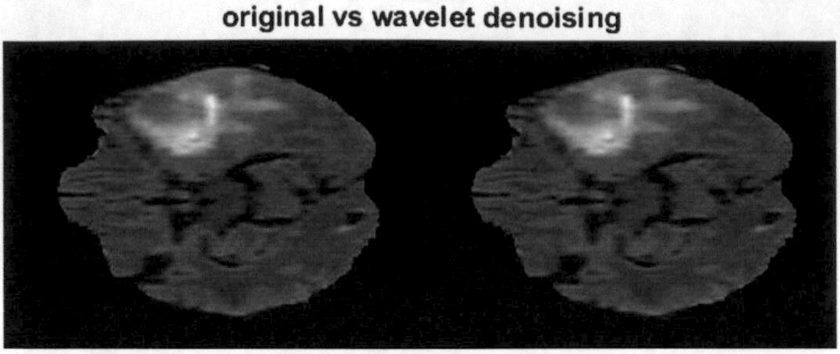

original vs wavelet denoising

Fig. 3 Wavelet denoising

detail sub-bands are composed of the wavelet coefficients. Several properties of the wavelet transform, which make this representation attractive for denoising, are easily recognized. For denoising a medical image, the wavelets are being widely used [8]. The resultant of wavelet denoising is shown in Fig. 3.

4 Segmentation

Brain tumor segmentation is used to classify the different tumor tissues such as active cells, necrotic core, and edema from normal brain tissues of white matter, gray matter, and cerebrospinal fluid. MRI brain tumor segmentation is drawing added attention in recent times because of its noninvasive imaging and good soft tissue distinction of MRI images [5].

Many methods are available which can produce good output for different image processing applications and are not always suitable for all types of applications and there is no unique segmentation algorithm which can give good results.

Extraction of required information from all types of multidimensional images is playing an important role in segmentation [9]. Manual separation of brain tissues for diagnosis purpose is more complex in terms of time and expertise. This leads to the development of automated segmentation techniques.

4.1 EM-Based Segmentation

Classification techniques are characterized as parametric and nonparametric. In parametric approach, the maximum likelihood (ML) or maximum a posteriori (MAP) approach is used to estimate the model parameters and its optimization is done by using the expectation–maximization (EM) algorithm [10]. The cancer extracted using this method is shown in Fig. 4.

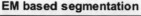

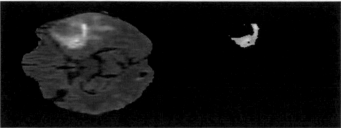

Fig. 4 EM segmented slice

The steps for implementation of EM algorithm are as follows:

1. Start: Let us consider the primary parameter set is (0).
2. E-step: After "t" iterations, parameter set is (t) and conditional expectation is calculated as

$$Q\left(\theta|\theta^{(t)}\right) = E\left[\ln P(x, y|\theta)|y, \theta^{(t)}\right]$$

$$Q\left(\theta|\theta^{(t)}\right) = \sum_{x \in X} P\left(x/y, \theta^{(t)}\right) \ln P(x, y|\theta) x \in X$$

where X indicates set of labels.

3. M-step: To obtain the next estimate, maximize $Q\left(\theta|\theta^{(t)}\right)$:

$$\theta^{(t+1)} = \arg\max_{\theta} Q(\theta|\theta^t)$$

then let $\theta^{(t+1)} \to \theta^{(t)}$ and repeat from the step 2.

In this algorithm, MAP estimation will be done to solve for x* which minimizes the total posterior energy

$$X^* = \arg\min_{X}\{U(Y|X, \theta) + U(X)\}$$

$$X* = \arg\max_{X}\{P(Y|X, \theta)P\{X\}\}$$

4.2 Histogram-Based Thresholding

Histograms of brain magnetic resonance imaging images indicate a classic form, wherein the thresholding is done by parameters which may not vary much from one patient to another. In fact, it appears that brain MRI histograms are frequently bimodal which stands for the intensity values such as the image background and all

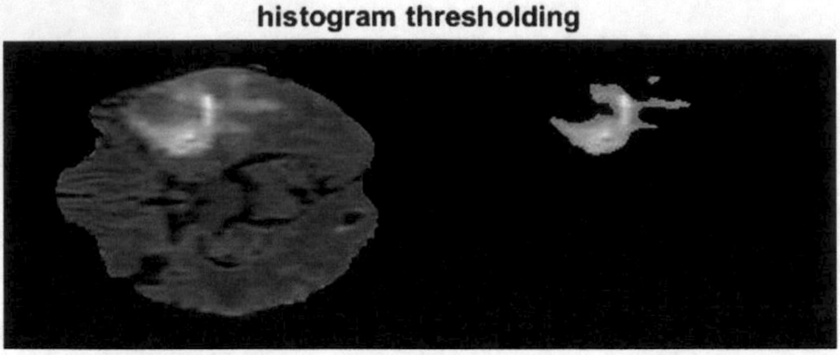

histogram thresholding

Fig. 5 Result of histogram thresholding

gray values. In our case, for every patient, two magnetic resonance imaging modalities (T1C and FLAIR) are used to extract the two corresponding histograms. Generally, edema region is brighter on FLAIR than on T1C, and therefore, the gadolinium-enhanced lesion has the alternative behavior [11]. The cancer extracted using this method is shown in Fig. 5.

Step 1: The histogram for MRI image is obtained after separating it into two similar parts about its central axis.
Step 2: Threshold is calculated by comparing to histograms.
Step 3: By using threshold, segmentation is done for two equal parts.
Step 4: Identified matter is sliced along its outline to get the dimension of the tumor.
Step 5: Construct an image of the original size. By checking the segmented image pixel value, 255 will be assigned if it is greater than threshold value or else 0 will be assigned.
Step 6: Segment the tumor area.

4.3 Object-Based Thresholding

The object-based thresholding technique is more suitable for images with different intensities. In this, the image is divided into various regions based on the intensity values [9]. For implementing thresholding technique, let us consider f(x, y) be the input image and "T" be the threshold value then the segmented image g(x, y) is given by

$$g(x,y) = \begin{cases} 0, & f(x,y) \le T \\ 1, & f(x,y) > T \end{cases} \qquad (1)$$

The image can be segmented into two groups using Eq. (1). The threshold value is applied to a selected region and segments the image objects that are different in intensity from their surroundings. Whereas the global thresholding is suitable only

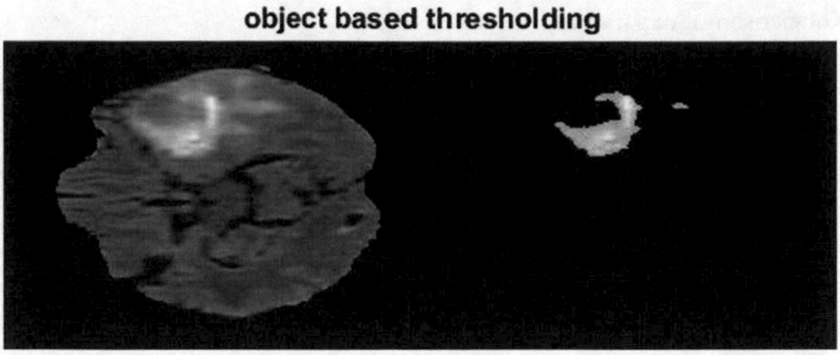

object based thresholding

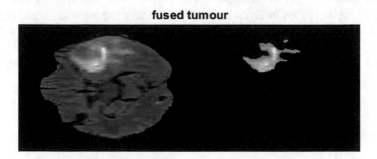

Fig. 6 Result of object-based thresholding

for images which contain objects with uniform intensity or the contrast between the objects and the background is high. The cancer extracted using this method is shown in Fig. 6.

4.4 Proposed Fusion Algorithm

The MRI brain image is subjected to abovementioned segmentation algorithms such as EM-based segmentation, histogram-based thresholding, and object-based thresholding individually, and then the results are combined (fused) pixel by pixel based on the maximum criterion to attain the absolute segmented image. The proposed fused technique exhibits better segmentation results comparing with individual segmentation algorithms. The cancer extracted using this method is shown in Fig. 7.

fused tumour

Fig. 7 Result of proposed fusion algorithm

Table 1 Comparison of existing and proposed methods

S. No	Method	Area (Pixels)	Perimeter (Pixels)
1.	Proposed	1059	306
2.	EM segmentation	440	175
3.	Histogram thresholding	1051	304
4.	Object-based thresholding	698	243

5 Results

First, the 2D MRI slices are extracted from mha file. These images are preprocessed for denoising. Now by applying existing segmentation methods and developed method, tumor is identified as shown in above results. All the information extracted is in 2D. In order to compare the performance of existing and proposed methods, area and perimeter of the identified tumor are considered as parametric measures. By comparing these metrics, it can be observed that proposed method is performed well in identifying all the edges of tumor without missing any information as shown in Table 1. This information can be used in constructing the 3D surface of tumor which is in the development stage.

6 Conclusion

A new hybrid algorithm based on expectation–maximization, histogram, and object-based thresholding methods is developed in this paper to identify the tumor in the MRI slices. For proper extraction of the features, initially, the images are denoised with wavelet methods. To identify the tumor in MRI slices, segmentation methods such as expectation–maximization, histogram, and object-based thresholding are applied individually. It is found that, in all three methods, some edges are missing. In order to overcome this, a new algorithm is proposed, which is achieved by fusing above three methods. To evaluate its performance, parametric measures like area and perimeter are consider, and it is observed that the proposed algorithm performed well in identifying all the edges of tumor.

References

1. Nasir, M., Baig, A., Khanum, A.: Brain tumor classification in MRI scans using sparse representation. In: Elmoataz, A. et al. (Eds.) ICISP 2014. LNCS 8509, pp. 629–637. Springer International Publishing Switzerland (2014)
2. Selvaraj, D., Dhanasekaran, R.: Mri brain image segmentation techniques—a review. Indian J. Comput. Sci. Eng. (IJCSE) **4**(5) (2013). ISSN: 0976-5166
3. The Whole Brain Atlas http://www.med.harvard.edu/aanlib/

4. SavajiPP, S., AroraP, P.: Denoising of MRI images using thresholding techniques through wavelet. Int. J. Sci. Eng. Technol. Res. (IJSETR) **1**(3) (2012)
5. Liu, J., et al.: A survey of MRI-based brain tumor segmentation methods. Tsinghua Sci. Technol. **19**(6), 578–595 (2014). ISSN l11007-0214ll04/10ll
6. Abdel-Maksoud, E., Elmogy, M., Al-Awadi, R.: Brain tumor segmentation based on a hybrid clustering technique. Egypt. Inform. J. (2015)
7. Agrawal, S., Sahu, R.: Wavelet based MRI image denoising using thresholding techniques. Int. J. Sci. Eng. Technol. Res. (IJSETR) **1**(3) (2012)
8. Pi_zurica, A., Wink, A.M., Vansteenkiste, E., Philips, W., Roerdink, J.B.T.M.: A review of wavelet denoising in MRI and ultrasound brain imaging. Current Med. Imaging Rev. 2005
9. Gordillo, N., Montseny, E., Sobrevilla, P.: State of the art survey on MRI brain tumor segmentation. Magn. Reson. Imaging (2013)
10. Shah, S., Chauhan, N.C.: An automated approach for segmentation of brain MR images using gaussian mixture model based Hidden Markov random field with expectation maximization, **2**(4),17th August 2015
11. Salah, M.B., Diaz, I., Greiner, R., Boulanger, P., Hoehn, B., Murtha, A.: Fully automated brain tumor segmentation using two MRI modalities. In: ISVC 2013, Part I, LNCS 8033, pp. 30–39. Springer, Berlin, Heidelberg (2013)

Adaptive Parameter Estimation-Based Drug Delivery System for Blood Pressure Regulation

Bharat Singh and Shabana Urooj

Abstract Controlled drug delivery system (DDS) is an electromechanical device that enables the injection of a therapeutic drug intravenously in the human body and improves its effectiveness and care by controlling the rate and time of drug release. Controlled operation of mean arterial blood pressure (MABP) and cardiac output (CO) is highly desired in clinical operation. Different methods have been proposed for controlling MABP; all methods have certain disadvantages according to patient model. In this paper, we have proposed blood pressure control using integral reinforcement learning-based fuzzy inference system (IRLFI) based on parameter estimation technique. To further increase the safety of the proposed method, a supervisory algorithm is implemented, which maintains the infusion rate within safety limit. MATLAB simulation depends the model of MABP, elucidate the ability of the suggested methodology in designing DDS and control postsurgical MABP.

Keywords Drug delivery system · Fuzzy inference system · Mean arterial blood pressure · Sodium nitroprusside · Maximum a posteriori estimators

1 Introduction

The nonstop advancement of innovative medication delivery structures is driven by the essential expand remedial action while limiting negative symptoms [1]. Controlling MABP of a postsurgical patient is a key variable to describe the system's operating point. MABP can be controlled by sodium nitroprusside (SNP), which is a drug that reduces the tension in the arteries walls; thus, blood pressure reduces [2].

B. Singh (✉)
Bharatividyapeeth's College of Engineering, New Delhi, India
e-mail: singh.bharat10@gmail.com

S. Urooj
School of Engineering, Gautam Buddha University, Greater Noida, Uttar Pradesh, India

© Springer Nature Singapore Pte Ltd. 2018   465
S. C. Satapathy et al. (eds.), *Information and Decision Sciences*,
Advances in Intelligent Systems and Computing 701,
https://doi.org/10.1007/978-981-10-7563-6_48

A closed-loop proportional–integral–derivative controller (PID) control has been widely applied in many real applications. Also, some other well-known methods are Ziegler–Nichols method, IMC method, and loop-shaping method [3]. Since single controller is not capable of achieving the desired control objective of clinical person and this problem has been acknowledged by some authors, this prompts a variety of adaptive control systems being proposed in the course of the most recent three decades; researches are based on self-tuning regulators, multiple-model adaptive control (MMAC), model-reference adaptive control and model-predictive control, as well as fuzzy control and rule-based nonlinear control [4, 5]. Among these methods, fuzzy base PID controller can be used in drug delivery system for guaranteed stability and accuracy. It has been proposed by many researchers that use of fuzzy logic controller (FLC) improves the PID controller response as far as taking care of progress in a working point for nonlinear systems and update the controller parameters the problem associated with the PID-based tuning methods is that it avoids the essential property of fuzzy PID controller, which may result in poor response [6]. To overcome this problem a fractional order fuzzy proportional–integral–derivative (FOFPID) can be implemented with a digital filter [7]. FOFPID are also used in speed regulation of machines such as turbines. According to Takagi–Sugeno (TS), it makes controller more robust because of the presence of fractional order integral and derivative function and can easily model the nonlinear system [8].

Blood pressure as being an instable one, and hence in order to reduce the fluctuation, digital filters can be used. All the research so far is focused on pre-determined response of patients. Somehow if we could determine the patient's parameter-based current surroundings, health, and physiological condition. In this way, controller can optimize and can provide best control. For this purpose, in this work, we have used the parameter estimation technique to predict patient's response. The oscillatory change in MABP through the administration of SNP can be decreased by using parameter estimation-based automatic drug delivery controller.

2 Problem Formulation

An intravenous drug delivery system intended here in such a way can only be more flexible and adaptive to any drug, since it is related to medical field and hence life of patients, a drug delivery system should be trustworthy with high accuracy and sensitivity. A patient's response to mean arterial pressure for SNP infusion can be described by following dynamic model, originally proposed by slate [9].

$$MAP(t) = p_o + \Delta p(t) + p_d(t) + n(t) \tag{1}$$

Here, MAP is actually mean arterial pressure, initial blood pressure is p_0, the change in blood pressure is $\Delta p(t)$, and $pd(t)$ is considered as rennin reflex action.

When vasodilator drug is infused to body, then body reaction to particular drug is known as renin reflex, and n(t) is a stochastic noise. A continuous-time deterministic model describing the transfer function relating the output as change in MAP with respect to input as administration rate of SNP is given in (2) as [10]

$$\Delta p(t) = \frac{ke^{-ds}(1 - \alpha e^{-ms})}{1 - \tau s} d(s) \qquad (2)$$

where $\Delta p(t)$ is the blood pressure change, drug sensitivity is k, d(s) is infusion rate of SNP, k is patient's sensitivity, α is the recirculation constant of drug, d is the initial transport delay, m is considered as the recirculation time delay, and τ is the system time constant. By Discretizing the continuous-time model shown in (2) with the sampling time of 15 s gives (3) the following discrete-time model [10]:

$$\Delta p(t) = \frac{q^{-d_1}(b_o + b_1 q^{-d_2})}{1 - aq^{-1}} I(t) \qquad (3)$$

where q − 1 indicates a unit delay operator, are the parameters of the numerator, and b0, b1 are the delays achieved from the sampling the continuous-time model. A refined automated drug delivery system is necessary to overcome the limitations of the past research works such as latency and instability. A scale factor is introduced in order to equalize the quantity or to give more weightage or less weightage to a function by which it was multiplied with.

3　Control Algorithms

Patient can have different sensitivities to particular drug, and response may vary with time, so closed-loop monitoring model has been presented to improve the patient's response with respect to intravenously infused drug. Parameter estimation technique has been used to detect the patient's response, based on response integral learning-based fuzzy system is implemented to regulate whole process. We provide an automated drug delivery system using parameter estimation with fuzzy inference system which is shown in Fig. 1.

3.1　Fuzzification

Fuzzy logic controller is implemented here by considering two inputs: one is error and another one is rate of change of error required to be fuzzified in addition to a single output. The entire process requires a sufficient number of membership functions (MFs), thereby causing in a larger rule base for two inputs. These membership functions have their own functional area in the input space as a

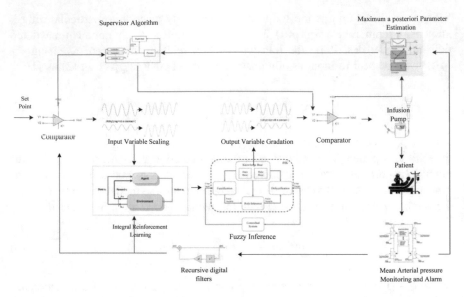

Fig. 1 Automated drug delivery model using parameter estimation-based learning with fuzzy inference system

membership value. Fuzzy rules depend on membership function and fuzzy rules, each MF is assigned an SNP infusion rates in (mcg/kg/min). Initially at normal blood pressure (90 mm Hg), the administration rate is zero, because the difference between the MABP and set point is zero. While the MABP increases through its membership curve, the infusion rate also increases with respect to the fuzzy rules. Another important thing is the infusion time with MABP of 160 mm Hg an infusion rate of 5 mcg/kg/min may reduce the MABP to normal level (90 mm Hg). The infusion time is adopted from IRL, which also has its effect on the decision of infusion rate as long as it updates the membership function by its learned things from the real time process, there upon it increases the robustness of the system in any uncertain conditions.

The all useful condition can be described by combination of these variables without doubt. After the selecting input variables, the shape and numbers of fuzzy sets have to be decided to sense them. A robust fuzzy partition has been proposed to keep the rule legibility (4)

$$\forall S_n \in S_n, \ \sum_{j=1}^{N_L(n)} \mu L_n^j(S_n) = 1 \tag{4}$$

where the input variable S_n is defined as $N_L(n)$ the fuzzy sets number. This type of separation suggests that there will be no additional fuzzy sets "activated" for an input variable (Fig. 2).

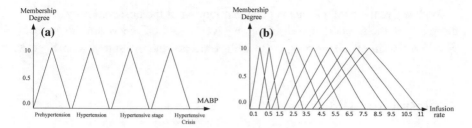

Fig. 2 Fuzzy partition of variable. **a** Input membership function. **b** Output membership function

With the help of input variables N_1 and number of labels help us determine the number of rules N of the database. They can be defined as fuzzy sets with a membership function of the following properties:

$$\mu o_m^i(Y_m) = \begin{cases} 1.0, & Y_m = o_m^i \\ 0.0, & otherwise \end{cases} \tag{5}$$

The function Y_m is approximated by O_m. Here, O_m is considered as conclusion vector that is related with rules. The actuated rule set represented by A is characterized by rule set that authenticates this condition. The maximum number of rules in A is equal to $2^N T$ according to fuzzy partition.

3.2 Estimation of Parameters by Maximum a Posteriori Estimators (MAPE)

Parameter estimation helps the system in updating the online variation, being an integral reinforcement learning method the control decisions also depend on the variations is the real-time ongoing process. A posterior distribution contains all the knowledge about the unknown quantity; therefore, we can use the posterior distribution to find point or interval estimates of. To obtain a parameter estimate is to choose the value of that maximizes the posterior probability density function (pdf).

The parameters of premises and consequents and structure identification will be identified by MAPE, which concerns partitioning the input space. Suppose that there are N observations and $y = [y(x_1)...y(x_N)]^T$ have been obtained with the design $X = (x_1,..., x_N)$ and assume $\pi(.)$ indicate a prior posterior density function for one or a subset of it. Bayesian estimation depends on the construction of the posterior (pdf) (6) for θ as [11],

$$\pi_{X,y}(\theta) = \frac{\varphi_{X,\Phi}(y)\pi(\theta)}{\varphi_X^*(y)} \tag{6}$$

With $\varphi_X^*(y)$ the p.d.f. of the marginal distribution of the observations y and $\varphi_{X,\Phi}$ the p.d.f. of their conditional distribution given θ. When the y_k are independent, $y_k = y(x_k)$ having the density $\varphi_{xk}, \overline{\theta}(y)$ with respect to the measure μ_{xK} on $Y_{xk} \subset R$.

4 Result and Discussion

Each pair of membership function, i.e., input as well as output membership function, is related with the fuzzy set rules; the relation is in the form of logic operator may be either "or"/"and". Our proposed IRLFI model is verified for adaptive control of the drug delivery system, the reference path shows a drop of MAP from 150 to 90 mmHg initially, and finally settled the level at 90 mmHg. The desired change of normal blood pressure is shown in Fig. 3.

The rate of change of SNP infusion depends on the variation in MABP level at a discrete-time period, the actual infusion rate of the SNP is shown in Fig. 4. At any time period, the infusion rate would be zero if the MABP level is within 90 mmHg and in case if the MABP level of a patient increases beyond the set point, the infusion rate also increases with respect to the increase in MABP level and the path followed by infusion rate for a certain time interval is given in graph below.

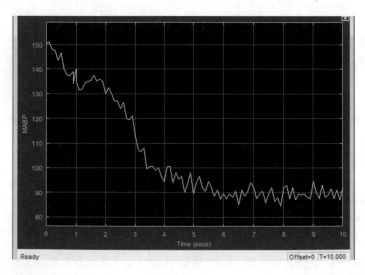

Fig. 3 Time versus normal mean arterial blood pressure

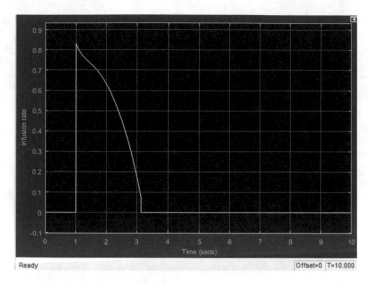

Fig. 4 Time versus SNP infusion rate

5 Conclusion

In an automated system which controls physiological variables, there are essentially three components sensors, a controller, and an actuator or an infusion pump. In this paper, the main focus was on control strategies using IRLFI system, by directly controlling the mechanical determinants of infusion pump. Our automated drug delivery system using IRLFI allows simultaneous control of MABP with stability and accuracy, and therefore, it can be used for the controlling of MABP and can also be applicable for any other infusion rate control system. The proposed method displays the ability of controller to improve the patient's response over less time, and also it improved the overall response of the closed-loop control of drug delivery system even in the presence of disturbances. In future, we can use this adaptive parameter estimation method for healing of different chronic diseases.

References

1. Rösler, A., Vandermeulen, G.W.M., Klok, H.-A.: Advanced drug delivery devices via self-assembly of amphiphilic block copolymers. Adv. Drug Deliv. Rev. **64**, 270–279 (2012)
2. Malagutti, N., Dehghani, A., Kennedy, R.A.: Robust control design for automatic regulation of blood pressure. IET Control Theory Appl. **7**(3), 387–396 (2012)
3. Neckebroek, M.M., De Smet, T., Struys, M.M.R.F.: Automated Drug Delivery in Anesthesia, vol. 3, pp. 18–26. Springer Science (2013)
4. Han, J., Zhu, Z., Jiang, Z., He, Y.: Simple PID parameter tuning method based on outputs of the closed loop system. Chin. Mech. Eng. Soc. **10**, 01–09 (2016)

5. Shabani, H., Vahidi, B., Ebrahimpour, M.: A robust PID controller based on imperialist competitive algorithm for load-frequency control of power systems. ISA Trans. (2012)
6. Sheng, W., Bao, Y.: Fruit Fly Optimization Algorithm Based Fractional Order Fuzzy-PID Controller for Electronic Throttle, vol. 10. Springer Science Business Media, Dordrecht (2013)
7. Tarasov, V.E.: Vector calculus in non-integer dimensional space and its applications to fractal media. Commun. Nonlinear Sci. Numer. Simul. **10**, 01–26 (2014)
8. Mici, A.D., Matauševkb, M.R.: Optimization of PID controller with higher-order noise filter. J. Process Control **24**(5), 694–700 (2014)
9. Slate, J.B., Sheppard, L.C.: Automatic control of blood pressure by drug infusion. IEE Proc. Phys. Sci. Meas. Instrum. Manage. Educ. Rev. **129**(9): 639 (1982)
10. Duan, X.-G., Deng, H., Li, H.-X.: A saturation-based tuning method for fuzzy PID controller. IEEE Trans. Ind. Electron. **60**(11), 5177–5185 (2013)
11. Huang, M., Wang, X., Wang, Z.: Multiple model adaptive control for a class of linear-bounded nonlinear systems. J. Latex Class Files **11**(4), 1–6 (2012)

Mathematical Modeling of Sensitivity and Specificity for Basal Cell Carcinoma (BCC) Images

Sudhakar Singh and Shabana Urooj

Abstract In this paper, mathematical modeling for malignant and non-malignant basal cell carcinoma is proposed. Image features are used for the modeling of growth rate, sensitivity, and specificity. Newton's law and Hooke's law play very important role in the modeling of growth, sensitivity, and specificity of basal carcinoma cell (BBC). Two features mean and entropy are used for two different positions. These two features are taken from the image data database of 550. Maximum growth rate of non-malignant and malignant BBC for the two different positions are 7.945, 10 and 19.76, 12, respectively. Maximum sensitivity and specificity calculated for malignant and non-malignant images are 0.8425, 0.3225 and 0.8512, 0.1992, respectively.

Keywords Growth of BCC · Sensitivity · Specificity · Newton's law

1 Introduction

Early location of doubtful skin wounds is basic to counteract skin spites, especially the melanoma, which is the most unsafe type of human skin disease. In a decade, imaging systems have been an undeniably critical instrument for early discovery and scientific models assume a pertinent part in plotting the movement of wounds.

Skin malignant is one of the diseases in the modern world which causes death to quite a few numbers of patients. Medical professionals and researchers are trying to find the root cause of the disease and produce effective medicines for the same. Most of the patients do not know about their skin malignant till it is in final stages. The basic diagnosis of a skin cancer is done using visual inspection by a general

S. Singh (✉) · S. Urooj
Department of Electrical Engineering, Gautam Buddha University,
Greater Noida, India
e-mail: sudhakarsingh86@gmail.com

S. Urooj
e-mail: shabanaurooj@ieee.org

© Springer Nature Singapore Pte Ltd. 2018
S. C. Satapathy et al. (eds.), *Information and Decision Sciences*,
Advances in Intelligent Systems and Computing 701,
https://doi.org/10.1007/978-981-10-7563-6_49

physician and if any further investigation needed the case is referred to a dermatologist. A biopsy and pathological analysis is carried out to know the type of cancer (basal cell carcinoma (BCC), squamous-cell skin carcinoma, and malignant melanoma). The deadliest among these three is malignant melanoma. It has an asymmetrical area, color variation, irregular border and diameter is often greater than 6 mm. Computer-aided detection (CAD) can be used on the skin disease images to assist the skin expert radiologist as a subsequent reader. CAD can find and differentiate the skin disease that a skin disease expert may not sprout. Once the CAD analysis has been done, the skin expert can take the decision based upon the classification results. Dermoscopy is a first and foremost extremely helpful standard technique for diagnosing the spiteful cells of skin disease [1]. The vital signs of dermoscopy are broadening in correctness matched with naked eye examination (up to 28% in the case of sensitivity and up to 14% in the case of specificity), thereby sinking the occurrence of pointless surgical removals of benign lesions [2, 3]. Skin infection is a large amount added unsafe if it is not found in the premature stages [4].

1.1 Related Work

Dermoscopy is a non-obtrusive imaging procedure used to acquire computerized imaginings on skin apparent by utilizing a gadget known as dermoscopy. It contains an amplifying focal point and a light source appended to a computerized camera. This gadget permits the representation of pigmented pattern in the epidermis and shallow dermis. Most dermoscopy structures are related to particular histopathological markers; in this manner, dermoscopy could be viewed as a connection between experimental (plainly visible) and histopathological (infinitesimal) morphology [5]. Most dermoscopy constructions are connected with definite histopathological signs; consequently, dermoscopy could be considered as a connection between medical test (macroscopic) and histopathological (microscopic) morphology [6].

Malignancy and related sicknesses remain a noteworthy issue to established researchers. In spite of being the fewest among all skin growths, the melanoma is a standout among the most destructive types of the illness [7]. Melanoma is due to the melanocytes, cells that create melanin, the skin tinge shade. From a study of disease transmission of tumor laid skin growth as the most widely recognized type of threat in the course of recent decades declared by USA [8]. Specifically, the melanoma is likewise the most well-known type of growth for youthful grown-ups (25–29 years of age) and the second most basic type of disease for youngsters (15–29 years of age) [9].

European examinations have recorded an expansion of melanoma rate over the most recent couple of decades. On account of Portugal, the assessed frequency for 2012 was 7.5 for every 100 000, mortality 1.6 for every 100 000, and predominance at 1, 3, and 5 years at 12.08, 33.99, and 53.93% separately. Because of the high harm capability of melanoma, early identification of doubtful skin damages is basic to anticipate threat and to expand the treatment adequacy.

2   Materials and Methods

In this work, image features have been used for the modeling of sensitivity and specificity of the BCC images. Dermoscopy images are acquired with regard to an experimental inspection, and can be impacted by some nonattendance of time constrained by money related and human obliges. These various edges can impact the idea of the picked up picture. Moreover, the proximity of artifacts (if fluid is associated in the interface between skin and dermoscopy) [10] needs a prepreparing dare to accumulate the photo to the division. The division, being the underlying stage in the image examination, requires an effective approach, since whatever remains of the strategy depends upon its correct yield. Segmentation of images is based on pixel-based separation, area-based separation, and edge ID. In this work multiscale local normalization (MLN) strategy is used for the calculation of area of suspicious part in images.

Genuine dermoscopy images used to test the approach were acquired with regard to clinical exercise by a dermatologist in an open doctor's facility, with a past educated agree to utilize these pictures for look into purposes.

2.1   The Mathematical Modeling

From a mechanical viewpoint, the BCC can be measured as a sealed thin-wall construction, kept up in strain by a force called turgor force. These BCC cell are further divided by Hooke's law [11]. Development is appeared by considering Newton's law. The running with course of action of conditions is understood for the position x and speed v and of the vertices of the bcc cells

$$F_{T,i} = m_i \frac{dv_i}{dt} \tag{1}$$

$$v_i = \frac{dx_i}{dt} \tag{2}$$

where m_i is the mass of the bcc cell vertex that discloses the rate of variation in speed (expanding speed) of the vertices comparable to the total force following up on the vertex; x_i and v_i are the position and speed of center topic i, exclusively, and the total force ($F_{T,\,i}$) is following up on this center of bcc cell.

The total pressure experienced by BCC cell's vertex is given by the equation below

$$F_T = \sum_{w \in W} F_w + F_d \qquad (3)$$

where F_w are total forces exerted by the set of walls w incident to the vertex, and

$$F_d = -bv_i \qquad (4)$$

is a damping force, and b is damping factor and the vertex speed v_i the force F_w is the resultant of the net turgor force accommodate Γ_{turgor} between the two bordering cells working average to the divider. Total wall force is given as

$$F_W = F_{turgor} + F_s \qquad (5)$$

The total turgor compel on the vertex is assumed by captivating the distinction in turgor weight P_{cell} of the two nearby cells increased considerably the distance of the divider as it is separated by the two occurrence peaks characterizing the divider.

In the proposed work, two distinct positions are considered for the investigation of carcinogenic and non-dangerous information.

2.2 Modeling of Sensitivity and Specificity

The logistic regression model is rapidly used for the analysis of sensitivity and specificity in the field of medical engineering. From the database of BCC images, a random data is selected for the diagnosis. Sensitivity and specificity is defined by a dependent variable V. If V = 1 then melanoma is malignant, if V = 0 then BCC is not malignant. In the proposed technique six variables are taken for the noninvasive diagnosis of BCC images. Predictive model is presented by the equation.

The prescient estimation of a true or false test may likewise be demonstrated utilizing this strategy. Be that as it may, the needy variable (V) must be determined to be the consequences of the "best quality level", then the sensitivity of the screening test v_i comprised as an informative factor (X,) [12]. The positive and negative prescient estimations of the screening test may then be assessed utilizing conditions (7) and (8) individually.

In the proposed work, a data set of extracted features is designed for malignant BCC and non-malignant BCC images, using these data set mathematical model is proposed and planned sensitivity and specificity of the image data set.

$$Logit \ \Pr(V = 1 \rightarrow X) = \alpha + \sum_{k=1}^{K} \beta_k X_k \qquad (6)$$

where K = 1, 2, 3, ..., k is positive integer, α and β are independent variables, and X_k is binary explanatory variable.

Fig. 1 Sensitivity, specificity
and predictive values of
diagnosis

N_{11}	N_{12}
N 21	N 22

Sensitivity of the diagnostic test for melanoma images was developed by Cornfield [8].

$$Sensitivity = \left\{ \frac{1}{1 + \exp\left[-\left(\alpha + \sum_{k=1}^{K} \beta_k X_k \right) \right]} \right\} \qquad (7)$$

where $X_1 = 1$ (Fig. 1).

For the specificity estimation of melanoma images is given below:

$$Specificity = 1 - \left\{ \frac{1}{1 + \exp[-(\alpha + \beta_1 X_1 + \beta_2 X_2)]} \right\} \qquad (8)$$

where $X_1 = 0$.

$$N_A = N_{11} + N_{12}, \quad N_B = N_{21} + N_{22}, \quad N_C = N_{11} + N_{21}, \quad N_D = N_{21} + N_{22} \qquad (9)$$

Sensitivity $= \frac{N_{11}}{N_B}$ and Specificity $= \frac{N_{22}}{N_D}$,
Predictive value for positive diagnosis $= \frac{N_{11}}{N_A}$,
Predictive value for positive Diagnosis $= \frac{N_{22}}{N_B}$.

The sensitivity of the diagnosis at covariate level W may be calculated using the variance–covariance matrix. If there are no additional covariates. Thus, a 95% confidence limit for the estimated sensitivity is given as below:

$$Sensitivity = \frac{1}{1 + \exp[-(\alpha + \beta_1 X_1 + \beta_2 X_2)]} \qquad (10)$$

And

$$Specificity = 1 - \left\{ \frac{1}{1 + \exp[-(\alpha + \beta_1 X_1 + \beta_2 X_2)]} \right\} \qquad (11)$$

Here, $X_1, X_2, \ldots, X_k$ all are equal to 1, and α is neglected.

3 Experimental Results

See Table 1 and Figs. 2, 3.

Table 1 Modeling of sensitivity and specificity of malignant and non-malignant images

S. No.	Time (s)	Non-malignant images		Malignant images		Non malignant images		Malignant images	
		Entropy	Mean	Entropy	Mean	Sensitivity	Specificity	Sensitivity	Specificity
1.	0.2	0.3178	0.4	0.7904	0.6	0.8294	0.1705	0.8007	0.1992
2.	0.4	0.6484	0.34	0.9748	0.64	0.6774	0.3225	0.8354	0.1645
3.	0.6	0.3391	0.36	0.9748	0.64	0.8425	0.1574	0.8354	0.1645
4.	0.8	0.5611	0.38	0.9239	0.82	0.6998	0.3001	0.8512	0.1487
5.	1.0	0.4774	0.4	0.8067	0.84	0.7324	0.2675	0.8385	0.1614

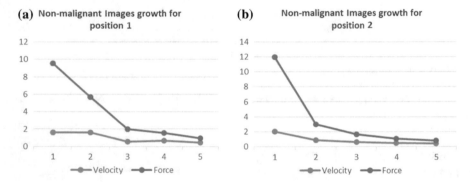

Fig. 2 Growth of the non-malignant images, **a** for the first position and **b** second position

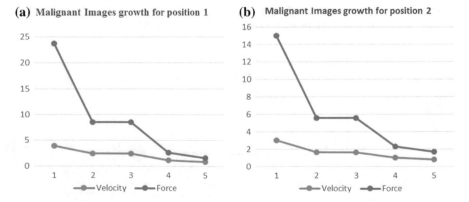

Fig. 3 Growth of the malignant images, **a** for the first position and **b** for second position

4 Discussion

In this work, modeling of growth, sensitivity, and specificity has been done using the features of the malignant and non-malignant area of the basal cell carcinoma. Table 1 depicted the feature of bcc image. Using the Newton's law, the growth of the malignant and non-malignant of the bcc is calculated. From Fig. 1, the growth of the non-malignant cells is investigated, and time interval from 0.2 to 1.00 s is taken for the modeling of growth of bcc cell. Figure 1a represents the growth of the non-malignant cell when position of the database is at entropy, and Fig. 1b presents the growth of the non-malignant bcc cell is at position mean. The rate of growth is maximum 7.945 at the entropy position 0.3178. When the position is sifted from entropy to mean, the growth is maximum 10 at mean position 0.4. The growth rates are minimum 0.4774 and 0.4 of the non-malignant images at two different positions 0.4774 and 0.4, respectively. In Fig. 2, the growth rate of the malignant images is shown. Growth rate is maximum 19.76 and 12 at positions 0.7904 and 0.6, respectively. The growth rate is minimum 0.8067 and 0.84 at position 0.8067 and 0.84, respectively.

Modeling of sensitivity and specificity has been done using features of the malignant and non-malignant images. For non-malignant images, maximum sensitivity 0.8425 occurs at the value of positions 0.3391, 0.36. For malignant images, maximum sensitivity 0.9239 is achieved at positions 0.82, 0.8512.

5 Conclusion

In this paper, the growth rate at different positions is investigated by mathematical modeling of malignant and non-malignant images. The growth rate is calculated by the modeling of Newton's third law. The maximum growth rate is at minimum time for the malignant and non-malignant images. Modeling of sensitivity and specificity is also has been done. Calculated maximum sensitivity and specificity is 0.8512 and 0.1992, respectively, for non-malignant images. For non-malignant images, maximum sensitivity and specificity is 0.8425 and 0.3225, respectively.

References

1. Yerushalmy, J.: Statistical problems in assessing methods of medical diagnosis, with special reference to X-ray techniques. Public Health Rep. (1896–1970) **62**, 1432–1449 (1947)
2. Greenberg, R.A., Jekel, J.F.: Some problems in the determination of the false positive and false negative rates of tuberculin tests. Am. Rev. Respir. Dis. **100**(5), 645–650 (1969)
3. Goldberg, J.D.: The effects of misclassification on the bias in the difference between two proportions and the relative odds in the fourfold table. J. Am. Stat. Assoc. **70**(351a), 561–567 (1975)

4. Eves, P., Layton, C., Hedley, S., Dawson, R.A., Wagner, M., Morandini, R., Ghanem, G., Mac Neil, S.: Characterization of an in vitro model of human melanoma invasion based on reconstructed human skin. Brit. J. Dermatol. **142**(2), 210–222 (2000)
5. Nissen-Meyer, S.: Evaluation of screening tests in medical diagnosis. Biometrics **20**, 730–755 (1964)
6. DeLong, E.R., Vernon, W.B., Bollinger, R.R.: Sensitivity and specificity of a monitoring test. Biometrics **41**, 947–958 (1985)
7. Ransohoff, D.F., Feinstein, A.R.: Problems of spectrum and bias in evaluating the efficacy of diagnostic tests. N. Engl. J. Med. **299**(17), 926–930 (1978)
8. Sudha, J., Aramudhan, M., Kannan, S.: Development of a mathematical model for skin disease prediction using response surface methodology. Biomed. Res. (2017)
9. Mendes, A.I., Nogueira, C., Pereira, J., Fonseca-Pinto, R.: On the geometric modulation of skin lesion growth: a mathematical model for melanoma. Rev. Bras. Eng. Biomed. **32**(1), 44–54 (2016)
10. Diamond, G.A., Rozanski, A., Forrester, J.S., Morris, D., Pollock, B.H., Staniloff, H.M., Berman, D.S., Swan, H.J.C.: A model for assessing the sensitivity and specificity of tests subject to selection bias: application to exercise radionuclide ventriculography for diagnosis of coronary artery disease. J. Chronic Dis. **39**(5), 343–355 (1986)
11. Pereira, J.M., Mendes, A., Nogueira, C., Baptista, D., Fonseca-Pinto, R.: An adaptive approach for skin lesion segmentation in dermoscopy images using a multiscale Local Normalization. In: Bourguignon, J.P., Jeltsch, R., Pinto, A.A., Viana, M. (eds.) Dynamics, Games and Science DGS II (CIM Series in Mathematical Sciences; 1), pp. 537–45. Springer (2015)
12. Liu, Z., Sun, J., Smith, L., Smith, M., Warr, R.: Distribution quantification on dermoscopy images for computer-assisted diagnosis of cutaneous melanomas. Med. Biol. Eng. Comput. **50**(5), 503–13 (2012). PMid: 22438064. http://dx.doi.org/10.1007/s11517-012-0895-7

Integer Representation and B-Tree for Classification of Text Documents: An Integrated Approach

S. N. Bharath Bhushan, Ajit Danti and Steven Lawrence Fernandes

Abstract Text document classification is creating more interest because of the availability of the information in the textual or electronic form. Generally, in conventional approaches, representation of text data and classification of text documents are considered as nondependent issues. In this research article, we have considered that overall efficiency of the text classification system depended on the effective representation of text data and efficient methodology for classification of the text documents. Here effective compressed representation for text documents is proposed for the text documents. Followed by a B-Tree-based classification methodology is adapted for classification. The proposed compressed representation and B-Tree methodologies are verified on the publically available large corpus to validate the effectiveness of the proposed models.

Keywords Text representation · B-Tree · Text classification

1 Introduction

Due to the increased popularity of the World Wide Web and Internet, text documents are becoming the one of the common kind of data (information) storehouse. Few most productive sources of text data are web pages, emails, discussion forums

S. N. Bharath Bhushan (✉)
Department of Computer Science & Engineering,
Sahyadri College of Engineering & Management, Adyar, Mangalore 575007, India
e-mail: sn.bharath@gmail.com

A. Danti
Department of Computer Applications, JNN College of Engineering, Shimoga 577204, India
e-mail: ajitdanti@yahoo.com

S. L. Fernandes
Department of Electronics and Communications,
Sahyadri College of Engineering & Management, Adyar, Mangalore 575007, India
e-mail: steven.ec@sahyadri.edu.in

© Springer Nature Singapore Pte Ltd. 2018   481
S. C. Satapathy et al. (eds.), *Information and Decision Sciences*,
Advances in Intelligent Systems and Computing 701,
https://doi.org/10.1007/978-981-10-7563-6_50

and newsgroup conversations, etc., contains billions of text documents [1]. These web portals are the main sources for the production of large text data which help many real-time applications like e-mail spam filtering, content-based document filtering to gain a lot of research interest and importance.

Generally, the theme of the document classification algorithm is to assign the label to the query document QD based on the features extracted from the training classes. Classification is a supervised learning task, where the algorithm is trained with a set of features extracted from the different classes. The process of text classification is to assign value say 0 or 1 to all query documents say QD to one of the predefined categories. The problem is to approximate the training function say ftrain: D and K → {0 or 1} by another function say testing function ftest: D and K → {0 or 1} such that the functions ftrain and ftest will coincide to the maximum level. Generally, the function ftest is considered as the classifier which will be trained scientifically by taking into consideration of the features extracted from the training samples. Though conventional text representation technique captures the textual information in an effective manner, it presents the number of issues like high dimensionality, fail to capture semanticity, volume, and sparsity. One of the solutions to this problem can be found at [2].

In this article, an efficient compressed representation for the text documents which will require small amount of computational time to save a term which will in turn minimizes the understanding cost is presented. The proposed model is based on the fact that integer values are easy to store and process rather than a complete word. The details of the efficient compressed representation of text data are presented in the respective section of this paper.

The remaining part of this article is planned as follows. In Sect. 2, a brief literature survey of the state-of-the-art techniques on the text classification is presented. In Sect. 3, a proposed model effective compressed representation for text data and classification using b-tree is presented. Section 4 discusses experimentation and comparative analysis of the proposed model with the existing state-of-the-art techniques. This paper will be concluded in Sect. 5.

2 Literature Review

The present section of the article presents existing works of text compression techniques considered for classification of text documents. But in literatures, few authors comment that text compression and classification are treated as two different issues. But this article presents an unconventional method for solving text classification problem using compression algorithms. By keeping this is in mind, our survey is more focusing on different compression and classification algorithms.

Generally compression algorithms are used to model the incoming text data and compress it based on statistical approaches. These are the algorithms requires more time since they require two passes on the input data. In literature, some of the works on compression-based classification can be seen. Mortan et al. [3] present a model-based approach for compression-based text classification. This research article presents three different models considered for compression. The three different models are standard minimum description length (MDL), approximate description length and BCN, best compression neighbor. In [4], a cross entropy-based approach for text classification is seen. This article is on the information that entropy is one of the best and appropriate methods to evaluate the information content in the text data. Text compression techniques, from information theory background, are generally based on the fact that language model is important for text classification issue. Also, authors have shown that, character level compression using PPM: prediction by partial matching is better compared to term level compression model. In [5], two different model-based approaches for text classification are seen. This paper illustrate two different models say modela and modelb for different classes of the training data. Query document will be compressed by making use of different abovementioned models and classification is achieved using the compression rate for the different considered models. Similar kind of word for low complexity devices can be seen in [6]. Here also authors present that have shown that, though PPM works better at character level, it requires more computational power because of this reason PPM is not practically advisable for low complexity devices say mobile phone or tablets. Because of this fact, the proposed compression algorithms for low complexity devices are based on static context models which will drastically reduce the storage space. Similar kind of technique for low complexity devices can be seen in [7]. Here it is based on the fact that the input data is divided into 16 bit followed by the application of Quine-McCluskey Boolean minimization function to determine the input data. Later well-known compression Huffman is used for compression of the data. Dvorski et al. [8] presents a different index-based approach for text compression. These kind works can be considered as the best approaches for content text retrieval systems. In [9], Khurana and Koul treat an English file as a combination terms (meaning full words) and nonterms (represents data or any other number). Term-based semi-static enhanced Huffman compression technique is seen in [10]. In this approach, algorithm captures the word features and constructs a byte-oriented tree. In [11], ETDCC: End tagged Dense code compression technique is presented. The ETDCC looks very similar to tagged Huffman compression technique, but the proposed ETDCC has more capacity in generating better compressed output. A different method for term-based compression technique can be seen in [12]. The present method tries to compress the text file at two different levels. The first level of compression is the word look table. It is based on the fact that generally word look-up table is located in the operating system, the first level of reduction is done at the operating system only. Compression will be achieved by replacing each word with its index number.

Now let us look toward classification approaches. Classification of text documents are nearest neighbor [13], Naïve Bayes [14, 15], support vector machines [16] regression-based approach [17, 18], and artificial neural network [19, 20].

3 Proposed Method

This section of the article explains the proposed compression-based integer representation-based approach for text documents for addressing classification problem. Classification of text documents is a supervised learning task, where the classification algorithm is trained with a set of features extracted from the different classes. The process of text classification is to assign value say 0 or 1 to all query documents say QD to one of the predefined categories. The proposed compression-based integer representation for text data is based on the fact that the memory required to store an integer is less than the memory required to store a character number. It would be better and less space is required if we store a text document as a collection of integer numbers. Once the text documents are represented in integer numbers, it will be very easy to handle integer numbers which contribute a lot to classification problem. The detailed explanation for this is presented in the corresponding subsections and the proposed model is diagramatically presented in Fig. 1.

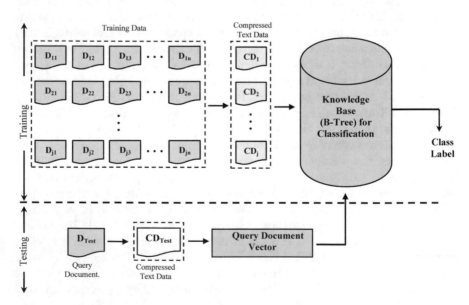

Fig. 1 Detailed diagram for the proposed technique

3.1 Text Representation

The proposed model is based on the fact that the sequence of characters requires more memory units compared to integer numbers. Based on this, a novel text representation algorithm is proposed, which has the facility of representing character string (a word) by unique integer numbers. It is known that words are the collection of alphabets, which represent a specific meaning. Similarly, a text document is a collection of such strings (word) which represent a specific domain. Words from the text documents are extracted and subjected for integer representation algorithm and represented by an integer number. The whole procedure is illustrated as below.

3.2 Classification Stage

Once the words from the text documents are extracted and subjected for integer representation algorithm, the text document can be viewed as a collection of integer numbers. As a result, text classification problem got transformed into integer number searching problem. Once the data is represented by an integer value, b-tree is used for searching and classification of text documents.

During the training stage, documents are subjected for integer representation algorithm and represented as integer numbers in the knowledge base of dimension k × tq, where k is the number of classes and tq represents the number of terms in that particular document. Once the construction of the knowledge base is completed, the concept of b-tree is exploited for classification of a query documents. Cells in the b-tree are designed in such way that word information and class information of the training documents can be easily preserved. After the construction of b-tree, query document is subjected for integer representation and compared with the b-tree for classification of a query document. The b-tree is accessed through in search of each term and the list of word indices. If the ith term Ti of the query document is present in the knowledge base of the class Cj, then the proximity score between the query and training class Cj is set to 1 otherwise set to 0.

Illustration 1: Compressed Integer Representation for each term in a document
Compressed Integer Representation to each term in the document will be using Eq. 1

$$Cw = \sum_{k=1}^{n} a_k b^k \qquad (1)$$

where

C_w **Cumulative sum of ASCII values,**
w   Collections of terms in each document,
k   Collection of alphabets in each term,
a   ASCII number of a character, and
B   Base.

ASCII values of "*apple*":
$$97 \times 2^4 + 112 \times 2^3 + 112 \times 2^2 + 108 \times 2^1 + 101 \times 2^0 = 3213$$

Input term	Memory required for regular compression	Memory required for after compression
apple	Five bytes	Two bytes

apple will be represented by an integer number **3213**.

$$Proximity_{ij} = \begin{cases} 1 & if\ T_i \in C_j \\ 0 & Otherwise \end{cases} \quad (2)$$

Based on the proximity score the query document will be classified to C_j class. The proposed model is presented in the following block diagram and the working procedure of the b-tree in illustration 2. As there are k classes and a query document contains t_q terms, $O(t_q log_r t)$ computational units are required for construction of knowledge base (b-tree).

4   Experimentation

To demonstrate the efficiency of the proposed compressed integer representation, four well-known datasets are considered for the experimentation.

dataset I: Vehicle Wikipedia dataset consists of vehicle information extracted from Wikipedia web portal. This data contains 440 documents from four different classes.

dataset II is Google Newsgroup dataset which consists of 1000 documents from ten different domains.

dataset III is the standard 20 mini newsgroup dataset, contains about 2000 documents from 20 different usenet discussion groups.

dataset IV is the 20 newsgroup dataset which contains 20,000 documents from 20 different usenet discussion groups. Dataset III is the subset of the dataset IV which contains randomly selected files from the original dataset. Two different kinds of experimentation are conducted to systematically verify the proposed

Table 1 Classification result for the proposed compressed integer representation

Datasets	40 percent: 60 percent			60 percent: 40 percent		
	Num of training	Num of testing	f-measure	Num of training	Num of testing	f-measure
Vehicle Wikipedia	44	66	0.8540	66	44	0.8962
Google Newsgroup	40	60	0.8660	60	40	0.9083
20 Mini Newsgroup	40	60	0.8659	60	40	0.9057
20 Newsgroup	400	600	0.8927	600	400	0.9153

algorithm. Two different kinds of experiments refer to division of the dataset into different ratios like 40% for training 60% testing and vice versa. Table 1 in the article presents the detail of the two different sets of experiments conducted to check the efficiency of the proposed models.

For the purpose of evaluation of results, f-measure, which is the harmonic mean of precision and recall, is calculated for each set of experiments using Eqs. (3), (4), and (5), respectively.

$$f - measure = 2PR/P + R \tag{3}$$

$$P(Precession) = a/(a + c) \tag{4}$$

$$R(Recall) = a/(a + d) \tag{5}$$

where *a, b, c,* and *d*, respectively, denote the number of correct positives, false negatives, false positives, and correct negatives.

Generally, increase in the value of P(precession) is often obtained at the cost of R(Recall) and vice versa. But conventional classifier will produce $P > R$ and $P < R$ is some special cases. This creates another issue in comparing the results of the classification algorithms. In order to balance precision and recall, higher order statistics such as f-measure is considered. F-measure is nothing but a harmonic mean of P and R. The higher value of f-measure denotes the efficiency of the proposed algorithm.

5 Conclusion

Current research article presents a compressed integer representation of text data for classification of text documents problem. Classification of compressed text data is accomplished using B-Tree in this article. The proposed article is based on the truth that memory required to store an integer is much lesser than storing a sequence of characters (a term). Here in this article, each term in the document will be represented using integer representation and a B-Tree is constructed to for effective indexing which is helpful for classification problem. The proposed compressed integer representation-based approach is evaluated by considering suitable publically available corpuses.

Illustration 2
Domain 1: Business is an art of making money.
   After stop word elimination: {business, make, money}
   Integer representation for D1 is {27021, 1575, 3345}

   * * *

Domain 2: Heart beats for 72 times per minute.
   After stop word elimination: {heart, beat}
   Integer representation D2 is {3204, 1498}

   * * *

B-Tree from the Training Documents from the domains D₁ and D₂.

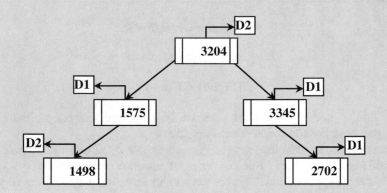

   * * *

Query Document: Heart is muscular tissue which beat 72 times.
   After stop word elimination: {heart, muscular, tissue, beat}
   Integer representation: {3204, 28380, 7107, 1498} ... (3)
   Now classification problem got reduced into searching for numbers of query document in b-tree.

Now,

Classification Stage:

Proximity score for Domain-1 and Domain-2

$$D1 = \{0,0,0,0,0\} \quad D2 = \{1,0,0,0,1\}$$

Now it is clear from the proximity score, query document belongs to **D2**.

References

1. Rigutini, L.: Automatic text processing: machine learning techniques. Ph.D. thesis, University of Siena (2004)
2. Bhushan Bharath S.N., Danti, A.: Classification of text documents based on score level fusion approach. Pattern Recogn. Lett. **94**, 118–126 (2017)
3. Marton, Y., Wu, N., Hellerstein, L.: On compression-based text classification. In: Proceedings of the European Colloquium on IR Research (ECIR), pp. 300–314 (2005)
4. Teahan, W., Harper, D.: Using compression based language models for text categorization. In: Proceedings of 2001 Workshop on Language Modeling and Information Retrieval (1998)
5. Frank, E., Cai, C., Witten, H.: Text Categorization using compression models. In: Proceedings of DCC-00, IEEE Data Compression Conference (2000)
6. Clemens, S., Frank, P.: Low complexity compression of short messages. In: Proceedings of IEEE Data Compression Conference, pp. 123–132 (2006)
7. Snel, V., Plato, J., Qawasmeh, E.: Compression of small text files. J. Adv. Eng. Inform. Inf. Achieve **20**, 410–417 (2008)
8. Dvorski, J., Pokorn, J., Snsel V.: Word-based compression methods and indexing for text retrieval systems. In: Proceeding Third East European Conference on Advances in Databases and Information Systems, pp. 75–84 (1999)
9. Khurana, U., Koul, A.: Text compression and superfast searching. In: Proceedings of the CoRR, 2005 (2005)
10. Moura, E., Ziviani, N., Navarro, G., Yates, R.B.: Fast searching on compressed text allowing errors. In: Proceedings of the 21st Annual International ACM Sigir Conference on Research and Development in Information Retrieval, pp. 298–306 (1998)
11. Nieves, G., Brisaboa, E.L., Param, J.: An efficient compression code for text databases. In: Proceedings of the 25th European Conference on IR Research, pp. 468–481 (2003)
12. Horspool, R.N., Cormack, G.V.: Constructing word based text compression of short messages. In: Proceedings of the IEEE Data Compression Conference, pp. 62–71 (1992)
13. Danti, A., Bhushan Bharath, S.N.: Document vector space representation model for automatic text classification. In: Proceedings of International Conference on Multimedia Processing, Communication and Information Technology, Shimoga, pp. 338–344 (2013)
14. Sebastiani, F.: Machine learning in automated text categorization. ACM Comput. Surv. **34**, 1–47 (2002)
15. Salton, G., Buckely, C.: Term weighting approaches in automatic text retrieval. J. Inf. Process. Manag. **24**(5), 513–523 (1988)
16. Joachims, T.: Text categorization with support vector machines: Learning with many relevant features. In: Proceedings of European Conference on Machine Learning (ECML), No. 1398, pp. 137–142 (2000)

17. Danti, A., Bhushan Bharath, S.N.: Classification of text documents using integer representation and regression: an integrated approach. Spec. Issue of The IIOAB Scopus Index. J. **7**(2), 45–50 (2016)
18. Bhushan Bharath, S.N., Danti, A., Fernandes, S.L.: A novel integer representation-based approach for classification of text documents. In: Proceedings of the International Conference on Data Engineering and Communication Technology, pp 557–564 (2017)
19. Hotho, A., Nurnberger, A., Paab, G.: A brief survey of text mining. J. Comput. Linguist. Lang. Technol. **20**, 19–62 (2005)
20. Mccallum, A.K., Nigam, K.: Employing EM in pool-based active learning for text classification. In: Proceedings of the 15th International Conference on Machine Learning, USA, pp. 350–358 (1998)

Switching Angle and Power Loss Calculation for THD Minimization in CHB-Multilevel Inverter Using DEA

Gayatri Mohapatra and Manas Ranjan Nayak

Abstract Voltage source multilevel inverter has acquired extensive applications in industries. They have the advantages of higher output voltage with lower switching frequencies, less THD for increased level. The evaluation of multilevel inverter loss is a trivial task due to the problem of unequal current sharing among the switches for different levels. This paper applies differential evolution algorithm (DEA) to evaluate the losses of switches of CHB-multilevel inverter including conduction and switching loss with IGBT as the switch. It optimizes the switching angle as well as the modulation index and evaluates the losses across the switch for minimum THD.

Keywords CHB-MLI · DEA · Power loss

1 Introduction

In recent year, the power requirement has increased at a faster rate. For a voltage requirement for medium rated drives and power level, it is difficult to operate with one semiconductor switch. Multilevel inverter draws enormous interest in the power industry due to the competency to develop a voltage from number of levels, received from capacitor voltage source [1]. Multilevel inverter generates a sinusoidal voltage from different voltage levels with reduced THD and dv/dt stress with respect to the increased level. It achieves high power along with empowering the use the renewable energy sources such as PV, wind and other sources. In the context of power system optimization and choice of multilevel inverter, a strict calculation of inverter efficiency with respect to power loss is mandatory. Reliability and system economical operation also has a factor of consideration. MLI has a

G. Mohapatra (✉) · M. R. Nayak
S 'O' A (Deemed to be University), Bhubaneswar, Odisha, India
e-mail: gayatrim79@gmail.com

M. R. Nayak
e-mail: manasnk72@gmail.com

© Springer Nature Singapore Pte Ltd. 2018
S. C. Satapathy et al. (eds.), *Information and Decision Sciences*,
Advances in Intelligent Systems and Computing 701,
https://doi.org/10.1007/978-981-10-7563-6_51

difficult process of loss evaluation due to unequal current sharing in each power switch of the inverter.

In this article, THD of different levels of inverter and optimized the same to IEEE standard using differential evolutionary algorithm (DEA) for a specific level. It also evaluates the losses of the inverter for the specific values of switching angle leading to minimum THD. A modeling technique for CHB-MLI (cascaded multi-level inverter) is applied to calculate the conduction and switching loss. Each IGBT is modeled by characteristics curve applying curve fitting exponential equation as a function of load current.

This paper is organized as follows: In Sect. 2, cascaded H-bridge multilevel inverter (CHB-MLI) is described, modeling of the system is done in Sect. 3, problem formulation is done in Sect. 4, and optimization technique is briefed in Sect. 5. Simulation and result are discussed in Sect. 6, and finally, the conclusion is given in Sect. 7.

2   Cascaded H-Bridge Multilevel Inverter (CHB-MLI)

The inverter with different levels is given in Fig. 1, which contains an array of power semiconductor and capacitor voltage sources. This contains a set of switches and voltage sources. The sequence of conduction of switches adds the voltages across the and finally, it is amplified at the output node. Hence, each switch has to resist only reduced voltage [1].

The multilevel converter starts from three levels, as the level increases, the output voltage increases the steps producing a waveform that will approach to a sine wave with less deformation. In the same context, a zero harmonic distortion can be obtained by infinite number of levels. More number of levels lead to the achievement of higher voltage levels with the issue due to voltage sharing. Furthermore, the maximum number of levels that can be allowed is limited due to voltage imbalance, need of voltage clamping, circuit complexities and packaging problem. The CHB inverter, as explained in Fig. 2, the inverter is comprised of a set of single-phase H_bridge inverter units connected in series.

Fig. 1 The one phase diagram of an inverter with different levels

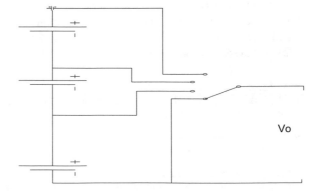

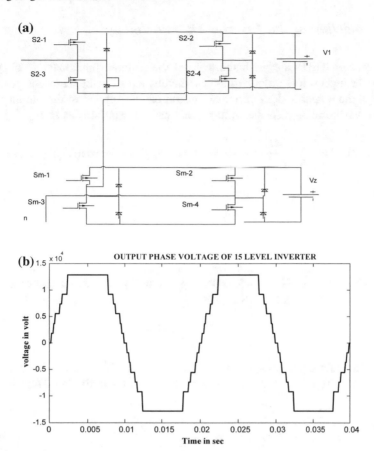

Fig. 2 **a** A single-phase n level cascaded inverter circuit, and **b** the 15 levels output voltage waveform

3 Modeling of the System

3.1 Modeling of the Inverter

If n is the number of steps of the voltage in phase for one half-cycle, then the number of levels in the phase voltage across the load is given by Eq. (1) [2]

$$k = 2s - 1 \tag{1}$$

For S as the number of voltage sources in dc (across capacitors), the number of voltage levels in Cascaded H-bridges inverter is given by $2 \times S + 1$ (phase) and they have $4 \times S + 1$ (line).

3.2 Modeling of the Inverter Output Voltage

The angles of IGBTs are predicted to form the required fundamental voltage and remove the highest low-order harmonics, thereby reducing the harmonic percentage (THD) at the output node. Fourier expansion can be applied to the output voltage stepped waveform, and (phase voltage) V_{an} can be expressed as in Eq. (2) [3].

$$V_{an}(wt) = \sum_{x=1,3,5}^{m} \frac{4E_{dc}}{x\pi} [\cos(h\beta_1) + \cos(h\beta_2) + \cdots + \cos(h\beta_s)] \sin(hwt) \quad (2)$$

where

S	signifies the number of H-bridge cells of the inverter,
E_{dc}	is the voltage of the dc input capacitor,
h	explains the no of harmonic order starting from $5-\infty$,
$\beta_1 - \beta_s$	are the switching angles depending upon the no of level, and
h	the number of harmonic order is given in Eq. (3). Here k takes the value 1 if S is even and 2 if S is odd.

$$h = 3S - k \quad (3)$$

The required output voltage (r m s) is the addition of the outputs of each bridge. The output voltage V_{an} is the summation of each of the individual H-bridges such that [4]

$$V_{an} = \sum_{z=1}^{S} V_{az} \quad (4)$$

where

V_{an} is expressed as the inverter phase voltage, and
V_{az} implies the individual source voltage for S number of sources.

A set of nonlinear equation is derived from Eq. (2) to get the value of switching angle (β). The nonlinear equations are given in Eq. (6) [5]

$$\begin{aligned}
\cos(\beta_1) + \cos(\beta_3) + \cdots + \cos(\beta_S) &= S \times M_1 \\
\cos(5\beta_1) + \cos(5\beta_3) + \cdots + \cos(5\beta_S) &= 0 \\
\cos(7\beta_1) + \cos(7\beta_3) + \cdots + \cos(7\beta_S) &= 0 \\
&\cdots \\
&\cdots \\
\cos(x\beta_1) + \cos(x\beta_3) + \cdots + \cos(x\beta_S) &= 0
\end{aligned} \quad (5)$$

where

M_1 is the index of modulation, and
x is the highest harmonic order which has to be removed.

These equations generate S number of nonlinear solutions [6]. Iterative techniques are used to provide initial assumed values of the delay angles to find a solution. DEA is a very strong tool to evaluate the problem using smartest approach. The main objective is to transform the challenges of selective harmonic elimination into an optimization problem, where the set of transcendental equations will be the matter of consideration for the optimization.

3.3 Power Loss Evaluation

Inverter losses are basically generated by IGBT/diode sets. The losses have two parts, conduction losses are calculated by taking a voltage V_{CE} interpreting the voltage drop and a resistor R_{ON} as in series with the ideal device. The nonlinear characteristic of the current–voltage is modeled.

3.3.1 Conduction Losses

The device conduction loss develops when the switch is on carrying current. This can be expressed as in Eq. 6 [5].

$$p_{conduction} = |I_C| \times V_{on} \tag{6}$$

3.3.2 Switching Losses

The switching loss occurs as the turn-on loss E_{on} and the turn-off loss E_{off} respectively. Especially, for the diode connected antiparallel with the switch, has the turn-off E_{rec} loss is considered only as the turn-on loss is very small or neglected. To calculate the switching losses, energy factor curve K is obtained by making a ratio of the switching energy to that of switching current as referred in the device datasheet. After that, curve fitting tool is used to approximate the energy factor curves by a polynomial equation [7].

The block diagram for conduction and switching losses are shown in Figs. 3 and 4.

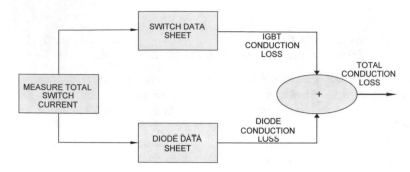

Fig. 3 The block diagram to calculate the conduction loss

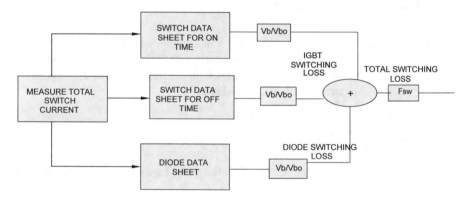

Fig. 4 The block diagram to calculate the switching loss

3.4 Modeling of IGBT

Mathematical modeling of IGBT is done by parabola interpolation method which can be applied to the characteristics graph between V_C and I_{CE} from the datasheet, and the curve fitting equations [6] are obtained as per Eqs. (7) and (8).

$$V_C = a1^2 I_{CE} + a2 I_{CE} + a3 \tag{7}$$

$$
\begin{aligned}
V_C &= -2 \times 10^{-6} I_{CE}^2 + 0.0073 I_{CE} + 1.173 \\
V_{PD} &= -3 \times 10^{-6} I_{PD}^2 + 0.0061 I_{PD} + 0.7479 \\
J_{ON} &= 8 \times 10^{-6} I_{CE}^2 + 0.0064 I_{CE} + 4.2836 \\
J_{OFF} &= 3 \times 10^{-6} I_{CE}^2 + 0.0031 I_{CE} + 2.2269 \\
K_{DIODE\_REC} &= 8 \times 10^{-6} I_D^2 + 0.0098 I_D + 3.8363
\end{aligned}
\tag{8}
$$

4 Formulation of Problem

4.1 Objective Function

Here the objective is to obtain low harmonic distortion and to get the output voltage close to sinusoidal. The objective function deals to minimize the THD, which is expressed as

$$Obj_F = (V_1 - SM_1)^4 + (V_5)^2 + (V_7)^2 + \cdots + (V_z)^2 + THD \tag{9}$$

$$THD = \sqrt{\sum_{n=2}^{\infty} \left(\frac{V_n}{V_1}\right)^2} \tag{10}$$

where

THD is the total harmonic distortion of the output voltage,
V_n responds to the output voltage harmonic for the order nth,
V_1 explains the output voltage (fundamental), and
M_1 corresponds to the index of modulation.

4.2 System Operational Constraints

(1) The harmonics of even order and the DC component are canceled. Only harmonics of odd are considered. Assuming a balanced three-phase system, all triplet order is zero in the output voltage equation in (2).
(2) The dc source voltage of all the capacitor and load is assumed to be constant to optimize the solution.
(3) Switching angle of IGBT, β takes the values between $0°$–$90°$.

5 Optimization Technique

DEA is an evolutionary computational algorithm first introduced by Storn Price (1995). It is a process which is carried out following the steps as mutation, crossover, and selection. Initialization leads to the process of mutation. (ϑ) generates random variables for 100 populations.

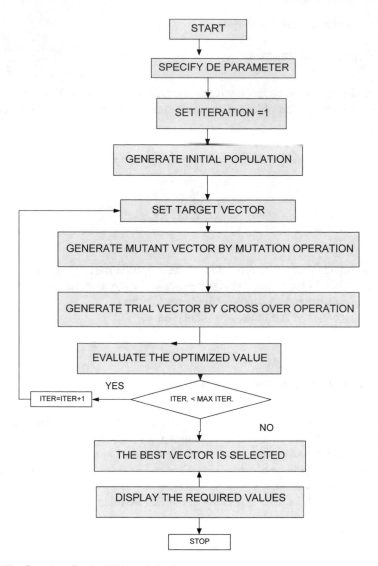

Fig. 5 The flowchart for the DEA optimization

5.1 Process of DEA Optimization

Initialization: Initialization is generating population within its prescribed limits of $0 - N$ where N represents population and D = variables. In this study, the switching angle is taken as a variable.

Mutation: The operator develops vectors a randomly selected vector. Vector indicates a1, a2, ..., an belong to {1, ..., Np} And a1 $\neq$ a2 $\neq$ a3 $\neq$ H. L is a user-defined constant known as the scaling mutation factor. This is typically chosen from within the range [0, 90].

Crossover: Crossover accelerates the diversity among the mutant parameter vectors and helps the algorithm to get rid of local optima. At the generation G, the crossover operation creates trial vectors (Hi) by mixing the parameters of the mutant vectors (H1I) with the target vectors (H2i) according to a selected probability distribution.

Selection: The selection chooses the vectors which are trying to drag the population in the next generation. This checks the fitness of the (Hi) and corresponding (H2i) and best solution is taken to the next generation.

The process is terminated whenever a maximum number of generations are reached or convergence is satisfied which is explained in Fig. 5.

6 Simulation and Result Analysis

The objective of this paper is to study the variation in the power loss with respect to load parameter for the optimized delay angle for a 15-level inverter.

For this, a model is developed in Simulink as in Fig. 7 modeling with the data given in Eq. (8) and the output voltage is calculated as per Fig. 2 for the optimized switching angle and modulation index, applying DEA. The optimized values are fed to the model by a pulse generator block and loss is evaluated. The loss calculation is performed as per the modeling given in Figs. 3 and 4 for a load of 60 Ω, 20 mH and found to be 970 W for the corresponding result.

The optimized THD values for different levels of inverter are evaluated and a graph is obtained as per Fig. 6. A Simulink model is developed as per Fig. 7, and the loss is calculated.

The variation in the load resistance and inductance has an effect on the loss which can be justified in Figs. 8 and 9 (Fig. 10).

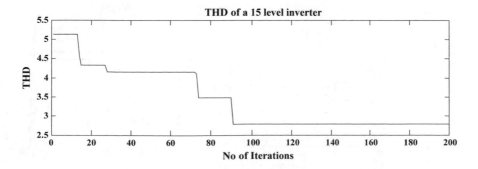

Fig. 6 The optimized graph of THD with respect to the number of iteration

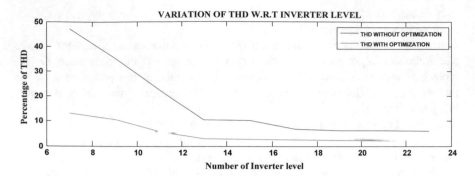

Fig. 7 Comparison of THD with respect to level of inverter is observed with and without optimization

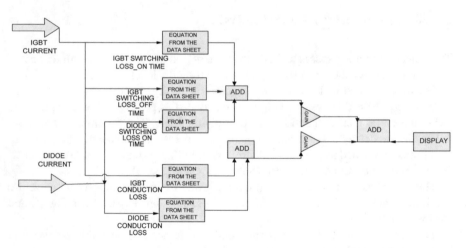

Fig. 8 The assembled diagram for loss calculation taking the switching and conduction loss into account for IGBT and DIODE

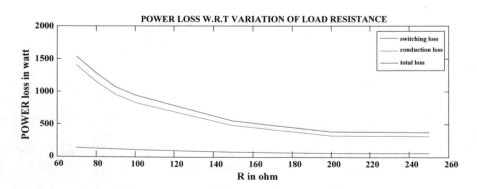

Fig. 9 The change of power loss with respect to load resistance keeping L constant

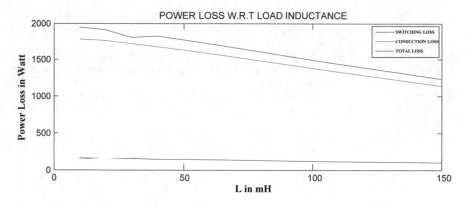

Fig. 10 The variation in power loss with respect to load inductance keeping R constant

7   Conclusion

In this article, differential evolution algorithm is applied to optimize the objective function where THD is one of the variables along with the number of harmonic voltages of a 15-level inverter. The optimized delay angles and modulation index are obtained. The loss of the switch corresponds to obtained angle is calculated and found to be 0.9 kW for a load of 60 Ω and 20 mH. The THD for different multi-level inverters is optimized starting from 7 to 27 level, and the switching loss is found to be around 7% of the total loss which can be minimized by controlling the switching frequency.

8   Future Scope

This work can be carried forward to find out the optimized inverter switching angle in order to have the minimum THD when connected to a grid through any renewable source integrated into the input side.

References

1. Rodriguez, J., Lai, J.S., Peng, F.Z.: Multilevel inverters: a survey of topologies, controls, and applications. IEEE Trans. Ind. Electron. (2002)
2. Chiasson, J., Tolbert, L.M., McKenzie, K., Du, Z.: Eliminating harmonics in a multilevel inverter using resultant theory. In: Proceedings of the IEEE Power Electronics Specialists Conference, pp. 503–508, June 2002. Cairns, Australia
3. Yao, W., Hu, H., Lu, Z.: Comparisons of SVM and carrier based modulation of multilevel inverter. IEEE Trans. Power Electron. (2008)

4. Mittal, N., Singh, B., Singh, S.P., Dixit, R., Komar, D.: Multi-level inverter: a literature survey on topologies and control strategies. In: 2nd International Conference on Power, Control and Embedded Systems (2012)
5. Aghdam, M.G.H., Fathi, S.S., Ghasemi, A.: The analysis of conduction and switching losses in OHSW multilevel inverter using switching functions. In: International Conference on Power Electronics and Drives Systems (2005)
6. Forzani, A., Nazarzadeh, J.: Precise loss calculation in cascaded multilevel inverters. In: Second International Conference on Computer and Electrical Engineering, 28–30 Dec 2009, vol. 2, pp. 563–568
7. Ramu, J., Parkash, S., Srinivasu, K., Ram, R., Prasad, M., Hussein, Md.: Reducing switching losses in cascaded multilevel inverters using hybrid-modulation techniques. Int. J. Eng. Sci. Invent. 2(4) (2012)
8. Govindaraju1, C., Baskaran, K.: Performance improvement of multiphase multilevel inverter using hybrid carrier based space vector modulation. Int. J. Electr. Eng. Inform. (2010)
9. Ramu, J., Parkash, S., Srinivasu, K., Ram, R., Prasad, M., Hussein, Md.: Comparison between symmetrical and asymmetrical single phase seven level cascaded h-bridge multilevel inverter with PWM topology. Int. J. Multidiscip. Sci. Eng. 3(4), 16–20 (2012)

Load Balancing of Unbalanced Matrix with Summation Method

Ranjan Kumar Mondal, Payel Ray, Enakshmi Nandi,
Biswajit Biswas, Manas Kumar Sanyal and Debabrata Sarddar

Abstract We know that cloud computing is an online-based servicing. So there are more than a million number of web servers, who are connected to online cloud computing to offer various types of online web services to cloud customers. Limited numbers of web servers connected to the cloud networks have to execute more than a million number of tasks at the same time. So, it is not simple to execute all tasks at a particular moment. Some machines execute all tasks, so there is a need to balance all loads at a time. Load balance minimizes the completion time as well as executes all tasks in a particular way. It is not possible to have an equal number of servers to execute equal tasks. Tasks to be completed in cloud environment system or environment will be greater than the connected components. Hence, a less number of servers have to execute a greater numbers of jobs. We propose a new algorithm in which some machines complete the jobs, where a number of jobs are greater than the number of machines and balance every machine to maximize the excellence of services in the cloud system.

R. K. Mondal (✉) · P. Ray · E. Nandi · D. Sarddar
Department of Computer Science and Engineering, University of Kalyani,
Kolkata, India
e-mail: ranjan@klyuniv.ac.in

P. Ray
e-mail: payelray009@gmail.com

E. Nandi
e-mail: pamelaroychowdhurikalyani@gmail.com

D. Sarddar
e-mail: dsarddar1@gmail.com

B. Biswas · M. K. Sanyal
Department of Business Administration, University of Kalyani, Kolkata, India
e-mail: biswajit.biswas0012@gmail.com

M. K. Sanyal
e-mail: manassanyal123@gmail.com

© Springer Nature Singapore Pte Ltd. 2018
S. C. Satapathy et al. (eds.), *Information and Decision Sciences*,
Advances in Intelligent Systems and Computing 701,
https://doi.org/10.1007/978-981-10-7563-6_52

Keywords Minimum completion time · Load balancing · Cloud computing

1 Introduction

We know cloud computing [1] is a web technology in support of the web or Internet and essential isolated servers to preserve records and applications. Cloud computing gives customers to make use of resources with no installation and right for using their resources from any PC with Internet. This technology gives incompetent computing by centralizing memory, bandwidth, and processing.

Cloud computing is a web form of internetwork where applications execute in more than one server on the web rather than on a PC or smartphone. Like the traditional client–server, a client connects to a server to perform a job. The main variation with cloud computing is that the computing process is run on more than one linked PCs to utilize the virtualization concept. With virtualization, more than one bodily servers can be configured and divided into multiple unattached "virtual machines" (VMs), all functioning autonomously and seem to the customer to be a distinct material machine. Such VMs do not really exist, so these could be moved. The Internet resources have been developed into grainy, giving customer and operator the benefits including open access to several devices, on-demand service, resource pooling, service surveying potential, and quick elasticity.

2 Load Balancing

In cloud computing, load balancing [2] is an online-based scheme for distributing total workloads across all web servers, network interfaces, and other resources. Typical Internet datacenter resources and applications rely on huge, dominant computing hardware, and network infrastructure which is an issue to the standard risks related to any device or PC, including hardware breakdown, lack of energy, and resource limitations in times of high demand.

This load balancing is utilized to assure that none of the currently available resources are idle or busy while others are being performed. To stabilize the distribution of loads, we could migrate the loads from the origin machines to the comparatively lightly loaded destination machines.

Cloud computing provides a variety of services to the customer, for example, apps sharing, online software, online gaming, and online storing. In the cloud environment, every single machine performs a task or a subtask [3].

Opportunistic Load Balancing algorithm intends to carry on each machine busy despite the available workloads of each machine [3–5]. Opportunistic load balancing algorithm allocates tasks to presented machines in arbitrary order. The Minimum Completion Time (MCT) [5] allocates tasks in the machines having the expected least execution time of this job over other machines [5]. The Min-Min algorithm assumes the similar scheduling approach as the MCT to allocate a job, and the machine has to complete this work with minimum completion time over other machines [6]. The Load Balance Min-Min (LBMM) [7] adopts the min-min approach and load balancing policy. It may stay away from the unnecessarily duplicated assignment.

3 Proposed Method

To find out the matrix cost of a collection of the machine(s) versus job(s) of the matrix, we would give a concentration to on a cost matrix problem consisting of machines (node) $M = \{M_1, M_2, \ldots, M_m\}$. An "n" number tasks or jobs $J = \{J_1, J_2, \ldots, J_n\}$ is considered to be allocated for implementation on the "m" existing machines and the completing cost S_{ij}, wherever $i = 1, 2, \ldots, m$ and $j = 1, 2, \ldots, n$ are referred to the matrix problem wherever $m > n$. Initially, we take the addition of each row; store the results in the array, named, Row-sum. Then we decide on the first n rows by row-sum, i.e., starting with smallest to next smallest to the array Row-sum and deleting rows related to the remaining tasks. Store the outcome in the new array that should be the array for the first subproblem. Repeat this progression until all tasks are assigned to their related machine until rows number are equal to the columns number. A left behind rows number makes another array named the second subproblem. We apply row-sum Mmethod at second subproblem to get a best possible outcome.

3.1 Algorithm

To discuss an algorithmic illustration of this technique, allow us to think a problem consisting of "m" Machines (nodes) $M = \{M_1, M_2, \ldots, M_m\}$. All jobs $J = \{J_1, J_2, \ldots, J_n\}$ is considered to be assigned for finishing on presented machines or node as well as the completing time S_{ij}, wherever $i = 1, 2, \ldots, m$ and $j = 1, 2, \ldots, n$, where $m > n$.

Step 1: *m*n matrix.*
Step 2 *This is to determine the* **Row Sum** *of each task, correspondingly.*
Step 3: *We split the matrix into two parts. First part will be n*n matrix based on maximum row-sum and remaining parts would be the n*m format. To calculate easy, we follow this rule. (The first part will be followed by one machine with one job with executes minimum execution time based on maximum row-sum and the second part will be followed by a machine that executes minimum execution time based on maximum row-sum.).*
Step 4: *In the second part, find a minimum cost of unassigned jobs J_i of all already assigned machines from maximum Row-sum. Next, it is to be allocateded to the consequent machine, and remove from the assigned job and the corresponding machine from the list. Then find next minimum cost of unassigned jobs from next maximum row-sum and allocate the job to its corresponding machine, and so on until all jobs assigned to their corresponding machine.*
Step 5: *So all jobs assigned at least to its corresponding machine and more the two jobs are not assigned to any machine.*
Step 8: *The End.*

4 Discussion with an Example

Consider a matrix having 5 nodes, i.e., $M = \{M_1, M_2, ..., M_5\}$, and 8 jobs, i.e., $J = \{J_1, J_2, ..., J_8\}$. This matrix has the execution time of each job to each node.

Step 1: Matrix

J_n/M_m	M_1	M_2	M_3	M_4	M_5
J_1	151	277	185	276	321
J_2	245	286	256	264	402
J_3	246	245	412	423	257
J_4	269	175	145	125	156
J_5	421	178	185	425	235
J_6	257	257	125	325	362
J_7	159	268	412	256	286
J_8	365	286	236	314	279

Step 2: It is to estimate the **Row-Sum** of each task, correspondingly.

J_n/M_m	M_1	M_2	M_3	M_4	M_5	Row-sum
J_1	151	277	185	276	321	**1210**
J_2	245	286	256	264	402	**1453**
J_3	246	245	412	423	257	**1583**

(continued)

(continued)

J_n/M_m	M_1	M_2	M_3	M_4	M_5	Row-sum
J_4	269	175	145	125	156	**870**
J_5	421	178	185	425	235	**1444**
J_6	257	257	125	325	362	**1326**
J_7	159	268	412	256	286	**1381**
J_8	365	286	236	314	279	**1480**

Step 3: We split the matrix into two parts. First part will be n*n matrix based on maximum row-sum and remaining parts would be the n*m format. The first part will be followed by one machine with one job that executes with a minimum execution time based on maximum row-sum and the second part will be followed by a machine that executes with a minimum execution time based on maximum row-sum.

J_n/M_m	M_1	M_2	M_3	M_4	M_5	Row-sum
J_2	**245**	286	256	264	402	**1453**
J_3	246	**245**	412	423	257	**1583**
J_5	421	178	185	425	**235**	**1444**
J_7	159	268	412	**256**	286	**1381**
J_8	365	286	**236**	314	279	**1480**

Step 4: As all machines assigned by jobs already, so we want to remain the minimum cost of unassigned jobs to its used machines. Find minimum cost of unassigned jobs J_i of all already assigned machines from maximum row-sum and it is to be assigned to the corresponding machine and remove from the assigned job and corresponding machine from the list and find the next minimum cost of unassigned jobs from next maximum row-sum and allocate the job to its corresponding machine and so on until all jobs assigned to their corresponding machine.

J_n/M_m	M_1	M_2	M_3	M_4	M_5	Row-sum
J_1	*151*	277	185	276	321	1210
J_4	269	175	145	*125*	156	870
J_6	257	257	*125*	325	362	**1326**

Step 5: So all jobs are assigned at least to its corresponding machine and more than two jobs are not assigned to any machine.

J_n/M_m	M_1	M_2	M_3	M_4	M_5
J_1	*151*	277	185	276	321
J_2	**245**	286	256	264	402
J_3	246	**245**	412	423	257
J_4	269	175	145	*125*	156
J_5	421	178	185	425	**235**
J_6	257	257	*125*	325	362
J_7	159	268	412	**256**	286
J_8	365	286	**236**	314	279

4.1 Final Result

The total execution time of each machine is as follows:
$M_1 \rightarrow J_1*J_2 \rightarrow 151 + 245 = 396$
$M_2 \rightarrow J_3 \rightarrow 245$
$M_3 \rightarrow J_6*J_8 \rightarrow 125 + 236 = 361$
$M_4 \rightarrow J_4*J_7 \rightarrow 125 + 256 = 381$
$M_5 \rightarrow J_5 \rightarrow 235$
And the final result is as follows:

M_m	M_1	M_2	M_3	M_4	M_5
Total execution time	396	245	361	381	235

5 Result Analysis

The following table displays the execution time (ms) of every task at dissimilar machines.

Figure 1 is showing the execution time for all tasks at dissimilar machines. The performance of our proposed work is compared with further methods by the case given in Fig. 1. This Figure displays the evaluation of the execution time of each machine. The completion times for completing all tasks by means of the proposed algorithm, HM [8], MM and LBMM are 396, 461, 555, and 423 ms, respectively. Our approach gets the lowest completion time with improved load balancing than another algorithm.

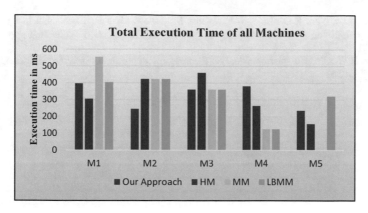

Fig. 1 Execution time (ms) of every task at dissimilar machines

6 Conclusion

The clarification of current unbalanced matrix, the matrix acquired with the support of summation technique is mentioned. Each subtask with minimum completion time is assigned to its corresponding machine so that result is best possible for the system. Although some machines execute more than one job, then completion time is best possible here, and loads are balanced with our proposed work.

This method is offered in an algorithmic shape and put into practice on the number of sets of input data to analyze the performance and efficiency of the algorithm. It is a straightforward technique to implement for load balancing.

References

1. Peter, M., Grance, T.: The NIST definition of cloud computing. 20–23 (2011)
2. Mondal, R.K., Ray, P., Sarddar, D.: Load Balancing. Int. J. Res. Comput. Appl. Inf. Technol. **4**(1), 01–21 (2016). ISSN Online: 2347-5099, Print: 2348-0009, DOA: 03012016
3. Armstrong, R., Hensgen, D., Kidd, T.: The relative performance of various mapping algorithms is independent of sizable variances in run-time predictions. In: 7th IEEE Heterogeneous Computing Workshop, pp. 79–87 (1998)
4. Freund, R., Gherrity, M., Ambrosius, S., Campbell, M., Halderman, M., Hensgen, D., Keith, E., Kidd, T., Kussow, M., Lima, J., Mirabile, F., Moore, L., Rust, B., Siegel, H.: Scheduling resources in multi-customer, heterogeneous, computing environments with SmartNet. In: 7th IEEE Heterogeneous Computing Workshop, pp. 184–199 (1998)
5. Ritchie, G., Levine, J.: A fast, effective local search for scheduling independent jobs in heterogeneous computing environments. J. Comput. Appl. **25**, 1190–1192 (2005)
6. Braun, T.D., Siegel, H.J., Beck, N., Bölöni, L.L., Maheswaran, M., Reuther, A.I., Robertson, J.P., Theys, M.D., Yao, B., Hensgen, D., Freund, R.F.: A comparison of eleven static heuristics for mapping a class of independent tasks onto heterogeneous distributed computing systems. J. Parallel Distrib. Comput. **61**, 810–837 (2001)

7. Wang, S.C., Yan, K.Q., Liao, W.P., Wang, S.S.: Towards a load balancing in a three-level cloud computing network. In: Computer Science and Information Technology, pp. 108–113 (2010)
8. Kuhn, H.W.: The Hungarian method for the assignment problem. Naval Res. Logist. Q. **2**(1–2), 83–97 (1955)

Detection of Outliers Using Interquartile Range Technique from Intrusion Dataset

H. P. Vinutha, B. Poornima and B. M. Sagar

Abstract Unpredictable usage of Internet adds more problems to the network. Protecting the system from the anomalous behavior plays a major issue in NIDS. Data mining approaches in the field of Intrusion Detection System (IDS) is becoming more popular. The outlier is a current problem faced by many data mining researches. Outliers are the patterns which are not in the range of normal behavior. Outliers in the dataset produce more false positive alarms, and this has to be reduced to increase the efficiency of IDS. We have used Interquartile Range technique to identify the outliers in the NSLKDD'99. In this, the continuous range of input is divided into quartiles and these quartiles are analyzed to target the range of outliers. Then the obtained outliers are removed by a filter called remove with value. The experiment is conducted using Weka data mining tool.

Keywords NIDS · Outlier · Interquartile range · Remove with value

1 Introduction

Intrusion Detection System has to be with more precision and stable in order to detect intruders. Usages of the internet are increased unpredictably and this adds more problems to the network. Protecting the network from this anomalous behavior is a major issue in Network Intrusion Detection System (NIDS). NIDS is one of the best security mechanisms which protect the network from different attacks. Keen

H. P. Vinutha (✉)
CS&E Department, Bapuji Institute of Education & Technology, Davangere, India
e-mail: vinuprasad.hp@gmail.com

B. Poornima
IS&E Department, Bapuji Institute of Education & Technology, Davangere, India
e-mail: poornimateju@gmail.com

B. M. Sagar
IS&E Department, RVCE, Bengaluru, India
e-mail: sagarbm@gmail.com

© Springer Nature Singapore Pte Ltd. 2018
S. C. Satapathy et al. (eds.), *Information and Decision Sciences*,
Advances in Intelligent Systems and Computing 701,
https://doi.org/10.1007/978-981-10-7563-6_53

observation of network is needed to get protected by attacks. The detection system has many stages to identify the anomalous behavior of the network. Data mining approaches like preprocessing, clustering, classification, and association rule play a major role in Network Intrusion Detection System to study network traffic pattern and to identify the attacks that take place in the network. Outlier detection is one of the approaches which help to reduce false alarm rate, and this reduction of false alarm improves the efficiency of the detection system. The outlier is the current problem faced by many data mining researches.

1.1   Outlier Detection

Outlier detection is a problem of finding patterns in data that are not in the range of normal behavior. These anomalous patterns are called outliers. Detection of such outliers helps to reduce the false alarm rate in NIDS [1].

Figure 1 shows the simple example of 2-dimensional outlier detection. In this, N1 and N2 are two normal data regions, O1 and O2 are the outlying instances, O3 is outlying region. Every dataset has outliers due to one or the other reasons, some of the common reasons are [1] the following:

(1) Malicious activity.
(2) Instrumentation Error.
(3) Change in the environment.
(4) Human error.

Three types of outliers are there and based on its relation to the data they are type 1, type 2, and type 3 outliers. Type 1 is based on individual outlying instances,

Fig. 1 Simple example of outlier detection

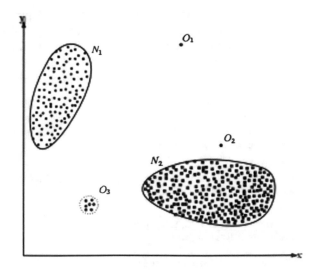

type 2 is due to individual data instances, and type 3 is a subset of data instances. Outlier detection has various application in different areas like intrusion detection, fraud detection, medical and public health data, industrial damage detection, image processing, novel detection in text data and other domains [1].

After the introduction, the paper is organized in the following sections. Section 2 gives brief literature survey, Sect. 3 describes the outlier detection approach, In Sect. 4 results are discussed with screenshots, Sect. 5 concludes the paper followed and by that references are given.

2 Background

In this [2] paper, author has studied about different outlier detection algorithms like statistical outlier detection, distance-based outlier detection, cluster-based outlier detection, density-based outlier detection, parametric and nonparametric outlier detection, classification-based outlier detection, sliding window-based outlier detection, and DSS and LDSS outlier detection algorithms. They have compared all the above-said algorithms using basic parameters like efficiency, computational cost, scalability, and applicability.

In paper [3], the author referred some real-time database to detect the outliers. The data mining techniques like clustering, classification, and association results are influenced by outliers. They have used mean, median, mode, and interquartile range in their experiment to detect outliers. The experiment is performed on 10 different datasets to prove all the instances are normal.

Paper [4] describes the unsupervised outlier detection for anomaly-based network detection. This unsupervised method overcomes the drawback of supervised anomaly detection. They have proposed a framework to detect outlier using random forest algorithm. This builds the patterns and detects the attacks in the datasets using outlier detection algorithm. The performance is compared and reported.

In paper [5], the author has proposed a system architecture which computes the distance and stores the information to improve the performance. Intrusion packets are received from the Internet and dataset is collected using SNORT. If the outlier values are greater than the threshold it generates false alarms. They have proposed neighborhood outlier factor (NOF), which is cable of capturing outliers by considering the density all around the points.

Paper [6] has implemented the intrusion detection system using various data mining techniques which helps to identify different attacks. They have used KDD dataset to analyze different techniques. In the proposed work algorithms like K-means, Genetic and SVM, Naïve Bayes, FPOP, and SCF. Finally, the paper is concluded that the use of hybrid approach of genetic and SVM to retrieve outlier performs well on the dataset.

3   Outlier Detection Approach: IQR

Interquartile range (IQR) is a technique that helps to find outliers in the data which are continually distributed. It is a difference between first quartile and third quartile; IQR = Q3 − Q1. The IQR is also known as midspread or middle 50% and technically H-spread. IQR can be used to divide the points of 25% and to build boxplot which is a simple graphical representation of interquartile range. In this method the dataset is divided into quartiles and orders the dataset into four equal parts. The divided ranges are Q1, Q2, and Q3 named as first, second, and third quartiles respectively. Figure 2 shows how box plot is built [7].

Algorithm Steps:

(1) Median is used to the calculate quartiles recursively.
(2) If 2n is the even number of entries

> Then,
>> Q1 is the first quartile = smallest entry median
>> Q3 is the third quartile = largest entry median,

(3) If 2n + 1 is the odd number of entries

> Then,
>> Q1 is the first quartile = smallest entry median
>> Q3 is the third quartile = largest entry median,

(4) Q2 is second quartile an ordinary median.

Example:
In this example, we have used some of the samples for odd entries with 13 rows. The samples have to be arranged is an ascending order as follows: −4, 4, 12, 14, 19, 22, 22, 24, 35, 44, 48, 83, 99.

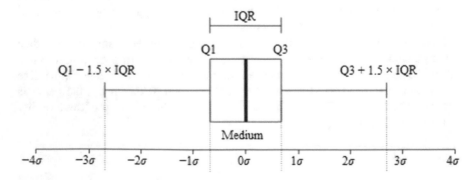

Fig. 2 Box plot

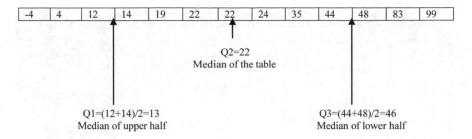

| -4 | 4 | 12 | 14 | 19 | 22 | 22 | 24 | 35 | 44 | 48 | 83 | 99 |

Q2=22
Median of the table

Q1=(12+14)/2=13
Median of upper half

Q3=(44+48)/2=46
Median of lower half

The first quartile is 13 and the third quartile is 46 then the **IQR = Q3 − Q1 = 46 − 13 = 33**. Now calculate the lower boundary and upper boundary to define the outlier calculation. The value which is less than and greater than these boundary values are considered as outliers.

$$\text{Lower boundary} = Q1 - (1.5*IQR) = 13 - (1.5*33) = -36.5$$
$$\text{Upper boundary} = Q3 - (1.5*IQR) = 46 + (1.5*33) = 95.5.$$

4 Results and Screenshots

The experiment is conducted on NSLKDD'99 dataset's training dataset, the original dataset contains 25192 instances and 42 attributes along with the class attribute. The usage of all 42 features in Intrusion Detection System is not a best practice, so it is necessary to find the relevant features. In order to find the best relevant features, feature extraction has to be performed on the dataset and some of the important relevant 29 features are extracted with one class feature. K-means clustering is applied on the reduced dataset to group the data into different clusters. Apply interquartile range technique on the clustered dataset to detect the outliers and extreme values. After performing IQR on the dataset, two more features are added to the dataset: Outlier, Extreme value. Last two columns are updated in the dataset with new values like yes and no. Yes indicated the outlier data which is out of range and no indicates the data within the range. Using remove with value method, the detected outliers can be removed from the dataset. This experiment is performed using Weka 3.7, which is a well-known data mining tool. Following screenshots give the experimental details.

Figure 1 shows the screenshot of the dataset before removing outliers, Fig. 2 shows the screenshot after identifying the outliers, and Fig. 3 is the screenshot after removing the outliers (Figs. 4 and 5).

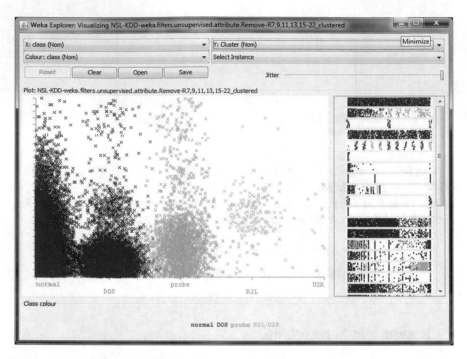

Fig. 3 Clustered data before applying IQR

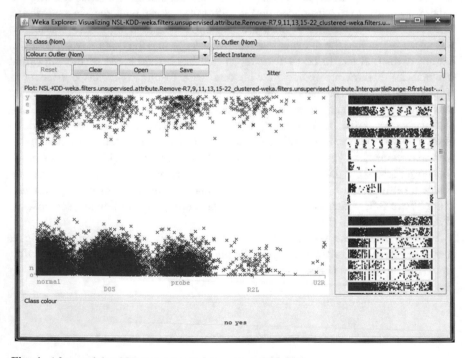

Fig. 4 After applying IQR (no: normal data, yes: outlaid data)

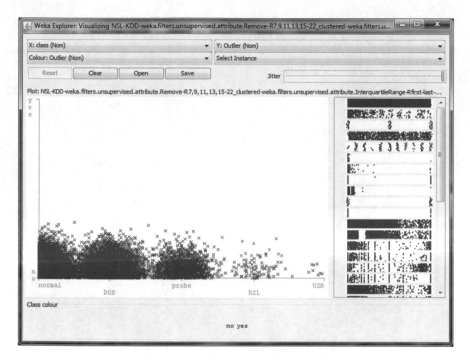

Fig. 5 After removing outliers

5 Conclusion

Detection of outliers is very important in Network Intrusion Detection System. Outliers in the dataset generate more false positive alarms, it is important to reduce the outliers from the dataset. Identifying outliers are also one of the biggest tasks. We have used interquartile range method to identify the number of outliers. This divides the continues data into three Quartiles like Q1, Q2, and Q3. IRQ is calculated using Q1 and Q3, finally using this IRQ upper boundary and lower boundary is identified. Screenshots are given to show the difference between before and after the identification of outliers.

References

1. Chandola, V., Banerjee, A., Kumar, V.: Outlier detection: a survey
2. Kaur, K., Garg, A.: Comparative study of outlier detection algorithms. IJCA **147**(9) (2016)
3. Sunitha, L., Balaraju, M., Sasikiran, J., Ramana, E.V.: Automatic outlier identification in data mining using IQR in real-time data. IJARCCE **3**(6) (2014). ISSN: 2278-1021
4. Zhang, J., Zulkernine, M.: Anomaly based network intrusion detection with unsupervised outlier detection

5. Jabez, J., Muthukumar, B.: Intrusion detection system: anomaly detection using outlier detection approach. ICCC, 338–346 (2015)
6. Mishara, M., Gupta, N.: Outlier detection and system analysis using mining technique over KDD. IJETTCS **4**(4) (2015)
7. Wikipedia

A Novel Arbitrary-Oriented Multilingual Text Detection in Images/Video

H. T. Basavaraju, V. N. Manjunath Aradhya and D. S. Guru

Abstract Text in images and videos plays a vital role to understand the events. The textual information is a prominent source and semantic information of a particular content of the respective image or video. Text detection is a primary stage for text recognition and text understanding. Still, text detection process is a challenging and interesting research work in the field of computer vision due to illumination, alignments, complex background and variation size, color, fonts of the text. The multilingual text consists of different geometrical structures of languages. In this paper, a simple and yet effective approach is presented to detect the text from arbitrary oriented multilingual images/video. The proposed method is based on Laplacian of Gaussian information and full connected component analysis. The proposed method is evaluated on four datasets such as Hua's dataset, arbitrarily oriented dataset, Multi-script Robust Reading Competition (MRRC) dataset and MSRA dataset with performance measures precision, recall and f-measure. The results show that the proposed method is promising and encouraging.

Keywords Multilingual text · Arbitrary-Oriented · Laplacian of Gaussian · Full connected component

1 Introduction

Text in an image or a video provides the semantic information and helps us to understand an event. Hence, the separation of textual information from its background is an important task. To do text detection task, initially, the gap between textual infor-

H. T. Basavaraju (✉) · V. N. Manjunath Aradhya
Department of Master of Computer Applications,
Sri Jayachamarajendra College of Engineering, Mysore, Karnataka, India
e-mail: basavaraju.com@gmail.com

D. S. Guru
Department of Studies in Computer Science, University of Mysore,
Mysore, Karnataka, India

© Springer Nature Singapore Pte Ltd. 2018
S. C. Satapathy et al. (eds.), *Information and Decision Sciences*,
Advances in Intelligent Systems and Computing 701,
https://doi.org/10.1007/978-981-10-7563-6_54

mation and background needs to be increased. The ample of works has been done on text detection and many of those methods considered particular properties and challenges to distinguish the text region from non-text region. Multilingual textual information consists of multiples of languages, with different geometrical shapes, multicolor, multisize, and multifonts. Hence, there is a demand for new approaches. Khare et al. [1] used k-means and gradient direction of pixels to obtain the textual information. The text candidates are extracted by computing median deviation of coefficient, K-means algorithm, and morphological functions in Minemura et al. [2]. Local block strength data is obtained to get accurate candidates by He et al. [3]. Wu et al. [4] developed an adaptive color scheme by observing image color histogram. Zhou et al. [5] used pipeline concepts to get the fast and accurate text. Ma et al. [6] introduce a rotation-based framework for arbitrarilyoriented text detection. A comprehensiveness of wavelet, Gabor, and K-means concepts for discovering the multilingual text is discussed by Pavithra and Aradhya [7]. Jeong and Jo [8] proposed a scheme for discovering the multilingual textual information on the basis of fast SWT. The text structure component detector has been developed by Ren et al. [9] to extract the text structure features. Liao et al. [10] introduced an integrated approach for detecting the multilingual scene text.

2   Proposed Methodology

The proposed method consists of two phases, in the first phase, Laplacian of Gaussian is used to extract text information from the background and in the second phase, full connected component concept is employed to determine the true text candidates.

2.1   Laplacian of Gaussian

The Laplacian of Gaussian detects edges as well as noise from the given input [i.e., Fig. 1a]. The rapid intensity change across the regions is highlighted by Laplacian method. This method uses Gaussian smoothing filter to smooth the regions and reduce the sensitivity of noise. The Laplacian of Gaussian computes the second order spatial derivatives. The Laplacian of Gaussian responses zero at long distance from the image, positive at one side of the edge, and negative at other side of the edge. The Laplacian of Gaussian edge detection method (refer Eq. 1) is processed on the gray level of given input to extract the two line structure of the text information. Figure 1 represents the extraction of Laplacian of Gaussian information from the input.

$$LoG(x, y) = -\frac{1}{\pi\sigma^4}\left[1 - \frac{x^2 + y^2}{2\sigma^2}\right]e^{-\frac{x^2 + y^2}{2\sigma^2}}, \tag{1}$$

where
σ is a Gaussian standard deviation.
x and y are Spatial coordinates.

2.2 Full Connected Component

We have carefully observed that text present in images or videos is made up of double line structures, as shown in Fig. 1c. It is quite evident to fill the double line structures in order to obtain the true text candidates. This structure of text has helped us to make edge information of each and every text in the form of full connected component. In this sequence, the given frame is transformed into gray level to increase the gap between text region and non-text region. After the application of Laplacian of Gaussian, holes are filled and these holes are considered as possible text candidates. Finally, true text candidates are detected by taking the difference between edge information (i.e., Fig. 1c) and hole filled regions (i.e., Fig. 2a). Figure 2 depicts the procedure of the full connected component on Laplacian of Gaussian information. Figure 2c shows the respective color components of the true text candidates (i.e., Fig. 2b).

| (a) Input Image | (b) Gray Image | (c) Two line structure |

Fig. 1 Laplacian of Gaussian information on input

| (a) Holes Filled | (b) Text Candidates | (c) Final Output |

Fig. 2 Extraction of full connected components

3   Experimental Results

The performance of the proposed method is evaluated by conducting an experimentation on standard dataset such as, Hau's dataset [11], arbitrary-oriented dataset used in [12], MRRC [13] and MSRA used in [14] datasets. These datasets consist of both graphic and scene text of multilingual languages like Kannada, English, Hindi, and Chinese. Precision (refer Eq. 3), recall (refer Eq. 2), f-measure (refer Eq. 4), and time parameters are used as performing measures of the proposed method and these parameters are calculated on the basis of text lines. The parameters of text detection are divided into three categories: (1) Actual Text Block (ATB) represents total number of text blocks present in single image or video frame, (2) Truly Detected text Block (TDB) is a text information detected by the proposed algorithm, and (3) Falsely Detected text Block (FDB) means non-text information detected by the proposed algorithm. The proposed method used a Laplacian of Gaussian technique effectively to produce the double line structure or full connected component of textual information. Arbitrary hole concept identifies the true text candidates by removing non-textual information. As well, time is a very important factor to take into consideration, thus we did our experiments on Intel CORE2Duo Processer system with 2GB of RAM. To illustrate the input frame, as shown in Fig. 2a, the execution time is estimated to 0.5 s. Time complexity has not described in the state of the art methods and hence we are not compared time with other techniques. The following subsections show the experimental results on different challenging datasets:

$$Recall(R) = \frac{TDB}{ATB},$$
(2)

$$Precision(P) = \frac{TDB}{TDB + FDB},$$
(3)

$$F - measure(F) = \frac{2RP}{R + P},$$
(4)

3.1   Experimental Results on Hua's Dataset

Hua's dataset consists of 45 horizontal text images with complex background collected from news sports events. The proposed method has correctly extracted the horizontal text information with very few false positives in case of low contrast images. Figure 3 shows the sample results of the proposed method. From Table 1, it is quite evident to notice that the proposed method is competitive when compared to the other state-of-the-art techniques with respect to recall.

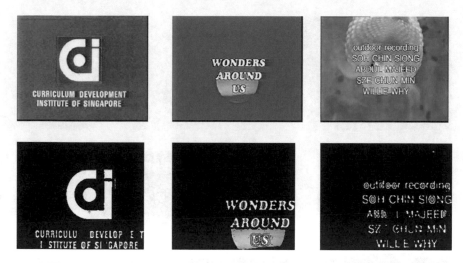

Fig. 3 Inputs and corresponding outputs of Hua's dataset

Table 1 Performance of the proposed and existing methods on Hua's data

Methods	R	P	F
Shivakumara et al. [15]	82	88	84
Sharma et al. [16]	88	77	82
Bayesian [17]	87	85	85
Laplacian [18]	93	81	87
Zhou et al. [19]	72	82	77
Fourier-RGB [20]	81	73	76
Lu et al. [21]	75	54	63
Wong and Chen [22]	51	75	61
Cai et al. [23]	69	43	53
Proposed method	86.02	62.43	72.35

3.2 Experimental Results on Arbitrary-Oriented Dataset

An arbitrary-oriented dataset has 142 arbitrary-oriented images with complex background, illumination, and low contrast. The proposed method accurately detects the text information region with very minute false positives in case of illumination effect. Figure 4 shows the sample results of the proposed method. Table 2 depicts the quantitative analysis of the proposed method on an arbitrary oriented dataset. The obtained results are very competitive when compared to that of other methods w.r.t recall and F-measure.

Fig. 4 Inputs and corresponding outputs of arbitrary-oriented dataset

Table 2 Performance of the proposed and existing methods on arbitrary-oriented dataset

Methods	R	P	F
Shivakumara et al. [15]	80	83	81
Sharma et al. [16]	73	88	79
Bayesian [17]	59	52	55
Laplacian [18]	55	68	60
Zhou et al. [19]	41	60	48
Fourier-RGB [20]	52	68	58
Lu et al. [21]	47	54	50
Wong and Chen [22]	34	90	49
Cai et al. [23]	45	62	52
Proposed method	72.54	57.58	64.20

3.3 Experimental Results on MRRC Dataset

Multi-script Robust Reading Competition (MRRC-334) dataset composed of 167 training images and another 167 testing images with all kinds of challenges of text detection. This dataset contains Kannada, English, Hindi, and Chinese with variety of degradation and challenges like curve, handwritten, multicolor, multiscript, illumination, artistic, emboss, night vision, depth, occlusion, low resolution, engrave, shear, glossy, slant, and shading. The proposed method correctly detects the text information region with very minute false positives. Figures 5 and 6 show the sam-

Fig. 5 Inputs and corresponding outputs of MRRC testing dataset

Fig. 6 Inputs and corresponding outputs of MRRC training dataset

Table 3 Performance of the proposed and existing methods on MRRC data

Methods	R	P	F
Yin et al. [13]	64	42	51
Proposed method	81.81	65.24	72.59

ple results of the proposed method. Table 3 depicts the experiment results of the proposed method on MRRC dataset. As well, the obtained results are compared with existing approach and outperform all the parameters considered.

3.4 Experimental Results on MSRA Dataset

The MSRA Text Detection 500 database (MSRA-TD500) is a publicly available benchmark dataset. This dataset contains 500 images, which are collected from indoor and outdoor scenes. The indoor images have signs, doorplates, and caution plates, while the outdoor images have guide boards and billboards in complex background. This dataset is very challenging due to the diversity of the texts and the complexity of the backgrounds in the images. The texts are in different lan guages like Chinese, English, or mixture of both with different fonts, sizes, colors,

Fig. 7 Inputs and corresponding outputs of MRSA testing dataset

Fig. 8 Inputs and corresponding outputs of MRSA training dataset

Table 4 Performance of the proposed and existing methods on MSRA data

Methods	R	P	F
Dey et al. [24]	85	52	65
Lu et al. [12]	63	29	40
Shivakumara et al. [25]	70	68	69
Epshtein et al. [26]	64	50	56
Rong et al. [27]	65	17	27
Mosleh et al. [28]	53	56	55
Li et al. [29]	65	26	37
Zhao et al. [30]	69	34	46
Yin et al. [31]	71	61	66
Risnumawan et al. [32]	68	70	69
Kang et al. [33]	62	71	66
Yao et al. [34]	62	64	61
Yin et al. [35]	63	81	71
Proposed method	84.79	68.79	75.95

and orientations. The proposed method is able to detect the text information region very effectively. Figures 7 and 8 show the sample results of the proposed method on this dataset. Table 4 depicts the quantitative analysis of the result obtained by the proposed method along with other existing contemporary methods on MSRA dataset. The previous methods used very complex and time-consuming techniques to detect the text present in MSRA dataset. But the proposed method explored a simple technique to extract the complex text. The f-measure outperforms the other methods. The obtained results are quite evident to claim that the proposed method is simple and effective in addressing the challenges of text detection in complex environment.

4 Conclusion and Perspectives

In this work, a very simple yet effective feature is introduced to detect the arbitrary multilingual text in images/video. The proposed method is so generic and not dependent on any specific language. The Laplacian of Gaussian effectively extracts full connected components. As well, full connected components are filled to extract the complete region of the text. The proposed method is evaluated on four challenging datasets considering all types of variations. In future work, the effective and efficient preprocessing method will be developed to increase the gap between text and non-text region.

References

1. Khare, V., Shivakumara, P., Raveendran, P.: Multi-oriented moving text detection. International Symposium on Intelligent Signal Processing and Communication Systems, pp. 347–352 (2014)
2. Minemura, K., Palaiahnakote, S., Wong, K.: Multi-oriented text detection for intra-frame in H. 264/AVC video. International Symposium on Intelligent Signal Processing and Communication Systems, pp. 330–335 (2014)
3. He, T., Huang, W., Qiao, Y., Yao, J.: Text-attentional convolutional neural network for scene text detection. IEEE Trans. Image Process. 25(6), 2529–2541 (2016)
4. Wu, H., Zou, B., Zhao, Y.Q., Guo, J.: Scene text detection using adaptive color reduction, adjacent character model and hybrid verification strategy. Vis. Comput. 33(1), 113–126 (2017)
5. Zhou, X., Yao, C., Wen, H., Wang, Y., Zhou, S., He, W., Liang, J.: EAST: an efficient and accurate scene text detector (2017). arXiv:1704.03155
6. Ma, J., Shao, W., Ye, H., Wang, L., Wang, H., Zheng, Y., Xue, X.: Arbitrary-oriented scene text detection via rotation proposals (2017). arXiv:1703.01086
7. Pavithra, M.S., Aradhya, V.N.M.: A comprehensive of transforms, Gabor filter and k-means clustering for text detection in images and video. Appl. Comput. Inform., 1–15 (2014)
8. Jeong, M., Jo, K.H.: Multi language text detection using fast stroke width transform. 21st Korea-Japan Joint Workshop on Frontiers of Computer Vision, pp. 1–4 (2015)
9. Ren, X., Zhou, Y., Huang, Z., Sun, J., Yang, X., Chen, K.: A novel text structure feature extractor for chinese scene text detection and recognition. IEEE Access 5, 3193–3204 (2017)
10. Liao, W.H., Wu, Y.C.: An integrated approach for multilingual scene text detection. Int. J. Comput. Inf. Syst. Ind. Manag. Appl. 8, 033–041 (2016)
11. Hua, X.S., Wenyin, L., Zhang, H.J.: An automatic performance evaluation protocol for video text detection algorithms. IEEE Trans. CSVT, 498–507 (2004)
12. Lu, C., Wang, C., Dai, R.: Text detection in images based on unsupervised classification of edge-based features. In: Proceedings of ICDAR, pp. 610–614 (2005)
13. Multi-script robust reading competition. http://mile.ee.iisc.ernet.in/mrrc/index.html
14. Yao, C., Bai, X., Liu, W., Ma, Y., Tu, Z.: Detecting texts of arbitrary orientations in natural images. In: Proceedings of CVPR, pp. 1083–1090 (2012)
15. Shivakumara, P., Basavaraju, H.T., Guru, D.S., Tan, C.L.: Detection of curved text in video: quad tree based method. In: 12th International Conference on Document Analysis and Recognition (ICDAR), pp. 594–598 (2013)
16. Sharma, N., Shivakumara, P., Pal, U., Blumenstein, M., Tan, C.L.: A new method for arbitrarily-oriented text detection in video. In: Proceedings of DAS, pp. 74–78 (2012)
17. Shivakumara, P., Sreedhar, R.P., Phan, T.Q., Lu, S., Tan, C.L.: Multi-oriented video scene text detection through Bayesian classification and boundary growing. IEEE Trans. CSVT, 1227–235 (2012)
18. Shivakumara, P., Phan, T. Q., Tan, C.L.: A laplacian approach to multi-oriented text detection in video. IEEE Trans. PAMI, 412–419 (2011)
19. Zhou, J., Xu, L., Xiao, B., Dai, R.: A robust system for text extraction in video. In: Proceedings of ICMV, pp. 119–124 (2007)
20. Shivakumara, P., Phan, T.Q., Tan, C.L.: New Fourier-statistical features in RGB space for video text detection. IEEE Trans. CSVT, 1520–1532 (2010)
21. Lu, C., Wang, C., Dai, R.: Text detection in images based on unsupervised classification of edge-based features. In: Proceedings of ICDAR, pp. 610–614 (2005)
22. Wong, E.K., Chen, M.: A new robust algorithm for video text extraction. Pattern Recognit., 1397–1406 (2003)
23. Cai, M., Song J., Lyu, M.R.: A new approach for video text detection. In: Proceedings of ICIP, pp. 117–120 (2002)
24. Dey, S., Shivakumara, P., Raghunandan, K.S., Pal, U., Lu, T., Kumar, G.H., Chan, C.S.: Script independent approach for multi-oriented text detection in scene image. Neurocomputing 242, 96–112 (2017)

25. Shivakumara, P., Phan, T.Q., Tan, C.L.: New wavelet and color features for text detection in video. In: Proceedings of ICPR, pp. 3996–3999 (2010)
26. Epshtein, B., Ofek, E., Wexler, Y.: Detecting text in natural scenes with stroke width transform. In: Computer Vision and Pattern Recognition (CVPR), pp. 2963–2970 (2010)
27. Rong, L., Suyu, W., Shi, Z.: A two level algorithm for text detection in natural scene images. In: 11th IAPR International Workshop on Document Analysis Systems (DAS), pp. 329–333 (2014)
28. Mosleh, A., Bouguila, N., Hamza, A.B.: Automatic inpainting scheme for video text detection and removal. IEEE Trans. Image Process. **32**(4), 460–472 (2013)
29. Li, Y., Jia, W., Shen, C., Hengel, A.V.D.: Characterness: an indicator of text in the wild. IEEE Trans. Image Process. **23**, 1666–1677 (2014)
30. Zhao, X., Lin, K.H., Fu, Y., IIu, Y., Liu, Y., Huang, T.S.: Text from corners: a novel approach to detect text and caption in videos. IEEE Trans. Image Proces. **20**, 790–799 (2011)
31. Yin, X.C., Yin, X., Huang, K., Hao, H.W.: Robust text detection in natural scene images. IEEE Trans. PAMI **36**, 970–983 (2014)
32. Risnumawan, A., Shivakumara, P., Chan, C.S., Tan, C.L.: A robust arbitrary text detection system for natural scene images. ESWA **41**, 8027–8048 (2014)
33. Kang, L., Li, Y., Doermann, D.: Orientation robust text line detection in natural images. In: Proceedings of CVPR, pp. 4034–4041 (2014)
34. Yao, C., Bai, X., Liu, W.: A unified framework for multioriented text detection and recognition. IEEE Trans. Image Process. **23**, 4737–4749 (2014)
35. Yin, X.C., Pei, W.Y., Zuang, J., Hao, H.W.: Multi-orientation scene text detection with adaptive clustering. IEEE Trans. PAMI **37**, 1930–1937 (2015)

Working with Cassandra Database

Saurabh Anand, Pallavi Singh and B. M. Sagar

Abstract Traditional databases cannot handle a huge amount of data. NoSQL database like Cassandra can handle large data with easier management and lower cost compared to SQL databases like Oracle and other relational databases. Cassandra is an open source technology. This paper explains how Cassandra helps in handling large data efficiently with almost 30–35% more efficient than the relational database like Oracle when data size increases to ten thousand and beyond. For this, the paper explains an experiment which was carried out by varying the size of number of data records and comparing the performance of Cassandra and Oracle. As the data size was continuously increased, most of the Cassandra queries took almost 30–35% less time as compared to Oracle. Though Oracle was more efficient when data size was less (up to 40×10^3 records), the performance dropped thereafter continuously for every 10×10^3 increase in data size thereafter.

1 Introduction

Databases are important components of information processing and it coexists with computer technology. Data storage and retrieval in appropriate manner require the use of apropos database. Relational databases have strongly dominated the modern computing applications for centuries. It is popular because of its consistency and functionality. Data management is efficient in management system because of ACID properties (Atomicity, Consistency, Isolation, and Durability). However, when it comes to the nonstructured and vast amount of data, this relational model

S. Anand (✉) · P. Singh
J.C. Penney, Bengaluru, India
e-mail: saurabh789789@gmail.com

P. Singh
e-mail: pallavisingh.jc@gmail.com

B. M. Sagar
R.V College of Engineering, Bengaluru, India
e-mail: sagar.bm@gmail.com

© Springer Nature Singapore Pte Ltd. 2018
S. C. Satapathy et al. (eds.), *Information and Decision Sciences*,
Advances in Intelligent Systems and Computing 701,
https://doi.org/10.1007/978-981-10-7563-6_55

has its own limitations. Relational databases are sufficient enough to work with the structured data but not when it comes to today's requirement for velocity.

For relational databases, it becomes a challenge to provide cost-effective and fast Create, Read, Update, and Delete (CRUD) operation as it has to deal with the overhead of joins and maintaining relationships among various data.

Hence, a new mechanism is required to deal with such data in an easy and efficient way. Here forth, NoSQL comes into the consideration to deal with unstructured BIG data in an effective way to provide maximum business value and customer satisfaction.

NoSQL databases are topically manipulating the architecture of the database and it predominately allows to increase the performance and availability of services. Thereby, it seems fruition when switching from conventional database to a NoSQL database.

1.1 Apache Cassandra

A database such as Cassandra [1] manages such huge amount of data because data is not stored in a single server and is distributed across different nodes in a cluster. The write operation goes to any random Cassandra node and this acts as the proxy node and writes data to the cluster. The data in is written to all nodes in the cluster using a replication placement strategy. There is no concept of a master node in Cassandra. Replication strategy determines how the data will be propagated across different nodes in the same or different cluster. The data to be written to the nodes is pushed into Memtable [1] with changes. Memtable is an in-memory table used by Cassandra. The full memtable is continuously written to SSTable [1] on disk and so there will be less in-memory data at a time. The use of memTable and SSTable for writes makes writes faster in Cassandra as there are no intensive synchronous writes blocking the disk as Cassandra does not update the data in disk in place. As the SSTables are continuously merged into one another and so the temporary SSTables are garbage collected. The writing of memtable to a disk-based structure called SSTable and merging of SSTables happen asynchronously and continuously and so the writes are very fast in Cassandra.

2 Background Study

Business is growing at a rapid pace and so is the data generated and data began to increase both in interconnectedness and volume. Maintaining this data efficiently with real-time data querying and updates has been challenging. According to a report published by IBM, 90% of world's data is created in recent 2 years. One reason for the recent boom of data is Internet. Data is created everyday web via web and business applications and large volume of data is becoming extremely difficult for relational databases to handle. To handle such large data which posed a challenge for relational database in terms of both storage and data fetch requirements, the concept of NoSQL database gained popularity especially years following 2009 where technologies such as Google Big Data and Amazon's Dynamo DB became areas of interest to people. The popularity of technologies which can handle large sets of data grew such as MongoDB, Cassandra and many research works were published which focused primarily on evaluation of its performance and focusing on areas such as scalability, availability, etc. One of the earliest performance evaluations for NoSQL database including Cassandra was done using Yahoo! Cloud Serving Benchmark (YCSB). YCSB [2] is still used today as the benchmark for measuring the performance of NoSQL [3] Database in terms of read/write operation, latency etc.

Data can be stored and fetched from Oracle, MySQL [4] for smaller data set in an efficient way but if data set is huge, relational database tends to work inefficiently and takes almost 90–95% more time compared NoSQL database like Cassandra and this was found by Rabl et al. [5].

An environment which provides data storage and real-time querying and update capabilities in acceptable time constraint is the suitable data storage and maintenance system. In recent years, software developers have been investigating ways of data fetching, manipulating, and real-time querying. Cassandra has evolved as one of the most efficient to provide such capabilities. It is an evolutionary change in data management environments as it provides efficient reads and writes by using memtable structure and partitions spread across SSTables.

3 Workflow Architecture

Our workflow architecture (Fig. 1) for the working of Cassandra consists of storing data (here "teaches" POJO-Plain Old Java Object) object in Cassandra and then retrieving data from Cassandra. In the process of saving data to Cassandra, the request initially goes to a coordinator node which replicates data across other nodes and finally returns success once the consistency level is met as set by the user or default consistency. Cassandra stores the data temporarily into a memory table called memtable. The size of memtable can be best adjusted according to size of data being dealt with. Also, Cassandra writes a log called commit log to recover data from memtable in case of any failures related to hardware. The data in memtable are flushed to disk whenever threshold exceeds and individual partitions are written across several SSTables on Disk.

The size of individual SSTables plays a significant role in determining the query time for a record if not present in memtable. The compaction strategy helps Cassandra to better store and fetch rows from partitions spread across several SSTables on Disk. In the process of a read request, the request is made to coordinator node and data is returned to the coordinator node from other nodes which wait on consistency level being met across nodes before returning the data.

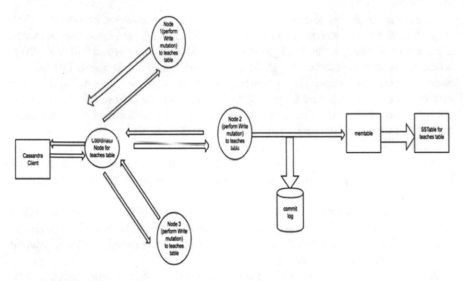

Fig. 1 Read and write Cassandra operation architecture

Once consistency level is met which can be ALL (all nodes must return the data) or Quorum (majority nodes out of a total number of nodes must return data) or some other consistency level as set by user or default is met.

4   Read and Writes in Cassandra

The read and writes are controlled by consistency levels across the Cassandra node as Cassandra offers eventual consistency. For a single data center, the reads will go to coordinator node and based on the read consistency level set, data returns from the nodes. For example, a read consistency of QUORUM in a single data center cluster with a replication factor of 3 then 2 of three replica nodes must respond to fulfill the request. If the contacted replicas have different data version, then the replica with most recent version returns the result. If the third replica is out of date with other two replicas, a read repair is initiated for out-of-date replicas. Consistency levels for read control how much consistency of data is required before the read operation returns the query result from replica nodes.

Same as read consistency, the write consistency level determines the number of replicas on which writes must be successful before the acknowledgement is returned to the client. For example, a write consistency of QUORUM in a single datacenter cluster with a replication factor of 3 then 2 of the three replica nodes must return success for write operation before the acknowledgement is returned to client application. For the analysis done in this paper, we have used a single-node Cassandra cluster with read and write consistency of one.

5   Data Modeling in Cassandra

Cassandra offers extensive data modeling technique through the use of clustering and partition key [6].

Cassandra stores data in different partitions based on partition key. Inside the partition key, the records are default sorted by cluster key.

Always make frequently queried fields as partition key as they can be easily put under different partitions and can be easily retrieved. Assume we have a system where students continuously get enrolled for a subject or course. Each subject or course is handled by a specific subject or course teacher. Now, we need to update student record for the subject for which the student enrolls for. A good data model will be able to make the subject id or teacher id teaching a particular subject as partition key and last updated timestamp as clustering key. By making teacher id or subject id as partition key, we can directly retrieve all student data enrolled for the same course at a given point of time from a single partition which would be have been not possible if student data for students enrolled in the same subject or taught by the same teacher would have lied in different partitions. Also, the recent most student to enroll in a particular subject can be found easily if we sort the student data by clustering key (last updated timestamp) in DESC order inside the course id or teacher id partition. Here, making the student id as partition key will be ineffi- cient as every new student enrolled for a subject will lie in a different partition with other details as course id enrolled for or teacher id, etc. As student id will be different for every incoming new student, to find the number or details of student enrolled under a course will be looking through all partitions and filtering out the require student data for a particular course as students are now distributed across partitions for the same subject in which they are enrolled.

6   Migrating Data to Cassandra

There are many ways to move data from RDBMS [7] to Cassandra. One of the common approaches to load large data into Cassandra is using SSTable loader utility of Cassandra. Cassandra provides SSTable Loader which can directly load the SSTables generated through custom the code in Cassandra. The generation of SSTables can be achieved through object of SimpleUnsortedWriter [8] which writes records into SSTables. These SSTables are then directly imported into Cassandra using SSTable importer [9]. SSTable loader can dump large data into Cassandra in short time. There are also other ways to read data from RDBMS and write to Cassandra such as using custom spring batch reader to read data from RDBMS and custom writer to write data to Cassandra. But using this approach requires reading data from a data source and writing to Cassandra. Whereas using the SSTable approach, we can generate the SSTables beforehand and dump into Cassandra in much lesser time. For example, consider we want to dump the student table and teacher table discussed.

7   Performance Evaluation on Basis of Demonstration

On analyzing the performance of both Cassandra and Oracle to Fetch information (student information of student taught by the same teacher), the performance of Cassandra is more when compared to that of Oracle by almost Cassandra taking 30% less time when working with records over 40000 and more. In an ideal scenario, Oracle is applicable only for small data set up to few thousands as in our case after dataset size exceeded 10^4 records Cassandra queries were more time efficient as compared to Oracle.

For evaluating the performance of Oracle and Cassandra, a table in Cassandra is created with a column (teacher_id) as partition key and last updated timestamp as clustering key. As more and more students are enrolled to be taught by the same teacher, all their information (name, student_id, etc.) will rest in the same partition as teacher_id is the partition key which will be same for all students taught by the same teacher. Similarly, two tables were created in Oracle namely student and teacher. The relationship between student and teacher table is established by teacher_id as primary key in teacher table and foreign key in student table linked to each other by foreign key constraint. Each Cassandra and Oracle table was populated continuously by increasing the data size and query to find the details of all students taught by the same teacher were executed against both the Cassandra and Oracle table (using a foreign key relationship between teacher and student table).

Oracle queries proved to be good initially for few records up to 40000 records. But as records increased, the efficiency of Cassandra was far better as shown (Figs. 2 and 3) by average of almost 20–30% more efficiency.

Below figure shows performance analysis:

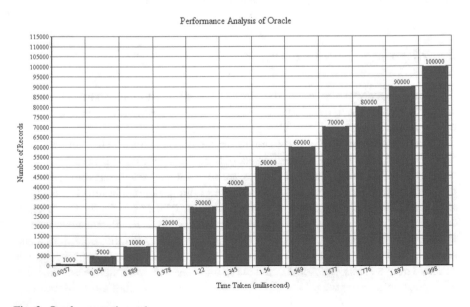

Fig. 2 Oracle query time taken

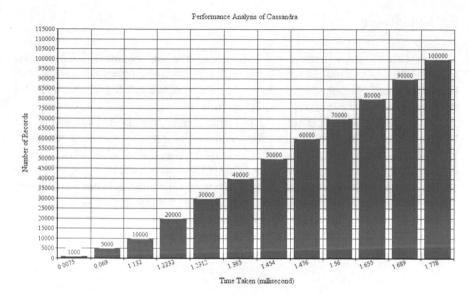

Fig. 3 Cassandra query time taken

<u>Oracle Table Schema</u> :

```
CREATE TABLE teacher
( teacher_id number(10) NOT NULL,
   teacher_name varchar2(50) NOT NULL,
   department varchar(30) NOT NULL,
   CONSTRAINT teacher_pk PRIMARY KEY (teacher_id)
);
create table student
( student_id number(10) NOT NULL,
   last_updated_ts timestamp,
   teacher_id  number NOT NULL,
   subject varchar (50) NOT NULL,
   CONSTRAINT student_pk PRIMARY KEY (student_id, last_updated_ts),
 FOREIGN KEY (teacher_id)    REFERENCES teacher(teacher_id)
);
```

```
Cassandra Table Schema:
CREATE KEYSPACE teach
   WITH REPLICATION = { 'class' : 'SimpleStrategy', 'replication_factor' :
1 };

   CREATE TABLE teach.teaches (
     teacher_id text,
     last_updated_ts timestamp,
     teacher_name text,
     subject text,
     dept text,
     student_id text,
        PRIMARY KEY (teacher_id,last_updated_ts)
     );
```

8   Conclusion

This paper undergoes performance evaluation between relational database like Oracle and NoSQL database like Cassandra to fetch student information from both the database and compare their performance. We created Fetch Student Information Application to get the retrieval time for records using Oracle and Cassandra.

On comparing overall performance of Cassandra, it exceeds the Oracle database by taking almost 30% less time to return the result of query operation as compared to Oracle approach to store and retrieve data. In conclusion, Cassandra is more efficient than the Oracle queries for large data set. As we can see from Figs. 2 and 3, as data size increases beyond 10^4, the efficiency of Cassandra is better as for every 10000 increase in record size Cassandra time taken follows a gradual increase in slope as compared to Oracle queries where time taken increased with a greater slope as compared to Cassandra and this trend continues as data increases by every 10000 up to 1000000.

References

1. Hewitt, E.: The Definitive Guide, O'Reilly Media, 2nd edn. Cassandra (2016)
2. Abdi, D.: Problems with CAP and Yahoo's little known NoSQL system http://dbmsmusings.blogspot.com/2010/04/problems-with-cap-and-yahoos-little.html, Yale University (2010)
3. Harrison, G.: Next Generation Databases, NoSQL, New SQL (2015)
4. McMurtry, D., Oakley, A., Sharp, J., Subramanian, M., Zhang, H.: Data Access for Highly Scalable Solutions: Using SQL, NoSQL and Polyglot persistence (2013)
5. Rabl, T., Gómez-Villamor, S., Sadoghi, M., Muntés-Mulero, V., Jacobsen, H.A., Mankovskii, S.: Solving big data challenges for enterprise application performance management. Proc. VLDB Endow. 5(12), 1724–1735 (2012)
6. Eure, I.: Looking to the future with Cassandra http://about.digg.com/blog/looking-future-cassandra, Sept (2009)
7. Date, C.J.: The New Relational Database Dictionary (2015)
8. Ellis, J., et. al.: Cassandra Gossiper Architecture http://wiki.apache.org/cassandra/Architecture Gossip (2010)
9. Coulouris, G., Dollimore, J., Kindberg, T.: Distributed Systems: Concepts and Design Addison Wesley (2005)

Learning to Solve Sudoku Problems with Computer Vision Aided Approaches

Tuan T. Nguyen, Sang T. T. Nguyen and Luu C. Nguyen

Abstract Sudoku puzzle is one of the most interesting logic based games with various levels which give our brains a good workout and make them active. However, it is difficult at a number of stages and easy to despondent players, especially for new players or people not enough confidence or endurance. The aim of this study is to develop a support tool for Soduku players, i.e. encouraging players to solve the hard Soduku puzzles or when seeking for help. It gives necessary steps as hints in order to solve the puzzles or check the correctness of their steps or even to solve the game completely. The input data, i.e. the Sudoku puzzle, come from the camera or even from the image/text files. Auto-detecting and recognising digits and the propagation constraint algorithms to generate the Sudoku hints or solutions are utilised. Methods and the tool itself are heavily tested to guarantee a stable and excellent program.

Keywords Computer vision · Sudoku · Propagation constraint

1 Introduction

Among logical games, Sudoku is an excellent brain game and one of the most popular great puzzle games of all time to play in many countries, though its origin is from Japan. Players must follow the rules to win. Solving Sudoku puzzles is a real challenge that uses much of logical thinking and combinatorial number placement. Playing Sudoku daily via infinite different stages will improve players' concentration, patience and developing logical thinking. This game does not demand any

T. T. Nguyen (✉)
University of Buckingham, Buckingham, UK
e-mail: tuan.nguyen@buckingham.ac.uk

S. T. T. Nguyen · L. C. Nguyen
International University, VNU-HCMC, Ho Chi Minh City, Vietnam
e-mail: nttsang@hcmiu.edu.vn

L. C. Nguyen
e-mail: ncluuititiu@gmail.com

© Springer Nature Singapore Pte Ltd. 2018
S. C. Satapathy et al. (eds.), *Information and Decision Sciences*,
Advances in Intelligent Systems and Computing 701,
https://doi.org/10.1007/978-981-10-7563-6_56

calculation or extraordinary maths skills. In spite of its benefits, Sudoku puzzles are still not accessible for everyone or easy to make mistakes. Eventually, players may get stuck somewhere in the difficult levels. Therefore, this study aims to build a software tool which is able to help players generate the correct solution for a certain Sudoku puzzle or to do a quick check with their own solution or to give a suggestion (hint) whilst playing it. To do that, a number of algorithms have been studied and the constraint propagation algorithm is the selection for the Sudoku solver implementation. The constraint propagation is an amazing efficient method to solve this kind of puzzle. In addition, players can also save the Sudoku puzzle in the text format to share with other players.

Moreover, image processing and computer vision algorithms are utilised to speed up the input process and provide a fast way to enter Sudoku puzzles by providing clear images or (physical) papers containing Sudoku puzzle, instead of manually typing number by number (up to 80 numbers) like other ordinary programs. This makes the interaction between users and the tool friendly and convenient. The users can seek helps or hints at any time and any stage during solving the puzzles. The hints or solutions appear right on the image, captured by the tool, and exact grid locations which make users easier and more friendly to follow than existing tools which only provide solutions in text.

This paper's structure is organised as follows. First, Sect. 2 discusses briefly algorithms used by computers to do the Sudoku solvers. This section also mentions computer vision based methods being used in this proposed tool. Second, Sect. 3 discusses the proposed computer vision aided framework with a learning algorithm to solve Sudoku problems. This shows why image processing actively involves in our application. Third, Sect. 4 conducts several experiments and provides discussions on results found. Lastly, Sect. 5 concludes this study and gives some future work for enhancement.

2 Related Work

Each Sudoku puzzle initialises numbers filled in some cells. Players have to use these seed numbers to depict other numbers in all boxes to form an unique solution. One important rule is that no number from 1 to 9 can be repeated in any row or column, although it can be repeatedly appeared along the diagonals. Figure 1c shows an unique Sudoku solution for the puzzle in Fig. 1b.

In difficult Sudoku puzzles, it requires many practices to win the game, e.g. listing all candidates in each square to gradually eliminate candidates to come up to the solution.

A1	A2	A3	A4	A5	A6	A7	A8	A9
B1	B2	B3	B4	B5	B6	B7	B8	B9
C1	C2	C3	C4	C5	C6	C7	C8	C9
D1	D2	D3	D4	D5	D6	D7	D8	D9
E1	E2	E3	E4	E5	E6	E7	E8	E9
F1	F2	F3	F4	F5	F6	F7	F8	F9
G1	G2	G3	G4	G5	G6	G7	G8	G9
H1	H2	H3	H4	H5	H6	H7	H8	H9
I1	I2	I3	I4	I5	I6	I7	I8	I9

(a) Name of each cell

6				1	3		9	
7	8					5	6	
	1						4	7
9				6		1	7	4
	7		1		9		5	
				7			2	3
		8	5	7	4			9
		7		2		6		
		4					3	2

(b) A Sudoku example

6	4	5	7	1	3	2	9	8
7	8	3	9	4	2	5	6	1
2	1	9	5	8	6	3	4	7
9	3	8	2	6	5	1	7	4
4	7	2	1	3	9	8	5	6
5	6	1	4	7	8	9	2	3
3	2	6	8	5	7	4	1	9
1	9	7	3	2	4	6	8	5
8	5	4	6	9	1	7	3	2

(c) Its full solution

Fig. 1 A Sudoku puzzle contains a 9 × 9 grid and seed numbers from 1 to 9

2.1 Computer Approaches Based Sudoku Solver

In computer programming, there are a variety of methods to solve Sudoku puzzles and some are briefly described as follows:

- Backtracking [1]: used in many applications such as eight queen puzzles. It visits unfilled cells in arbitrary order then fills in digits sequentially from possible choices, and backtrack (i.e. discard unsuccessful choices) when deadlocks are met. At each backtracking time, it changes the digit in the cell most recently filled before the deadlock took place. If that specific cell is attempted with every single possible digit, the algorithm goes back to the second prior cell filled before the last deadlock and iterates that cell's digit. Backtracking algorithm is simple to implement, however, it needs a large amount of memory space because of the recursive technique.
- Brute Force algorithm: the basic idea is to go through all possible solutions extensively. It does many iterations to look for all possible solutions for Sudoku puzzles. If found solutions cannot solve the problem, the algorithm removes them and rolls back to the original solutions then try again. It does not require a huge of memory space, but it requires a lot of processing time.
- Stochastic search [2]: this approach first assigns digits randomly to the empty cells in the grid. Then, it calculates the number of errors, and rearranges these filled digits around the grid until the number of errors is reduced to zero. Finally, the solution to the Sudoku puzzle is found. The process of stochastic search is pretty quick and consumes less memory space, but it is difficult to implement.
- Constraint propagation [3, 4]: this method applies a set of criteria to the possible candidates to find the solution. A candidate that satisfies all the criteria will be the solution to the puzzle. This typical algorithm is applied in this study to build the application because of its fast and efficient. Some appropriate constraints are used to eliminate the candidates to reduce the complexity whilst finding solutions. The detail is discussed in Sect. 3.

2.2   Computer Vision Aided Approaches

k-Nearest Neighbours (kNN) algorithm is one of the simple classification algorithms but highly competitive. Providing a new data, the algorithm tries to predict its label by searching for the nearest match (closest distance) among the predefined classes, i.e. tell me who your neighbours are, and I will tell you who you are. The distance measure can be any metric function and Euclidean distance is the most common use.

The kNN classification algorithm is used to implement Optimal Character Recognition (OCR), which performs reading characters in the captured image. A novel approach in this study is to employ computer vision techniques to help players much when they only have papers-based Sudoku puzzles and do not know or have a less experience in Sudoku. How to apply kNN to do OCR will be discussed in the following sections.

3   Method and Implementation

To build the Sudoku solver application in this study, a computer vision aided framework is proposed with two main parts: (1) Sudoku image processing, and (2) machine learning to solve Sudoku problems, as shown in Fig. 2.

Initially providing a paper-based Sudoku puzzle, the tool automatically captures, by an attached camera, its image. Next, a series of image processing and machine learning methods are utilised to extract all necessary information, i.e. the digits and their correspondent cells to form a Sudoku puzzle ready to solve. All above processes reduce considerably amount of time, comparing to manually input digits into the Sudoku grid. Then, the puzzle is passed through the solver to find an unique solution or produce hints for users when they consult the tool. The following sections provide each step in details of the framework.

3.1   Sudoku Image Processing and Machine Learning

This component's key task is to enable receiving Sudoku puzzles from the camera. It has two main steps: preprocessing to enhance and standardise input images allowing the information extraction step to obtain data more accurately.

Preprocessing is a necessary and crucial step. In this step, a captured puzzle image is processed through a series of preprocessing methods including greyscale, blurring, thresholding [5] on the Sudoku grid area needed to filtrate.

After preprocessing (Fig. 3a), all the object contours in the image are now clearer and more accurate. Moreover, it is compulsory that the Sudoku grid is at the centre of the camera. Therefore, finding the Sudoku grid in the image can be done by looking for the biggest contour which has 4 vertices, named as top-left, top-right,

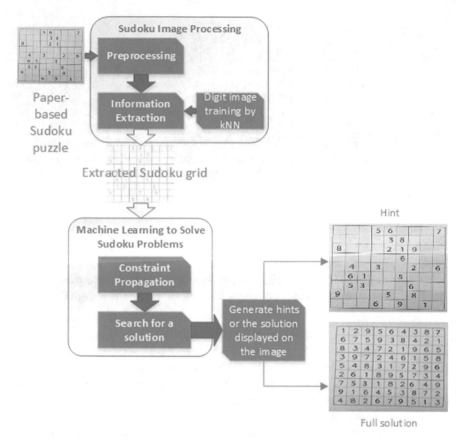

Fig. 2 The proposed framework

bottom-right, bottom-left. The top-left point has the smallest value of x and y-coordinate, whilst the bottom-right point has the largest one. These points are also used to draw all grid cells.[1]

After the whole Sudoku grid is positioned, the perspective transformation will be performed. A morphological transformation, with the ellipse kernel (11×11),

---

[1] All cells in the Sudoku grid have to be drawn out exactly. The Sudoku grid has 81 cells formed by 9 rows and 9 columns. They are equivalent to 10 horizontal and 10 vertical lines. Therefore, determining all the cells in Sudoku grid is quite simple since 4 key points have been determined (Fig. 3b). First, the coordinates of points in vertical line are calculated by the top-left and bottom-left points, so 8 points are expected to be determined. The interval between 2 points is the distance between top-left and bottom-left divided by 9. It is similar for the points in last vertical line. Then, the coordinates of every point in all horizontal lines are easily defined.

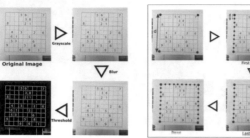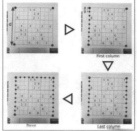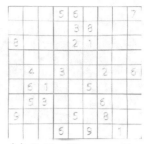

(a) Preprocessing image containing Sudoku grid.

(b) Process of creating cells of Sudoku grid.

(c) Grid with perspective transformation.

Fig. 3 Process to extract cells

dilates then erodes the image. Then, it is normalised to obtain a better processing result.[2] Figure 3c describes the transformation of the extracted Sudoku grid.

The next step is to locate the digits in the Sudoku grid and to recognise them. A cell of the corresponding digit is determined by coordinates of that digit divided by 50.[3]

Information Extraction: There are several ways to perform the optimal digit recognition (ODR) such as [6, 7]. However, kNN is applied because it is simple and requires only a small data to recognise digits. Training data is a step to obtain the database for the kNN algorithm. The training samples are digits from 0–9 in 10 popular fonts. During the training phase, the data of digit contours and the learned digits are stored as a database.

After training, extracting or reading digits from a pre-processed Sudoku image can be achieved.[4] Each contour will be compared with the database to identify the correct number. Its position in which cell is also determined.

3.2 Sudoku Solver

The constraint propagation [3, 4] is adopted to solve the Sudoku puzzle. It is a process of finding a solution to a set of constraints imposing conditions, and a solution is a set of values for the variables that satisfy all constraints, i.e. Sudoku rules. Moreover, it might be combined with search to produce the complete solution to every Sudoku puzzle. First, define **Unit**: a column, a row or a 3 × 3 subgrid of the Sudoku

---

[2]Particularly, the Sudoku grid in the original image is transformed into a result image of 450 × 450 pixels in size. To increase the precision, the perspective transformation is applied to each cell from the original image.

[3]Because the extracted Sudoku image's size is 450 × 450.

[4]Because all contours are determined.

grid, a collection of 9 squares, **Peers**: the squares that share a unit. Then, the *Process steps* to solve Sudoku problems as follows:

1. Set and name the Sudoku squares in the grid as Fig. 1a. The columns are labelled with numbers from 1 to 9 and rows with letters from *A* to *I*.
2. Input a Sudoku puzzle into the grid.
3. Define digits in all the squares by two following strategies:

   - If a square has only one possible digit, then eliminate that digit from the squares peers.
   - If a unit has only one possible place for a digit, put the digit there.

In case of failure, an incomplete solution is returned, i.e. some squares remain uncertain. To handle this problem, we do searching better solutions, as follows:

1. Choose one unfilled square and consider all its possible digits.
2. Repeat assigning the square each digit and searching for a solution until success. This process is a depth-first search.

4   Experimental Results and Discussion

Analysing the Solver: To compare the efficiency of the constraint propagation algorithm, two other methods, backtracking and exact cover using Algorithm X [8] are included. Table 1 shows the average time of 10 runs for each game level of each algorithm. Numbers in bold show the smaller the number, the better the performance. Figure 4a visualises Table 1 to show the relationship between the algorithms and their solving times for different levels of difficulty.[5]

The more difficult the puzzle, the more time it needs to be solved. Backtracking method is efficient to very easy puzzles because it requires the least time to solve them, whilst constraint propagation and exact cover demand a little more time. However, as the difficulty of the puzzle is rising, backtracking method required a huge time to solve extreme problem (i.e. extreme/very easy ratio is 766.5!). Meanwhile, the difficulty level does not really affect the efficiency of constraint propagation and exact cover (i.e. extreme/very easy ratios are 1.25 and 1.9 respectively).[6] Overall,

---

[5]What defines the difficulty of the Sudoku puzzle is still debatable. The important fact is that the number of given digits is not really related to the difficulty of the Sudoku puzzle, the positions of digits are more decisive. One difficult Sudoku can contain more given digits than an easy one. Generally, a Sudoku puzzle, requiring more difficult techniques to solve, can be considered more difficult. Moreover, a Sudoku requiring more repetitions, i.e. many times beginning over again when a certain technique does not work or many uses of a difficult technique, can also be considered more difficult. These two factors determine the difficulty of the game.

[6]The time solving a difficult puzzle of backtracking method is more than 100 s times that one of the constraint propagation and exact cover methods. Because in the hard puzzle, it has to go forth and back in loops many times to look for the correct solution while constraint propagation and exact cover just need to eliminate some determined candidates. Constraint propagation method uses simpler and less constraints than the exact cover method, so it is faster to solve a puzzle.

Table 1 Average time (s) to solve a Sudoku puzzle corresponding to difficulty level of three different algorithms. Numbers in bold are the best

Level	Constraint propagation	Backtracking	Exact cover
Very easy	0.003593513	**0.003304386**	0.004111327
Easy	**0.003686002**	0.008127502	0.00426706
Medium	**0.00368348**	0.011563614	0.004484119
Hard	**0.003677662**	0.058336855	0.004449537
Very hard	**0.00396528**	1.1457886	0.00590327
Extreme	**0.004481116**	2.53272209	0.007799193

Table 2 Average time (s) of each function in camera Sudoku solver

Capture	Form cell	Perspective transformation	OCR	Solve	Display	Sum
0.0273445	0.007667	1.105628167	0.048173	0.004661	0.017316	1.2108

the constraint propagation algorithm is the best method for the Sudoku game problem because it quickly produces correct solutions and is easy to understand and implement.

Analysing image processing steps: Many images of Sudoku puzzles are tested to evaluate the tool's efficiency and speed. To have a clear and critical view, Table 2 and its visualisation (Fig. 4b) show the average runtime of each component function from capturing Sudoku puzzle to display the solution. It shows performing perspective transformation takes more than 90% of the whole running time.[7] Other functions, OCR, capture, display, form Cell and solving a puzzle, take 4%, 2.25%, 1.43%, 0.63% and 0.385% of the whole process, respectively. Figure 5 shows the input by camera, and the tool provides the solution or hint immediately on the captured image.

User interface examples: Besides inputting Sudoku puzzles by a camera, the tool allows users to enter puzzles manually or load them from the saved text file. Then, the tool provides the hint (Fig. 5a, c) and solution (Fig. 5b, d) for camera input and text input respectively.

Although the tool works well with some kinds of image, there are some drawbacks easily pointed out. First, it efficiently works with clear images and easy-to-read fonts, i.e. not be complicated and contain many brushes or strokes, because the training data are not covered all cases. Another is that if the image is containing many considerable different size rectangles, it might wrongly detect cells. Last is the large rotation angles such as more than 90° from the vertical axis. It can cause an unreliable digit recognition.

---

[7]The reason is that every of 81 cells is transformed in turn. Though it takes almost all the time to determine the input Sudoku, this application is much better than the manual input. Users' mistakes and tedious tasks are reduced significantly.

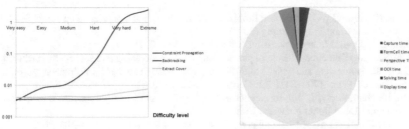

(a) Algorithm vs. time to solve the Sudoku puzzles at different levels.

(b) Percentage of each function in Camera Sudoku Solver

Fig. 4 Comparing processing time

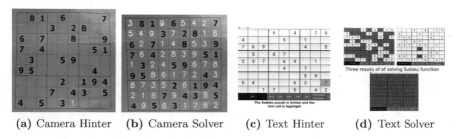

(a) Camera Hinter   (b) Camera Solver   (c) Text Hinter   (d) Text Solver

Fig. 5 Sudoku examples and their solution or hints

5 Conclusions

The experimental results have shown that the proposed framework with constraint propagation algorithm is efficient to solve Sudoku puzzles from easy to difficult. The algorithm is outstanding compared with others (i.e. backtracking, exact cover) because of giving correct solutions within the shortest time.

In addition, the computer vision based Sudoku Solver developed is much useful for new players. Players can obtain hints or solutions by using text/image files or directly from camera without wasting time to type digits. The application provides variety of useful functions, e.g. giving hints, solving, saving and checking the correctness or detecting input errors of Sudoku puzzles. Moreover, those drawbacks addressed in the previous section are the coming enhancements.

References

1. Golomb, S., Baumert, L.: Backtrack programming. J. ACM **12**, 516–524 (1965)
2. Lewis, R.: Metaheuristics can solve sudoku puzzles. J. Heurist. **13**(4), 387–401 (2007)
3. Bessiere, C.: Constraint propagation. Found. Artif. Intel. **2**, 29–36 (2006)

4. Simonis, H.: Sudoku as a Constraint Problem, 11th International Conference on Principles and Practice of Constraint Programming (2005)
5. Rafael, C.G.: Digital Image Processing, 3rd edn. Pearson (2007)
6. Kamal, S., et al.: Detection of Sudoku puzzle using image processing and solving by Backtracking, simulated annealing and genetic algorithms: a comparative analysis. In: Proceedings of the 2015 Third ICIIP, pp. 179–184. IEEE Computer Society (2015)
7. Wicht, B., Henneberty, J.: Mixed handwritten and printed digit recognition. In: Sudoku with Convolutional Deep Belief Network, 13th ICDAR (2015)
8. Knuth, D.: Dancing links, Millennial perspectives in computer science, 187–214 (2000)

Combined Effect of Cohort Selection and Decision Level Fusion in a Face Biometric System

Jogendra Garain, Ravi Kant Kumar, Dipak Kumar, Dakshina Ranjan Kisku and Goutam Sanyal

Abstract There are different parameters which degrade the performance of a face biometric system due to their variations. The baseline biometric systems can get relief to some extent from this kind of negative effect by utilizing the information of the cohort images and fusion methods. But to achieve the set of suitable cohorts for each and every enrolled person is a task of great challenge. Determining the cohort subset using k-means clustering cohort selection based on the matching proximity is presented in this paper. SIFT and SURF are used as facial features to represent each face image and to calculate the similarity score between two face images. The clusters having highest and lowest centroid value are fused using union rule to form the target, user dependent cohort subset. The query-claimed matching scores are normalized with the help of T-norm cohort normalization technique. The scores after normalization are used in recognition separately for SIFT as well as SURF. Finally, the responses from the classifier for these two different features are fused at decision level to cover up the shortcomings of the cohort selection method if any. The experimental execution is done on FEI face database. This integrated face biometric system gains a significant hike in performance that evidences its effectiveness over baseline.

J. Garain (✉) · R. K. Kumar · D. Kumar · D. R. Kisku · G. Sanyal
Department of Computer Science and Engineering, National Institute of Technology
Durgapur, Durgapur 713209, West Bengal, India
e-mail: jogs.cse@gmail.com

R. K. Kumar
e-mail: vit.ravikant@gmail.com

D. Kumar
e-mail: dipakcsi@gmail.com

D. R. Kisku
e-mail: drkisku@gmail.com

G. Sanyal
e-mail: nitgsanyal@gmail.com

© Springer Nature Singapore Pte Ltd. 2018   549
S. C. Satapathy et al. (eds.), *Information and Decision Sciences*,
Advances in Intelligent Systems and Computing 701,
https://doi.org/10.1007/978-981-10-7563-6_57

Keywords Face biometric system · SIFT · SURF · K-means clustering
Cohort score · Cohort subset · Score normalization technique
Fusion

1 Introduction

The main adversary of a face biometric system for person authentication is the
uncountable variations of face images like illuminations, expressions, rotation, head
movement, occlusion, makeup, etc., which cause the performance of the system to
fall down. Therefore, to repulse the roles of these unwanted variations, cohort's
properties help out the researchers to some extent in the domain of computer vision
and biometrics. It can be improved further if fusions are added up with cohort
selection. In some of the applications of biometric authentication, the false
acceptance cannot be negotiated to be high at any cost rather the complexity of
system may be allowed to increase. For that very reason, utilization of cohort
selection is spreading into the research domain. It helps the traditional systems to
furnish them further. Fusion is another mode to increase the robustness. The perception of cohort images is nothing but to extract and utilize the meaningful features available in imposter template. Auckenthaler et al. [1] first proposed to
augment the system's performance used to verify a speaker with the help of Bayes'
theorem to normalize cohort scores. Thereafter, the concept of cohorts comes to be
used with the other biometric traits also. Merati et al. [2] applied on multimodal
biometric system consisted of face and fingerprint. They state in their literature that
in case of sorted cohorts, there is a distinguishable pattern between genuine score
and imposter scores. The plot of mean and variance versus cohort model [3] show
that only the high and low scores are having this discriminative feature but the
average scores look identical for genuine and imposters both. Garain et al. [4]
proposed Bezier curve cohort selection because the Bezier curve passes through the
dense points which are having average scores. So the target is to select the points
scattered far from the curve which are bearing discriminative features. This is the
main motivation to use k-means clustering for cohort selection where the clusters
having the templates of high scores and low scores are fused to form cohort subset
excluding the templates with zero score. Without using cohorts, a biometric system
may run faster but the literature Tistarelli et al. [5], Aggarwal et al. [6] say that the
improvement in performance is remarkable by adopting cohort features. Tulyakov
et al. [7] state that if two different techniques of score normalization are applied, it
may further increase the accuracy of a system although only one normalization
technique is used in this work.

The proposed work applies SIFT [4, 8] and SURF [9, 10] to get the feature
points on each face images and further to calculate the matching score between any
two face images, a matching algorithm, best-fit [10] is used. These matching scores
are used to pick the cohort scores and put into the cohort subset. The experimental

evaluation of the work is done on FEI face database [11] and achieves a significant gain in recognition rate.

The paper is structured as follows. Section 2 presents cohort selection method and overall framework of the proposed work. Results are placed in Sect. 3. The conclusion is in Sect. 4.

2   Cohort Selection

Since the initial cohort set in this work is predefined to half of the total enrolled subjects to reduce over computation, there are N-1 cohort scores for each of the 2 N enrolled subjects. The first N subjects have N-1 cohort templates because there is one image of their own which is excluded. But for the other N subjects, the number of cohort templates is also N. So the last cohort score is excluded to make the size of initial cohort set equal for all enrolled subjects. It is not necessary to exclude the last one only but it can be anyone. Let $S = \{\delta_1, \delta_2, \ldots, \delta_{n-1}\}$ is a set of cohort score of a particular subject registered in the database. Using this set of scores the reference subset of cohort for that person is fixed as pre-verification process. This is executed for all the enrolled subjects. The steps are stated in Algorithm 1.

Algorithm 1 k-means Clustering Cohort Selection
Input: Set of cohort scores $S = \{\delta_1, \delta_2, \ldots, \delta_{n-1}\}, 2n \rightarrow$ number of enrolled template, $\delta_i \rightarrow$ cohort score with ith subject.
Output: The target subset of cohorts $\{C_S\}$

1. Apply k-means clustering over all cohort scores δ to build k number of cluster. Here, k is considered as 5.
2. Find the centroid of all k clusters.
3. Fuse the clusters C_{max} and C_{min} using union rule to constitute the final cohort subset C_S.

$$C_S = \{C_{max} \cup \{C_{min} - X_0\}\}, \; X_0 \; indicates \; \text{the tempaltes having zero score.}$$

$$where \; C_{max} = C_m | \psi_m = \max(\psi_i) \quad and \quad C_{min} = C_n | \psi_n = \min(\psi_i), \; i = 1 \; to \; k$$

The complete framework of the person verification system integrated with the proposed cohort selection method is depicted in Fig. 1 describing that the test samples are matched with the claimed template as well as the corresponding cohort templates and the raw matching score is normalized. This is done for SIFT feature and SURF feature separately. The decisions of the classifier based on SIFT as well as SURF are fused using OR rule to get the final decision.

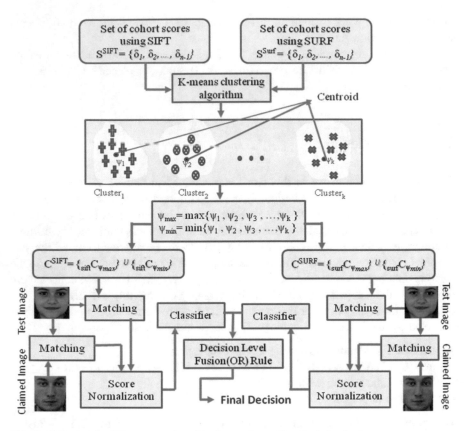

Fig. 1 Framework of the proposed verification system with selected cohorts

Figure 2 plots the number of cohorts selected for each subject using k-means clustering cohort selection method. For example, it is showing that the number of cohort in the reference set for first person is 25 and for the last person is 29. Figure 3 specifies the strength of this cohort selection mechanism by displaying the range of the number of selected cohorts. The average number is neither very large nor very small. This value may change if 'k' changes in k-means clustering. It is experimentally analyzed that the proposed method gives most suitable cohort subset while k = 5 with the initial cohort set length 100. The raw similarity score between query and claimed image is normalized using T-norm score normalization technique as written in Eq. (1).

$$S_{Tnorm} = \frac{g - \mu}{\sigma}, \tag{1}$$

where g stands for raw matching score, μ is mean and σ is the standard deviation of all cohort scores.

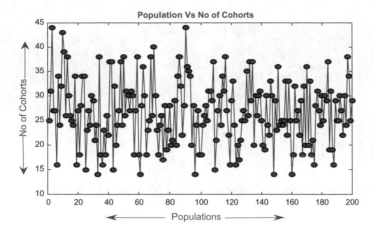

Fig. 2 No of cohorts selected for each subject using k-means clustering

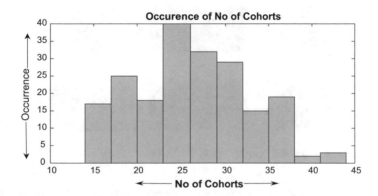

Fig. 3 Frequency of 'number of selected cohorts'

3 Experimental Setup and Result Analysis

FEI database [11] consists of 2800 face images of 200 Brazilian peoples with equal contribution. The variations available in this database are head rotation up to 180 degrees, facial expression (smiling), and illumination changes. But the background for all images is uniform. All the images are color and originally the size is 640 × 480 pixels other than a folder separately provided which contains 200 neutral and 200 smiling face images of size 360 × 260 pixels. These smiling and neutral face images are used for cohort set and enrolled set, respectively. Samples are shown in Fig. 4a, b. Among these, 200 smiling face images only first 100 samples are taken into the cohort pool to avoid the unnecessary computation. Cohort selection is offline process that is why the size of the images is kept as it is but the size of the test samples is reduced to 160 × 110 pixels to speed up the

(a)

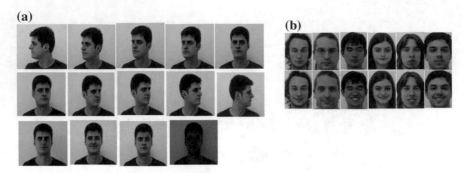

Fig. 4 **a** Sample faces from FEI face database. **b** Sample of enrolled face (1st row) and the cohort faces (second row)

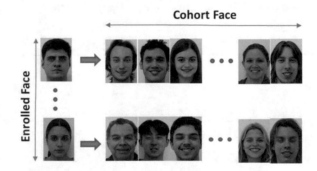

Fig. 5 Enrolled faces and their corresponding cohort faces

testing process. All the remaining 12 samples per subject are taken as test sample. Total 2360 samples are tested excluding the images where no feature (SIFT/SURF points) or very less feature points are extracted, specially the templates with very poor light (14th instances). The samples of selected cohorts per subject are also depicted in Fig. 5.

The comparison of performance is shown in Fig. 6 in which ROC (FAR in X-axis and GAR in Y-axis) is plotted for baseline (SIFT and SURF), after modification with cohort selection and finally after applying decision level fusion (OR rule) for 'with cohort' and 'without cohort' both. The quantitative measures are showcased in Table 1. The parameters used in the table like FAR, FRR, GAR, EER, and Acc are defined in Appendix. The proposed method has dominating accuracy, marked as bold, over all other cases placed in the table.

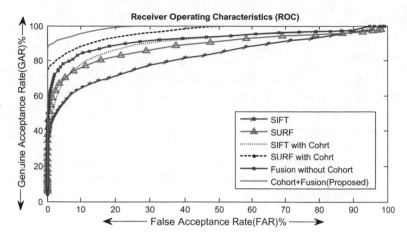

Fig. 6 Receiver operating characteristics (ROC) curve

Table 1 Comparison table

Cohort selection	Feature	FAR (%)	FRR (%)	EER (%)	GAR (%)	Acc (%)
Before applying	SIFT	10.5209	35.1154	22.8181	64.8846	77.1819
After applying		11.2831	20.4231	15.8531	79.5769	84.1469
Before applying	SURF	7.8427	26.0769	16.9598	73.9231	83. 0402
After applying		7.5485	13.8462	10.6973	86.1538	89.3027
Without cohort	Fusion	8.9030	17.0769	12.9899	82.9231	87.0101
With cohort (Proposed)		3.0734	8.5000	5.7867	91.5000	**94.2133**

4 Conclusion

This work illustrates how to utilize the non-matched templates available in the gallery. Besides, it also states how to improve the accuracy of a biometric system using fusion at decision level. Unlike the existing literatures, this work does not fix the number of cohorts as a constant rather it varies from 14 to 44 shown in Fig. 3. Although there is an overhead of extra computation of 25 (on an average) face pair matching along with the fusion process at decision level but it is tolerable in comparison with the accuracy hike. In respect of computational cost also it is better than the system where all non-matched templates are used as cohort without bothering about cohort selection. However, to reduce the overhead of unnecessary computations, an efficient cohort selection method is very much essential. This work can also deal with other than face biometric traits. So, there is a future score to enhance the work for other traits as well as for multimodal. There is another scope to make the system more robust by fusing the information in different levels.

Appendix

Nomenclature

SIFT	Scale-Invariant Feature Transform
SURF	Speeded up Robust Feature
δ	Cohort Score
ς	Set of cohort scores
C_s	Final Subset of Cohorts
C_{max}	Cluster of cohort scores having maximum centroid value
C_{min}	Cluster of cohort scores having minimum centroid value
ψ_m	Centroid of the cluster C_m
ψ_n	Centroid of the cluster C_n
S_{Tnorm}	Normalized score using T-norm cohort score normalization technique
μ	Mean of all cohort scores in a set
σ	Standard deviation of all cohort scores in a set
FAR	False Acceptance Rate $= \frac{No\ of\ imposter\ samples\ accepted}{Total\ number\ of\ imposter\ sample\ tested} \times 100\%$
FRR	False Rejection Rate $= \frac{No\ of\ genuine\ samples\ rejected}{Total\ number\ of\ genuine\ sample\ tested} \times 100\%$
GAR	Genuine Acceptance Rate $= (100 - FRR)\%$
EER	Equal Error Rate $= \frac{FAR + FRR}{2}\%$
Acc	Accuracy $= (100 - EER)\%$

References

1. Auckenthaler, R., Carey, M., Thomas, H.L.: Score normalization for text-independent speaker verification systems. Digit. Signal Proc. **10**, 42–54 (2000)
2. Merati, A., Poh, N., Kittler, J.: User-specific cohort selection and score normalization for biometric systems. IEEE Trans. Info. Forensics Secur. **7**(4) (2012)
3. Merati, A., Poh, N., Kittler, J.: Extracting discriminative information from cohort models. In: Proceedings of 4th IEEE International Conference on Biometrics: Theory Applications and Systems (BTAS), 1–6 (2010)
4. Garain, J., Kumar, R.K., Sanyal, G., Kisku, D.R.: Selection of user-dependent cohorts using Bezier curve for person identification. In: International Conference Image Analysis and Recognition, pp. 566–572. Springer International Publishing (2016)
5. Tistarelli, M., Sun, Y., Poh, N.: On the use of discriminative cohort score normalization for unconstrained face recognition. IEEE Trans. Inf. Forensics Secur. **9**(12), 2063–2075 (2014)
6. Aggarwal, G., Ratha, N.K., Bolle, R.M.: Biometric verification: looking beyond raw similarity scores. In: Proceedings of Computer Vision and Pattern Recognition Workshop, pp. 31–31 (2006)
7. Tulyakov, S., Zhang, Z., Govindaraju, V.: Comparison of combination methods utilizing T-normalization and second best score model. In: Proceedings of the IEEE Conference on Computer Vision and Pattern Recognition Workshop, pp. 1–5 (2008)
8. Lowe, D.G.: Distinctive image features from scale-invariant key points. Int. J. Comput. Vis. **60**(2), 91–110 (2004)

9. Bay, H., Tuytelaars, T., Van Gool, L.: May. surf: speeded up robust features. In: European conference on computer vision (pp. 404–417). Springer, Berlin, Heidelberg
10. Garain, J., Shah, A., Kumar, R.K., Kisku, D.R., Sanyal, G.: BCP-BCS: best-fit cascaded matching paradigm with cohort selection using Bezier curve for individual recognition. In: Asian Conference on Computer Vision, pp. 377–390. Springer, Cham (2016)
11. Thomaz, C.E., Giraldi, G.A.: A new ranking method for Principal Components Analysis and its application to face image analysis. Image Vis. Comput. **28**(6), 902–913 (2010)

Global Stability of Harvested Prey–Predator Model with Infection in Predator Species

Nishant Juneja and Kulbhushan Agnihotri

Abstract In the present work, a model on predator–prey system in which only the predator population is infected by a disease is proposed and analyzed. It is considered that both the prey and predator species are harvested at different harvesting rates and differential predation rates for susceptible, and infected predators are also taken. Conditions for the existence and stability of all feasible equilibrium points are obtained. It is observed that nonzero equilibrium point is always globally stable whenever it is locally stable. It has been noticed that the reduced predation rate of infected predator and controlled harvesting of predator species is helpful in controlling the spread of disease.

Keywords Carrying capacity · Harvesting · Predator · Prey
Predation rate

1 Introduction

The population dynamics is one of the widely studied branches of mathematical biology and recently the study of infectious diseases is gaining momentum. The study of effects of diseases on the dynamics of ecological systems is biologically relevant [1–3]. Most of the eco-epidemiological studies deal with the situations where disease floats in prey species only [4–6], whereas some studies considered the case with disease in predator species [7–9]. The role of infection in the dynamics of the system is studied. Currently, researchers are showing a great interest in studying the effect of harvesting on the dynamical behavior of the ecological system. Some of them considered the case where prey species is subjected to constant harvesting

N. Juneja (✉)
I. K. Gujral Punjab Technical University, Kapurthala, India
e-mail: rtmct2016@gmail.com

K. Agnihotri
Shaheed Bhagat Singh State Technical Campus, Ferozepur, India
e-mail: agnihotri69@gmail.com

© Springer Nature Singapore Pte Ltd. 2018
S. C. Satapathy et al. (eds.), *Information and Decision Sciences*,
Advances in Intelligent Systems and Computing 701,
https://doi.org/10.1007/978-981-10-7563-6_58

whereas others considered the harvesting of predator species [10–15]. But as far as our knowledge is concerned, a very little work has been carried out for eco-epidemiology model in which both the prey–predator populations are subjected to harvested. We considered a situation where only the predator population is invaded by a disease and the disease does not cross the species barrier. F. Gulland considered several examples of such cases in Table 2 of [16]. The most common example is common seal (*Phoca vitulina*) and striped dolphin (*Stenella coeruleoalba*) where they are infected by virulent disease named phocine distemper virus (PDV). In 1988, the epidemic of PDV infections is found among harbor seals. This virus becomes the main cause of death of 18,000 harbor seals (*P. vitulina*) and 300 gray seals (*Halichoerus grypus*) along the Northern European coast. In this case, fish are considered as prey. Therefore, we make the strong assumption that the disease does not cross the species barrier. The present paper deals with a harvested predator–prey model in which the disease is spreading from infected predator species to susceptible predator species. It is assumed that disease in predator species reduces the mobility of infected predator. So we consider the differential predation rate for susceptible and infected predators. The main purpose of the present paper is to study the dynamical behavior of the system where both the prey–predator species are subjected to harvesting.

2   Formulation of the Model

Let us consider a prey–predator model in which the disease is transmitted in predator species only. As a result, the predator species is categorized into two classes, namely, susceptible and infected predator. The population of susceptible and infected predator at any time "t" is given as y and z. Hence, the total population at any time "t" is given as $P = y + z$. The prey population follows logistic growth with "r" as intrinsic growth rate and "k" as the carrying capacity of the environment. The disease is transmitted in predator species with law of mass action having λ as transmission parameter. The susceptible and infected predators have μ and μ' as their respective mortality rates with $\mu' > \mu$. The predator and prey species are also subjected to harvesting having harvesting attempt E and E_1 correspondingly. Let γ and γ' be the catch ability coefficients of susceptible and infected predators with $\gamma' > \gamma$ as clearly understood. The susceptible and infected predators meet the prey with α and α' as their respective capture coefficients $(\alpha' < \alpha)$.

Based on the above assumptions, we write the following equations for our eco-epidemiological model:

$$\frac{dx}{dt} = rx\left(1 - \frac{x}{k}\right) - \alpha xy - \alpha' xz - \gamma_1 E_1 x$$

$$\frac{dy}{dt} = \alpha\beta xy - \mu y - \gamma Ey - \lambda yz \tag{1}$$

$$\frac{dz}{dt} = \alpha'\beta xz - \mu' z - \gamma' Ez + \lambda yz \quad (0 \le x(0) \le k, 0 \le y(0), 0 \le z(0))$$

3 Equilibrium Points

The system can have following different equilibriums:

(a) The trivial equilibrium point $A(0,0,0)$, where all the populations extinct.
(b) A predator-free equilibrium point $B(\hat{x},0,0)$, where $\hat{x} = k(1 - \gamma_1 E_1/r)$ which exist

$$\text{if } E_1 < \frac{r}{\gamma_1} \tag{2}$$

(c) A disease-free equilibrium (DFE) $C(x',y',0)$, where

$$x' = \frac{k}{R_0}, y' = \frac{r}{\alpha}\left(1 - \frac{1}{L}\right) - \frac{\gamma_1}{\alpha}E_1, \text{ which exist if } E_1 < \frac{1}{\gamma_1}\left(r - \frac{r}{L}\right) \tag{3}$$

where $L = \frac{\alpha\beta k}{\mu + \gamma E}$

(d) An endemic semi-positive equilibrium $D(\bar{x},0,\bar{z})$, where

$$\bar{x} = \frac{k}{L'}, \bar{z} = \frac{r}{\alpha}\left(1 - \frac{1}{L'}\right) - \frac{\gamma_1}{\alpha}E_1, \text{ which exist if } E_1 < \frac{1}{\gamma_1}\left(r - \frac{r}{L'}\right) \tag{4}$$

where $L' = \frac{\alpha'\beta k}{\mu' + \gamma' E}$

Conditions (3) and (4) can be combined as

$$E_1 < \frac{1}{\gamma_1}\left(r - \frac{r}{L'}\right) \tag{5}$$

(e) An endemic positive equilibrium E (x^*, y^*, z^*), where

$$x^* = K\left[1 - \frac{(\alpha\mu' - \alpha'\mu) + (\alpha\gamma' - \alpha'\gamma)E + \gamma_1 E_1}{r}\right] \text{ and } y^* = -\left[\frac{\alpha'\beta x^* - \mu' - \gamma'E}{\lambda}\right], z^* = \frac{\alpha\beta x^* - \mu - \gamma E}{\lambda}$$

exists if $r - \gamma_1 E_1 > (\alpha\mu' - \alpha'\mu) + (\alpha\gamma' - \alpha'\gamma)E$ and $\frac{\alpha'\beta x^* - \mu'}{\gamma'} < E < \frac{\alpha\beta x^* - \mu}{\gamma}$

$$(6)$$

4   Uniform Boundedness

Lemma *All the solutions of the system (1) will be in the region* $R = \left[(x, y, z) \in R_+^3 : 0 \leq x + y + z \leq l/m\right]$ *as* $t \to \infty$ *for all positive initial values* $(x(0), y(0), z(0)) \in R_+^3$ *where* $m = \min\left(\gamma_1 E_1, \gamma E, \gamma'E\right)$ *and* $l = rx + \alpha\beta xy + \alpha'\beta xz$.

Proof The proof of this lemma follows directly from comparison theorem.

5   Stability Analysis

The following theorems are direct consequences of linear stability analysis of the system (1):

Theorem 1 *The trivial equilibrium* $A(0, 0, 0)$ *is locally stable if* $E_1 > \frac{r}{\gamma_1}$.

Proof The eigenvalues for the equilibrium $A(0, 0, 0)$ are given by

$$\xi_1 = r - \gamma_1 E_1, \ \xi_2 = -\mu - \gamma E, \ \xi_3 = -\mu' - \gamma'E$$

Clearly $\xi_2, \xi_3 < 0$, Now $\xi_1 < 0$, Iff $E_1 > \frac{r}{\gamma_1}$ (7)

Theorem 2 *The predator-free equilibrium* $B(\hat{x}, 0, 0)$ *if exist, is locally asymptotically stable for* $E > (\alpha\beta\hat{x} - \mu)/\gamma$, *where* $\hat{x} = k\left(1 - \frac{\gamma_1}{r}E_1\right)$.

Proof The eigenvalues for the equilibrium $B(\hat{x}, 0, 0)$ are given by

$$\xi_1 = -\frac{r\hat{x}}{k}, \ \xi_2 = \alpha\beta\hat{x} - \mu - \gamma E, \ \xi_3 = \alpha'\beta\hat{x} - \mu' - \gamma'E$$

Now $\xi_2, \xi_3 < 0$ iff $E > (\alpha\beta\hat{x} - \mu)/\gamma$ (8)

So, excess harvesting can lead to predator extinction which should be avoided for biodiversity.

Theorem 3 *The disease-free equilibrium (DFE) $C(x',y',0)$ if exist is locally asymptotically stable if $\frac{\lambda y'}{\mu'+\gamma'E-\alpha'\beta x'}<1$.*

Proof For $C(x',y',0)$, the characteristic roots are given by the equation

$$\left(\xi-\alpha'\beta x'+\mu'+\gamma'E-\lambda y'\right)\left(\xi^2+\frac{rx'}{k}\xi+\alpha^2\beta x'y'\right)=0$$

Here $\xi_1=\alpha'\beta x'-\mu'-\gamma'E+\lambda y'$. ξ_2,ξ_3 are clearly negative. For $\xi_1<0$, we have

$$\frac{\lambda y'}{\mu'+\gamma'E-\alpha'\beta x'}<1 \tag{9}$$

Theorem 4 *The endemic semi-positive equilibrium $D(\bar{x},0,\bar{z})$ if exist is locally asymptotically stable if $\frac{\lambda\bar{z}}{\alpha\beta\bar{x}-\mu-\gamma E}>1$,*

Proof The characteristic roots of the equilibrium $D(\bar{x},0,\bar{z})$ are given by the equation $\left(\xi+\frac{r\bar{x}}{k}\right)\left[\xi(\xi-\alpha\beta\bar{x}+\mu+\gamma E+\lambda\bar{z})\right]+\alpha'\bar{x}\left[\alpha'\beta\bar{z}(\xi-\alpha\beta\bar{x}+\mu+\gamma E+\lambda\bar{z})\right]=0$
Clearly, this equation will have negative roots if $\alpha\beta\bar{x}-\mu-\gamma E-\lambda\bar{z}<0$

$$\text{i.e.} \quad \frac{\lambda\bar{z}}{\alpha\beta\bar{x}-\mu-\gamma E}>1 \tag{10}$$

Theorem 5 *The positive endemic equilibrium $E(x^*,y^*,z^*)$ is locally asymptotically stable for all parametric values.*

Proof The eigenvalues for $E(x^*,y^*,z^*)$ are given by the equation $\xi^3+\frac{rx^*}{k}\xi^2+\left(\lambda^2y^*z^*+\alpha^2\beta x^*y^*+\alpha'^2\beta x^*z^*\right)\xi+\left(\frac{\lambda^2rx^*y^*z^*}{k}+2\lambda\alpha\alpha'\beta x^*y^*z^*\right)=0.$

Here ξ_1,ξ_2,ξ_3 are roots of above equation having the entire coefficients positive. So ξ_1,ξ_2,ξ_3 are clearly negative. Hence, $E(x^*,y^*,z^*)$ is always locally stable.

Theorem 6 *A locally asymptotically stable nonzero equilibrium point $E(x^*,y^*,z^*)$ is always a globally stable.*

Proof For global stability, define a function a positive definite function

$$\text{Let } V=A_1\left(x-x^*-x^*\log\frac{x}{x^*}\right)+A_2\left(y-y^*-y^*\log\frac{y}{y^*}\right)+A_3\left(z-z^*-z^*\log\frac{z}{z^*}\right)$$

$$\frac{dV}{dt}=A_1\left(\frac{x-x^*}{x}\right)\frac{dx}{dt}+A_2\left(\frac{y-y^*}{y}\right)\frac{dy}{dt}+A_3\left(\frac{z-z^*}{z}\right)\frac{dz}{dt}$$

By taking $A_1=A_2\beta, A_2=A_3$ and little simplification yields

$$\frac{dV}{dt}=-\frac{A_1r}{K}\left(x-x^*\right)^2<0$$

So, the equilibrium point $E(x^*,y^*,z^*)$ is globally asymptotically stable.

6 Numerical Simulations

Numerical simulations have been carried out to study the dynamics of the proposed 3-D model (1). Consider the following set of parametric values:

$$r=0.6, \ k=50, \ \alpha_1=0.015, \ \alpha_2=0.012, \ E=0.11, \ E_1=1.9, \ \lambda=0.6$$
$$\mu_1=0.25, \ \mu_2=0.3, \ \beta=0.3, \ \gamma_1=0.35, \ \gamma=0.2, \ \gamma'=0.6, \tag{11}$$

The system (1) has equilibrium point $A(0,0,0)$ for the data set (11). It is locally asymptotically stable by Theorem 2, as the computed value of E_1 is so large that Eq. (7) is satisfied (see Fig. 1).

Again, consider the following set of parametric values:

$$r=0.6, \ k=50, \ \alpha_1=0.015, \ \alpha_2=0.012, \ E=0.11, \ E_1=0.6, \ \lambda=0.6$$
$$\mu_1=0.25, \ \mu_2=0.3, \ \beta=0.3, \ \gamma_1=0.35, \ \gamma=0.2, \ \gamma'=0.6, \tag{12}$$

We had taken all the parametric values same as in data set (11) except E_1, and the harvesting effort for prey is taken so small that existence condition (2) is satisfied. System (1) has equilibrium point $B(32.5, 0, 0)$. It is locally asymptotically stable by Theorem 2, as the computed value of E is so large that Eq. (8) is satisfied. So the excess harvesting of predator can lead to predator extinction (see Fig. 2).

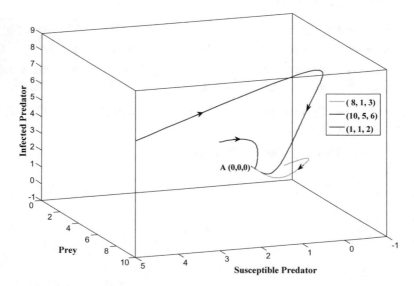

Fig. 1 Phase diagram depicting the behavior of the model equation (1) for the stability of equilibrium point $A(0, 0, 0)$

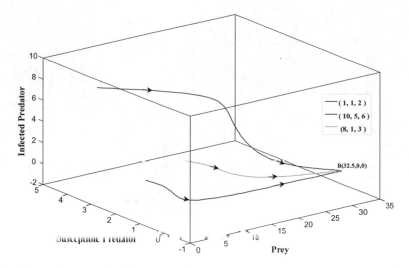

Fig. 2 Phase diagram depicting the behavior of the model equation (1) for the stability of equilibrium point $B(32.5, 0, 0)$

Now if we consider the parametric set of values

$$r = 0.6, \ k = 50, \ \alpha_1 = 0.515, \ \alpha_2 = 0.012, \ E = 0.5, \ E_1 = 0.6, \ \lambda = 0.6$$
$$\mu_1 = 0.25, \ \mu_2 = 0.3, \ \beta = 0.3, \ \gamma_1 = 0.35, \ \gamma = 0.2, \ \gamma' = 0.6, \tag{13}$$

System (1) has disease-free equilibrium point $C(2.2654, 0.7045, 0)$ for the data set (13). This equilibrium point exists as the harvesting effort of prey species is so small that it satisfies Eq. (3). Moreover, the equilibrium point $C(2.2654, 0.7045, 0)$ is locally asymptotically stable by Theorem 3. Here the lesser value of predation rate of infected predator is helping in the local stability of the disease-free equilibrium point (see Fig. 3).

Now consider the parametric set of values

$$r = 0.6, \ k = 50, \ \alpha_1 = 1.15, \ \alpha_2 = 1.012, \ E = 0.1, \ E_1 = 0.6, \ \lambda = 0.6$$
$$\mu_1 = 0.25, \ \mu_2 = 0.3, \ \beta = 0.3, \ \gamma_1 = 0.35, \ \gamma = 0.2, \ \gamma' = 0.6, \tag{14}$$

System (1) has disease-free equilibrium point $D(1.1858, 0, 0.3713)$ for the data set (14). Also, the equilibrium point $D(1.1858, 0, 0.3713)$ is locally asymptotically stable by Theorem 4 (see Fig. 4).

Finally, consider the following parametric set of values:

$$r = 0.6, \ k = 50, \ \alpha_1 = 0.115, \ \alpha_2 = 0.012, \ E = 0.11, \ E_1 = 1, \ \lambda = 0.6$$
$$\mu_1 = 0.25, \ \mu_2 = 0.3, \ \beta = 0.3, \ \gamma_1 = 0.35, \ \gamma = 0.2, \ \gamma' = 0.6, \tag{15}$$

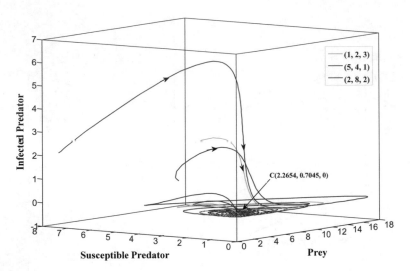

Fig. 3 Phase diagram depicting the behavior of the model equation (1) for the stability of disease-free equilibrium point C(2.2654, 0.7045, 0)

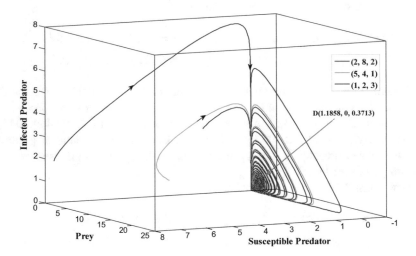

Fig. 4 Phase diagram depicting the behavior of the model equation (1) for the stability of equilibrium point D(1.1858, 0, 0.3713)

System (1) has disease-free equilibrium point E(15.4408, 0.5174, 0.4345) for the data set (15). This equilibrium point exists as reasonable harvesting efforts of prey and predator species are applied so as to satisfy Eq. (6) (see Fig. 5)

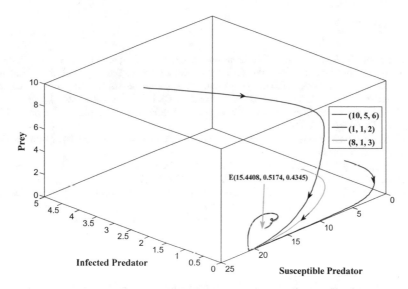

Fig. 5 Phase diagram depicting stability of equilibrium point D(1.1858, 0, 0.3713)

7 Conclusion

In this work, a prey–predator model having disease in predator species is proposed and analyzed. Harvesting of both the prey and predator species is done. Predation rate for diseased predator is considered less as compared with the healthy predator due to low mobility of diseased ones. It has been shown that the lesser predation rate of the infected predator is helping in making the system disease free. The infected predator population $\bar{z}$ decreases due to less predation rate of infected predator α'. The prey harvesting also reduces the size of $\bar{z}$ which reduces the chance of stability of equilibrium point $D(\bar{x}, 0, \bar{z})$ which is good for our biological environment. It is shown theoretically as well as numerically that if harvesting effort for prey, i.e., E_1 is greater than r/γ_1, then all the populations will be wiped out from the system. The predator species will extinct if harvesting effort is applied beyond certain threshold, i.e., $E > (\alpha\beta\hat{x} - \mu)/\gamma$. So excess harvesting can lead to predator elimination. So, it is desirable to do reasonable harvesting of the species for their coexistence.

Acknowledgements The first author Nishant Juneja, being a research scholar at I.K.G. Punjab Technical University, Kapurthala highly acknowledges this university for its valuable inputs on present research.

References

1. Anderson, R.M., May, R.M.: The invasion, persistence and spread of infectious diseases in animal and plant communities. Philos. Trans. R. Soc. Lond. B Biol. Sci. **314**, 533–570 (1982)
2. Anderson, R.M., May, R.M.: The population dynamics of micro-parasite and their invertebrate hosts. Philos. Trans. R. Soc. Lond. B Biol. Sci. **291**, 451–524 (1982)
3. Hadeler, K.P., Freedman, H.I.: Predator-prey populations with parasitic infection. J. Math. Biol. **27**, 609–631 (1989)
4. Chattopadhay, J., Arino, O.: A predator-prey model with disease in the prey. Nonlinear Anal. **36**, 747–766 (1999)
5. Chattopadhay, J., Arino, O., Bairagi, N.: Pelicans at risk in Salton sea—an eco-epidemiological study. Ecol. Model. **136**, 103–112 (1999)
6. Greenhalgh, D., Haque, M.: A predator-prey model with the disease in prey species only. Math. Methods Appl. Sci. **30**, 911–929 (2006)
7. Lenbury, Y., Rattanmongkonkul, S., Tumrasvin, N., Amornsamankul, S.: Predator-prey interaction coupled by parasitic infection: limit cycles and chaotic behavior. Math. Comput. Model. **30**, 131–146 (1999)
8. Agnihotri, K., Juneja, N.: An eco-epidemic model with disease in both prey and predator. IJAEEE **4**(3), 50–54 (2015)
9. Agnihotri, K., Juneja, N.: Predator-prey system with infection in predator only. Elixir Appl. Math. **83**, 32909–32917 (2015)
10. Chattopadhay, J., Bairagi, N., Chaudhuri, S.: Harvesting as a disease control measure in an eco-epidemiological system—a theoretical study. Math. Biosci. **217**(2), 134–144 (2009)
11. Kar, T.K., Ghorai, A.: Dynamic behavior of a delayed predator-prey model with harvesting. Appl. Math. Comput. **217**, 9085–9104 (2011)
12. Guria, S.: Effect of harvesting and infection on predator in a prey-predator system. Nonlinear Dyn. **81**, 917–930 (2015)
13. Wang, X., Liu, H., Xu, C.: Hopf bifurcations in a predator-prey system of population allelopathy with a discrete delay and a distributed delay. Nonlinear Dyn. **69**(4), 2155–2167 (2012)
14. Lenzini, P., Rebaza, J.: Non-constant predator harvesting on ratio-dependent predator-prey models. Appl. Math. Sci. **4**(16), 791–803 (2016)
15. Gupta, R.P., Banerjee, M., Chandra, P.: Bifurcation analysis and control of Leslie-Gower Predator–Prey model with Michaelis-Menten type prey-harvesting. Differ. Equ. Dyn. Syst. **20**, 339–366 (2016)
16. Gulland, F.: The impact of infectious diseases on wild animal populations: a review. Ecol. Infect. Dis. Nat. Popul. **1**, 20–51 (1995)

Flying Ad hoc Networks:
A Comprehensive Survey

Amartya Mukherjee, Vaibhav Keshary, Karan Pandya, Nilanjan Dey
and Suresh Chandra Satapathy

Abstract An ad hoc network is the cooperative disposition of a collection of dynamic (mobile) nodes without the necessity of existing infrastructure or any centralized access points. Recently, ad hoc networks have aroused great scientific curiosity and have led to wide-scale research works into this field. In this paper, we provide a complete survey on Flying Ad hoc Networks (FANETS) as an emerging field among Mobile Ad hoc Networks (MANETS) and Vehicular Ad hoc Networks (VANETS). FANET implies creating an ad hoc network between multi-UAV systems, which is connected to the base station. The base station can be remotely ground based or an aircraft. In FANET, communication between UAVs is dependent on node mobility and topological changes. In this paper, we provide a comprehensive survey of the design issues, communication methodologies, and routing protocols of UAVs with open research issues.

Keywords Ad hoc network · FANETS · MANETS · VANETS
Routing · Mobility · Swarm

A. Mukherjee (✉) · V. Keshary · K. Pandya
IEM Unmanned Aerial Vehicle R&D, Institute of Engineering & Management,
Salt Lake, Kolkata, India
e-mail: mamartyacse1@gmail.com

V. Keshary
e-mail: vaibhavjee10@gmail.com

K. Pandya
e-mail: kp249646@gmail.com

N. Dey
Techno India College of Technology, Rajarhat, Kolkata, India
e-mail: neelanjan.dey@gmail.com

S. C. Satapathy
PVP Siddhartha Institute of Technology, Vijaywada, Andhra Pradesh, India
e-mail: sureshsatapathy@gmail.com

© Springer Nature Singapore Pte Ltd. 2018   569
S. C. Satapathy et al. (eds.), *Information and Decision Sciences*,
Advances in Intelligent Systems and Computing 701,
https://doi.org/10.1007/978-981-10-7563-6_59

1   Introduction

Unmanned Aerial Vehicle (UAV) is an aerial vehicle that does not carry a human on board, which is powered by a jet or reciprocating engine, and can be piloted remotely or flown autonomously based on preprogrammed flight plans [1, 2]. The application scenario of UAVs has evolved in the past few decades, from specific military applications to civilian domain [3, 4]. In recent times, ad hoc networks deploying UAVs have gained scientific importance.

An ad hoc network is a network which does not have a fixed infrastructure and each node in the network is dynamic and can move from one place to another limited by the coverage area of the network [5]. The possibility of extending the wireless coverage, improving the overall capacity and enabling network auto-configuration with no infrastructure support has sparked the idea of multi-hop wireless networks such as the flying ad hoc networks. FANET (Illustrated in Fig. 1) is a novel mobile ad hoc network type, where the communicating nodes are UAVs [6, 7]. In FANETS, one of the UAVs is directly linked to the infrastructure, whereas the rest of the UAVs in the swarm has multi-hop communication in which each node acts as hop count or a relay [8]. During a FANET flight scenario, not necessarily all of the UAVs may communicate to a ground control station, only one of the cluster head communicating should be adequate as proposed in [9]. The various challenges in the communication can be illustrated as

Node Mobility: The speed of UAVs ranges from 30 to 45 m/s, which pose a serious challenge in terms of maintaining data link and coverage quality. The internode distance in FANETs is much large compared to VANETs, MANETs. The topology change infuses key constraint on communication on the account of high node mobility.

Geographical and Environmental Constraints: Radio signal depends on the terrestrial environment such as the presence of high-rise buildings, mountains, or ravines [4]. Environmental factors play an important role [10] such as high winds can cause drifting of nodes from previous coordinates, thus affecting the radio signal sent from the transmitter node to the present receiver node.

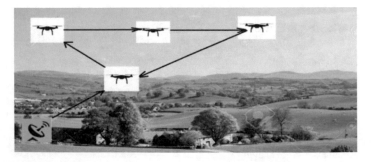

Fig. 1 FANET application scenario

Synchronization Among the UAV Swarms: In [11], the decision is taken onboard imposing full autonomy to the nodes. The greatest challenge lies in correct analysis and implementation of the data available through the various sensors deployed on board the UAVs [4].

Low Latency and Sufficient Bandwidth Availability: Latency is the delay in the packet data transmission among the dynamic nodes. FANET protocol must satisfy bandwidth capacity requirement so that latency is low- and high-resolution real-time images and videos are communicated [4].

The paper comprises of 7 subsections initiating from Introduction to References. In Sect. 2, we provide Related Research Work done in this domain while Sect. 3 deals with Design Considerations. Communication Methodologies and Extensive Research Areas are highlighted in Sects. 4 and 5 respectively. Open Research Areas is outlined in Sect. 5. In Sect. 6, a detailed Conclusion is provided followed by References.

2 Related Research Work

Bekmezci et al. [4] clearly outline the difference between MANETs, VANETs, and FANETs. The multi-hop communication between multi-UAV systems is depicted along with the survey of communication protocols in FANETs. In [12], the mobility of nodes in flying ad hoc networks is discussed and various models have been surveyed. Paparazzi mobility model is a stochastic model that mimics the paparazzi UAV response based on the state machine.

Omnidirectional antennas and Distributed Coordinated Function (DCF) MAC protocols were deployed earlier. Directional antennas and DMAC protocols for flying ad hoc networks are alternatives in application [9]. Routing poses immense challenges in flying ad hoc networks. In [7], Location-oriented directional medium access control protocol is proposed which is more promising than DMAC protocol in terms of throughput, utilization, average network latency, and fairness. The proposed LODMAC protocol characteristics is coded and adapted for deployment in 3D scenarios. In [12], four routing protocols are discussed and Muzaffar et al. propose route switching algorithm, which tracks all routes from source to destination and switches to another route if the present link is going to undergo failure for flying ad hoc networks. This algorithm takes into account the topology change effect on the relay nodes and thereby reduces the decadence of network performance. Rosati et al. propose Predictive Optimum Link State Routing (P-OLSR) algorithm which takes cognizance of GPS information of the unmanned aerial vehicles to predict how the quality of the wireless links evolve. It is the only FANET-specific routing mechanism that has an available LINUX implementation [13]. Berkley Aerobot Team (BEAR) is outdoor testbed implementation of FANETs that allows effective communication between relay nodes [4]. Real-time indoor Autonomous Vehicle (RAVEN) developed at Aerospace Control

Laboratory, MIT employs a motion capture system to enable prototyping of aerobatic flight controllers for helicopters and airplanes; robust coordination algorithms for multiple nodes and vision-based sensing algorithm for indoor flight [4].

3 FANET Design Issues

FANET is a specialized subset of VANET with distinctive design considerations. VANET constitutes as a subsection of MANET as specified in the figure. The design parameters are highlighted as follows:

- **Node Mobility**: As already discussed in Introduction, node mobility plays a pivotal role in designing the FANET architecture. Maintenance of robust and effective communication is a serious design issue [4].
- **Mobility Models**: Various mobility models [10] have been proposed to overcome node mobility constraint. Mobility models provide a description of speed and position variation of a multi-UAV system. A brief overview of the models is provided below.

  - **Random Way Point Mobility Model**: In this model the UAV nodes does a selection of random speed position after the pause time taken it selects another random speed position [14] that is suitable for the simulated environment.
  - **Pheromone-Based Model**: The basic unit of this model is based on a pheromone map. The pheromones are the guiding source for the UAVs. The model focuses on maximizing the area of surveillance.
  - **Paparazzi Mobility Model (PPRZM)**: The paparazzi model is a probability model which is based on comparing the protocols available in the ad hoc networking of UAVs. It relates to real-life mobility issues subjected to environmental conditions more than other models [14].

- **Node Density**: The number of relay nodes present per unit area is defined as node density. The node density is lesser in FANET in contrast to VANET or MANET architecture,
- **Topology Changes**: FANET architecture depends on node mobility and node density. The variations in internodal distances introduce topology change in the system which results in a topological update. Whenever a UAV is removed or added to the existing system, a topological update is carried out [6]. Link outages and communication failures are key research issues.
- **Radio Propagation Model**: FANET architecture supports Line-of-Sight (LoS) communication with little provisions for NLoS communication between sender and receiver [4]. NLoS highlights the uniqueness of FANET.
- **Power Consumption and Refueling**: Present-day communication protocols are energy efficient with the focus on network lifetime [15]. UAV is a Li-ion-powered vehicle with limited battery life and small payload capacity

which is a key constraint in the continuous execution of mission. A novel mechanism of refueling the batteries mid-air using LASERs is discussed in [16]. Extension of flight time and payload capacity offer research opportunities.

- **Localization**: The nodes in FANET are dynamic and require efficient and effective tracking through Global Positioning System (GPS). In various communication protocols GPS update interval of 1 s is ineffective and specialized AGPS and DGPS is deployed [4, 15]. LODMAC proposed by Temel et al. operates on a GPS interval of 1 s.

4 Communication Methodologies

For effective communication, UAV systems must consider the communication requirements and commercial applications, different data and Quality-of-Service (QoS) requirements [17]. Communication in multi-UAV system can have four categories as shown in Fig. 2.

4.1 Communication in FANET

In FANET, the relay nodes must have preplanned information about the location of the nearest node and a brief outline of the whereabouts of other relay nodes. In any wireless network where the source node communicates with the destination node through a combination of dynamic nodes, the problem of link failure and connectivity is common [18].

4.1.1 Physical Layer (PHY)

Physical layer of FANETs deals with the signal transmission from UAV-2-UAV, UAV-2-GCS, and the antenna architecture affecting communication due to variable distance between the nodes [4, 19]. Antenna architecture reflects the link stability and efficiency toward link outages. Omnidirectional antennas suffer from lack of security as they disseminate the signal in all directions. A novel directional antenna

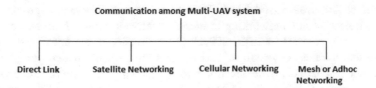

Fig. 2 Communication methodology for multi-UAV system

Table 1 Comparison between omnidirectional and directional antennas in FANET

Specification	Omnidirectional antenna	Directional antenna
Security	Low	Relatively high
Communication range	Small	Large
Spatial reuse	Low	High
Node location information	Not needed	Needed
Network capacity	Less	More
Latency of data packets	High	Low

emanates the signal from the transmitter to receiver without disseminating in all other directions [4]. Directional antenna requires the information of next hop or UAV to transmit the signal [7]. A comparative study of omnidirectional and directional antennas is shown in Table 1.

The transmission range of directional antennas is higher on the account of higher directivity of radio waves. Real-time application demand low latency of data packets at relay node to enhance the communication range and maintain the link sanctity.

4.1.2 Medium Access Control Layer (MAC)

To ensure communication effectively, a stringent and regulatory framework is needed [6]. The communication medium is open to all the nodes and can be accessed by any of the nodes. If this is uncontrolled, then collision can occur between the UAVs [6]. To prevent such instances wireless medium needs regulatory guidelines or MAC protocols. In [6], MAC protocols for wireless networks is classified as

- **Contention-Based MAC Protocol**: The communicating nodes compete to access the shared medium through random access.
- **Contention-Free MAC Protocol**: The access to the medium is not governed by random access but is synchronized and governed by novel application scenarios.

Based on the unique features of FANET such as node mobility and node density, various MAC protocols have been proposed such as Directional Antenna-based MAC protocol (DMAC), Location-Oriented MAC protocol (LODMAC), and full duplex radio circuits multi-packet reception [4, 7, 20, 21].

Directional Antenna-Based MAC Protocol: Directional antennas provide leverage in wireless ad hoc networking to increase network connectivity and overcome link failures in contrast to IEEE 802.11 (omnidirectional antennas). Robert et al. propose that wireless environment has its own impairments which complicate the design such as channel errors, location carrier sensing [22]. A novel Time Division Multiple Access (TDMA)-based MAC protocol is proposed in [20] for a large collection of multi-UAV systems ranging at 50–80 in number for enhanced ad hoc networking through a single multi-meshed tree algorithm by deploying four-phased directional antenna arrays. The constraint in MAC using directional antennas is discussed in detail as follows:

- **Node Mobility**: As discussed in the introduction, node mobility dominates the design issues. Alshabatat et al. also highlight wind and other environmental factors affecting the end-to-end delay of data packets and throughput of the ad hoc network [21].

- **Direction Carrier Sensing**: In recent MAC protocols, the carrier sensing technique is used because when a signal is transmitted into space, it limits the carrier sensing or neighbor sensing. Carrier sensing is used to increase the spatial reuse of radio resources. This impairment exists in wireless networks.

- One of the primary problems in utilization of directional antennas lies in utilizing directional MAC protocols, which are affected by deafness, Head-Of-Line blocking (HOL), and prominent hidden node problem [6, 7].

Recently, a new MAC protocol, LODMAC, is proposed with switched beam directional antennas which promise to overcome the mentioned constraints and disseminate node location information per second, viz., global GPS update time [7]. Temel et al. propose a novel protocol in which a new Busy-to-Send data packet (BTS) is deployed along with the traditional RTS/CTS data packets to overcome deafness and avoid collisions.

4.2 FANET Networking Protocols

Earlier efforts in designing of networking protocols encircled around MANET architecture. With specific design issues of FANET, the routing protocols underwent modification (Fig. 3). Bellur et al. conducted various flight experiments deploying *Topology Broadcast Based on Reverse Path Forwarding Routing Protocol (TBRPF)* on FANET architecture for intra-team communication. TBRPF is a proactive network protocol based on the broadcasting of link state updates aimed at reducing overhead [4, 23]. *Directional Optimized Link State Routing Protocol (DOLSR)* is a proactive protocol based on Optimized Link State Routing Protocol (OLSR) which uses directional antennas and aimed at reducing end-to-end delay [24]. DOLSR with directional antennas use less number of Multipoint Relay Node (MPR) in contrast to OLSR leading to enhanced performance [21].

Reactive protocols like *Dynamic Source Routing (DSR)* suffer from inherent data packet latency due to repetitive route discovery causing exhaustion [25]. *Ad hoc-On-Demand Distance Vector Protocol (AODV)* is a reactive protocol that broadcasts route request packets in the network to find a path to the destination. AODV differs from DSR in routing table maintenance [26]. *Time-Slotted On-Demand Routing (TSODR)* proposed by Forsmann et al. uses dedicated time

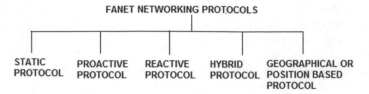

Fig. 3 Taxonomy of FANET networking protocols

slots in which only one node can send data packet. Although it increases the use of network bandwidth but reduces the packet collisions and ensure packet delivery [18].

Route Switching Algorithm proposed by Yanmaz et al. is a novel routing mechanism which tracks all possible routes from the sender node to receiver besides switching to another route if the present route is at risk of failure [12]. It is a trajectory-aware routing protocol aimed at overcoming mission failure in hostile conditions.

Hybrid and geographical position-based routing protocols employed on FANET architecture overcome the inherent route discovery delay of reactive protocols and huge overhead of control messages in proactive protocols. *Zone Routing Protocol (ZRP)* considers a congregation of relay nodes as "zone". Communication within the zone is proactive whereas intrazonal communication involves reactive protocol [15]. *Temporary Ordered Routing Algorithm (TORA)* is a hybrid protocol proposed by Corson et al. which is aimed at reducing network overhead [27]. *Geographic Position Mobility Oriented Routing protocol (GPMOR)* proposed by Lin et al. invokes Gaussian–Markov mobility model to predict the movement of UAVs [28]. It is a position-based routing protocol, which relies on information of next hop node and minimizes data packet delay, congestion, and overhead [4]. Kung et al. proposed *Greedy Perimeter Stateless Routing protocol (GPSR)* for FANET route discovery [29]. It is a position-based routing protocol that use beacon signals to broadcast the information of nodes in the network.

Reactive and proactive protocols coupled in a hierarchical manner constitute the hierarchical protocols. Complexity and traffic congestion limit the use of hierarchical protocols for FANET. *Clustering algorithm for UAV networking* proposed by Jun et al. is a hierarchical protocol that forms clusters on the ground which are executed in the missions [30]. Scalability of coverage area is a distinct advantage of cluster formation.

Static routing protocols are those protocols in which a predefined routing table is present before the execution of a mission. *Data-Centric Routing Protocol* and *Load Carry and Deliver Routing* are static routing protocol [15]. Multi-UAV system is designed for specific application. It is difficult to adapt multi-UAV system for other applications [4, 15]. Data-centric routing solutions can be used in FANETs for different types of applications on the same multi-UAV system [4]. It is a static protocol based on publisher–subscriber mode as proposed by Mahajan et al. in [31]. If UAV loaded with data ferries it to the destination, then this routing mechanism is

Table 2 Brief summary of FANET networking protocol

Protocol	Type	Feature	Advantage(s)
TBRPF	Proactive	Reverse path forwarding	Reduced overhead
DOLSR	Proactive	Multipoint Relay Node (MTR), directional antennas	Lower end-to-end delay
DSR	Reactive	Mobility awareness	Adaptability to node mobility
AODV	Reactive	Broadcasts route request data packets	
TSODR	Reactive	Dedicated slot to send data packet	Reduced data packet collisions and better data delivery
ZRP	Hybrid	Congregation of UAV nodes is assigned as "zones"	Reduced overhead and maximized throughput
GPMOR	Geographic	Gaussian–Markov model	Better data packet ratio, improved data forwarding in terms of latency [4]
Clustering algorithm	Hierarchical	Highest weight to decide cluster head [16]	Scalability of mission
Load carry and deliver protocol	Static	Nonstop ferrying of loaded data from source to destination affected by coverage area	Increases throughput and security

Load Carry and Design Routing; it increases throughput, security, however, increasing coverage area adversely affects the performance [15]. A brief comparative overview of FANET Networking Protocol is given in Table 2.

5 Extensive Research Areas

Communication in FANET is a serious design consideration. In [4], the author highlights the need to design and implement protocols considering UAV nodes in three-dimensional spaces; implementing Free Space Optics (FSO) in data transmission enhance performance, thereby reducing network latency. Security in FSO is increased while it remains less prone to jamming. The various issues are highlighted as follows:

- **UAV-to-UAV Communication**: Node-to-node communication is difficult to maintain in FANET on the account of high node mobility. Maintaining the sanctity of data link is an open research issue. Data-centric routing algorithm offers new research avenues [4, 15]. Communication among UAVs considering nodes in 3D remains largely unexplored.
- **Minimal Network Latency and Managing Bandwidth Constraints**: Dissemination of audio–video data in FANET application scenario imposes a key

constraint on bandwidth requirement. Increasing bandwidth of data transfer introduces noise into the transmission. Minimizing network latency in dense FANET deployment remains an important research issue. Free Space Optics (FSO) provides a novel mechanism for using the optical link to transmit data. Recent findings have found FSO to be independent of fading [32].

- **Standardization of FANET**: FANET uses various wireless communication bands, such as, VHF, UHF, L-Band, C-band, Ku-Band, etc., [33], which is also used in different application areas, such as GSM networks, satellite communications. For reducing congestion problem, FANET requires standardization. Civilian UAVs standardization remains a challenging task beforehand.
- **FANET Algorithms**: Robust algorithms [34] provide stability to the in-flight hostile environment. Research into framing strong algorithms is a hot topic for researchers and scientific community worldwide.
- **Swarms**: FANET [35, 36] architecture-based swarms of UAV leads to ample new application scenarios. Forest fire monitoring, search and rescue, disaster management is few to name.

6   Conclusion

Unmanned aerial systems have gained prominence among the scientific community over the years. Deployment of multi-UAV system in noisy hostile and erratic environment with appreciable performance offers a plethora of unexplored avenues. In this paper, we surveyed on flying ad hoc networks emphasizing on various design challenges, communication methodologies, and issues relating to the transport layer. FANET provides a novel mechanism to overcome scalability issue present in direct link networking.

FANET architecture used in non-line-of-sight communication remains largely unexplored. Communication remains one of the most serious constraints encircling FANETs. Data-centric routing algorithm and multi-level hierarchical routing protocols are promising new routing algorithms. We provide extensive research areas to encourage researchers in this domain.

References

1. Alshbatat, A.I., Dong, L.: Cross-layer design for mobile ad-hoc unmanned aerial vehicle communication networks. In: 2010 International Conference on Networking, Sensing and Control (ICNSC). IEEE (2010)
2. Dey, N., Mukherjee, A.: Embedded Systems and Robotics with Open Source Tools. CRC Press (2016)

3. Sun, Z., Wang, P., Vuran, M.C., Al-Rodhaan, M., Al-Dhelaan, A., Akyildiz, I.F.: BorderSense: border patrol through advanced wireless sensor networks. Ad Hoc Netw. **9** (3), 468–477 (2011)
4. Bekmezci, I., Sahingoz, O.K., Samil, T.: Flying ad-hoc networks (FANETs): a survey. Ad Hoc Netw. **11**(3), 1254–1270 (2013)
5. Samara, G., Al-Salihy, W.A.H., Sures, R.: Security analysis of vehicular ad hoc networks (VANET). In: 2010 Second International Conference on Network Applications Protocols and Services (NETAPPS). IEEE (2010)
6. Bazan, O., Jaseemuddin, M.: A survey on MAC Protocols for wireless adhoc networks with beamforming antennas. IEEE Commun. Surv. Tutor. **14**(2), 216–239 (2012)
7. Temel, Samil, Bekmezci, Ilker: LODMAC: location oriented directional MAC protocol for FANETS. Comput. Netw. **83**, 76–84 (2015)
8. Jiang, F., Lee Swindlehurst, A.: Dynamic UAV relay positioning for the ground-to-air uplink. 2010 IEEE GLOBECOM Workshops (GC Wkshps). IEEE (2010)
9. Samil, T., Bekmezci, I.: On the performance of flying adhoc networks (fanets) utilizing near space high altitude platforms (haps). In: 2013 6th International Conference on Recent Advances in Space Technologies (RAST). IEEE (2013)
10. Mukherjee, A., et al.: A disaster management specific mobility model for flying ad-hoc network. Int. J. Rough Sets Data Anal. (IJRSDA) **3.3**, 72–103 (2016)
11. Chaumette, S., Laplace, R., Mazel, C., Mirault, R., Dunand, A., Lecoutre, Y., Perbet, J.-N.: CARUS, an operational retasking application for a swarm of autonomous UAVs: first return on experience. In: Military Communication Conference—MILCOM 2011, pp. 2003–2010 (2011)
12. Muzaffar, R., Yanmaz, E.: Trajectory aware adhoc routing protocols for micro aerial vehicle networks. In: IMAV 2014: International Micro Air Vehicle Conference and Competition 2014. Delft University of Technology, Delft, The Netherlands, 12–15 Aug 2014
13. Rosati, S., et al.: Dynamic Routing for Flying Ad Hoc Networks. No. EPFL-ARTICLE-207491 (2015)
14. Bouachir, O., et al.: A mobility model for UAV Ad Hoc network. In: 2014 International Conference on Unmanned Aircraft Systems (ICUAS). IEEE (2014)
15. Tareque, M.H., Atiquzzaman, M.: On the routing in flying ad hoc networks. In: Proceedings of the Federated Conference on Computer Science and Information Systems, vol. 5, pp. 1–9. ACSIS. https://doi.org/10.15439/2015f002
16. Singh, S.K., et al.: A comprehensive survey on Fanet: challenges and advancements. Int. J. Comput. Sci. Inf. Technol. (IJCSIT) **6**(3), 2010–2013 (2015)
17. Mohammed, F.: Efficient data communication in unmanned aerial vehicles. Theses, Paper 35 (2015)
18. Forsmann, J.H., Hiromoto, R.E., Svoboda, J.: A time-slotted on-demand routing protocol for mobile ad hoc unmanned vehicle systems. SPIE **6561**, 65611P (2007)
19. Shah, B., Kim, K.I.: A survey on three dimensional wireless ad hoc and sensor networks. Int. J. Distr. Sensor Netw. (2014)
20. Huba, W., Shenoy, N.: Airborne surveillance networks with directional antennas. In: ICNS 2012, The Eighth International Conference on Networking and Services, pp. 1–7 (2012)
21. Alshabatat, A.I., Dong, L.: Adaptive MAC protocol for UAV communication networks using directional antennas. In: Proceedings of International Conference on Networking, Sensing and Control (ICNSC), pp. 598–603 (2010)
22. Vilzmann, R., Bettstetter, C.: A survey on MAC protocols for Adhoc Networks with directional antennas. In: EUNICE 2005: Networks and Applications Towards a Ubiquitously Connected World, 187–200 (2006)
23. Bellur, B., Ogier, R.G.: A reliable, efficient topology broadcast protocol for dynamic networks. In: Proceedings of Eighteenth Annual Joint Conference of the IEEE Computer and Communications Societies, INFOCOM' 99, vol. 1, pp. 178–186. IEEE (1999)
24. Clausen, T., Jacquet, P.: Optimized link state routing protocol (OLSR). RFC 3626 (Experimental), Oct 2003

25. Brown, T.X., Argrow, B., Dixon, C., Doshi, S., Thekkekunel, R.G., Henkel, D.: Ad-hoc UAV ground network (AUGNet). In: 3rd AIAA Un-manned Unlimited Technical Conference, pp. 29–39 (2004)
26. Murthy, S., Aceves, J.L.: An efficient routing protocol for wireless networks. ACM Mobile Netw. Appl. 183–197 (1996)
27. Park, V., Corson, S.: Temporarily-ordered routing algorithm (TORA). In: Version 1. Internet draft: IETF MANET working group. http://tools.ietf.org/html/draft-ietf-manet-tora-spec-04, vol. 3, Aug 2013
28. Lin, L., Sun, Q., Li, J., Yang, F.: A novel geographic position mobility oriented routing strategy for UAVs. J. Comput. Inf. Syst. 8, 709–716 (2012)
29. Karp, B., Kung, H.T.: GPSR: greedy perimeter stateless routing for wireless networks. In: Proceedings of the 6th Annual International Conference on Mobile Computing and Networking, MobiCom' 00. ACM, New York, USA, pp. 243–254 (2000)
30. Kesheng, L., Jun, Z., Tao, Z.: The clustering algorithm of UAV networking in near space. In: 8th International Symposium on Antennas, Propagation and EM Theory, 2008 (ISAPE 2008), pp. 1550–1553 (2008)
31. Ko, J., Mahajan, A., Sengupta, R.: A network-centric UAV organization for search and pursuit operations. In: Aerospace Conference Proceedings, 2002, vol. 6, pp. 2697–2713. IEEE (2002)
32. Chlestil, C.H., Leitgeb, E., Sheikh Muhammad, S., Friedl, A., Zettl, K., Schmitt, N.P., Rehm, W., Perlot, N.: Optical Wireless on Swarm UAVs for High Bit Rate Applications
33. Sahingoz, O.K.: (FANETs): concepts and challenges. Springer J. Intel. Robot Syst. 74, 513–527 (2014)
34. Mukherjee, A., et al.: Unmanned aerial system for post disaster identification. In: 2014 International Conference on Circuits, Communication, Control and Computing (I4C) IEEE (2014)
35. Murthy, C.S.R., Manoj, B.: Ad Hoc Wireless Networks: Architectures and Protocols. Prentice Hall PTR, Upper Saddle River, NJ, USA (2004)
36. A. 4b Network Environmental Committee, JAUS/SDP Transport Specification. http://standards.sae.org/as5669a. Accessed 28 Aug 2016

Author Index

© Springer Nature Singapore Pte Ltd. 2018
S. C. Satapathy et al. (eds.), *Information and Decision Sciences*,
Advances in Intelligent Systems and Computing 701,
https://doi.org/10.1007/978-981-10-7563-6

Printed in the United States
By Bookmasters